—Foreword by
BRUCE BABBITT

ECOSYSTEM MANAGEMENT *for* SUSTAINABILITY

*Principles and Practices Illustrated
by a Regional Biosphere
Reserve Cooperative*

Edited by

JOHN D. PEINE, PH.D.

*United States Geological Survey
Biological Resources Division
Southern Appalachian Field Laboratory
The University of Tennessee
Knoxville, Tennessee*

LEWIS PUBLISHERS

Boca Raton Boston London New York Washington, D.C.

Cover photo by Mary Anne Peine.

Library of Congress Cataloging-in-Publication Data

Ecosystem management for sustainability : principles and practices
    illustrated by a regional biosphere reserve cooperative / edited by
    John D. Peine.
          p. cm.
     Includes bibliographical references and index.
     ISBN 1–57444–053–5 (alk. paper)
     1. Ecosystem management.   2. Sustainable development.
   3. Ecosystem management--Case studies.   4. Sustainable development-
  -Case studies.   I. Peine, John D. (John Douglas), 1944–   .
   QH75.E323   1998
   333.7--dc21   98-15035
                                                                                                CIP

The views expressed in this volume are solely those of the authors and do not necessarily reflect the policy of the organization with which they are affiliated.

No claim to original U.S. Government works
International Standard Book Number 0-57444-053-5
Library of Congress Card Number 98-15035
Printed in the United States of America 1 2 3 4 5 6 7 8 9 0
Printed on acid-free paper

Foreword

The President of the United States' Council on Sustainable Development recently stated that human survival may well depend on widespread acceptance of principles of sustainable living and ecosystem management. When we act locally, we must think globally. We must also think holistically, considering the relationships between Nature and the economic, cultural, and spiritual life of our communities and societies. As the 21st century and the next millennium approach, the need to put these principles into practice has never been more obvious.

We are all on a steep learning curve as we struggle to apply idealistic principles to real world situations. The road map is still mostly theoretical at this point, but consultation and cooperation clearly must be common denominators of our efforts. To empower communities to work toward a sustainable future, we need specific applications from which to learn. This book is an important contribution toward meeting this need.

The Man And Biosphere Program (MAB) was initiated 25 years ago under the auspices of the United Nations Educational, Scientific, and Cultural Organization. MAB was the first international effort to develop the knowledge, skills, and cooperation required to sustain harmonious relationships between people and Nature. The USMAB Program offers a framework for building the broad public–private partnerships needed to achieve this purpose domestically and share our experiences globally. At last count, 115 countries had established autonomous national organizations to coordinate their participation in MAB.

A cornerstone of MAB is the growing World Network of Biosphere Reserves, which today includes more than 329 areas in more than 83 countries. The biosphere reserves are MAB's geographic hubs for cooperation among government agencies, the private sector, and local communities to find practical solutions to complex problems of conservation and development, and for sharing information and experience globally.

More than 20 years of quiet cooperation have made the southern Appalachians a standard-bearer for biosphere reserves in the United States and the world. Without fanfare, dedicated individuals have created a remarkable framework for gathering, sharing, and applying information to help people work together to solve problems in a region comprising 135 counties in parts of 6 states. Their most recent effort was a regional assessment of the status and trends of natural resources and human activities — one of three prototype projects of its kind in the country. The assessment has developed an extraordinary compilation of information which can be used to anticipate problems and seek reasonable solutions. The hundreds of data sets available on the Internet include a wide range of subject matter associated with the natural environment and the human dimensions of the landscape.

In the southern Appalachians, quiet cooperation has truly paid off. There are no major environmental train wrecks here between those concerned with protecting the environment and those representing economic development interests. This may be in part attributed to the extraordinary efforts documented in this volume. The lack of confrontation has allowed for the development of broad alliances that might not have otherwise so readily occurred.

I congratulate the many men, women, and children of southern Appalachia for their participation in this visionary effort. They are teaching us all about ways to implement the principles of sustainable living and ecosystem management. Their message has global relevance as we all continue the struggle for balance between conserving native species and the processes of the natural world and meeting the needs of human communities for natural resources and beneficial uses of ecosystems.

Bruce Babbitt

Dedicated to

Anita
Bill
Marie
Mary Anne
Emelie
Josedale Renowned

Proceeds from sale of this book
are donated to the
Southern Appalachian Man And Biosphere Foundation

Preface

What is the use of a house if we don't have a habitable planet to put it on?

Henry David Thoreau

The inspiration for this book came from my 10 year experience (1982 to 1992) as research administrator at Great Smoky Mountains National Park. During that assignment, I managed a research program that included a wide spectrum of social and natural science disciplines addressing a variety of ecosystem management issues. I was a member of the senior management team for the Park and as such was exposed to and participated in numerous conversations and debates associated with a full spectrum of management policies and program operations. In addition, I was very involved with the development of the UNESCO Man And Biosphere program in the southern Appalachians.

This book is about ecosystem management. The term reflects current philosophy of resource management. It represents a point in time on the evolutionary spectrum of policy related to natural resources management. Catch-phrases reflecting philosophical benchmarks from previously accepted norms include multiple use management and integrated resource management. Over the last 35+ years, there has been a steadily growing awareness of the complexity of ecosystems and the human dimensions of "natural" ecosystems. This growing awareness of the interrelationships of ecosystem components has led to more holistic perspectives on management practices. This in turn has led to a greater recognition of the need for interdisciplinary research concerning the environment.

The term "ecosystem management" suggests that this evolution of philosophy has brought us to yet a higher level of awareness. The primary distinction of this level is the awareness of the interdependency of the processes and functions associated with ecosystems and the incorporation of the human dimensions related to those processes. Although many of my colleagues in the "natural" sciences have been reluctant to fully embrace this shift in the philosophy of resource management, progressive ecosystem managers are actively practicing it, as documented in this volume.

Academicians and resource managers have published profusely on the theory of ecosystem management and sustainable development. Few have documented a comprehensive spectrum of the practice of these principles. The primary contribution of this publication is the provision of linkage between theory and practice of the emerging philosophies of ecosystem management and sustainable development. The next level in the evolution of philosophy of ecosystem management will be shaped through lessons learned from practical experience. The development of theory in the 1990s will yield to accepted practices in the next century.

Section I of this book reviews the debate associated with definition of terms and how the principles have evolved in the region and been applied internationally. An overview of the system of biosphere reserves in the U.S. is described.

In Section II of the book, the physical setting and institutional framework of the case study in the southern Appalachians is described. An overview is provided of the natural and cultural resources of the region. The framework for integrated ecosystem management established in the southern Appalachians is documented. A comparison is made between a biosphere reserve in the southern Appalachians and central Russia to place the case study of the southern Appalachians in a more global perspective.

Each chapter in Section III describes a key dimension of ecosystem management and features a case study in the southern Appalachians. The litany described in Chapters 7 through 21 may initially appear as a disparate set of topics, but they collectively represent the range of issues typically addressed by the progressive ecosystem manager. My most striking observation when reviewing the assemblage of programs described is the extraordinary spectrum of complex issues facing stewards of the natural environment. The assemblage of disciplines required to address the myriad of the challenging issues facing ecosystem managers is truly daunting.

The initial chapter deals with the conduct of environmental assessments, followed by a chapter on the role of environmental monitoring. Ecosystem monitoring is inextricably linked to ecosystem assessments. More specific topics follow in subsequent chapters.

For instance, the next chapter is about neotropical migratory birds, which represent charismatic indicator species sensitive to the dynamics of forest structure and fragmentation. The chapter on black bears illustrates management of a large charismatic animal that ranges beyond the boundaries of protected areas into harm's way. Next, there is a discussion of red wolves, illustrating the principles of species repatriation. Then, the story of the southern Appalachian strain of brook trout, representing an isolated species whose range has become extremely limited, is discussed. In this case, the genetic variation within species is an issue. The next topic concerns an exotic species, the European wild boar. This mammal is very destructive, and attempts to control the population have been most difficult.

The chapter on pests and pathogens documents one of greatest concerns of ecosystem managers. The topic of air pollution is extremely complex, requiring cooperation among industry, science, ecosystem managers, community, and government regulatory agencies. In the next chapter, the role of fire in ecosystem management is discussed. It is one of the most powerful tools to manipulate ecosystems available to managers. The next topic is land use planning and how it can be used to mitigate potential adverse effects from development on privately owned lands adjacent to protected areas. The chapter on high elevation grassy balds illustrates some of the difficult choices when managing a historic landscape within the context of a natural area. The chapter on climate change illustrates actions needed now to prepare for this most potentially catastrophic force threatening ecosystem processes in the next century.

The last two chapters in Section III illustrate the complexity in dealing with highly threatened ecosystems. The Clinch–Powell river basin illustrates ecosystem restoration practice at a watershed scale. The high elevation spruce–fir ecosystem is arguably the most disturbed ecosystem in the southern Appalachians.

The link among chapters in Section III is the expression of principles of ecosystem management associated with each topic and the *Risk Assessment Report Card,* presented as a concluding statement for each case study. The Report Card includes a rather long list of factors associated with making decisions within the extremely complex world of ecosystem management. Elements of the Report Card and a brief explanation of the scale of values are as follows:

- **Vision**
  A = Clear and concisely defined vision and/or purpose
  F = Total lack of vision

- **Resource risk**
  A = Extremely low risk to the natural environment
  F = Extremely high risk to the natural environment

- **Resource conflicts**
  A = Low likelihood for conflicts among elements of ecosystem
  F = High likelihood for conflicts among elements of ecosystem

- **Socioeconomic conflicts**
  A = Low likelihood of conflicts with industry or society
  F = High likelihood of conflicts with industry or society

- **Procedural protocols**
  A = Procedures are well defined and accepted
  F = Procedures are not at all defined nor accepted

- **Scientific validity**
  A = Topic is well documented and procedures are well defined by refereed scientific research
  F = Topic is not documented nor procedures defined by scientific research

- **Legal jeopardy**
  A = Minimal chance of legal challenge
  F = High degree of probability for legal challenge

- **Public support**
  A = High degree of public acceptance of management policy and practice
  F = Significant public hostility toward management policy and practice

- **Adequacy of funding**
  A = Fully adequate financial support
  F = Woeful lack of funding to support management initiative

- **Policy precedent**
  A = Public policy precedent well established
  F = Current policy does not support management initiative

- **Administrative support**
  A = Agency provides all required in-kind services and facilities
  F = Agency does not support management initiative

- **Transferability**
  A = Issue and actions are relevant to other protected areas in similar biogeographic regions
  F = Issue and actions are limited in relevance to one ecoregion

In the concluding section of the book, Section IV, there is first a chapter discussing the role of reforming institutions in order to reach the goals of ecosystem management and sustainable development. The last chapter provides a vision to transform the regional biosphere cooperative to a more activist, operative role of leadership in ecosystem management in the southern Appalachians.

Enjoy the book. I hope you find it as stimulating as I have working with all the 51 contributing authors, each of whom are leaders in the advancement of ecosystem management and champions of the natural environment. Here's hoping the message of this volume contributes to raising awareness of the critical need to alter our behavior to insure that the next millennium brings improvement in our ability to perpetuate the sustainable presence of our fellow 4 to 30 million or so species on this our Planet Earth.

John D. Peine, Ph.D.

Contributors*

Caffilene Allen, Ph.D.
U.S. EPA, Region IV
Atlanta, GA

Robert L. Anderson
U.S.D.A. Forest Service
Atlanta, GA

Cory W. Berish, Ph.D.
U.S. EPA, Region IV
Atlanta, GA

Edward R. Buckner, Ph.D.
The University of Tennessee
Knoxville, TN

David A. Buehler, Ph.D.
The University of Tennessee
Knoxville, TN

Faith Thompson-Campbell
National Coalition of Exotic Plant Pest Councils
Washington, D.C.

Joseph D. Clark, Ph.D.
U.S.G.S.
Southern Appalachian Field Laboratory
The University of Tennessee
Knoxville, TN

Jaime A. Collazo, Ph.D.
North Carolina State University
Raleigh, NC

Barron A. Crawford
U.S. Fish and Wildlife Service
Great Smoky Mountains National Park
Townsend, TN

Margarita A. Davydova
Moscow State Pedagogical University
Moscow Russia

B. Richard Durbrow
U.S. EPA, Region IV
Atlanta, GA

Christopher Eagar, Ph.D.
U.S.D.A. Forest Service
Durham, NC

Kathleen E. Franzreb, Ph.D.
Southern Research Station, U.S.D.A.
  Forest Service
Clemson University
Clemson, SC

Donald W. Gowan
The Nature Conservancy
Abbington, VA

William P. Gregg, Ph.D.
U.S.G.S., Biological Resources
  Division
Reston, VA

Stanley Z. Guffy, Ph.D.
The University of Tennessee
Knoxville, TN

James E. Harrison
U.S. EPA, Region IV
Atlanta, GA

Hubert Hinote
Southern Appalachian Man and
  Biosphere Program
Great Smoky Mountains National
  Park
Gatlinburg, TN

William A Jackson
National Forest in North Carolina
U.S.D.A. Forest Service
Asheville, NC

---

* The views expressed in this volume are solely those of the authors and do not necessarily reflect the policy of the organization with which they are affiliated.

Vladimir A. Koshevoi, Ph.D.
Moscow State Pedagogical University
Moscow, Russia

Richard A. Lancia, Ph.D.
North Carolina State University
Raleigh, NC

Chris F. Lucash
U.S. Fish and Wildlife Service
Great Smoky Mountains National Park
Townsend, TN

Karen A. Malkin, Esq.
National Park Service
Washington, D.C.

Frank J. McCormick, Ph.D.
The University of Tennessee
Knoxville, TN

Gary F. McCracken, Ph.D.
The University of Tennessee
Knoxville, TN

Michael L. McKinney, Ph.D.
The University of Tennessee
Knoxville, TN

William R. Miller III, Ph.D.
Saturn Corporation
Spring Hill, TN

Leslie Cox Montgomery
U.S. EPA, Region IV
Atlanta, GA

Stephen E. Moore
Great Smoky Mountains National Park
Gatlinbug, TN

Brian J. Morton, Ph.D.
North Carolina Environmental Defense Fund
Raleigh, NC

Niki Stephanie Nicholas, Ph.D.
Tennessee Valley Authority
Norris, TN

John E. Nolt, Ph.D.
The University of Tennessee
Knoxville, TN

David M. Ostermeier, Ph.D.
The University of Tennessee
Knoxville, TN

Charles R. Parker, Ph.D.
U.S.G.S.
Great Smoky Mountains National Park
Gatlinburg, TN

Michael R. Pelton, Ph.D.
The Univeristy of Tennessee
Knoxville, TN

John D. Peine, Ph.D.
U.S.G.S.
Southern Appalachian Field Laboratory
The University of Tennessee
Knoxville, TN

Kerry N. Rabenold, Ph.D.
Purdue University
West Lafayette, IN

James C. Randolph
Indiana University
Bloomington, IN

Kert H. Riitters
U.S.G.S.
Southern Appalachian Field Laboratory
University of Tennessee
Knoxville, TN

Scott E. Schlarbaum, Ph.D.
The University of Tennessee
Knoxville, TN

Theodore R. Simmons, Ph.D.
North Carolina State University
Raleigh, NC

Elizabeth R. Smith, Ph.D.
Tennessee Valley Authority
Norris, TN

Barry R. Stephens
Division of Air Pollution Control
State of Tennessee
Nashville, TN

Robert Sutter
The Nature Conservancy
Chapel Hill, NC

Jack D. Tuberville
Tennessee Valley Authority
Norris, TN

Nichole L. Turrill
Indiana University
Bloomington, IN

Karen P. Wade
Great Smoky Mountains National Park
Gatlinburg, TN

Peter S. White, Ph.D.
University of North Carolina
Chapel Hill, NC

Dennis H. Yankee
Tennessee Valley Authority
Norris, TN

Igor A. Zhigarev
Prioksko-Terrasny Biosphere
  Reserve
Serpukov, Russia

Contents

Foreword
Bruce Babbitt

Preface
John Peine

SECTION I: The Concepts and their Relevance to the Man and Biosphere Program

Chapter 1
Principles of Ecosystem Management and Sustainable Development 3
F. McCormick

Chapter 2
Environmental Policy, Sustainable Societies and Biosphere Reserves..23
W. Gregg

Chapter 3
The Evolution of Land Use Ethics and Resource Management: Coming Full Circle...................41
J. Nolt and J. Peine

SECTION II: The Case Study Setting and Institutional Framework

Chapter 4
The Southern Appalachians: The Setting for the Case Study 63
J. Randolph, M. McKinney, C. Allen and J. Peine

Chapter 5
Framework for Integrated Ecosystem Management: The Southern Appalachian Man
and the Biosphere Cooperative..81
H. Hinote

Chapter 6
A Comparison in the Management of Two Biosphere Reserves: Prioksko-Terrasny Reserve,
Russia and Great Smoky Mountains National Park, U.S. 99
J. Peine, V. Koshevoi, K. Wade, I. Zhigarev and M. Davydova

SECTION III: Components of Ecosystem Management

Chapter 7

The Environmental Assessments: The Southern Appalachian Experience.....................................117
C. Berish, R. Durbrow, J. Harrison, W. Jackson, and K. Riitters

Chapter 8

Environmental Monitoring...167
E. Smith, C. Parker, and J. Peine

Chapter 9

The Role of Indicator Species: Neotropical Migratory Song Birds...187
T. Simons, K. Rabenold, D. Buehler, J. Collazo, and K. Franzreb

Chapter 10

Management of a Large Carnivore: Black Bear 209
J. Clark and M. Pelton

Chapter 11

Species Repatriation: Red Wolf..225
C. Lucash, B. Crawford, and J. Clark

Chapter 12

Management of Isolated Populations: Southern Strain Brook Trout..247
S. Guffy, G. McCracken, S. Moore, and C. Parker

Chapter 13

Control of Exotic Species: European Wild Boar 267
J. Peine and R. Lancia

Chapter 14

Control of Pests and Pathogens...291
S. Schlarbaum, R. Anderson, and F. Tompson-Campell

Chapter 15

Air Quality Management: A Policy Perspective 307
J. Peine, L. Cox Montgomery, B. Stephens, W. Miller III, B. Morton, and K. Malkin

Chapter 16

Fire Management 329
E. Buckner and N. Turrill

Chapter 17

Land Use Planning: Sustainable Tourism...349
J. Peine

Chapter 18

Managing Biodiversity in Historic Habitats: A Case History of the Southern
Appalachian Grassy Balds 375
P. White and R. Sutter

Chapter 19

Climate Change: Potential Effects in the Southern Appalachians 397
J. Peine and C. Berish

Chapter 20

Ecosystem Stabilization and Restoration: The Clinch–Powell River Basin Initiative 417
D. Yankee, J. Peine, J. Tuberville, and D. Gowan

Chapter 21

Threatened Ecosystem: High elevation Spruce–Fir Forest 431
N. Nicholas, C. Eagar, and J. Peine

SECTION IV: The Future

Chapter 22

The Role of Institutions in Ecosystem Management 457
D. Ostermeier

Chapter 23

Moving to an Operational Level: A Call for Leadership From the Southern Appalachian
Man and the Biosphere Cooperative 475
J. Peine

Index 485

Section I

The Concepts and their Relevance to the Man and Biosphere Program

1 Principles of Ecosystem Management and Sustainable Development

Frank J. McCormick

CONTENTS

Introduction 3
Sustainability 3
Ecosystem Management 6
Examples of Growing Acceptance of Sustainability and Esosystem Management 10
    Internal Acceptance 10
    National Acceptance..10
    Regional and Local Acceptance..10
Principles of Sustainability and Ecosystem Management 12
References 19

INTRODUCTION

This discussion of sustainability and ecosystem management will focus upon four points: (1) What is sustainability?; (2) What is ecosystem management and its relevance to sustainability?; (3) Examples of the growing acceptance of sustainability and ecosystem management; and (4) Principles of sustainability and ecosystem management.

SUSTAINABILITY

Sustainability is a goal, a goal too seldom achieved. Sustainability has become the primary goal of both economic development and natural resource management. In addition to similar goals, development planning and resource management planning employ similar procedures and methods. In both endeavors we evaluate alternative action plans in terms of their short-term and long-term consequences. Planning and managing for sustainability require that we focus more intently upon the long-term consequences (World Commission on Environment and Development 1987). **Sustainable development is development which: integrates economic, environmental, and social values during planning; distributes benefits equitably across socioeconomic strata and gender upon implementation; and ensures that opportunities for continuing development remain undiminished to future generations**.

Greatest success in implementing sustainable development has been in developing countries. Expenditure of U.S. tax dollars in developing countries is accompanied by intensive and extensive oversight by congressional committees and nongovernment organizations (NGOs). There is no comparable oversight activity or public pressure associated with development in the southern Appalachians or elsewhere in the U.S. Although implementation of sustainable development (EESI 1991; World Bank 1993) is more advanced in developing countries than in the U.S., noteworthy examples are to be found within the U.S., including the Southern Appalachian Man and the Biosphere (SAMAB, 1994) program.

If economic development or resource management are to be sustainable, evaluation of alternative action plans must be in terms of integrated economic, environmental, and social values. Our greatest challenge is to identify the optimal balance among these three sets of values. It is the integration of these three value systems which, in part, distinguishes sustainability from other strategies of resource management or economic development. The following paradigm of environmental sociology provides the rationale for integrating these three value systems: "The physical-biological environment influences, and is influenced by human behavior and societies" (Taylor et al. 1990). Although the validity of this fundamental paradigm is difficult to challenge, contemporary assessment procedures (Council on Environmental Quality 1987; USAID 1989) routinely disregard it and thereby jeopardize efforts to achieve sustainability. Interdependencies among physical, biological, and social components of the human environment are not discernable when economic, environmental, and social assessments are conducted independently of one another, as is the custom.

Sustainability is further defined in terms of equitability. **Sustainability is characterized by equitable distribution of benefits and opportunities across socioeconomic groups, gender, and generations**. Economic development and resource management have shared a common evolutionary history as they progressed toward intergenerational equitability. Until the mid-20th century, both shared the same single minded goal of maximum economic yield. By the 1960s an additional goal confronted development planners and resource managers. Policy reforms required that benefits be distributed equitably across socioeconomic strata. By the late 1970s, planners and managers were further challenged by policies requiring that benefits and opportunities be distributed equitably by gender. Most recently, and most germane to sustainability, policy reforms require that benefits and opportunities be distributed equitably across generations. Extension of the time dimension to several generations places new demands upon perceptions and skills of planners and managers. It forces them to be increasingly dependent upon scientific principles and methods, for within good science lies the potential for prediction and projection across expanded time horizons. Many of the scientific principles and methods needed to achieve sustainability are to be found within ecological science. During the formative years of ecosystem science, no agency or department of government contributed as much to the development of ecosystem principles and methods as did the Department of Energys (DOE) predecessor, the Atomic Energy Commission (AEC). Dr. Stanley Auerbach, (the first Director of the Environmental Science Division, Oak Ridge National Laboratory), described the contributions of AEC programs to ecosystem science as follows:

> One cannot ignore the contributions to basic ecology. The combination of systems analysis and radioisotope techniques to delineate and quantify ecosystem processes has had a profound impact on that part of ecology dealing with ecosystem analysis. Not only the techniques themselves but the insights, understandings, and ideas derived from the application of these radioecological concepts have served to stimulate wide interest in ecosystem analysis and have provided new insights into the structure and functioning of ecosystems (Auerbach 1971).

It is totally appropriate that a quarter century later these principles and methods are being used to guide ecosystem management at the Oak Ridge Reservation (ORR) and other sites participating in SAMAB. Ecological science is guided by a holistic philosophy (Gause 1934), which is totally compatible with the integration of economic, environmental, and social values into a comprehensive approach to planning and management. Once it became obvious that sustained economic productivity

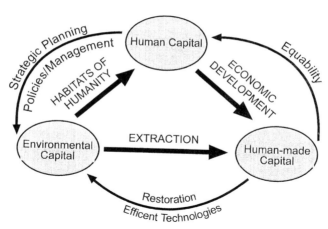

FIGURE 1 A dynamic model of sustainable development. (From McCormick, J.F., Institute for International Training in Sustainable Development Info. Doc. E01-1035-001-95, University of Tennessee, Knoxville, 1994.)

is dependent upon sustained ecological productivity (McCormick 1991), and once society began to recognize significant economic values of ecosystem goods and services (Ahmad et al. 1989), it took little time for society to demand that these values be shared equitably. **Just as pro-life advocates argue that the unborn have a right to life, advocates of sustainability argue the unborn have a right to an environment capable of sustaining life, and sustaining it at a level of quality enjoyed by the present generation.** When sustainability is defined in terms of integrated value systems and intergenerational equitability, it becomes obvious that sustainability involves issues of empowerment and ethics as well as issues of environment and economics. There is no prerequisite to achieving sustainability more crucial than the integration of social values and local peoples into economic development and natural resource planning (Gagnon et al. 1993). Local peoples are survivors. Knowledge acquired from generations of survivors provides pragmatic lessons regarding sustainability.

Managing for sustainability requires that there be no net reduction in the sum of environmental capital, human resource capital, or human-made capital (Goodland 1993) available to future generations (Figure 1) (McCormick 1994). Any one of these three forms of capital can be expected to increase or diminish from time to time. It is the availability of "total capital," representing all three value systems, which must remain undiminished to future generations. In this dynamic model of sustainability, human ingenuity provides the means of compensating for periodic imbalances among the various forms of capital. In this model it is the feedback loops representing various forms of human ingenuity which distinguish sustainable development and management from traditional nonsustainable policies and practices.

At a more operational level, what capabilities, procedures, skills, and tools do we need in order to achieve sustainability? A quarter century of experience working with policy makers in our government and those of numerous developing countries has convinced me of one thing; we already have enormous scientific and human resources we are not using effectively. The most urgent need is not more fundamental scientific research, but rather, the transfer of already existing information to policy makers, planners, and managers. Do not mistake this emphasis upon information transfer as a call for cessation or reduction of basic research; it is a plea to adjust our priorities to the realities of planning and managing for sustainability. For example, the Sustainable Biosphere Institute (SBI) of the Ecological Society of America (Lubchenco et al. 1991) identifies 55 fundamental research topics which cut across three priority areas of research. During the past 3 years, more than 300 professionals in sustainable development planning and implementation (ASSET 1994) have been asked to respond to the SBI in the context of professional training courses on sustainable development. Responses were consistently as follows:

1. It is amazing and admirable that a group of scientists could come to a consensus regarding a prioritized research agenda
2. Some, but very little, of the research agenda would be helpful. However, the research agenda reflects academic research interests more than what we, as practitioners of sustainable development, need in order to do our job
3. Perhaps sustainability is a smoke screen being used to acquire funds for basic research
4. Environmental decision making; the third component of the SBI should be given higher priority

The point of this example is that those who are responsible for implementing policies of sustainability do not view lack of basic research as their greatest constraint. Science and technology have provided us with a variety of rapid and accurate procedures for the inventory, monitoring, and evaluation of natural resource capital. Satellite imagery, geographic information systems, and concepts of landscape ecology are conspicuous examples. Technologies of information transfer have undergone hitherto unimaginable advances. Internet and interactive satellite communication provide exciting opportunities to transfer information and share experiences.

There are scientifically based planning procedures which have been tested, and which, when properly implemented, are exceedingly effective in predicting and projecting consequences of alternative development or management actions. Among these procedures are environmental impact assessment (EIA) (Wathern 1988) and various forms of risk assessment (RA) (Suter 1993) (Figure 2; Hakonson 1993). Present limitations upon the efficacy of these procedures lie as much in the transfer of results to decision makers as in the quantity or quality of information being transferred. Policy reform too often exceeds institutional capabilities to implement new policies. One solution to these constraints is training of information donors and receivers. Personnel must receive professional training in order to assimilate new procedures and skills into new strategies and management activities (McCormick 1994). Otherwise they will, quite naturally, ignore or circumvent new initiatives. In many instances institutional reform may be equally as important as information transfer. Globally, but especially in developing countries, well-informed scientists and managers express frustration regarding institutional barriers to implementation of reforms. Common barriers include outdated procedures, traditions, lack of intra- and interagency cooperation, political interference, corruption, inadequate planning, lack of accountability, and an untrusting public.

The "absolute priority" (World Bank 1993) that principles of sustainability guide economic development and resource management was explained most succinctly by the late President of Rwanda, "If in the struggle for economic development we jeopardize the sustained productivity of our natural resources, the struggle will have been in vain" (McCormick 1983).

ECOSYSTEM MANAGEMENT

Like sustainability, ecosystem management also reflects the holistic philosophy of ecological science (Slocombe 1993). Of all forms of stewardship, ecosystem management embodies strategies and methods best suited to achieving sustainability. Ecosystem management is an attempt to manage entire ecological systems rather than individual and fragmented components such as timber, fish, or wildlife species. Automobile service departments understand the importance of providing whole vehicle maintenance as opposed to servicing selected parts. Physicians understand the value of holistic medicine as opposed to treating spatially or temporally isolated symptoms. Likewise, ecologists have long understood the value of managing entire ecosystems rather than scattered and

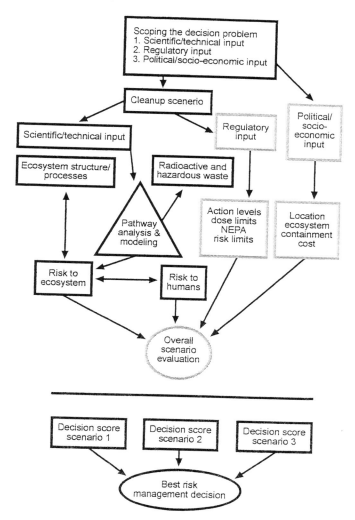

FIGURE 2   A decision process to manage risks. (Courtesy of T. Hakonson, Colorado State University, Fort Collins.)

diverse components. The rationale for holistic ecosystem management is nowhere better explained than in John Wolfe's foreword to *A Tropical Rain Forest* (Wolfe 1970):

> No environmental factor in an ecological system operates in isolation. And because the environment is holocoenotic and undergoing changes minute to minute, hour to hour, day to day, season to season, and year to year and because a change in one factor affects the rates of others and the rates of all life processes, ecological systems are not to be understood by the study of each factor, process or condition separately.

Boundaries, sizes, and shapes of ecosystems are determined by natural and anthropogenic discontinuities in our physical-chemical environment. Species populations are the fundamental biological components of ecosystems. Within permeable ecosystem boundaries, ecosystem structure and function are characterized by spatial and temporal interactions among these species. Ecology is the scientific analysis of environmental relationships of plants, animals, microbes, and man. This includes complex interactions with their physical-chemical environment as well as with one another. One advantage of holistic systems management is that it may focus upon dynamic processes rather than upon relatively static structural components. Flows of energy and cycling of elements provide more powerful and useful insights into limits to productivity (Lugo and McCormick 1981) and sustainability than do instantaneous measurements of standing crop of timber, fish, wildlife, or people. Continuing advances in molecular biology enable us to monitor biological activity closer and closer to deterministic loci. This capability provides us with mechanistic explanations of biological processes. Likewise, advances in ecosystem science (Kormondy and McCormick 1981) enable us to monitor dynamic processes which provide mechanistic explanations of productivity and sustainability (Risser 1985). Enhanced understanding of causality gives resource managers greater ability to predict positive and negative impacts of alternative management plans. The scientific basis for ecosystem management is well documented (Ecological Society of America 1995). Ecosystem science has revealed the dynamic and complex nature of ecosystems and the nonlinearity of interdependencies among components. Ecosystems constantly change, our understanding of them constantly changes, and management goals are always subject to change. Consequently, ecosystem management must be adaptive (Holling 1978). Following chapters provide case studies of spatial and temporal interactions between species populations and their environments. These case studies reflect our understanding of the dynamic processes which should guide ecosystem management.

Holistic management of ecological systems acknowledges man, and habitats of humanity, as being integral components of ecological systems. The following quote from Aldo Leopold reflects the significant difference between a "man in nature" vs. a "man and nature" philosophy. This quote also serves as a guiding principle for the Chattooga River ecosystem management project (Cawrse 1994). "When we see the land as a community to which we belong, we may begin to use it with love and respect." Ecosystem management provides a framework for assimilating human-social values into management policies and strategies traditionally dominated by economic values, and to a lesser extent by environmental values. **To achieve sustainability we must have resource management policies and implementation strategies which promote economic, environmental and social equitability across generations. Holistic ecosystem management meets these criteria.** In terms of fundamental principles, sustainability and ecosystem management involve planning and management across greatly expanded dimensions of the human environment. The dimensions of environment are space and time. Ecosystem management requires management of entire ecosystems at a geographic scale of watersheds and river basins. Ecosystem management often requires integrated management of public and private lands. This is especially challenging in the context of emotionally sensitive issues such as range reform, reduced timber harvests from federal lands, or enforcement of environmental regulations on private lands. These are significant expansions of traditional spatial dimensions of environment. Sustainability requires extending planning across an intergenerational time scale. Intergenerational planning is not so great a challenge to ecologists as it is to economists. Few, if any, concepts or methods of economic assessment have been applied to intergenerational timescales. On the other hand, ecological assessments have traditionally addressed landscape changes requiring centuries. We have well-tested procedures such as EIA (Wathern 1988) and RA (Suter 1993) and analytical methods such as Sorenson networks (Sorenson 1971), Leopold matrices (Leopold et al. 1971) and integrated value functions (U.S. Fish and Wildlife Service 1980) to assist us in projecting or predicting consequences of alternative action plans across intergenerational time scales.

Few ecosystems in the world are unaffected by impacts of human activities. Most anthropogenic impacts have their origins beyond the boundaries of affected ecosystems. Relatively simple ecosystem input–output models (van Dyne 1981) provide insights into interdependencies of on-site ecosystems, ecological buffer zones, and regional or global sources of stressors. Holistic ecosystem management is essential when we cannot escape integrating diverse economic, environmental, and social values across geographic boundaries or traditional boundaries between private and public lands. Ecosystem management provides a framework within which we can derive the optimal balance of multiple values, as perceived by multiple and diverse resource users. A growing problem confronting natural resource managers is that they must manage multiple interdependent resources, each of which has multiple uses and multiple users with multiple value systems. Under such complex demands any management strategy that is less than holistic is doomed to failure. Optimized management for one species or habitat within an ecosystem will inevitably be damaging to the sustainability of others. An ecosystem perspective requires that we strive for the greatest good accompanied by the least harm. You may recognize this as the goal of a planning process called environmental impact assessment. The Tennessee Valley Authority (TVA) recently implemented a Robust Multiple Objective Linear Programming Model (Wu 1991) and user interface to assist them in achieving the optimal solution when confronted with multiple objective decision making. Management goals are derived from agency policies and from stakeholder suggestions. Weights (prioritization) are determined on the basis of scientific information and stakeholder attitudes. Solutions are expressed as the percentage of the landscape which should be represented by each type of habitat. When linked to geographic information systems (GIS), solutions identify specific sites best suited for management as a specific habitat type and which type of habitats should be juxtaposed to one another. Solutions justify management prescriptions and simultaneously provide information required in associated environmental impact statements. Attainment of a defendable optimal solution is not only a prerequisite to sustainability, it is also prudent policy for government stewards of public lands at a time when their stewardship is subject to severe criticism and lack of public trust. Ecosystem management is no panacea, but it offers broader opportunities than more restricted and myoptic species management.

Ecosystem management is inherently cost effective because it is, in large part, passive management (INTECOL 1984). Once fundamental structural and functional relationships within ecosystems are understood, the major responsibility of management is to passively monitor limits of acceptable change (Stankey et al. 1985) in ecosystem processes and components. The guiding principle is, if it works don't spend time and money trying to fix it. When limits of acceptable change are exceeded, the corrective action most often required is regulation and restoration of human interventions.

Another contribution of ecosystem management to the achievement of sustainability is that it assists us in overcoming the arrogance of our own ignorance. In most instances, if ecosystem management provides high quality unfragmented habitat, species populations can do quite well on their own, without further management interventions. Too often, in efforts to manage species populations, we arrogantly assume that we know better than they the conditions under which they best live, grow, reproduce, die, and decay. Our less presumptive responsibility should be limited to providing species of concern with opportunities to do their own thing. We can best do this through ecosystem management and habitat protection.

Finally, it is important to remember that ecosystem management is not a new concept, it is merely newly advertised. It is a humbling experience to read classics such as *The Earth as Modified by Human Action* (Marsh 1874) and realize that much of contemporary ecological theory and many of the current buzz words are merely new jargon for time tested wisdom. Biodiversity, conservation biology, landscape ecology, ecosystem management and sustainability are hardly new ideas. If you doubt this contention, *The Background of Ecology* (McIntosh 1985) should convince you otherwise. What *is* new is the vast array of planning procedures and technologies available to implement good stewardship. What is also new is increased pressure from an informed public to insure that we implement good stewardship. The President's Council on Sustainable Development and the White House Federal Ecosystem Management Task Force are more a product of public pressure than they are the product of enlightened administrators.

The goal of sustainability, and ecosystem management as a strategy for achieving this goal, are totally defendable in terms of good science, good ethics, and good economics. It would appear that, to do otherwise, is quite indefensible.

EXAMPLES OF THE GROWING ACCEPTANCE OF SUSTAINABILITY AND ECOSYSTEM MANAGEMENT

Acceptance of sustainability and ecosystem management paradigms has been a slow but continuous process. As described earlier, initial concern for socioeconomic equitability was later extended to race, gender, and finally to intergenerational equitability. The complex and subversive nature of these paradigms undoubtedly causes some readers to think, "get real"; this isn't going to happen. The following review of international, national, and local acceptance presents surprising evidence to the contrary. Perhaps most surprising are success stories from developing countries where conflicts between exponential population growth and diminishing natural resources would seemingly make implementation of the paradigms most difficult. In reality, those conflicts may be less of a barrier to achieving sustainability than irresponsible and excessive resource consumption by developed nations. Clearly, most irresponsible are those individuals and institutions that argue against family planning and birth control in the context of human suffering, and resource depletion. The consequential apocalypse of poverty, suffering and death presents a formidable barrier to achieving global sustainability.

INTERNATIONAL ACCEPTANCE

Two decades of steadily growing acceptance of sustainable development among international development agencies is evidenced by significant milestones within the past 3 years. In December 1993, the President of the World Bank (WB), Lewis T. Preston (World Bank 1993), urged policy makers to focus on the "absolute priority" of implementing sustainable development. Appointment of Ismael Serageldin as WB Vice President for Sustainable Development is evidence of the sincerity of this plea. Similarly, in January 1994, the U.S. Agency for International Development (USAID) implemented its Strategies for Sustainable Development. In recognition of post-Cold War realities, USAID's policy of sustainable development focuses on four challenges: (1) Protecting the environment, (2) building democracy, (3) stabilizing population growth and protecting human health, and (4) encouraging broad-based economic growth.

The International Biological Program (IBP) and Man and Biosphere Program (MAB), including projects at International Biosphere Reserves, have provided over a quarter century of international research and management experience necessary to support recent initiatives in sustainable development and ecosystem management.

The International Association for Ecology (INTECOL 1984) provided useful and early insights into ecosystem management in its report on "Managing Parks and Reserves as Ecosystems." This report recommended that throughout the world: "parks be managed as ecosystems"; that management be based upon research which provides understanding of the natural dynamics — and the limits to recovery"; and that when changes in the park ecosystem are within limits of acceptable change, "management should be passive." In the international arena, sustainable development and ecosystem management have been underway long enough to provide several success stories.

Success stories include USAID's sustainable agricultural development and ecotourism projects in Rwanda (McCormick 1983); development of sustainable tropical forestry in the Pal Cazu valley of Peru (Hartshorn et al. 1987); and sustainable ecotourism and ecosystem management of Amboseli National Park in Kenya. In each of these projects, inclusion of local peoples, their values, and needs, contributed significantly to project success. In each of these projects, environmental assessments identified alternatives which were, or were not, sustainable. In each of these projects, research, assessment, and planning focused upon entire ecosystems, including habitats of humanity.

NATIONAL ACCEPTANCE

For decades, national parks and large reserves such as those secured by the Nature Conservancy have been subject to ecosystem management in order to achieve sustainability. Large tracts of land with reasonable security and mandates to preserve and protect are especially conducive to ecosystem management. In 1992 the U.S. Forest Service (USFS) underwent policy reform in response to public pressure and adopted a policy of ecosystem management which states: "An ecological approach will be used to achieve the multiple use management of the National Forests and Grasslands. It means we must blend the needs of people and environmental values in such a way that the National Forests and Grasslands represent diverse, healthy, productive and sustainable ecosystems" (Crow 1994). Other federal agencies have followed the example of the USFS, the most recent example being the DOE. December 19, 1994, the Secretary of Energy released the following policy statement:

> It is Department of Energy policy to manage all of its land and facilities as valuable national resources. Our stewardship will be based on the principles of ecosystem management and sustainable development. We will integrate mission, ecologic, economic, cultural and social factors in a comprehensive plan for each site that will guide land and facility use decisions. Each comprehensive plan will consider the site's larger regional context and be developed with stakeholder participation. This policy will result in land and facility uses which support the Department's critical missions, stimulate the economy, and protect the environment.

Implementation of sustainability and ecosystem management by DOE is especially significant. For the past half-century DOE has been the steward of over 2,400,000 acres of land, much of which served as buffer zones behind security fences patrolled by security guards at 30 laboratories throughout North America (DOE 1994a). As a consequence, these little used and passively managed buffer zones contain high quality unfragmented examples of many of the ecosystems of North America. These protected ecosystems lie in the midst of an unprotected landscape which has been degraded and fragmented throughout the past half-century (Mann et al. 1995). It is not coincidental that scientists at the Savannah River Plant and Oak Ridge Reservation were among the leaders in developing ecosystem science and management. Both sites have been designated National Environmental Research Parks (DOE 1994b) for the high quality of indigenous ecosystems and ecosystem research. Based upon the economic principle of rarity, these well-protected buffer zones may be the most valuable lands under DOE stewardship, or for that matter, under federal stewardship anywhere in North America. For example, Upper Three Runs Creek at the Savannah River Plant has been described as, "a time capsule — a preserved specimen of a southern blackwater stream," perhaps having the highest biodiversity of any ecosystem on the planet (Seabrook 1994). At ORR, in the midst of surrounding urban development, 70% of the land is now forested. This includes ecosystem types of sufficient value to be designated State Natural Areas. ORR serves as a refugium of species and a source of species recruitment into lands being restored within the region (Mann et al. 1995). DOE buffer zones are similar to retirement funds. They have been set aside and allowed to grow without withdrawals of capital. In either case it would be foolhardy to cash them in to meet short-term, near-sighted needs and forfeit sustained high quality productivity.

REGIONAL AND LOCAL ACCEPTANCE

The Southern Appalachian Man and the Biosphere (SAMAB) program was recently cited as, "a national model in ecosystem management" (SAMAB NEWS 1994). SAMAB success in implementing ecosystem management is due in part to its role as a coordinator of overlapping responsibilities and concerns of 11 federal agencies, 3 state governments, 8 cooperating universities, and 14 affiliate companies and NGOs. Success in pursuing sustainable management of natural resources is due in part to its mission statement, which is: "to foster harmonious relationships between humans and their environment through programs and projects that *integrate the social, physical, and biological sciences* to address actual problems." Three recent success stories contribute to citation of SAMAB as a national model (Hinote 1994). Throughout the world, degraded air quality has become a transboundary issue. Air pollution may originate off site and its impacts may transcend traditional political, management, or geographic boundaries. While impacts may be significant and produce widespread public concern, responsibilities and opportunities confronting individual federal or state agencies are often ill defined and constrain responsiveness of those agencies. In March 1992, SAMAB brought together over 125 representatives of federal and state agencies responsible for monitoring air quality in the Great Smoky Mountains National Park (GSMNP). As a direct result of the SAMAB conference, these agencies formed a partnership (Southern Appalachian Mountain Initiative) to coordinate air monitoring programs and projects. This initiative has received over $600,000 in support and has resulted in coordination of four monitoring stations, dissemination of air quality information to the public, and initiation of an air quality management plan for southern Appalachia. A second example of ecosystem management is the role played by SAMAB in addressing limits of acceptable change in the GSMNP. In the absence of natural predators, the distribution and abundance of several indigenous and exotic herbivore species have exceeded limits of acceptable change. SAMAB played a key role in developing public receptivity to the controversial reintroduction of a traditional predator, the red wolf. The third example is an effort to integrate economic, environmental, and social values into a community development plan. The community of Pittman Center is a gateway to the GSMNP. Residents hope to avoid the aesthetic blight and cultural degradation associated with other gateways. The fierce independence characteristic of southern Appalachian peoples provided a potential constraint to sustainable development planning. SAMAB, in cooperation with the University of Tennessee and the Economic Development Administration, assisted Pittman Center in producing a strategic development plan consistent with concepts of sustainability and methods of ecosystem management.

There is growing acceptance of sustainability and ecosystem management among participating agencies within SAMAB (SAMAB 1994). The GSNMP was the first U.S. National Park to adopt ecosystem management as a management strategy. As cited previously, ORR has been a leader in developing concepts and methods of ecosystem science. The USFS Coweeta Hydrologic Laboratory is world renowned for research on applications of ecosystem research to ecosystem management, especially as pertains to sustained productivity of water and related resources (Hornbeck and Swank 1992). The USFS Chattooga River Project (Cawrse 1994) is a model of ecosystem management being followed closely by resource managers throughout the country. This project, which includes over 120,000 acres of USFS land extending across three states, has as its goals to: work with the public to reach a shared vision of resource management; to integrate resource management across state boundaries; and to meet public demands for forest uses within the context of *sustaining* diverse, healthy, and beautiful ecosystems.

Clearly, there is widespread and growing acceptance of sustainable development and ecosystem management by international, national and regional peoples and institutions.

PRINCIPLES OF SUSTAINABILITY AND ECOSYSTEM MANAGEMENT

A set of principles should be available to policy makers and resource managers as they attempt to design and implement ecosystem management. In reviewing the scattered and diverse literature on sustainability and ecosystem management, ten sets of guiding principles were identified (Tables 1 through 10). Table 1 presents principles for achieving sustainability. Tables 2 through 10 provide principles of ecosystem management. These tables are presented as lists to assist the reader in developing her/his own set of principles and to identify commonality, contradictions, or omissions.

Fifteen principles for achieving "Our Vision of a Sustainable United States" (Table 1) are expressed in terms of policy reforms necessary to achieve sustainability. One third of these principles focus on scientific issues, such as environmental quality, carrying capacity, stability, conservation, and technology. Another one third of the principles address socioeconomic issues, such as market strategies, poverty, equitability, and public participation. The final one third specifically address issues of interdependence and integration of environmental, social and economic value systems. Clearly, the two most fundamental principles of sustainability are: (1) integration of economic, environmental and social value systems and (2) intergenerational equitability. One principle transcends the other fourteen in that it is blatantly subversive. It calls for fundamental changes in the conduct of government, institutions, and peoples in order to achieve the vision of sustainability. This is not surprising, since ecology, the science which underlies principles of sustainability, has long been acknowledged as a subversive science (White 1967).

Principles of ecosystem management fall into two categories: those which address scientific issues and those which address policy, planning, and management. Ecosystem management should be guided by ecological principles of: biodiversity, ecosystem complexity, scales of analysis, hierarchies of components, ecosystem boundaries, baselines and monitoring, stability and resiliency, carrying capacity, and most importantly, holistic analysis. Planning and management should be guided by principles of: interdisciplinary and intergenerational planning, integration of economic/environmental/social value systems, information transfer, public participation, human ecosystems, equitability, adaptive management, and recognition of constraints and limitations. While none of these principles are mutually exclusive, the issue of uncertainty is a significant concern to both the scientific and planning aspects of ecosystem management.

It is worth noting that eight of the nine sets of ecosystem management principles are concordant. The "Seven Pillars of Ecosystem Management" (Table 5) is discordant, in that all principles are expressed in terms of social benefit to only one species. The other eight sets of principles are less anthropogenic. The preceeding narrative and literature citations provide opportunities to expands ones' understanding of these fundamental principles.

Continuing debate and refinement of principles is an essential part of the process leading to increased acceptance of sustainability and ecosystem management. Ultimately, the acceptance and implementation of these paradigms is critical to the survival of the 5 million or more species sharing this planet. Acceptance will require a global common vision and commitment. Obviously, we have not yet reached that point, but the debate is well advanced, implementation has begun, and acceptance is growing. As you read the following chapters, especially those in Section IV, it may be instructive to assess the degree to which those programs of ecosystem management reflect fundamental principles of sustainability and ecosystem management.

TABLE 1
Principles for "Our Vision of a Sustainable United States"

Our Vision of a Sustainable United States:

Our vision is of a life-sustaining earth. We are committed to the achievement of a dignified, peaceful, and equitable existence. We believe a sustainable United States will have an economy that equitably provides opportunity for satisfying livelihoods and a safe, healthy, high quality life for current future generations. Our nation will protect its environment, its natural resource base, and the functions and viability of natural systems on which all life depends.

Principles: (Fifteen)

- We must preserve and, where possible, restore the integrity of natural systems — soils, water, air, and biological diversity — which sustain both economic prosperity and life itself
- Economic growth, environmental protection, and social equity should be interdependent, mutually reinforcing national goals, and policies to achieve these should be integrated
- Along with appropriate protective measures, market strategies should be used to harness private energies and capital to protect and improve the environment
- Population must be stabilized at a level consistent with the capacity of the earth to support its inhabitants
- Protection of natural systems requires changed patterns of consumption consistent with a steady improvement in the efficiency with which society uses natural resources

TABLE 1 (continued)
Principles for "Our Vision of a Sustainable United States"

- Progress toward the elimination of poverty is essential for economic progress, equity, and environmental quality
- All segments of society should equitably share environmental benefits and burdens
- All economic and environmental decision-making should consider the well-being of future generations and preserve for them the widest possible range of choices
- Where public health may be adversely affected, or environmental damage may be serious or irreversible, prudent action is required, even in the face of scientific uncertainty
- Sustainable development requires fundamental changes in the conduct of government, private institutions, and individuals
- Environmental and economic concerns are central to our national and global security
- Sustainable development is best attained in a society in which free institutions flourish
- Decisions affecting sustainable development should be open and permit informed participation by affected and interested parties; that requires a knowledgeable public, a free flow of information, and fair and equitable opportunities for review and redress
- Advances in science and technology are beneficial, increasing both our understanding and range of choices about how humanity and the environment relate. We must seek constant improvements in both science and technology in order to achieve eco-efficiency, protect and restore natural systems, and change consumption patterns
- Sustainability in the U.S. is closely tied to global sustainability. Our policies for trade, economic development, aid, and environmental protection must be considered in the context of the international implications of these policies

From President's Council on Sustainable Development, 1994.

TABLE 2
An Overview of Ecosystem Management Principle

Ecosystem management requires the maintenance of sustainable ecosystems, while providing for a wider array of uses, values, products, and services from the land to an increasingly diverse public (Overbay 1992). In many respects, ecosystem management represents a refocusing by the USDA Forest Service, on the "sustainable" part of the Multiple-Use, Sustained-Yield Act of 1960. Decisions that emphasize a view to the future, promote sustained production over the long run, and maintain all the pieces of ecosystems are characteristic of this emerging management philosophy (Risbrudt 1992).

Overbay (1992) proposes that the following **six principles** be used to describe the initial components of ecosystem management:

- Multiple-use, sustained-yield management of lands and resources depends on sustaining the diversity and productivity of ecosystems at many geographic scales
- The natural dynamics and complexity of ecosystems mean that conditions are not perfectly predictable and that any ecosystem offers many options for uses, values, products, and services, which can change over time
- Descriptions of desired conditions for ecosystems at various geographic scales should integrate ecological, economic, and social considerations into practical statements that can guide management activities
- Ecosystem connections at various scales and across ownerships make coordination of goals and plans for certain resources essential to success
- Ecological classifications, inventories, data management, and analysis tools should be integrated to support integrated management of lands and resources
- Monitoring and research should be integrated with management to continually improve the scientific basis of ecosystem management

From Jensen and Everett, 1994.

TABLE 3
Key Principles of Ecosystem Management

In the aggregate, the principles discussed below can contribute to achieving the new NPS ecosystem goal of preserving, protecting, and/or restoring ecosystem integrity (composition, structure, and function) and also maintaining sustainable societies and economies. The principles are either objectives of this goal or tactics to help achieve the goal and are intended to guide management policy and action. The **nine principles** (in no specific order) are multiple boundaries and scale; natural resources, biodiversity, and conservation biology; cultural resources and traditions; social, cultural, economic, and political factors; information management/scientific basis for decisions; partnerships; interdisciplinary approach to management; long-term ecosystem management focus; and adaptive and flexible management.

- **Multiple boundaries and scale** — Ecosystems do not have permanent or absolute boundaries. Rather, multiple factors considered in multiple scales with multiple boundaries are necessary for ecosystem management
- **Natural resources, biodiversity, and conservation biology** — It is imperative that the NPS work to restore and/or maintain biological diversity (species, genetic, and ecosystem) and the ecological patterns and processes that maintain that diversity
- **Natural resources and traditions** — This entails preserving and maintaining significant resources and advocating or assisting others to protect important archeological, historical, and ethnographic resources in their historic contest
- **Social, cultural, economic, and political factors** — NPS resources are not separate and removed from society. Rather, they are an integral part of society. Social, economic, and political reality must be understood by park management. Economic and social needs of surrounding communities may be supported without compromising NPS values. Political actions help determine NPS activities and the NPS should use its expertise to educate elected officials at all levels
- **Information management/scientific basis for decisions** — NPS management decisions should be grounded in the best scientific natural, cultural, economic, and social data available in order to gauge effectively the full impact of policy alternatives and to help choose the course of action that will best achieve ecosystem management goals
- **Partnerships** — Ecosystem management is best understood as shared responsibility, and the NPS should collaborate, communicate, cooperate, and coordinate with partners
- **Interdisciplinary approach to management** — Rather than separating employees by discipline, varied disciplines should work together in teams toward specific objectives
- **Long-term ecosystem management focus** — What: Managers of NPS resources common to an ecosystem should cooperatively develop a long-term ecosystem vision and specific management objectives in conjunction with partners
- **Adaptive and flexible management** — Ecosystem management can be best served by allowing innovative management approaches to be tailored to specific ecosystems

Derived from National Park Service, 1994.

TABLE 4
The Main Components of an Ecosystem Approach

- Describe parts, systems, environments and their interactions
- Are holistic, comprehensive, and transdisciplinary
- Include people and their activities in the ecosystem
- Describe system dynamics through concepts such as stability and feedback
- Define the ecosystem naturally, for example, bioregionally instead of arbitrarily
- Look at different levels and/or scales of system structure, process, and function
- Recognize goals and take an active, management orientation
- Incorporate stakeholder and institutional factors in the analysis
- Use an anticipatory, flexible research and planning process
- Entail an ethics of quality, well-being, and integrity
- Recognize systemic limits to action — defining and seeking sustainability

From Slocombe, D.S. 1993. *Bioscience,* 43, 9, 612–622.

TABLE 5
Seven Pillars (Principles) of Ecosystem Management

- **The first pillar of ecosystem management is Values and Priorities** — Ecosystem management reflects a stage in the continuing evolution of social values and priorities; it is neither a beginning nor and end
- **The second pillar of ecosystem management is Boundaries** — Ecosystem management is place-based, and the boundaries of the place of concern must be clearly and formally defined
- **The third pillar of ecosystem management is Health** — Ecosystem management should maintain ecosystems in the appropriate condition to achieve desired social benefits; the desired social benefits are defined by society, not scientists
- **The fourth pillar of ecosystem management is Stability** — Ecosystem management can take advantage of the ability of ecosystems to respond to a variety of stressors, natural and man-made, but there is a limit in the ability of all ecosystems to accommodate stressors and maintain a desired state
- **The fifth pillar of ecosystem management is Diversity** — Ecosystem management may or may not result in emphasis on biological diversity as a desired social benefit
- **The sixth pillar of ecosystem management is Sustainability** — The term *sustainability,* if used at all in ecosystem management, should be clearly defined — specifically, the time frame of concern, the benefits and costs of concern, and the relative priority of the benefits and costs
- **The seventh pillar of ecosystem management is Scientific Information** — Scientific information is important for effective ecosystem management, but is only one element in the decision-making process that is fundamentally one of public or private choice

Derived from Lackey, R.T. 1994.

TABLE 6
An Overview of Ecological Principles for Ecosystem Management

The process of formulating management guidelines for sustaining ecosystems
should be guided by **Eight central principles:**

- Management goals must be defined precisely
- Ecological hierarchies must be defined according to management goals
- Ecological patterns and diversity must be understood in terms of processes and constraints generating them, as well as in terms of their possible impact on other components of ecosystems
- The implications of management practices on patterns and processes must be understood at all scales of the hierarchies
- Management for sustainability of ecological patterns and diversity must include maintenance of all ecosystem attributes across their natural ranges of spatial–temporal scales
- Ecosystem management must be concerned with the sustainability of patterns and processes together rather than merely the maintenance of existing patterns
- The historical range of natural variability across a range of spatial-temporal scales must be defined if patterns and processes are to be maintained at all appropriate scales of organization (e.g., ecological and evolutionary). The role of natural variability should be recognized in the development of management plans
- Monitoring schemes must be designed that explicitly recognize the hierarchical nature of ecological systems. Monitoring multiple attributes at all appropriate ecological scales can provide a basis to assess ecosystem change

From Bourgeron, P.S. and Jensen, M.E., 1994.

TABLE 7
An Ecological Basis for Ecosystem Management

Guiding Principles (six)

- Humans are an integral part of today's ecosystems and depend on natural ecosystems for survival and welfare; ecosystems must be sustained for the long-term well-being of humans and other forms of life
- In ecosystems, the potential exists for all biotic and abiotic elements to be present with sufficient redundancy at appropriate spatial and temporal scales across the landscape
- Across adequately large areas, ecosystem processes (such as disturbance; succession; evolution; natural extinction; recolonization; fluxes of materials; and other stochastic, deterministic, and chaotic events) that characterize the variability found in natural ecosystems should be present and functioning
- Human intervention should not impact ecosystem sustainability by destroying or significantly degrading components that affect ecosystem capabilities
- The cumulative effects of human influences, including the production of commodities and services, should maintain resilient ecosystems capable of returning to the natural range of variability if left alone
- Management activities should conserve or restore natural ecosystem disturbance patterns

From Kaufmann, M.R. et al., 1994.

TABLE 8
Ecosystem Principles and Their Implications for Management

Basic Ecological Principles (five)

- **Define ecosystems** — Ecological systems are groups of interacting, interdependent parts (e.g., species, resources) linked to each other by exchange of energy, matter, and information. Ecological systems are considered complex because they are characterized by strong interactions between components; complex feedback loops; and significant time and space lags, discontinuities, *thresholds and limits*
- **Dynamic systems** — Ecological systems exhibit temporal changes along various developmental pathways that result in different types of organization
- **Hierarchial organization** — Complex ecosystem patterns, landscapes, and the multitude of processes that form them exist within a hierarchical framework. Hierarchy theory is concerned with multiscaled systems that can be viewed as systems of constraints in which a higher level of organization provides to some extent the environment in which lower levels evolve
- **System limits** — In the course of undergoing change, ecological systems may follow different pathways. Eventually, the processes that result in changes balance the processes that lead to ecological organization. Such a state is called an optimal operating point. An optimal operating point defines the limit of ecosystem development along one pathway
- **Public involvement** — Key principles (ten) in building a successful public involvement program are
  1. Clearly state goals and objectives
  2. Plan for effective dialogue
  3. Build cooperation constructively and deliberately
  4. Emphasize education and social learning
  5. Diversify the public involvement framework
  6. Be open, honest, and responsive
  7. Understand the multiplicity of publics and techniques
  8. Use a third party intervener when necessary
  9. Establish a framework for continuing dialogue
  10. Analyze, evaluate and monitor

Derived from Interagency Eco-Regional Assessment Team, 1994.

TABLE 9
Guiding Principles (five) for the U.S. Forest Service Land Management Planning Process

- People are a part of ecosystems; meeting people's needs and desires within the capacities of natural systems is a primary role of resource decision making. In truth, planning in this context is more about people than anything else
- Involving the public in National Forest System planning and decision making on an ongoing, open, and equitable basis is essential
- The scientific community should play a vital role in gathering and analyzing information for resource decision making
- The forest planning process should provide for efficient adjustment of forest plans in response to changing conditions and new information
- Ecosystems cross land ownerships, jurisdictions, and administrative boundaries; therefore, planning efforts should be coordinated with other landowners, Federal agencies, and State, local, and tribal governments in a manner that respects private property rights and the jurisdictions of other government entities

Derived from Thomas, J.W., 1995.

TABLE 10
Ecosystem Principles and Their Implications for Management

The implementation of ecosystem management requires consideration of at least **four principles:**

- **Ecosystems are dynamic and evolutionary** — Natural resource managers should be able to predict the consequences of management activities and consider their influence on ecosystem development, including possible change in developmental pathways. Measurement variables and methods should be selected to evaluate changes in ecosystem structures and functions
- **It is useful to view ecosystems as being organized within a hierarchy, with each level having a variety of time and space scales** — A critical characteristic of a hierarchical system is that every level is a discrete functional entity and at the same time is part of a larger whole
- **Ecosystems have biophysical and social limits** — In all ecosystems, there are limits to the rate and amount of accumulation of biomass (plant, animal, and human). The actual limit, or optimum operating condition, is only one of several possible, because of multiple developmental pathways. This point is only temporary because the environment is constantly in flux and ecosystems change accordingly. Furthermore, these limits determine the capability of the system to provide goods and services. Decisions in a ecosystem management context should consider intergenerational equity and tradeoffs. Investments in ecosystem restoration made by this generation would provide benefits and options for future generations
- **There are limits to the predictability of ecosystem patterns and processes; conditions and events may be predictable at some scales but not at others**
  - Adaptive management strategies improve our ability to predict
  - Implementation of ecosystem principles is guided by six approaches and a three-step process. Although identified as "approaches" and "steps" they provide valuable insight into "principles" of implementation
  - The main features of using an ecosystem management approach include:
    Clear description of components, ecosystems, environments, and interactions
    A holistic, comprehensive, interdisciplinary, and integrated process
    A system that includes people, their values, and activities
    A clear description and understanding of ecosystem dynamics that considers system patterns, processes, structures, and functions
    Consideration of different scales (temporal, spatial, social organizational) of system structures and functions
    Planning and management unit delineation considering ecological (biophysical and social) boundaries
- **There are three primary steps to implementing ecosystem management:** (1) management unit delineation, (2) understanding ecosystem functions, and (3) developing and implementing a management plan. The first two steps point to the need to (Seven Principles):
  - Determine the kind of information needed to define management units
  - Explore the implications for planning and management using different data and methods to define ecosystems and management units
  - Design an integrated multidisciplinary data collection scheme, including monitoring of past, present, and future ecosystem states, behaviors, and functions
  - Explore methods to organize, display, and illustrate interrelations of data collected
  - Design methods of integrated, multidisciplinary synthesis and interpretation of data
  - Understand how processes at different scales interact
  - Describe linkages between biological and social system functions and activities

TABLE 10 (continued)
Ecosystem Principles and Their Implications for Management

The third step requires developing a socially acceptable system of organizations for administering ecosystem management units. Developing the administrative system requires knowledge and learning in several areas that (Six Principles):

- Reveal the kind and quantity of human demands for ecosystem products
- Reveal the human values that prompt the expectation for ecosystem products
- Design efficient methods of acquiring information about human values and expectations
- Design methods to resolve conflicts arising from differences in expectations for ecosystem products
- Design means to inform people of the consequences of alternate ecosystem product choices
- Allow organizations the adaptability to develop the appropriate mix of skills, performance incentives, and organizational flexibility to implement ecosystem management

Derived from U.S. Forest Service, 1994.

REFERENCES

Ahmad, Y.J., Serafy, S.E., and Lutz, E. 1989. Environmental Accounting for Sustainable Development.The World Bank. Washington, D.C.

ASSET. 1994. U.S. Agency for International Development Training Division and the Institute for International Research. Arlington, VA.

Auerbach, 1971. Contributions of Radioecology to AEC Mission Programs. In Proc. Third National Symp. Radioecology., Ed. D.J. Nelson. Conf. 710501 — P1. National Technical Information Service. U.S. Dept. of Commerce, Springfield, VA. pp. 3-8.

Bourgeron, P.S. and Jensen, M.E. 1994. An overview of ecological principles for ecosystem management. In Ecosystem Management: Principles and Applications. USDA. U.S. Forest Service. Pacific Northwest Research Station PNW-GTR-36. pp 45–57.

Cawrse, D. 1994. Chattooga River Project. Ecosystem Management. Developing Ecosystem Management. Issue 1, Summer 1994. Chattooga Project. Clemson University. Clemson, SC.

Council on Environmental Quality. 1987. Regulations for Implementing the Procedural Provisions of the National Environmental Policy Act. 40 *CFR*, parts 1500–1508.

Crow, T.R. 1994. Ecosystem management. *Bull. Ecol. Soc. Am.* 75 (1), 33–35.

D.O.E. 1994a. Department of Energy. Stewards of a National Resource. DOE/FM-0002.

D.O.E. 1994b. Department of Energy. National Environmental Research Parks. DOE/ER-0615 p.

EESI, 1991. Environmental and Energy Study Institute. Partnership for Sustainable Development: A New U.S. Agenda for International Development and Environmental Security. Washington, D.C.

Ecological Society of America, 1995. The scientific bases for ecosystem management. Ad hoc Committee on Ecosystem Management. Washington, D.C. 63 pp.

Gagnon, C., Hirsch, P., and Howitt, R. 1993. Can SIA Empower Communities? Environ. Impact Assess. Rev. 13 (4), 229–253.

Gause, G.F. 1934. *The Struggle for Existence*. Baltimore: Williams & Wilkins.

Goodland, R. 1993. Definition of environmental sustainability. *Int. Assoc. Impact Assess. Newsl.* 5 (2), 2.

Hakonson, T. 1993. Center for Ecological Risk Assessment and Management. College of Natural Resources, Colorado State University, Fort Collins.

Hartshorn, G., Simeone, R., and Tosi, J., Jr. 1987. Manejo para rendimiento sostenido de bosques naturales: Un sinopsis del proyecto de desarrollo del Palcazu en la Selva Central de la Amazonia. Peruana. In *Management of the Forestry of Tropical Americas: Prospects and Technologies.* Proceedings of a Conference, Sept. 22 to 27, 1986. Institute of Tropical Forestry, Rio Piedras, Puerto Rico.

Hinote H. 1994. Testimony on the Southern Appalachian Man and Biosphere Program for The United States Senate Subcommittee on Agricultural Research, Conservation, Forestry, and General Legislation. April 14, 1994. SAMAB. Great Smoky Mountains National Park. Gatlinburg, TN.

Holling, C.S. 1978. *Adaptive Environmental Assessment and Management.* New York: John Wiley & Sons.

Hornbeck, J. and Swank, W.T. 1992. Watershed ecosystem analysis as a basis for multiple-use management of eastern forests. *Ecol. Appl.,* 2 (3), 238–247.

INTECOL, 1984. Managing Parks and Reserves as Ecosystems. A report from a workshop organized by The International Association for Ecology for the U.S. National Park Service. June 1984. Calloway Gardens, GA.

Interagency Eco-Regional Assessment Team, 1994. A national framework for integrated ecological assessment and USDA. U.S. Forest Service. 64 pp.

Jensen, M.E. and Everett, R. 1994. An overview of ecosystem management principles. In: Ecosystem Management: Principles and Applications. USDA. U.S. Forest Service. Pacific Northwest Research Station. PNW-GTR-318. pp. 6–15.

Kaufmann, M.R., Graham, R.T., Boyce Jr., D.A., Moir, W.H., Perry, L., Reynolds, R.T., Bassett, R.L., Mehlhop, P., Edminster, C.B., Block, W.M., and Corn, P.S. 1994. An ecological basis for ecosystem management. USDA. U.S. Forest Service. Rocky Mountain Forest and Range Experiment Station GTR-RM-24. 22 pp.

Kormondy, E.J. and McCormick, J.F. 1981. Rise of Modern Ecology. In *Handbook of Contemporary Developments in World Ecology.*, Eds. E.J. Kormondy and J.F. McCormick. Introduction. Greenwood Press, CT. p. 776.

Lackey, R.T. 1994. Seven pillars of ecosystem management. Modification of a presentation at the Symposium on Ecosystem Health and Medicine: Integrating Science, Policy, and Management, Ottawa, Ontario, Canada, June 19–23, 1994.

Leopold, L.B., Clarke, F.B., Hanshaw, B.B., and Balsley, J.R. 1971. A procedure for evaluating environmental impact. Geological Survey Circular 645. U.S. Geological Survey. Denver, CO. p. 13.

Lubchenco, J., Olson, A.M., Brubaker, L.B., Carpenter, S.R., Holland, M.M., Hubbell, S.P., Levin, S.A., MacMakon, J.A., Matson, P.A., Melillo, J.M., Mooney, H.A., Peterson, C.H., Pulliam, H.R., Real, L.A., Regal, P.J., and Risser, P.G. 1991. Sustainable Biosphere Initiative. *Ecology* 72 (2), 371–412.

Lugo, A.E. and McCormick, J.F. 1981. Influence of environmental stressors upon energy flow in a natural terrestrial ecosystem. In *Stress Effects on Natural Ecosystems.*, Eds. G.W. Barrett and R. Rosenberg. New York: Wiley & Sons. pp. 79–102.

Mann, L., Parr, P., Pounds, L., and Graham, R. 1995. Protection of biota on nonpark public lands. Examples from the Department of Energy Oak Ridge Reservation. Draft. Environmental Sciences Division. ORR. p. 23.

Marsh, G.P. 1874. *The Earth as Modified by Human Action.* New York: Charles Scribners and Sons. p. 629.

McCormick, J.F. 1983. A cooperative regional demonstration project jointly sponsored by the Government of Rwanda, the U.S. Agency for International Development, the University of Tennessee, and the Southeast Consortium for International Development. Ecology Department, University of Tennessee, Knoxville. p. 220.

McCormick, J.F. 1991. Ecological prerequisites for sustainable economic development and conservation of natural resources. In *J. Plant Resources Environ.* 1 (2), 43–49. (Peoples Republic of China.)

McCormick, J.F. 1994. Institute for International Training in Sustainable Development. Info. Doc. E01-1035-001-95. University of Tennessee, Knoxville.

McIntosh, R.P. 1985. *The Background of Ecology: Concept and Theory.* New York: Cambridge University Press. p. 383.

National Park Service, U.S. Dept. of the Interior, 1994. Key Principles of Ecosystem Management. In Ecosystem Management in the National Park Service. pp. 11–22. Washington, D.C.

Overbay, J.C. 1992. Ecosystem management. In Proc. National workshop: Taking an Ecological Approach to Management. April 27 to 30; Salt Lake City, UT. WO-WSA-3: Washington, D.C., U.S. Department of Agriculture Forest Service, Watershed and Air Management. pp. 3-15.

President's Council on Sustainable Development, 1994. Draft Vision and Principles of Sustainable Development (June 9, 1994). Washington, D.C.

Risbrudt, C. 1992. Sustaining ecological systems in the Pacific Northwest Research Station. In: Proc. of the National Workshop: Taking an Ecological Approach to Management. April 27 to 30. Salt Lake City, UT. WO-WSA-3. Washington, D.C., U.S. Department of Agriculture Forest Service, Watershed and Air Management: pp. 3–15.

Risser, P.G. 1985. Toward a holistic management perspective. *Bioscience* 35 (7), 414–418.

SAMAB. 1994. Southern Appalachian Man and the Biosphere Action Plan, October 1994.

SAMAB NEWS. 1994. Conference Keynote Calls SAMAB a Model. SAMAB NEWS December, 1994.

Seabrook, C. 1994. Where pollution meets pristine. Saturday Reader. *Atlanta Journal, Atlanta Constitution.* Dec. 17, 1994.

Slocombe, D.S. 1993. Implementing ecosystem based management. *Bioscience* 43 (9), 612–622.

Sorenson, J.C. 1971. *A framework for Identification and Control of Resource Degradation and Conflict in the Multiple Use Coastal Zone.* Berkeley: University of California, Department of Landscape Architecture.

Stankey, G.H., Cole, D.N., Lucas, R.C., Peterson, M.E., and Frissell, S.S. 1985. The Limits of Acceptable Change (LAC) System for Wilderness Planning. USDA, U.S. Forest Service Intermountain Forest and Range Experiment Station. Ogden, UT. Gen. Tech. Rep. INT-176. p. 39.

Suter, G.W., II. 1993. *Ecological Risk Assessment.* Ann Arbor: Lewis. p. 538.

Taylor, C.N., Bryan, C.H., and Goodrich, C.G. 1990. *Social Assessment — Theory, Process and Techniques.* Center for Resource Management. Lincoln University, Canterbury, New Zealand. p. 232.

Thomas, J.W. 1995. Forest Service Land Management Planning Process. Presentation of Chief, U.S. Forest Services to U.S. Senate Subcommittee on Forests and Public Land Management, Committee on Energy and Natural Resources. April 5, 1995. 2 pp.

USAID 1989. United States Agency for International Development. Environmental Procedures Handbook III. Annex 1. 22 *CFR,* Part 216.

U.S. Forest Service. 1994. Scientific framework for ecosystem management in the interior Columbia River basin. Version 2. pp. 80. Eastside Ecosystems Management Project, Walla Walla, WA.

U.S. Fish and Wildlife Service 1980. *Habitat Evaluation Procedures* (HEP). 102 ESM. U.S. Dept. of Interior Fish and Wildlife Service, Division of Ecological Services N.P. Washington, D.C.

van Dyne, G.M. 1981. Response of a short grass prairie to man induced stresses as determined from modeling experiments. In *Stress Effects on Natural Ecosystems.*, Eds.: G.W. Barrett and R. Rosenberg. Wiley & Sons. New York: pp. 57–70.

Wathern, P. 1988. *Environmental Impact Assessment. Theory and Practice.* Winchester, MA: Unwin Hyman. p. 332.

White, L., Jr. 1967. The historical roots of our ecological crisis. *Science,* 155, 1203–1206.

Wolfe, J. 1970. Foreword. In *A Tropical Rainforest.*, Eds.: H.T. Odum and R.F. Pigeon. Division of Technical Information. U.S. Atomic Energy Commission. T 10-24270 (PRNC-138).

World Bank. 1993. Environment bulletin. A Newsletter of the World Bank Environment Community. Volume 5, No. 1.

World Commission on Environment and Development. 1987. *Our Common Future.* New York: Oxford University Press.

Wu, X. 1991. Strategic Planning Models for Wildlife Habitat Management and Environmental Assessment of Forest Landscapes. University of Tennessee, Knoxville. Ph.D. dissertation. p. 129.

2 Environmental Policy, Sustainable Societies, and Biosphere Reserves

William P. Gregg

CONTENTS

Origins of MAB and the Biosphere Reserve Concept..24
   Biosphere Reserves Defined 25
Laying a Foundation of Designated Sites, 1974 to 1980 26
Establishing Multisite Linkages, 1981 to 1989...29
Supporting the Harmonization of Conservation and Development: the 1990s............34
   Establishing a Strategic Framework 35
   Improving Access to Information 35
   Building Local Partnerships..36
Future Directions..37
References 38

A distinguishing characteristic of the closing years of the 20th century is the universal recognition among the peoples of the world of the need for an ethic of sustainability. For example, the Rio Declaration, adopted by the member states of the United Nations at the 1992 United Nations Conference on Environment and Development, reflects broad agreement on the principles that must guide efforts to achieve harmony between human activities and the earth's environment.[1] The Declaration recognizes the ecological unity of the planet, the global interdependence of human activities and natural processes, the relationship between long-term economic progress and environmental protection, and the right of all people to "a healthy and productive life in harmony with nature." Widespread acceptance of such principles has fostered international cooperation in formulating and implementing treaties on biological diversity, forests, climate change, desertification, stratospheric ozone, marine pollution, and other environmental issues. At the national level, this ethic is reflected in policies that encourage multisector cooperation in balancing environmental, economic, and cultural considerations in achieving sustainable development. Government agencies and other interests concerned with natural resources have initiated cooperative programs and expanded ongoing efforts to manage resources holistically on an ecosystem basis. At regional scales, effective cooperation among diverse interests is now being demonstrated in large geographic areas, such as air sheds, watersheds, regional planning districts, ecoregions, and landscapes. The extent of these areas depends on the spatial and temporal scales of the particular issues of concern and the goals of the cooperating agencies and organizations. Locally, the ethic is motivating the cooperative efforts of neighborhoods and communities in managing their small areas of the planet to sustain a healthy environment that meets the needs of local people.

For more than two decades, UNESCO's Man and the Biosphere Program (MAB) has been facilitating cooperation among policy makers, resource users, and natural and social scientists to

0-57444-053-5/99/$0.00+$.50

foster sustainability at scales from local to global. To demonstrate the benefits of this cooperation, MAB is building a global network of biosphere reserves. Biosphere reserves focus attention on geographic areas that can serve as models for integrating conservation and sustainable development locally, while providing relevant information, technologies, and experience to help solve regional and global environmental problems.[2,3]

ORIGINS OF MAB AND THE BIOSPHERE RESERVE CONCEPT

The Man and the Biosphere Program was launched by the United Nations Educational, Scientific, and Cultural Organization (UNESCO) in 1970 to facilitate intergovernmental cooperation in fostering harmonious relationships between humans and the biosphere. MAB was the first deliberate international initiative to find ways to achieve sustainable development.[4] The Program's broad goal was to "develop the basis within the natural and social sciences for the rational use and conservation of the resources of the biosphere and for the improvement of the global relationship between man and the environment: to predict the consequences of today's actions on tomorrow's world and thereby to increase man's ability to manage efficiently the natural resources of the biosphere."[5]

The establishment of MAB was a manifestation of increasing public concern in the developed countries about the cumulative impacts of human activities on the environment. During the 1960s and early 1970s, the U.S. launched serious and wide-ranging efforts to protect environmental values. Federal laws were enacted to safeguard air, water, endangered species, wilderness, coastal environments, and other natural resources. New agencies and programs were established to monitor, assess, regulate, and maintain environmental quality. Private organizations were formed to provide channels for citizen advocacy and action. New policies and programs, and an increasing share of agency budgets, were directed toward preventing, mitigating, and redressing environmental impacts. The National Environmental Policy Act of 1969 (NEPA) established a national policy on the role of the federal government in sustaining the environment. This visionary law commits the national government to "use all practicable means … to create and maintain conditions under which man and nature can exist in productive harmony, and fulfill the social, economic, and other requirements of present and future generations." To carry out this policy, NEPA requires the federal government to "utilize a systematic, interdisciplinary approach which will insure the integrated use of the natural and social sciences and the environmental design arts in planning and decision making which may have an impact on man's environment." This policy, and the required integrative approach, enabled the U.S. to aggressively support the establishment of the international MAB program within UNESCO. It subsequently became one of the first nations, in 1974, to establish a national MAB organization to facilitate the participation of many agencies and institutions as partners in implementing MAB goals.

At its first meeting in 1971, MAB's International Coordinating Council identified 13 international scientific projects as the initial framework for coordinating the activities of countries, intergovernmental bodies, and nongovernmental organizations participating in MAB.[5] These projects focused on human impacts on forests, grasslands, arid lands, mountains and islands and other widely distributed ecosystems; the management of croplands, engineering works, urban ecosystems, and demographic change; and human perceptions of environmental quality. In each of these areas, MAB sought to encourage interdisciplinary research through the integration of the natural and social sciences. MAB's architects recognized the value of protected areas in meeting the scientific, educational, recreational, cultural, and economic needs of human societies, and foresaw an important role for these areas in facilitating the overall research program. They therefore included a project on conservation of natural areas, through which MAB would seek to support the establishment of a coordinated world-wide network of reserves, protected and managed in various ways. These sites would facilitate ecosystem research, the monitoring of environmental change, education and training, and the "means for maintaining the gene pools of plants, animals, and micro-organisms in all of their diversity." The project called for the designation of such areas as biosphere reserves in order to recognize their importance to humankind and foster international concern for their long-term conservation. The concept of a global network of protected areas for these purposes was endorsed in 1972 at the first global environmental conference ("Stockholm Conference on the Human Environment ").[6]

In an annex to the Nixon-Brezhnev Summit Agreement of 1974, the U.S. and the former Soviet Union gave momentum to building the network by agreeing to establish biosphere reserves on their respective territories. Later that year, the U.S. and several other countries did, in fact, unilaterally establish biosphere reserves, including 16 in the U.S. UNESCO subsequently convened a special task force, which set forth general criteria for use by national MAB organizations in nominating areas for designation by UNESCO as biosphere reserves.[7] The U.S. and the U.S.S.R. further honored their summit commitments by convening a joint symposium on biosphere reserves, which launched a bilateral program of ecological monitoring in paired biosphere reserves which, with minor interruptions due to the changing climate of the countries' relationship, has facilitated cooperation for 20 years.[8-10]

BIOSPHERE RESERVES DEFINED

The UNESCO criteria, as developed in 1974 and subsequently refined, encourage the nomination of areas that include a mosaic of ecological systems representative of a major biogeographical region, including a gradation of human interventions; that are significant for biodiversity conservation; and that offer opportunities to explore and demonstrate approaches to sustainable development on a regional scale.[11-14] Its size and configuration should be sufficient to carry out three basic functions of biosphere reserves:

- To contribute to the *conservation* of landscapes, ecosystems, species, and genetic variation
- To foster ecologically, socially, and culturally *sustainable development*
- To provide *logistic support* for environmental education, training, demonstration projects, monitoring, and research relating to local, regional, national, and global issues of conservation and sustainable development

To carry out these functions, the biosphere reserve should include one or more legally protected *core areas* managed to sustain indigenous biota and natural processes according to established conservation objectives, and of sufficient size and appropriate configuration to achieve these objectives. The core area provides important opportunities for conservation, long-term observational studies, and environmental education and serves as a regional benchmark of ecological health. The biosphere reserve should also include one or more legally or administratively established *buffer zones* (also referred to in the U.S. as *managed use areas*) that typically adjoin or surround the core area (s). These areas are managed for uses and activities consistent with conservation objectives. An important purpose of these zones is to develop and demonstrate approaches for managing and using natural resources while maintaining natural processes and biota. Buffer zones frequently provide areas for experimental research, education, demonstrations of sustainable production systems and appropriate technologies, and public recreation. The biosphere reserve should also recognize a *transition area* (also referred to in the U.S. as a *cooperative area*). The transition area surrounds the core area(s) and buffer zone(s) and supports a variety of resource uses and human activities characteristic of the larger region. The transition area serves as a zone of influence where the parties involved in the core areas and buffer zones of the biosphere reserve work with surrounding communities and other regional interests to communicate the lessons learned and help promote sustainable resource management. The transition area is open-ended and usually not delineated on a map. Its functional dimensions may vary in space and time depending on the scale of the issues being addressed cooperatively by biosphere reserve stakeholders. Within this spatial configuration, UNESCO's latest criteria specifically encourage organizational arrangements to involve public agencies, private organizations, local communities, and other interests in planning and implementing the biosphere reserve functions.[14]

The zonation categories were developed to describe, in general terms, the functional roles of the existing protected conservation areas and other legally established management units, which comprise the internationally designated units of the biosphere reserve, and the surrounding "bio-geocultural" area in which cooperative activities are especially encouraged. *In the U.S. and most other countries, the categories involve no legal or administrative requirements. They merely serve to aid the parties in their voluntary efforts to implement the biosphere reserve concept.*

By encouraging research, education, and demonstration activities within internationally recognized areas containing various ownerships, resource uses, and management strategies, UNESCO offered biosphere reserves as a new approach for demonstrating conservation and sustainable uses of ecosystems. However, in the early 1970s, there was little experience to draw upon for implementing such an approach. Recognizing this, UNESCO encouraged national MAB organizations to use flexibility in adapting UNESCO's general guidance to their particular situations. The implementation of the biosphere reserve concept would rely on the existing legal and administrative authorities, frameworks, and processes in each country. UNESCO's role would be to facilitate the sharing of information and experience from biosphere reserves, provide guidance and technical assistance, encourage international cooperation, and encourage financial and technical support for biosphere reserves from governments, other United Nations agencies and international organizations.

Since he began his career with the Department of the Interior in 1971, the author has followed the development of MAB's biosphere reserve network, initially as an interested observer and, since 1980, as a member of the USMAB (USMAB) organization, representing first the National Park Service and later the National Biological Service, which in 1996 became the Biological Resources Division of the U.S. Geological Survey. During this time, many factors, events, and trends have influenced the development of U.S. biosphere reserve network, which, as of early 1997, included 47 units (Figure 1). The remainder of this chapter highlights several stages in the evolution of the U.S. network, and the challenges in making biosphere reserves standard-bearers for demonstrating the ecosystem approach.

LAYING A FOUNDATION OF DESIGNATED SITES, 1974 TO 1980

In view of MAB's mission to facilitate intergovernmental scientific cooperation on natural resource issues, the responsibility for designing the initial U.S. biosphere reserve network fell to scientists in federal land managing agencies, namely, the U.S. Department of Agriculture (USDA) and the Department of the Interior (DOI) . These agencies administered securely protected sites with long histories of research in most of the country's biogeographical provinces. It is therefore logical that the first 27 U.S. biosphere reserves designated through 1978 were either protected conservation areas — mostly national parks — managed by DOI or experimental research areas managed by the Forest Service (FS) or Agricultural Research Service (ARS) within USDA However, the architects of the initial U.S. nominations realized the limitations of these federally protected areas in fulfilling the purposes of biosphere reserves. National parks provided sites of global importance for conservation, baseline studies of natural ecosystems, and improving public awareness of environmental issues . However, they were not suitable sites for manipulative research and development of sustainable economic uses of ecosystems. On the other hand, the USDA sites often had outstanding long-term records in implementing these functions, but were of limited significance as conservation areas. In biogeographical provinces containing both types of sites, the architects decided to pair

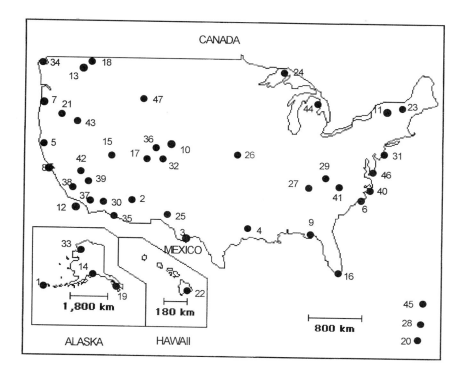

1. Aleutian Islands	17. Fraser	
2. Beaver Creek	18. Glacier	33. Noatak
3. Big Bend	19. Glacier Bay-Admiralty Island	34. Olympic
4. Big Thicket	20. Guanica	35. Organ Pipe Cactus
5. California Coast Ranges	21. H.J. Andrews	36. Rocky Mountain
6. Carolinian-South Atlantic	22. Hawaiian Islands	37. San Dimas
7. Cascade Head	23. Hubbard Brook	38. San Joaquin
8. Central California Coast	24. Isle Royale	39. Sequoia-Kings Canyon
9. Central Gulf Coastal Plain	25. Jornada	40. South Atlantic Coastal Plain
10. Central Plains	26. Konza Prairie	41. Southern Appalachian
11. Champlain-Adirondack	27. Land Between the Lakes	42. Stanislaus-Tuolumne
12. Channel Islands	28. Luquillo	43. Three Sisters
13. Coram	29. Mammoth Cave Area	44. University of Michigan
14. Denali	30. Mojave and Colorado Deserts	45. Virgin Islands
15. Desert	31. New Jersey Pinelands	46. Virginia Coast
16. Everglades	32. Niwot Ridge	47. Yellowstone

FIGURE 1 Biosphere reserves in the U.S.

separately designated biosphere reserves in the hope of facilitating cooperation in carrying out biosphere reserve functions within the province.[15]

Designation of the early biosphere reserves was noncontroversial. Administering federal agencies viewed the international recognition as a means to encourage greater scientific use of the designated sites and to increase scientific cooperation on important natural resource issues. Designation involved no legal or financial obligations, did not require changes in planning or local management practices, and encouraged activities of a strictly scientific, technical, and educational nature. Early designations therefore generated little public interest, and there were no pressures for public involvement in the nomination process.

In 1974, the USMAB National Committee established directorates of natural and social scientists from the federal agencies and the private sector to plan and implement interdisciplinary research

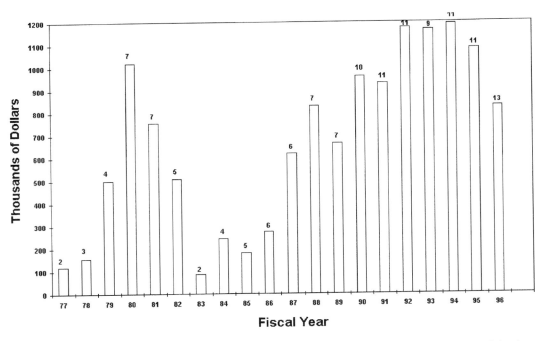

FIGURE 2   USMAB budget: 1977 to 1996. Numbers on graph represent number of agencies participating.

on particular topics (e.g., environmental pollution, human perception of environmental quality) and ecosystems (e.g., mountains, aridlands, islands, temperate forests, tropical forests). A separate directorate, with representatives from land managing agencies and organizations, was established to facilitate and support activities involving biosphere reserves. Between 1976 and 1978, USMAB sponsored workshops in five regions containing a number of paired biosphere reserves (the Eastern Forest, Pacific Northwest, the Southwest, the Pacific Southwest, and the Rocky Mountains). Important resource issues in each region and opportunities for future cooperation involving the region's biosphere reserves were identified. Beginning in 1978, USMAB funded extramural grants to support the directorates' programs. During the next 5 years, USMAB's Consortium for the Study of Man's Relationship with the Global Environment supported many cooperative research projects in biosphere reserves, some involving paired sites, on such regional resource issues as acidic deposition in the southern Appalachians, fire management in the northern Rockies, and integrated management of white-tailed deer in Sonoran oak–pine woodlands of paired biosphere reserves in the U.S. and Mexico. In 1978, a national USMAB forum developed a framework for long-term ecological monitoring in biosphere reserves.[16] Subsequently, USMAB, in cooperation with the U.N. Environment Programme, launched a multilateral program for monitoring background pollution levels in air, water, and soil involving biosphere reserves in the U.S., Russia, and Chile.

USMAB's initiatives increased interest among protected area administrators in filling gaps in the U.S. network. By the late 1970s, nonfederal organizations began to petition for biosphere reserve nomination. In 1979, the Nature Conservancy's Konza Prairie Research Natural Area in Kansas and the University of Michigan's Biological Station became the first nonfederal areas to be designated as biosphere reserves. By 1980, the U.S. network comprised 36 biosphere reserves, located in 16 of the country's 20 biogeographical provinces.

During the Carter administration (1977–1981), the number of federal agencies participating in USMAB and their funding contributions increased significantly (Figure 2). By the end of the decade, USMAB had become a recognized, respected, and modestly funded framework for applied interdisciplinary research. In 1979, the Office of Management and Budget (OMB) and the Office of Science and Technology Policy in the Executive Office of the President issued a joint memorandum underscoring the importance of MAB in international cooperation and as a focus for coordinating domestic programs aimed at improving the management of natural resources and the environment. Federal agencies were directed to participate fully in the program. In 1980, USMAB approved a national program plan, which included support for cooperative regional demonstration projects.[17] The partnership projects — the conceptual forerunners of ecosystem initiatives 15 years later — proposed to use MAB's interdisciplinary research capabilities and biosphere reserves to help address regional resource problems. The plan was favorably received at OMB, which referred the plan to MAB's participating agencies for funding.

As the interest of federal agencies in MAB increased, concern about threats to protected areas helped focus specific attention on biosphere reserves. In 1980, the National Park Service (NPS) prepared a report to the Congress on the state of the national parks.[18] The report summarized the perceptions of park managers on 53 categories of threats, the majority of which were reported to originate outside park boundaries. Parks designated as biosphere reserves reported particularly large numbers of threats — perhaps reflecting the strong history of research in these areas which enabled these threats to be documented. The "threats report" reinforced NPS interest in the role of biosphere reserves in obtaining information to document and assess threats, particularly external threats requiring expanded cooperation. As the steward of the majority of U.S. biosphere reserves at the time, this interest proved to be an important factor in the development of the U.S. Biosphere Reserve Program during the next decade.

At the international level, increasing global interest in the relationship between conservation and development influenced efforts to expand the unique niche of biosphere reserves. In 1980, the World Conservation Union (IUCN), in cooperation with the World Wildlife Fund and the U.N. Environment Program, issued the World Conservation Strategy, which provided an early conceptual blueprint for integrating conservation and development.[19] The strategy set forth an integrated and cooperative approach for maintaining ecological processes and biological diversity within the context of sustainable utilization of ecosystems and species. Although the Strategy makes only a brief reference to biosphere reserves, it helped foster international interest in exploring the role of biosphere reserves in demonstrating the ecosystem approach.

ESTABLISHING MULTISITE LINKAGES, 1981 TO 1989

In 1981, UNESCO held an international conference to review MAB's first 10 years and recommend future directions.[20] With respect to biosphere reserves, activities generally reflected the objectives of the national parks and research reserves which comprised most of the network at the time. The U.S. reported the many accomplishments of its 36 biosphere reserves in developing information, technology, and scientific capabilities to address resource issues of concern to the managers of the designated sites. However, there was little evidence of effective communication with the public about MAB, biosphere reserves, and interrelationships between human activities and the biosphere, and only limited progress in developing regional and international linkages involving the designated sites.[21] The U.S.S.R. emphasized environmental monitoring in existing protected areas, but indicated little emphasis on developing most other biosphere reserve functions.[22] The conference evoked a general concern regarding the need for more emphasis on development and integration of the functions of biosphere reserves, rather than the designation of sites per se — i.e., for quality over quantity. The 1981 conference paved the way for the first International Biosphere Reserve Congress, which was held in Minsk, Belarus, in 1983. This forum brought together specialists from scores of countries to produce the first detailed compendium of biosphere reserve case studies, along with numerous recommendations for developing a program framework.[23] UNESCO subsequently used these recommendations to develop the Action Plan for Biosphere Reserves, which was adopted in December 1984.[13]

More clearly than any previous guidance from UNESCO, the Action Plan conceptualizes biosphere reserves as models for linking the conservation of ecosystems and biological diversity with regional social and economic development. Although the concepts of "ecosystem management" and "sustainable development" had yet to become widely recognized, the Plan clearly envisions biosphere reserves as cooperative frameworks for demonstrating creative ways to integrate these concepts. The Action Plan stresses the importance of involving local interests, and that biosphere reserves should demonstrate economic and social benefits for local people as well as the relationship between conservation and sustainable development of the wider region. Seen in this context, the internationally designated protected core and buffer areas of the biosphere reserve play key roles in developing the knowledge, skills, and attitudes required to enable many cooperators to integrate ecosystem management and sustainable development functions. The Plan's proposed program framework included objectives and a 5-year program of recommended actions for filling gaps in the network; improving the management of designated sites; conserving ecosystems and key species; strengthening monitoring, research, education, and training; facilitating local participation and regional planning; and systems for international sharing of information from biosphere reserves. Unfortunately, voluntary contributions from international organizations and governments fell short of the requirements for implementing the ambitious Plan.

In the U.S., the change of national administrations in 1981 significantly affected the fortunes of the USMAB Program. The new administration's concerns over the policies and administrative inefficiencies of United Nations agencies quickly focused on UNESCO. Although these concerns had little to do with MAB and other programs in UNESCO's science sector, the administration's disaffection with UNESCO made Federal agencies reluctant to contribute to USMAB. The national program plan was tabled indefinitely. By 1983, USMAB no longer had sufficient funds to award research grants through the MAB Consortium. The administration signaled its intention to withdraw from UNESCO and reduced funds appropriated to the State Department for UNESCO activities. The National MAB Committee ceased to meet on a regular basis. Only the Forest Service and the State Department were contributing. The proposed U.S. withdrawal from UNESCO sparked congressional hearings on the effects of the withdrawal on U.S. science. The hearings documented the many benefits of MAB and other UNESCO science programs, and led to discussions within the Government on ways to maintain these benefit. Although the U.S. withdrew from UNESCO at the end of 1984, the administration subsequently requested and received a legislative appropriation to the Department of State to enable the U.S. to participate in these programs on a country-to-country basis, beginning in FY 1986.[24] The appropriation, in effect, gave these programs a political "clean bill of health." Additional domestic agencies began to join USMAB (Figure 2). Based on an independent review of USMAB, the National Committee moved to strengthen MAB's role in integrating the natural and social sciences to address resource policy and management issues. By 1990, the levels of agency participation and funding had returned to nearly those of 10 years earlier.

During the 1980s, the National Park Service remained a stalwart supporter of the Biosphere Reserves Directorate, which continued to meet regularly. NPS funded a series of comprehensive bibliographies and histories of scientific studies in NPS biosphere reserves, and a multidisciplinary ecosystem research program on the Virgin Islands Biosphere Reserve. NPS funding and in-kind support enabled the Directorate to begin a systematic review of the nation's biogeographical provinces to fill gaps in biosphere reserves. Under USMAB guidelines,[25,26] the U.S. became the first country to launch separate biosphere reserve selection programs for terrestrial and coastal/marine provinces. Nominations focused on sites of particular importance for conserving the characteristic ecosystems of the province, including major ecological communities and the associated gradients, processes, and physical factors. The site's history and potential for observational studies, experimental research, and education were also important considerations. As in the 1970s, only sites that were legally protected and managed for conservation or research objectives were considered. However, the guidelines added an important new provision to allow for nomination of multiple sites as units of the same biosphere reserve. Beginning in the early 1980s, multiple-site biosphere reserves — often including sites under many different administrators — became the U.S biosphere reserve model.

Between 1981 and 1986, USMAB panels of scientists reviewed eight terrestrial and three coastal biogeographical provinces. The panels subdivided the provinces based on biogeographical and ecological considerations, and identified the qualified conservation and research areas for nomination in each subprovince. The overall effort recommended 36 multisite biosphere reserves, containing more than 200 sites. By 1986, 13 new biosphere reserves, containing as many as 13 sites each under federal, state, local, and private administrators had been designated. The designated configurations of sides included a pair of national parks on adjacent islands in Hawaii (Hawaiian Islands BR); a group of sites managed by NPS as the Big Thicket National Preserve in southeast Texas; several clusters of national and state parks and research sites in the coast redwood forests of central and northern California; a five-site complex of federal- and state-protected areas in the California deserts, and clusters of sites under federal, state, and private ownerships in three coastal embayments of the southeast coast. Others included a single site, with additional sites to be added when conditions would permit. Still others included combinations of large areas under state regulation and protected areas managed to conserve particular ecosystems.

The MAB panels identified highly qualified conservation and research sites. However, they were not asked to assess whether the identified sites would actually cooperate in implementing the biosphere reserve concept. Although the linkages made sense from biogeographic and ecological points of view, sites were sometimes located in areas with very different cultures, histories of resource use, economic conditions, politics, priorities and institutional relationships. Cooperation in such areas following designation proved difficult.

For example, the South Atlantic Coastal Plain Biosphere Reserve, originally designated in 1983, consisted of the state of New Jersey's 438,000-hectare Pinelands National Reserve — a model for conservation and sustainable development ecosystems at the northern end of a large subprovince — and the Congaree Swamp National Monument, an outstanding brown water swamp ecosystem in South Carolina, at its southern end. Several protected areas filling gaps in the middle of the subprovince could not be nominated at the time. In 1988, the state of New Jersey, believing the Pinelands unit to fulfill biosphere reserve criteria in its own right and seeing little incentive for cooperation with sites hundreds of miles away, requested and received UNESCO redesignation as the New Jersey Pineland Biosphere Reserve, leaving Congaree Swamp as the sole unit in the South Atlantic Coastal Plain Biosphere Reserve.

Even concerted efforts to foster cooperation among designated sites sometimes proved unsuccessful. In the Carolinian–South Atlantic Province, which extends roughly from Cape Hatteras to Cape Canaveral, the MAB panel recommended a single biosphere reserve containing three clusters of sites. The Carolinian–South Atlantic Biosphere Reserve, designated in 1986, contains 13 sites — two in the North Carolina's Outer Banks, six in the South Carolina's Santee Delta, and five in Georgia's Sea Islands. The sites include most of the characteristic ecosystems of the Province. Many have substantial histories of research and monitoring. Following designation, the state of South Carolina designated a coordinator to help plan biosphere reserve activities. South Carolina's unit included nearly all the contiguous management units in the Santee River delta — a former rice-growing area containing large plantations. The federal, state, and private managers of the delta cluster were receptive to using the biosphere reserve as a framework for cooperation. However, the state of Georgia was concerned about the possible effect of designation on its prerogatives for managing forests, wetlands, recreation, and development in the Sea Islands. The state's lukewarm interest in the biosphere reserve concept, and its refusal to nominate any of its extensive holdings, sapped enthusiasm for participating in the biosphere reserve among managers and other potential partners in Georgia. The small Outer Banks cluster, consisting of a national seashore and an unmanaged marine area some distance offshore, had no previous history of cooperation and was not in a position to contribute substantially at the outset. The three areas had vastly different histories, contemporary cultures, economies, and interests. Each proved to be at very different stages of readiness to participate in a biosphere reserve. Two MAB-sponsored workshops failed to produce consensus on a focus or mechanism for cooperation. South Carolina eventually withdrew its coordinator. Had the South Carolina cluster been nominated in its own right as a separate biosphere reserve, the outcome might have been different.

As biosphere reserves began to link protected areas, the first instances of opposition began to be heard from local groups fearing government infringement on traditional resource uses and private property rights. In 1986, a U.S.–Canadian MAB panel recommended a biosphere reserve linking the existing Isle Royale National Park Biosphere Reserve in Lake Superior, Voyageurs National Park and the Forest Service's Boundary Waters Canoe Area in northern Minnesota, and the Quetico Provincial Park in Ontario. Some local groups saw the proposed "North Woods Biosphere Reserve" as yet another effort by managing agencies to limit traditional recreational uses of federal lands. Although biosphere reserve designation would not have affected these uses, unconvinced local groups continued to oppose the nomination. The Forest Service withdrew their site from further consideration. However, the NPS submitted the nomination of the two national parks. By the time the nomination reached the U.S. National MAB Committee, the controversy threatened to become an issue in a statewide political campaign. The Committee terminated further USMAB participation in the nomination and recommended public education on MAB goals and benefits. The role of local interests in biosphere reserves would become an increasingly important consideration in the years ahead.

Although some efforts encountered difficulties, others succeeded beyond expectations. A MAB Panel for the Eastern Forest Province recommended that three cooperating sites (the Great Smoky Mountains National Park and Coweeta Biosphere Reserves and the Oak Ridge National Environmental Research Park) in the Southern Appalachian region be redesignated together as the Southern Appalachian Biosphere Reserve, and that other candidate sites be considered for inclusion. After site managers endorsed the concept, a feasibility study was prepared by a local citizen familiar with biosphere reserves. The study identified resource issues of regional concern, as well as many agencies and organizations interested in seeing how MAB and biosphere reserves might help address them. It also recommended establishment of a coordinating mechanism for a regional MAB program in the Southern Appalachians. A series of interagency planning meetings eventually led to the establishment of the Southern Appalachian Man and the Biosphere (SAMAB) Cooperative, with federal and state members, in 1988. The next year, the SAMAB Foundation was chartered to facilitate participation of private organizations and individuals. Among the cooperative's first actions was to nominate five sites as the nucleus of the Southern Appalachian Biosphere Reserve and to describe a region containing parts of five states as SAMAB's principal area of cooperation and expansion within which additional sites could be considered in the future as units of the biosphere reserve. Chapter 5 (Hinote) documents SAMAB's many successes as an expanding multisector partnership for addressing issues of conservation and development.

Several factors appear to have influenced SAMAB's success. Local interests were in charge. The emphasis was on solving problems identified by local partners, rather than on designating a biosphere reserve. The stakeholders were able to reach consensus on the limits of the southern Appalachian region as the principal area of concern and cooperation, based on physical, ecological, historical, cultural, political, and institutional considerations. The region made sense to those with access to the intellectual, technical, and financial resources required to enable stakeholders to address a wide range of problems. The stakeholders organized their activities around the broad goals of MAB, rather than on the objective of establishing a biosphere reserve. The biosphere reserve was seen as a tool to help stakeholders achieve their goals, rather than as an end in itself. The linkage of sites was not nominated until the coordinating organization was in place, and stakeholders took time to consider how the candidate sites would contribute to SAMAB's cooperative activities.

SAMAB was the first partnership to take advantage of the synergy between building a cooperative program and the international designation of sites on the basis of their contributions to the program. In this case, existing biosphere reserves were linked to form a regional biosphere reserve, and additional biosphere reserve sites were subsequently nominated to the regional biosphere reserve as conditions permitted. The designation of the additional sites is seen as a way to recognize accomplishments, increase the stake of the sites in the cooperative program, facilitate cooperative management, and increase the support of stakeholders, including the sites' neighboring communities.[27]

Toward the end of the decade, the SAMAB experience was being widely shared with groups in other regions and helped foster and encourage many grassroots efforts to implement biosphere reserve concepts. USMAB provided seed funds to local groups in the Colorado Rockies, the Sonoran Desert, the Central California Coast, the Ozark Highlands, and the Champlain–Adirondack region to assess the feasibility of forming new cooperative programs or expanding cooperation involving existing biosphere reserves. Groups (e.g., in the Catskills) began to explore opportunities on their own initiative.

In 1989, the Biosphere Reserve Directorate prepared revised guidelines for selection and management of biosphere reserves, reflecting the experience of the previous decade.[28] Emphasis was placed on the role of biosphere reserves in facilitating cooperation in "biogeocultural" regions, described on the basis of some combination of ecological, physical, economic, cultural, and political factors.

In 1989, the National Committee restructured the USMAB Program. Approval of the guidelines was deferred, pending development of a program strategy for U.S. Biosphere Reserves. The new USMAB consisted of five interdisciplinary research directorates responsible for developing multi-year research projects on policy and management issues in high latitude, temperate, tropical, coastal/marine, and human-dominated ecosystems, respectively. The priority areas for the directorates' projects were: global change, arid lands and desertification, aquatic areas and wetlands, biodiversity, cultural diversity, sustainable development, and biosphere reserves. Each directorate included a balanced membership of natural and social scientists from government and academia. A new Coordinating Committee on Biosphere Reserves, replacing the former Biosphere Reserve Directorate, was established to strengthen the role of biosphere reserves within the overall program.

No discussion of U.S. biosphere reserves in the 1980s would be complete without mention of two major conferences which helped stimulate interest in biosphere reserves among the managers of U.S. Biosphere Reserves. In 1984, U.S. agencies, UNESCO, and MAB-Canada cosponsored the first biosphere reserve managers' workshop in conjunction with the 50th anniversary of Great Smoky Mountains National Park — one of the initial U.S. Biosphere Reserves. U.S. managers reviewed the international recommendations on the roles and functions of biosphere reserves, considered the merits of the multisite approach, brainstormed possibilities for using biosphere reserves to encourage cooperation in addressing regional issues, and recommended ways for individual sites to participate more fully in implementing the biosphere reserve concept.[29] In 1987, UNESCO, USMAB, MAB-Mexico, and various U.S. agencies cosponsored a symposium on biosphere reserves at the World Wilderness Congress in Colorado, which was attended by the managers of the majority of U.S. biosphere reserves. The symposium included general concept papers, case studies from developed and developing countries, including applications in coastal and marine areas. The proceedings from these conferences provided a window on the implementation of biosphere reserves and continue to served as basic references.[30]

SUPPORTING THE HARMONIZATION OF CONSERVATION AND DEVELOPMENT: THE 1990s

During the 1990s, the interrelated concepts of biodiversity, ecosystem management, and sustainable development have led to significant changes in the principles, policies, and practices of conservation and socioeconomic development. Each concept now has an extensive literature, and numerous organizations working to apply the concept at local, regional, national, and international levels. These efforts create important opportunities for using biosphere reserves as model areas for integrating these concepts.

The Convention on Biological Diversity (CBD), opened for signature by national governments in 1992, was an important milestone. The Convention focused world attention on the role of biological diversity to human survival and progress. It defines biodiversity to include all the levels of biological organization (ecosystems, habitats, species, populations, and genetics), establishes broad policy goals for its conservation and sustainable use, and provides a framework for international cooperation. Although President Clinton has signed the Convention, it has not been ratified by the Senate (as of early 1997). In the U.S., the Convention is a catalyst for raising public awareness, improving access to information, and encouraging cooperative management of biodiversity — particularly at the ecosystem level. The CBD does not specifically mention biosphere reserves. However, a world network of fully functional Biosphere Reserves could become a principal means for achieving many of the Convention's goals.

In 1993, the Clinton administration established the Interagency Ecosystem Management Task Force in response to a recommendation in the Vice President's National Performance Review, which called upon federal agencies to implement a proactive approach in managing ecosystems to ensure a sustainable economy and a sustainable environment. The approach is a goal-driven method for sustaining or restoring natural ecosystems, based on a shared vision of desired future conditions that integrates ecological, economic, and social factors. The emphasis is on sustaining ecosystem functions, including the continuing evolution of biodiversity. (The concept of ecosystem management is now widely accepted, and, in 1996, was endorsed by the parties to the CBD as the principal framework for implementing their activities under the Convention.)

To implement the national policy on ecosystem management, U.S. agencies are coordinating and adapting their activities to respond to the issues facing particular ecosystems, the people who live in them, and the economies based upon them.[31] Within these ecosystems, agencies are working to improve and integrate the information they obtain; to make this information widely accessible and available; and to foster communication among scientists, information managers, and information users. Because of their research capabilities and conservation values, many U.S. biosphere reserves could become important demonstration areas for ecosystem management, provided appropriate mechanisms for linking the designated sites with the surrounding cooperative area are implemented. Several cooperative areas containing biosphere reserves, such as the southern Appalachians, the Everglades, and the Greater Yellowstone Area, are already recognized for their contributions to ecosystem management.

Biodiversity and ecosystem management concepts have attracted the attention of agencies, organizations, and public interests concerned with conservation and use of natural resources. The concept of sustainable development is similarly attractive to entities concerned with socioeconomic development and related issues, such as human health and welfare, equitable access to resources, environmental justice, and the preservation of cultural traditions.

During the past several years, international forums — such as the U.N. Conference on Environment and Development (1992) and the World Summit for Social Development (1995) — have focused attention on various aspects of sustainable development. The Commission on Sustainable Development, established by the White House in 1993, provides a focal point within the federal government for coordinating U.S. policies and initiatives. In the U.S., scores of cities and communities are developing their own indicators and initiatives to implement the concept of sustainable development.

Many U.S. biosphere reserves continue to reflect the sectorial management objectives of the protected (core) areas and sustainable use areas designated years ago, rather than the broad goals of biosphere reserves, which require extensive interaction with stakeholders in the surrounding cooperative area. Such U.S. biosphere reserves generally do not yet incorporate sustainable social and economic development projects involving local communities in their cooperative programs. However, inclusion of such projects has enormous potential for generating goodwill and public support for biosphere reserve objectives.

ESTABLISHING A STRATEGIC FRAMEWORK

In 1994, USMAB approved a Strategic Plan for the U.S. Biosphere Program, based on the recommendations of biosphere reserve managers and stakeholders. The program's mission is to establish and support a network of designated biosphere reserves representative of the biogeographical areas of the U.S. and to promote " a sustainable balance among the conservation of biological diversity, compatible economic use, and cultural values, through public and private partnerships, interdisciplinary research, education, and communication.".[32] The plan set forth goals, objectives, and recommendations for action by USMAB and the biosphere reserves themselves during the next decade. The plan covers policy and program operations, network development, local participation, research, education, and communication. Within USMAB, the plan restored momentum for biosphere reserves. The National Committee moved quickly to establish a new Biosphere Reserve Directorate and increase its support for local partnerships.

In 1995, UNESCO sponsored a landmark conference of experts to share experience with biosphere reserves, and develop recommendations which UNESCO incorporated into a strategy for developing effective biosphere reserves. Utilizing a similar format to the U.S. plan, The "Seville Strategy" set forth goals, objectives, and actions for implementation of the biosphere reserve concept at the international, national, and biosphere reserve levels. The Strategy includes an extensive checklist of indicators for use in evaluating progress in implementation. The Conference also enabled UNESCO to finalize the "statutory framework of the World Network of Biosphere Reserves" which sets forth definitions, functions, selection criteria, and designation and review procedures, as well as guidelines on publicity, participation in the World Network (and regional and thematic subnetworks) and the responsibilities of UNESCO.[33] An important new provision in the UNESCO guidance requires states to forward a review of the status of each biosphere reserve every 10 years for comments and recommendations by MAB's International Coordinating Council, which may remove areas from the World Network if they do not satisfy the designation criteria.

Together, the conceptual plans and guidance for the U.S. Biosphere Reserve Program and the World Network provide a flexible framework for developing the roles of biosphere reserves in achieving sustainable societies.

IMPROVING ACCESS TO INFORMATION

In the 1990s, USMAB has played a leadership role in improving access to information on biosphere reserves. In cooperation with EuroMAB — an organization of European and North American MAB organizations, USMAB compiled a directory of contacts, resource maps, environmental databases, research activities, and existing infrastructure in 175 biosphere reserves in 32 countries.[34] The directory is available on the World Wide Web. Using the same protocol, UNESCO subsequently expanded the Directory to included the remaining sites in the World Network. ACCESS 96, a supplement containing metadata on permanent monitoring plots in EuroMAB biosphere reserves, was recently prepared in cooperation with the National MAB Committee of Germany.[35] Environ-mental data continue to be collected in paired biosphere reserves in Russia and U.S., through the bilateral small watershed program.

Through the Biosphere Reserve Integrated Monitoring Program (BRIM), launched by EuroMAB in 1991, a methodology for systematically recording information on records of flora and flora (MABFlora and MABFauna) was widely reviewed and tested in European biosphere reserves.[36] The methodology — originally developed for U.S. national parks — was improved and used to develop a self-contained computer program and manual, which is now in use in many EuroMAB and Latin American Biosphere Reserves (through MABNetAmericas, an electronic subnetwork of biosphere reserves launched following the 1994 Summit of the Americas). Initial efforts have focused on birds, mammals, and vascular plants. Data for many biosphere reserves are posted on the Internet. OBSERVE, a flexible computer program for recording observations of flora and fauna, supplements the MABFlora and MABFauna databases and is being tested in a number of biosphere reserves.

BioMon, a database from permanent monitoring plots, was developed through the Smithso-nian/MAB Biodiversity Program.[37,38] A data collection protocol and self-contained computer pro-gram for recording and analyzing data in the field were developed and tested in tropical forest biosphere reserves and similar protected areas in Latin America. The methodology was subsequently tested in tropical forests in Africa, Asia, and the Caribbean; and was recently adopted for use in protected forest ecosystems in Canada. A demonstration plot has also been established at Great Smoky Mountains National Park in the Southern Appalachian Biosphere Reserve. The SI/MAB Program provides training courses for foreign specialists, both in-country and in the U.S., on conservation practices and the use of the methodology.

The managers of the MAB's biodiversity databases are exploring ways to link the various methodologies and coordinate training efforts, under the broad umbrella of BRIM. The goal is to gradually incorporate other data sets on natural and human systems, to support interdisciplinary assessment and modeling in biosphere reserves.

BUILDING LOCAL PARTNERSHIPS

The restructuring of USMAB in 1990 reduced the emphasis on designating new biosphere reserves, pending completion of the Strategic Plan. Although sites have been added to several existing reserves, only two new units have been designated in the 1990s — i.e., the Land Between the Lakes Area in 1991 and the Mammoth Cave Area in 1990. Many local efforts to establish new biosphere reserves — some initiated as far back as the late 1980s — have encountered opposition from well organized, but misinformed, groups concerned about future infringement on property rights and traditional uses by government agencies or the U.N. The opposition prompted a hearing in the 104th Congress on the proposed American Land Sovereignty Protection Act, which includes a sunset provision calling for future termination of most U.S. biosphere reserves unless they are specifically authorized by the Congress. Although national policies have never been more favorable to the biosphere reserve concept, the opposition indicates the need for greater public education on the practical benefits of biosphere reserves to local people and in the nomination process.

Given the current situation, emphasis has been on developing the functions of existing biosphere reserves, including the role of MAB research through the five research directorates. In a national workshop in late 1995, biosphere reserve managers stressed the "value added" from cooperation on regional issues through the unifying aegis of MAB and the biosphere reserve concept. A survey just before the workshop indicated that biosphere reserves involved in cooperative programs that identify explicitly with biosphere reserve concepts and goals report more significant management benefits in more areas than those involved in cooperative programs not identifying with these concepts and goals. The reported benefits were much greater in such areas as ecosystem manage-ment, local political support, availability of information on both natural and human systems, international cooperation, environmental awareness and fostering an ethic of sustainability. On average, the former group reported greater levels of participation by 17 of 18 entities in their cooperative activities (there was no difference in participation of Native Americans). Differences between the groups were negligible for the principal cooperators — universities, and federal and state agencies. However, the former group was much more likely to report cooperation with international organizations, and various entities concerned with economic development and local community (e.g., local governments, schools, and citizen volunteers) goals. The survey appears to confirm the "value added" from linking cooperative activities explicitly with biosphere reserve concepts and goals.[39]

In the 1990s, U.S. biosphere reserves became partners in various cooperative mechanisms for implementing biosphere reserve concepts.[40] These include a trinational (U.S., Mexico, Tohono O'odham Nation) nonprofit community-based alliance in western Sonoran Desert, a biosphere reserve coordinating committee within a regional economic development authority in the Mammoth Cave Area, a 14-member MAB-affiliated regional cooperative in the Colorado Rockies representing 4 biosphere reserves, informal binational exchanges among biosphere reserves in the Chihuahuan Desert (U.S.–Mexico), and a nonprofit biosphere reserve association with associated science, management, and education councils for the 13-unit Central California Coast Biosphere Reserve. Each of these cooperatives provides a mechanism for the partners to plan and implement a wide range of activities relating to biodiversity conservation, ecosystem management, and sustainable development.

Finally, there is the important role of the interdisciplinary projects of the USMAB's ecosystem-based research directorates. Each of these projects addresses policy and management issues of paramount importance in biosphere reserves (e.g, comanagement of caribou populations in Alaska, design and management of coastal/marine biosphere reserves, multinational sharing of data and information among biosphere reserves in transborder ecosystems, development of GIS-based approaches for using interdisciplinary information in ecosystem management and land use analysis). Each involves one or more biosphere reserves or potential biosphere reserves; and each is developing practical tools for integrating data and information from the natural and social sciences to support cooperative ecosystem management. As the projects are completed, there remains some uncertainty regarding how best to assure that the methods and approaches developed continue to help the stakeholders in regional ecosystems and biosphere reserves address complex resource problems. Integration of the biosphere reserve and research components of USMAB is an important challenge that still remains to be fully addressed.[41]

FUTURE DIRECTIONS

Ronald Engel, in his various papers on the symbolic and ethical significance of biosphere reserves, suggests that biosphere reserves embody the "language of community" as distinguished from the "language of resource management," which characterizes the traditional management practices of agencies and institutions.[42,43] The former is concerned with fostering the participation, well-being, cultural values, and self awareness of human communities in their relationships with Nature; the latter, with managing nature as a physical resource for human progress and material well-being. Biosphere reserves have associations with both ethical languages, but Engel sees the ethic of community as the more important in motivating action to put the concept into practice.

Biosphere reserves offer an elegant tool for communities to obtain, share, and apply information and technologies. They offer a means to organize cooperation among many stakeholders and help them identify a cooperative area for conservation and development that makes sense ecologically, culturally, and politically. They offer a framework for cooperation in monitoring and research on natural and human systems, and linking local conditions with regional and global influences. In particular, they help build ties between the internationally designated biosphere reserve sites and the surrounding area, and help discover ways to integrate conservation and community development.

They offer the benefits of international cooperation with communities in other areas of the world having similar interests and problems. Realization of these benefits depends upon continuous communication among communities of biosphere reserve stakeholders at scales from local to global.

During the past few years, hundreds — perhaps thousands — of partnerships have been formed in the U.S. to implement activities consistent with biosphere reserve goals. In fact, many areas not designated biosphere reserves reflect biosphere reserve goals more fully than existing designated areas.

A significant effort is needed to inform potential stakeholders, including the U.S. Congress, of the accomplishments of U.S. biosphere reserves and the benefits of working toward a U.S. network that is fully implementing biosphere reserve concepts. Through USMAB, successful ecosystem-based partnerships involving existing biosphere reserves and local groups working toward sustainable development in the vicinity of existing biosphere reserves should be recognized and encouraged to consider the "value added" of identifying with biosphere reserve goals. In nominating new biosphere reserves, opportunities for cooperation in a "biogeocultural" area that makes sense to local people should be assessed. The nomination of particular sites for inclusion in a biosphere reserve should based on their capabilities and interests in pursuing these opportunities for cooperation. Local partnerships should be encouraged to recommend ways to improve the functionality of existing biosphere reserves, including the option of restructuring existing biosphere reserves through the addition, deletion, or consolidation of designated sites.

Finally, there is a need to strengthen the scientific functions of biosphere reserves as demonstration areas for integrating interdisciplinary research, existing networks for monitoring natural and human systems, and predictive assessment. In this regard, existing and future biosphere reserves could provide an important subregional coordinating framework for the National Science and Technology Council's efforts to integrate the nation's environmental monitoring and research networks and programs,[44] and help link this initiative to the efforts of stakeholders in a particular biogeocultural region to plan an ecologically, culturally and economically sustainable future.

The Southern Appalachian Biosphere Reserve is one of the world's longest-running efforts to implement the biosphere reserve concept. Over the past 20 years, the concept has catalyzed numerous projects that have helped the region's stakeholders address shared problems. These activities have led to a multisector framework for cooperation in demonstrating ways to conserve the region's significant ecosystems, protect its cultural heritage, and assist its communities in achieving sustainable development. The contributions to this book describe the many components of this continuously evolving demonstration, which exemplifies the global leadership of a unique region of the United States.[45] Such efforts should be recognized and encouraged. The lessons learned in the southern Appalachians should be considered by stakeholders in any region in evaluating the practical benefits of the biosphere reserve concept in helping them develop their own goals and strategies.

REFERENCES

1. Center for Our Common Future. 1993. The Earth Summit's Agenda for Change. The Center for Our Common Future, Geneva, Switzerland. 70 pp.
2. Batisse, Michel. 1986. Developing and focusing the biosphere reserve concept. *Nat. Resour.* 22(3):1–10.
3. Lasserre, P., M. Hadley, and J. Robertson. 1993. The International Network of Biosphere Reserves: narrowing the gap between reality and potential. Paper presented at Conference of Managers of U.S. Biosphere Reserves, Estes Park, Colorado, December 6 to 10, 1993. UNESCO, Paris.
4. Batisse, M. 1993. Biosphere reserves: an overview. *Nat. Resour.* 29(1-4):3-5.
5. United Nations Educational, Scientific and Cultural Organization. 1971. International Co-ordinating Council of the Programme on Man and the Biosphere (MAB), First Session, November 9 to 19, 1971: Final Report. UNESCO, Paris. 65 pp.

6. United Nations Conference on the Human Environment. 1973. Report of the United Nations Conference on the Human Environment, Stockholm, June 5 to 16, 1972. United Nations, New York.

7. United Nations Educational, Scientific and Cultural Organization. 1974. Report of the Task Force on Criteria for the Choice and Establishment of Biosphere Reserves. MAB Report Series No.22. UNESCO, Paris. 46 pp.

8. Franklin, J.F. and S.L. Krugman, Ed. 1979. Selection, Management and Utilization of Biosphere Reserves: Proc. U.S.-U.S.S.R. Symp. Biosphere Reserves, Moscow, U.S.S.R., May 1976. U.S. Department of Agriculture, Forest Service, Pacific Northwest Forest and Range Experiment Station. Gen. Tech. Rep. PNW-82. 307 pp.

9. Hemstrom, M.A. and J.F. Franklin, Ed. 1981. Successional Research and Environmental Monitoring Associated with Biosphere Reserves. Proc. Second U.S.-U.S.S.R. Symp. Biosphere Reserves, March 10 to 15, 1980, Everglades National Park, FL. U.S. National Park Service Publ. No. 1799. 271 pp.

10. Izrael, Y.A. 1984. The concept of ecological monitoring in biosphere reserves. Annex 2 *in* UNESCO-UNEP, Conservation, Science and Society. UNESCO, Paris. 2 vols. 612 pp., plus annexes.

11. United Nations Educational, Scientific and Cultural Organization. 1974. Report of the task force on criteria for the choice and establishment of biosphere reserves. MAB Rep. Ser. No. 22. UNESCO, Paris. 46 pp.

12. United Nations Educational, Scientific and Cultural Organization. 1987. A practical guide to MAB. UNESCO, Paris. 40 pp.

13. United Nations Educational, Scientific and Cultural Organization. 1984. Action plan for biosphere reserves. *Nat. Resour.* 20(4):1-12.

14. United Nations Educational, Scientific and Cultural Organization. 1996. Biosphere Reserves: the Seville Strategy and the Statutory Framework of the World Network. UNESCO, Paris. 18 pp.

15. Franklin, J.F. 1977. The biosphere reserve program in the United States. *Science* 195:262–267.

16. United States Man and the Biosphere Program. 1979. Long-term Ecological Monitoring in Biosphere Reserves. International Workshop on Long-term Ecological Monitoring in Biosphere Reserves, October 20 to 28, 1978, Washington, D.C. 31 pp., plus appendixes.

17. United States Man and the Biosphere Program. 1980. Program plan for the U.S. Man and the Biosphere Program. Unpublished report. U.S. Department of State, Washington, D.C.

18. U.S. Department of the Interior, National Park Service. 1980. State of the Parks — 1980: a Report to the Congress. National Park Service, Office of Science and Technology, Washington, D.C.

19. International Union for the Conservation of Nature and Natural Resources. 1980. World Conservation Strategy. IUCN, Gland, Switzerland.

20. DiCastri, F., F.W.G. Baker, and M. Hadley, Ed. 1984. *Ecology in Practice.* Tycooly International Publishers: Dublin, Ireland. 2 vols: 1:*Ecosystem Management,* 524 pp, and 2:*The Social Response,* 382 pp.

21. Gregg, W.P. and M.M. Goigel. 1984. Putting the biosphere reserve concept into practice: the United States experience. *In* DiCastri, F., F.W.G. Baker, and M. Hadley, Eds. 1984. *Ecology in Practice,* Part I: *Ecosystem Management.* Tycooly International Publishers; Dublin, Ireland.

22. Sokolov, V. 1984. The system of biosphere reserves in the U.S.S.R.: status and prospects. Pp. 492-499 *in* DiCastri, F., F.W.G. Baker, and M. Hadley, Eds. 1984. *Ecology in Practice,* Part I: *Ecosystem Management.* Tycooly International Publishers, Dublin, Ireland.

23. United Nations Educational, Scientific and Cultural Organization and United Nations Environment Programme. 1984. Conservation, Science, and Society. Contributions to the First International Biosphere Reserve Congress, Minsk, Byelorussia, U.S.S.R., September 26 to October 2, 1983. UNESCO, Paris. 2 vols. 612 pp, plus annexes.

24. Subcommittee on Natural Resources, Agriculture Research and Environment of the House Committee on Science and Technology. 1983. Hearings on U.S. Man and the Biosphere Program relating to proposed U.S. withdrawal from UNESCO. April 5, 1983.

25. United States Man and the Biosphere Program. 1981. Interim Guidelines for the Identification and Selection of Coastal Biosphere Reserves: a Report to the Directorate on Biosphere Reserves. Unpublished report, USMAB Secretariat, Department of State, Washington, D.C. 30 pp.

26. United States Man and the Biosphere Program. 1983. Guidelines for Identification, Evaluation and Selection of Biosphere Reserves in the United States. USMAB Report No.1 (First Revision). Department of State, Washington, D.C. 38 pp.

27. Gregg, William P. and Hubert H. Hinote. 1995. Toward a U.S. modality of biosphere reserves: the Southern Appalachian Biosphere Reserve. Presentation at the International Conference on Biosphere Reserves, Seville, Spain, March 20 to 25, 1995. Available from Southern Appalachian Man and the Biosphere Program, Gatlinburg, TN 8 pp.

28. United States Man and the Biosphere Program, Directorate on Biosphere Reserves. 1989. Guidelines for selection and coordination of U.S. biosphere reserves (draft report). U.S. Man and the Biosphere Program, Directorate on Biosphere Reserves, Washington, D.C.

29. Peine, J.D., Ed. 1985. Proc. Conf. Management of Biosphere Reserves, November 27 to 29, 1984, Great Smoky Mountains National Park, Gatlinburg, TN. U.S. Department of the Interior, National Park Service, Gatlinburg. 207 pp.

30. Gregg, W.P., Jr., S.L. Krugman, and J.D. Wood, Jr., Ed. Proc. Symp. Biosphere Reserves, Fourth World Wilderness Congress, September 14 to 17, 1987, YMCA of the Rockies, Estes Park, CO, U.S. Department of the Interior, National Park Service, Atlanta, GA. 291 pp.

31. Interagency Ecosystem Management Task Force. 1995. The ecosystem approach: healthy ecosystems and sustainable economies. Vol.1: Overview. Interagency Ecosystem Management Task Force, Washington, D.C. 54 pp.

32. U.S. Man and the Biosphere Program, Biosphere Reserve Directorate. 1994. Strategic Plan for the U.S. Biosphere Program. Department of State Publication No. 10186. Washington, D.C. 28 pp.

33. UNESCO. 1996. Biosphere Reserves: the Seville Strategy and the Statutory Framework of the World Network. UNESCO, Paris. 18 pp.

34. EuroMAB. 1993. ACCESS: a Directory of Contacts, Environmental Data Bases, and Scientific Infrastructure on 175 Biosphere Reserves in 32 Countries. U.S. Department of State Publ. No. 10059. Washington, D.C.

35. EuroMAB. 1996. ACCESS 1996: a Directory of Permanent Plots which Monitor Flora, Fauna, Climate, Hydrology, Soil, Geology, and the Effects of Anthropogenic Changes at 132 Biosphere Reserves in 27 Countries. Department of State Publ. No. 10322. Washington, D.C. 342 pp.

36. Quinn, J. F., R.J. Meese, D.C. Hudson, T.C. Lebeck, and J.A. Gaines. 1995. MABFauna (including Observe): A handbook for users of the MAB biological inventory system. University of California at Davis, Davis, CA. 51 pp.

37. Dallmeier, F., Ed. 1992. Long-term monitoring of biological diversity in tropical forest areas, MAB Digest 11. UNESCO, Paris. 72 pp.

38. Comiskey, J. A., G.E. Ayzanoa, and F. Dallmeier. 1995. A data management system for monitoring forest dynamics. J. Trop. For. Sci. 7{3}:419–427.

39. U.S. Man and the Biosphere Program, Biosphere Reserve Directorate. 1995. Biosphere reserve managers' survey 1995. (Analysis of survey results.) Available from USMAB Secretariat, Department of State, Washington, D.C. 16 pp.

40. United States Man and the Biosphere Program 1995. Biosphere reserves in action: case studies of the American experience. U.S. Department of State Publ. No. 10241, Washington, D.C. 86 pp.

41. Constable Commission. 1995. Final report to the United States Man and the Biosphere Program. Available from the USMAB Secretariat, Department of State, Washington, D.C. 15 pp.

42. Engel, J. Ronald. 1989. The symbolic and ethical dimensions of the biosphere reserve concept; pp. 21–31 In Gregg, W.P., Jr., S.L. Krugman, and J.D. Wood, Jr., Eds. Proc. Symp. Biosphere Reserves, September 14 to 19, 1987. U.S. Department of the Interior, National Park Service, Atlanta, GA. 1989. 190 pp.

43. Engel, J.R. 1985. Renewing the bond of mankind and nature: biosphere reserves as sacred space. *Orion* 4(3):52-59.

44. National Science and Technology Council. 1996. Integrating the nation's environmental monitoring and research networks and programs: a proposed framework. National Science and Technology Council, Washington, D.C.

45. Hinote, H.H. and W. P. Gregg. 1995. Toward a U.S. modality of biosphere reserves: the Southern Appalachian Biosphere Reserve. Paper presented at the UNESCO International Conference on Biosphere Reserves, Seville, Spain. March 20 to 25 1995.

3 The Evolution of Land Use Ethics and Resource Management: Coming Full Circle

John E. Nolt and John D. Peine

CONTENTS

Prehistoric Period...
Cherokee Nation (1540 to 1750)...42
Early European Settlement (1750 to 1838)..43
Large-Scale Resource Extraction (1838 to 1930)..45
    Cattle Ranching...47
    Timber Industry...48
    Beginning of the Tourist Industry..48
Creation of the National Park (1930 to 1975)...49
    Developing the Park Experience around the Automobile...................................50
    Creating a Sport Fishery on Abrams Creek...50
    Agriculture in Cades Cove..50
    Managing a Historic Landscape in Cades Cove..51
Natural and Cultural Resources Protection and Management (1976 to 1987).............52
    Black Bear Management..53
    Wild Hog Control...54
    The Deer Herd in Cades Cove...55
    Fire Management...55
Sustainable Development Strategies for the Southern Appalachians (1988 to Present)...............55
    The Southern Appalachian Man and the Biosphere Cooperative (SAMAB)............55
    The Southern Appalachian Mountain Initiative...56
    Pitman Center, a Future Gateway Community..56
    Red Wolf Reintroduction 56
    The Automobile and the Park 57
Future Challenges: How Deep the Spiral? (2000 and Beyond) 57
References 58

It is frequently claimed that aboriginal peoples have practiced something like an environmental ethic — so frequently, in fact, that a backlash has resulted. Many authors are now busy cataloguing the ecological *sins* of ancient and aboriginal cultures.* The history of land use ethics is complex, but it remains generally true that aboriginal cultures included the natural world in their ethical

---

* See, for example, "Environmental History Challenges the Myth of a Primordial Eden," *The Chronicle of Higher Education.* May 4, 1994. A56.

deliberations more deeply than civilized, and especially industrialized, cultures typically have — at least, until very recently. In this chapter, we sketch the history of land use ethics in a particular geographical region, the southern Appalachian highlands. We shall see that this history exemplifies the general pattern of a departure from and return to moral consideration of nature and that the evolution of land use ethics from prehistory to modern times is closely tied to social behavior highly influenced by economic trends.

In speaking of land use ethics, we use the term "land" broadly, as in Leopold (1966), to include not only the soil, but also plants, animals, rivers, mountains, and ecological systems in general. Thus the terms "land use ethic" and "environmental ethic" are for us synonymous.

The history presented here is the context from which an ethic for ecosystem management has emerged in the environs of the Great Smoky Mountains National Park, hereafter referred to as "the Park."

PREHISTORIC PERIOD (PRIOR TO 1540)

The earliest known inhabitants of the southern Appalachian region were here by the end of the most recent ice age, somewhere around 10,000 B.C. (Dickens 1976, introduction). Known as the Paleo-Indians, they were nomadic hunters of large mammals. Though some evidence remains of their behavior toward the land, virtually nothing is known about their beliefs. Whatever their ethic — or, better, mythos — was, it seems not to have prevented them from overexploiting their food sources. The mastodon, the American horse, the camel, and *bison antiques*, for example, all of which inhabited large portions of North America until 10,000 B.C., were extinct by 7000 B.C., apparently victims of climate change and/or overhunting (Hudson 1976, p. 41).

Around 8000 B.C., with the retreat of the glaciers and the demise the of large mammals, the Paleo-Indian culture gave way to the Archaic Tradition. Diet was now more diverse: fish, woodland animals, especially whitetail deer (*Odocoileus virginianus*), and acorns (*Quercus* spp.) and hickory (*Carya* spp.) nuts were staples. Bands became smaller, more sedentary, and increasingly efficient in exploiting local sources of food (Hudson 1976, pp. 44–47). There is some evidence from as early as 5000 to 8000 B.C. of the presence of fire in the landscape, presumably as a means to maintain open lands to enhance wildlife habitat and perhaps to facilitate primitive agricultural activity (Delcourt and Delcourt 1996).

Around 1000 B.C. there emerged a new tradition, distinctive to eastern North America. The Woodland tradition intensified the tendency toward increasing settlement and exploitation of local food sources. Sunflowers (*Helianthus* spp., *Rudbeckia* spp.), sumpweed (*Iva annua*), and possibly goosefoots (*Chenopodium* spp.) appear to have been cultivated, and many varieties of seeds and nuts were stored for winter and spring use. The Woodland peoples wove fabric from twilled vegetable fiber and made extensive use of pottery, whose design varied from place to place, suggesting localization and diversity. They also created large mounds and earthworks throughout the East. Some archaeologists point to these structures as evidence of a highly developed agriculture, but others argue that the woodlands were so rich in wild foods that the surplus required to supply the leisure to build large structures was available even with a very primitive agriculture. Trade was extensive during the Woodland period; lavish burial sites contain shark's teeth from the Atlantic and Gulf coasts and grizzly bear teeth and obsidian from the Rocky Mountains (Hudson 1976, pp. 55–77).

Between 700 and 900 B.C., beginning in the central Mississippi Valley, the Woodland tradition began to give way to the final prehistoric period, the Mississippian. Population increased, towns became larger and more permanent, and elaborate burials ceased. The Mississippians erected wooden buildings on large flat-topped pyramidal mounds and seem to have made considerable use of palisades, defensive towers, ditches, and other means of fortification.

The chief agricultural crops were now beans, corn, and squash, the latter two being plants of tropical origin whose cultivation had spread into North America during the Woodland period. Beans probably originated in the Southeast. These crops were now becoming staples, but hunting and gathering remained important.

The Mississippian culture lasted into historic times. Hernando de Soto observed Mississippian villages when he explored the Southeast in 1539 to 1543. But by the time extensive contact between the whites and the Southeastern Indians began a century and a half later, the old Mississippian social order was gone and the old religion was no longer practiced (Hudson 1976, pp. 77–97; Josephy 1993, 17). The causes of this decline remain obscure, but among them may be devastation from introduced European diseases (Frome 1980, p. 48).

To what extent did these prehistoric peoples alter their larger environment? Almost certainly they played a role in the extinction of large mammals, but apart from that little can be said. The landscape did change dramatically following the end of the ice age — boreal forests, for example, receded in the upper elevations in the Southeast, and oak–hickory forests advanced from the south — but the change was due primarily to global warming following the ice age (Delcourt and Delcourt 1996). As a result, it is difficult to define normalcy of ecosystem processes over the period of prehistoric human habitation. The still controversial role of deliberately set fires in altering the landscape exacerbates this difficulty.

CHEROKEE NATION (1540 TO 1750)

The Cherokee Nation, which evolved slowly out of the Mississippian tradition in the southern Appalachians, enters recorded history in the year 1540, when the DeSoto expedition, crossing the Blue Ridge Mountains, first encountered the Cherokees. By the time the European settlers arrived in the southern Appalachians in the 18th century, the Cherokee Nation was arguably the most civilized culture in North America. Cherokee territory encompassed portions of what are now Virginia, North Carolina, South Carolina, Tennessee, Georgia, and Alabama. Their population numbered 10,000 to 20,000 in some 40 to 60 settlements clustered along streams and rivers (Frome 1980, p. 21). The center of the Cherokee nation was Cowee on the Little Tennessee River approximately 12 air miles south of the Park.

The Cherokee were agriculturists, living in relatively permanent river bottom settlements. Over time, their agricultural activity became quite extensive, encompassing large tracts of bottomland. As with their Mississippian forebears, their chief crops were corn, beans, and squash, which they supplemented by hunting, fishing, and gathering. The Cherokees were also active participants in a transcontinental network of trade.

It is with the Cherokees that our knowledge of land use ethics in the Great Smoky Mountains region begins. The Cherokees saw themselves as part of a natural balance, the disturbance of which brought forth evil. This notion of balance is evident in the Cherokee myth of the origin of disease and medicine. In ancient times, according to this myth, excessive hunting by humans threatened the animals with extinction. (Is this an echo of the Paleo-Indian experience with large prehistoric mammals?) The animals responded by inventing diseases, which they magically inflicted upon humans. The plants, however, remained friendly to the humans and countered this magical attack by providing an herbal cure for each disease (Mooney 1982, pp. 250–252; Hudson 1970). Yet as a result it became necessary for the hunter of certain animals to apologize to his kill and take ritual precautions to prevent the dead animal's spirit from avenging itself by inflicting disease upon the hunter or his household (Hudson 1976, pp. 158 and 346).

The Cherokees regarded animals as not essentially different from human beings (Mooney 1982, 261–262). They believed, for example, that bears were a tribe of humans who had been transformed by dwelling in wild places. Some bears, they thought, retained the power of speech (Mooney 1982, pp. 264 and 325–327).

Plants, though not so closely related to humans, were nevertheless spiritual beings. James Mooney describes the Cherokee conception of the ginseng plant:

> The beliefs and ceremonies in connection with its gathering are very numerous. The doctor [shaman] speaks constantly of it as of a sentient being, and it is believed to be able to make itself invisible to those unworthy to gather it. In hunting it, the first three plants found are passed by. The fourth is taken, after a preliminary prayer, in which the doctor addresses it as the "Great Ada'wehi," and humbly asks permission to take a small piece of its flesh. On digging it from the ground, he drops into the hole a bead and covers it over, leaving it there, by way of payment to the plant spirit. After that he takes them as they come without further ceremony (Mooney 1982, p. 425).

The Cherokees respected not only animals and plants, but even nonliving natural objects as magically potent and spiritually alive. This is evident, for example, in their personification of rivers:

> In Cherokee ritual, the river is the Long Man, *Yunwi Gunahita*, a giant with his head in the foothills of the mountains and his foot far down in the lowland, pressing always, restless and without stop, to a certain goal, speaking ever in murmurs which only the priest may interpret. In the words of the sacred formulas, he holds all things in his hands and bears down all before him. His aid is invoked with prayer and fasting on every important occasion of life, from the very birth of the infant, in health and sickness, in war and love, in hunting and fishing, to ward off evil spells and to win success in friendly rivalries. Purification in the running stream is a part of every tribal function, for which reason the town-house, in the old days, was erected close to the river bank (Mooney 1900, p. 1–2).

As was common practice among native American tribes throughout the eastern U.S., the Cherokees routinely burned large expanses of bottomland to create rangeland for white tail deer and open space for the cultivation of maize, squashes, and tobacco. Hudson describes the burning as "light" and attributes several practical and ecological benefits to it:

> Not only did it reduce the threat of serious forest fires by reducing the accumulation of dead wood and litter on the forest floor, but it also laid down a bed of ashes, a soil nutrient, and it kept the forest open by clearing out underbrush, tree seedlings, and saplings. This stimulated the growth of open meadows and plant life on which deer could browse, it probably made acorns and chestnuts easier to find, and this in turn would have increased both the deer and turkey populations (Hudson 1976, 276–277).

Others, however, have argued that the impact of the burning was substantial, particularly prior to the coming of the Europeans.

The influence of native Americans on the landscape may have greatly diminished shortly after the Europeans arrived, as a result of a catastrophic loss of population due to the introduction of infectious diseases by the early explorers. The pre-Columbian native North American population estimates range from 1 to 18 million. It has been estimated that the native American population dropped by over 90% as a result of these epidemics, against which there was no immunity (Thomas 1993, p. 105). Some old-growth forest stands, such as the highly acclaimed Joyce Kilmer Forest and Wilderness Area, may be remnants of this release from frequent humanly kindled fires. This hypothesis would explain the fact that Joyce Kilmer harbors an even-aged stand of early succession tulip poplar trees (*Liriodendron tulipifera*) that all began growing 400 or 500 years ago (Ed Buckner, pers. commun.).

Burning was, no doubt, the Cherokees' most dramatic effect on the land. But even the burning was practiced in what might be regarded as a sustainable manner; there is no evidence of loss of ecosystems or extinction due to fire. In other respects, the Cherokee impact on the land was unobtrusive.

EARLY EUROPEAN SETTLEMENT (1750 TO 1838)

Though trade with the Cherokees began around 1700, white settlers did not arrive in the Smokies until around 1750. Thereafter, their numbers increased rapidly. It was in this period that the most dramatic shift in land use ethic occurred, both among the Cherokees themselves and with the arrival of the newcomers.

The Cherokees, ravaged by war, disease, and whiskey, experienced the rapid dissolution of their culture. Despite their tradition of appeasing and balancing the forces of nature, during the 18th century the Cherokees, in concert with other tribes and the settlers, hunted the white-tailed deer almost to extinction in a frenzy comparable to the slaughter of the buffalo on the Great Plains (Hudson 1976, 436–437). This behavior is sometimes cited as evidence that the Cherokees' respect for nature was a sham and that they would have been just as exploitative as the settlers if they had access to the same technology.

But such a conclusion is premature. The Cherokees needed guns to defend themselves against military assaults, both by the Europeans and by neighboring tribes, and they needed iron tools to work the marginal land to which they were increasingly forced to retreat. These items could be obtained only through trade, and the only marketable commodity the Cherokees had to offer was deerskin. Whether they would have slaughtered the deer if their own survival had not been at stake is a question that cannot be answered with confidence (Hudson 1976, pp. 427–443).

Their behavior may also be explained, at least in part, by the interaction of their beliefs with the changing circumstances. Diseases brought by whites had afflicted the Southeastern Indians ever since the DeSoto expedition. In 1738 a smallpox epidemic killed nearly half the remaining Cherokees (Hudson 1976, p. 173). Given their account of the origin of disease, it would have been natural for the Cherokees to conclude that the animals had concocted new maladies and were engaged in a magical aggression against them. Such reasoning could easily have eroded the original reluctance to overhunt and perhaps even sparked retaliation against the deer. There is evidence that such mythological thinking did, in fact, justify the roughly simultaneous slaughter of fur-bearing animals among tribes further north (Martin 1978). Thus the very ideology that supported sustainable hunting practices before the arrival of the Europeans may have served to rationalize overhunting as the plight of the Cherokee became increasingly desperate.

A very different land use ethic guided the settlers. We can trace it to two philosophical roots: (1) fundamentalist Protestantism, and (2) the Enlightenment ethic of property rights, stemming largely from the work of the British philosopher John Locke (1632 to 1704).

The Smoky Mountain settlers were overwhelmingly Protestant and fundamentalist: Baptists, Presbyterians, and Methodists. A measure of the difference between their land use ethic and that of the Cherokee can be gleaned from the respective attitudes of these two peoples toward the afterlife, for these attitudes reflect the relative importance assigned to life on Earth. For the traditional Cherokee, the souls of the dead may linger for a while as ghosts but eventually find their way to the Darkening Land in the West, a place less appealing than the here and now. For the settlers, Heaven was the true and glorious life; the earth was only a temporary habitation, stained with corruption and evil. This view was powerfully expressed in the theology of John Calvin. "If Heaven is our country," wrote Calvin, "what is the earth but a place of exile? — and if the departure out of this world is an entrance into life, what is the world but a sepulcher?"(Calvin 1928). Though only a fraction of the settlers were Calvinists in the strict predestinarian sense, this Calvinist denigration of the earth was probably fairly widespread among them (Peacock and Tyson 1989, pp. 6–7). Given the likewise popular belief in the imminence of the apocalypse, and the exhortation in Genesis 1:28 to "Be fruitful and increase and fill the earth and subdue it," their ethic, unlike the Cherokees', had little room for the notion of a sustainable natural balance.*

---

* Cf. Hudson 1970. The general influence of traditional Christian ideas on land use ethics is the subject of a wide-ranging debate. (White 1967; Whitney 1993).

Yet according to that same Genesis story, God found the creation good, and he gave all living things to humanity. These ideas, too, shaped the attitudes of the settlers, most of whom probably regarded earthly life as good, despite their glorification of Heaven.*

The second root of the settlers' land use ethic is the philosophy of the Enlightenment, particularly that of Locke, whose concepts of natural rights, liberty, and property ultimately spurred the American Revolution. Citing biblical authority, Locke held that the land was a gift of God to humanity. In spite of this divine origin, the materials of nature are, according to Locke, almost worthless in themselves, acquiring value chiefly through human labor. This justifies individual ownership:

> As much Land as a man tills, plants, improves, cultivates, and can use the product of, so much is his property. He by his labor does, as it were, inclose it from the common (Locke 1980, 21)".

In this way previously "unimproved" land becomes the property of any individual who "mixes his labor" with it. Since the native Americans did little to "improve" the land (i.e., to increase its economic value), Locke's theory rationalized the European seizure of American lands.

This philosophy was highly influential in colonial America. It is prominent, for example, in the works of Thomas Jefferson,** and was undoubtedly popular among the Smoky Mountain settlers. In sum, the settlers regarded the land ambivalently as both a divine gift and a place of exile; but they were firmly convinced of their right to subdue, civilize, and possess it by their labor.

And possess it they did. Large land holdings obtained by treaty from the Cherokees were distributed for settlement by the government. In addition, there was much illegal settlement on Cherokee land. Subsistence farming was the norm, but settlement was so rapid and intense that the best bottomland was quickly occupied; and hillsides, never used by the Cherokees for agriculture, were cleared, plowed, and cultivated. Erosion and overuse rapidly depleted the soil.

Farmers allowed their cattle and hogs to forage freely in the woods and routinely set fires, particularly along ridge tops, to enhance the growth of grasses on the forest floor. With their guns, the settlers rapidly depleted the game. The eastern bison (*Bison*), red wolf (*Canis rufus*), passenger pigeon (*Ectopistes migratorius*), American elk (*Cervus elaphus*), and eastern mountain lion (*Felis concolor*) were eventually eliminated.

Whereas the Cherokees saw the land (in the broad sense defined at the outset) as a community of spiritual beings, the Europeans saw its value almost exclusively in anthropocentric terms; the land existed for human use. This contrast was starkly reflected in their respective uses of ginseng, a plant regarded as a spiritual being by the Cherokees and as a source of income by the Europeans. (The settlers exported ginseng through markets from Philadelphia to China where it was esteemed for magical qualities.) North Carolina historian Foster A. Sondley, writing in the 1920s, describes the exploitation of ginseng by whites:

> This very peculiar and most interesting plant is now rarely, if ever met with anywhere in the country, and its extinction is a striking example of uncontrolled greed, especially on the part of the least industrious members of new communities, who regard all wild products as common property to be appropriated and even wantonly abused without right or restriction.***

---

* That has apparently been the case, at least, among their descendants. See Dorgan 1987, 203–207.

** Jefferson's view of property was more specific than Locke's in one particular: Jefferson held that property was an "adventitious" or "alienable" right, a right that could voluntarily be given up, as opposed to an "unalienable right," which could not. Locke did not make this distinction. Thus, whereas for Locke the fundamental human rights are "life, liberty, and property," Jefferson, seeking in Declaration of Independence to justify the American revolution in terms of the British violation of "unalienable rights," changes this triad to "life, liberty, and the pursuit of happiness." Still, Jefferson, like Locke and the American revolutionaries generally, explicitly maintained Locke's view that the individual has a natural, moral right to property. See, for example, White 1978, 213–228.

*** Quoted in Dykeman and Stokely 1978, p. 55.

Throughout the settlement period encroachment on Cherokee land intensified, despite its prohibition by numerous treaties. By the late 1820s, the Georgia legislature, clamoring for the elimination of Cherokees from the state, enacted laws to appropriate Cherokee lands, dissolve the laws of the Cherokee Nation, and require that any Cherokee actively resisting the sale of Cherokee lands serve not less than 4 years in the penitentiary. White missionaries who protested these injustices were beaten and jailed. The case of one, Samuel Worcester, eventually reached the Supreme Court, which declared some of the Georgia laws unconstitutional and ordered his release. President Andrew Jackson, however, refused to enforce the ruling (Hudson 1976, pp. 462–463). In 1830 Jackson sponsored and signed into law a bill mandating removal of all the Southeastern tribes, including the Cherokees. The Cherokees resisted longest and were the last to be exiled. Yet in 1838 they too were rounded up and, refusing to ride in government wagons, were forced to walk — young and old, sick and well — to desolate reservations in Oklahoma. Thousands died along this bitter trek, which has come to be known as the Trail of Tears. A few hundred, hiding in the mountains, survived to form the core of the Eastern Band, which now resides on a reservation on the south side of the Park. Yet the Cherokee religion and way of life were crushed, and dominion over the Smokies passed exclusively to the whites.

LARGE-SCALE RESOURCE EXTRACTION (1838 TO 1930)

As the white population increased, large-scale industries began to move into the southern Appalachians. Corporate interest in the region accelerated dramatically with the introduction of the railroad in the late 19th century. The major change in land use ethic as we enter the period of large-scale resource extraction is the shift from individual to corporate rights. Property rights, which Locke saw as accruing to individuals by their labor, are in this period granted by custom and law to corporations. At the heart of the pioneer ethic were independence and individual self-determination. In this increasingly corporate-centered ethic, the rights of the individual give way to the values of efficiency and profit. The result is a *laissez faire* economics of the kind envisioned by Adam Smith (1723 to 1790) (Smith 1930).

This time period has been referred to as the "colonization period" of the southern Appalachians (Lewis et al. 1978). While still seen as property, the land was now understood less as a source of individual livelihood than as an ensemble of resources for corporate economic development. The land was to generate, not merely livelihood or wealth, but great quantities of *surplus* wealth: capital. Immediate financial gain, epitomized in the then-popular maxim "get it while the gettin's good," took precedence over longer term economic return. This large-scale corporate exploitation of resources for short-term gain without regard for ecosystem sustainability produced a situation comparable to the current destruction of the rainforest to create short-term pastures for the cattle industry in the Brazilian state of Amazonia.

In accord with this *laissez faire* ethic, individuals and corporations obtained large blocks of land in the southern Appalachians by original land grants. Speculation was rampant on the acceleration in land value associated with such forms of development as tenant homesteading and livestock operations. In fact, much of the culture commonly associated with the "cowboys" of the Western U.S. in the late 19th century was present by midcentury in the Great Smoky Mountains. Toward the end of the century, logging became the dominant industrial activity. Tourism began to grow, as well.

CATTLE RANCHING

Typical of these ranching operations were the holdings of Drury Armstrong. In the early 1840s, Armstrong held several large tracts of land totaling over 50,000 acres in Sevier County, TN. In order to develop the land in the most cost effective means, he leased parcels to homesteaders for an average of 10 years. After 10 years, the improved property was returned under arrangements like the following:

> Said David Howser agrees to put up a comfortable, hewed log cabin, kitchen and smoke house and stable… He also agrees to open, fence, wall and put into cultivation, during the lease, not less than fifteen acres of land, and as much as 25 acres….All the cleared land to be kept and left under a good ten rail fence….At the expiration of this lease, … David Howser agrees and obliges himself to surrender to Armstrong the whole of the premises, in good repair (Armstrong 1851).

Armstrong organized large cattle drives for his leaseholders to move cattle from the mountains to nearby markets located, for example, in Maryville, TN.

TIMBER INDUSTRY

Logging was another early industry. Two rather distinct phases of logging occurred in the Smokies. The earlier period ended about the beginning of the 20th century. Characterized by small enterprises managed by local people and financed by local capital, these operations selectively cut the most valuable timber of the day, particularly tulip poplar (*Liriodendron tulipifera*), black cherry (*Prunus serotina*) and ash (*Fraxinus* spp.). Although they went deep into the forest for prime specimens, they generally stayed near streams and other accessible places (Lambert 1958). The Middle and West Prong of the Little River Basins were selectively logged and repeatedly burned. Ridgetops were also burned to provide grazing land. The loggers built splash dams to expedite floating the logs down river. An estimated 50% of the softwoods had been removed from these watersheds by the turn of the century (Ayres and Ashe 1905).

The later period saw the introduction of large-scale timber operations. The growing demand for lumber and paper and the development of improved methods for obtaining, transporting, and manufacturing wood products brought outside capital and management. The largest such enterprise operating on the Tennessee side of what was to become Great Smoky Mountains National Park was the Little River Lumber Company. That firm, under the guidance of its founder W. B. Townsend, built its mill outside the park in the community now bearing his name. The firm logged on three major watersheds, using a railroad that extended 18 miles along the Little River to the then logging camp of Elkmont. It is estimated that the Little River Lumber Company removed 560 million board feet of timber from the Park, representing roughly 28% of the estimated 2 billion board feet of timber removed by large and small operators (Lambert 1958).

By the 1920s, with the advent of steam powered skidders and logging railroads, harvest practices had become extremely destructive. Clearcutting was the norm. The clearing of lanes for inclines and overhead skidders laid bare large areas where no trees of log size existed. This hindered forest regeneration, and the disrupted soils eroded, harming stream biota.

Forest fires were the most destructive concomitant of logging. The loggers cut over all types of terrain and left large areas bare, allowing the remaining trees, shrubs, and soils to dry out — along with an enormous tangle of discarded limbs and tops of trees, that often accumulated to a depth of 7 feet or more. Sparks from wood burning locomotives and skidding machines frequently ignited fires, the most serious of which occurred in the 1920s. One series on Blanket Mountain reportedly burned for 2 months (Lambert 1958).

BEGINNING OF THE TOURIST INDUSTRY

Generally the mountaineers remained poor, despite this influx of industry. Some, however, began to make a living from tourism. A small hotel known as LeConte Lodge operated in the vicinity of the Greenbriar community within what is now the park boundary. The establishment entertained early tourists to the Smoky Mountains. At that time, it was common practice for local residents to trap black bears, keep them in cages, and fatten them with American chestnuts (*Castanea dentata*). When the keeper was ready to butcher the bear, lodge guests occasionally paid significant fees (up to $75) to shoot it while their picture was being taken. The tourist got the picture and the hide and the local resident got the meat and money (Glenn Cardwell, pers. commun.).

CREATION OF THE NATIONAL PARK (1930 TO 1975)

The idea for a park in the southern Appalachians arose before the turn of the century. In 1900, the Appalachian Park Association sent a message from the Secretary of Agriculture to Theodore Roosevelt advocating a national park. But the idea took hold only with the progressivism of the New Deal. This progressivism brought yet another dramatic shift in land use ethic. Since the European influx, human relations to the land had been dominated by the Lockean notion of individual and, later, corporate property rights. But in the 1930s property rights were challenged and frequently overridden by condemnation proceedings.

The underlying assumption of park advocates was no longer the Lockean idea that the right action must respect individual human rights, nor the *laissez faire* capitalist notion that the right action is the one that maximizes economic efficiency, but the utilitarian idea that the right action was the one which yields the greatest happiness for the greatest number. ("Utility," in its philosophical sense, means usefulness for the production of happiness.) Happiness, for the progressive utilitarians was most emphatically not defined merely as economic gain. It included as well such intangible values as aesthetic enjoyment and the rejuvenative benefits of the wilderness experience. As a moral theory, utilitarianism was first articulated by two British philosophers, Jeremy Bentham (1748 to 1832) and John Stuart Mill (1806 to 1873). Like *laissez faire* economics, it is essentially secular in motivation, positing as the goal of ethics neither the salvation of souls nor respect for natural or God-given rights, but the maximization of human happiness.

It is noteworthy that in the original formulations of both Mill and Bentham, allowance was made as well for the happiness of animals. Utilitarianism has thus had a large influence on modern humane and animal rights movements. But the early Park advocates were not concerned with the happiness or pain of animals. Their philosophy is succinctly expressed on the bronze plaque in the masonry wall at Newfound Gap, where the Park's dedication ceremony was held on September 2, 1940. In bold capital letters, the plaque reads: "*for the permanent enjoyment of the people.*"

A more explicit, yet still typical, expression of the utilitarian park ethic is to be found in the words of Assistant Park Superintendent David deL. Condon, who wrote in 1963:

> Each individual tree is serving mankind to its fullest capacity; it is giving him far greater values than the economics of so many board feet of timber, so many tons of paper pulp, or a given number of dollars as a medium of exchange. Such economic values as these are fleeting and are with man but a short while: he soon spends the money, he moves from the house, he discards the paper products, and he soon burns the logs in his fire. Yes, he needs all of these as part of life, but to be content and happy and to live fully, he needs more. It is this additional something … that he gets from the resources perpetuated in their natural state in the National Parks.
>
> These natural resources … will, if preserved, continue to give mankind the values of mental relaxation, physical recreation and spiritual stimulation so vital to a happy and full life — those values which are so fleeting and intangible that we can never assess their true value or meaning. (quoted in Campbell 1960, p. 58)

The values Condon expresses transcend the *laissez faire* economics of the previous period but remain well within the orbit of humanistic utilitarianism. Ultimately, the only value in view here is human happiness.

Where the goal of general human happiness conflicted with individual property rights or corporate profits, the progressive utilitarians were prepared to sacrifice these earlier ideals. Though some of the land for the Park was purchased from more or less willing sellers, there was extensive condemnation of both private and corporate land.

DEVELOPING THE PARK EXPERIENCE AROUND THE AUTOMOBILE

The national park movement began in the West, where it received considerable support from the railroad industry. In the East, the automobile industry played a similar role. Two complementary values were at work here: utilitarian enjoyment for the tourists and economic gain for the automobile corporations and the tourism industry.

Among the strongest lobbyists for park initiatives in the 1930s were the Automobile Association of America and the automotive industry, which wanted to provide a recreational destination for private automobiles. Moreover, there was keen local interest to create a tourist industry for Knoxville, TN and Asheville, NC.

The Park was developed as part of the Civilian Conservation Corp (CCC) activities during the Great Depression. The overall intent was to provide an infrastructure for experiencing the scenery from the comfort of the private automobile. Scenic highways, overlooks, trail heads, picnic and camp grounds, and visitor centers were all designed with the vehicle-bound visitor in mind. Over 50 years later, that tradition has not changed. The majority of the Park's $10 million annual budget is dedicated to the maintenance of facilities and services related to the vehicle-oriented visitor experience. Communities situated at the "gateways" to the Park have flourished as a result of the orientation of the Park development, dramatically demonstrating the economic importance of location to traffic patterns leading to the Park.

CREATING A SPORT FISHERY ON ABRAMS CREEK

Probably the most dramatic testament to the value link made between human happiness and consumptive activities in national parks occurred on Abrams Creek, with its headwaters in Cades Cove. In the summer of 1957, Abrams Creek was poisoned with rotenone for 14 miles below Abrams Falls to its then confluence with the Little Tennessee River in order to create a premium rainbow trout fishery (Etnier and Starnes 1993, p. 331). The poisoning took place on the day of the closing of the dam for Chilhowee Lake — yet another impoundment along the beleaguered Little Tennessee River — which was built to provide cheap power for the ever-increasing demands of the nearby Aluminum Company of America (Alcoa) plant. The Abrams Creek poisoning was a joint effort by the National Park Service, U.S. Fish and Wildlife Service, Tennessee Valley Authority, Tennessee Game and Fish Commission, and the Aluminum Company of America. The intent was to remove all the extraneous "rough fish," to avoid competition for the preferred "game species," the exotic brown (salmo trutta) and rainbow trout (Onchorhynchus mykiss).

It is difficult to comprehend the logic in this strategy, since an entire food chain was destroyed. As a result of the effort, the only population of smoky madtom (*Noturus baileyi*) known at the time was extirpated from the stream. Even today, 20 species of fish have never returned to Abrams Creek. Some of these species can never be reintroduced successfully, since they utilized the Little Tennessee River as their primary habitat (Steve Moore, pers. commun.). This act serves as a poignant example of the anthropocentric land use ethic harbored in natural resource managing agencies in the 1950s and 1960s.

AGRICULTURE IN CADES COVE

Cades Cove is a wide, flat-bottomed valley in the western end of the Park, long used for agricultural purposes by native Americans and early European settlers. The earliest European settlers moved into the Cove in the 1820s. By the 1850s hundreds more had arrived. This number dropped dramatically within a decade, as the resources in the Cove could not support such a dense population (Shields 1977, 10). Following the Civil War, building construction changed from log to framed structures. Surplus agricultural products were hauled to markets outside of the isolated Cove. The population peaked at 700 around 1900. By that time, the valley was divided into numerous small parcels. Most of the flat land in the Cove was reserved for grain farming. Cattle and pigs ranged freely in woodlots and the forests occurring on the slopes surrounding the Cove.

Today, agricultural permittees maintain roughly 1250 acres of pastures, approximately 65% of the Cove. The Park Service compromised the natural condition of the Cove in order to accommodate agricultural interests. Pastures are maintained in exotic fescue instead of native grasses. An estimated 83,000 board feet of timber were removed by agricultural permittees from woodlots in the Cove through the 1970s. Job Corps workers cut saw timber for the permittees in 1968. In the late 1960s the Soil Conservation Service directed extensive channeling of the water courses, exceeding 25,000 feet by one estimate (Ed Trout, pers. commun.). Wetlands were drained in order to enhance the pastures for cattle. The following are typical entries documenting the work to drain the pastures in Cades Cove during the 1960s and 1970s (Soil Conservation Service 1968, 1973):

August 1968 — Reopened main channel and cut two new channels to drain swampy area. Total channel work — 300 feet, one dozer day.

November 1968 — Hauled 3 loads of rock to Abrams to aid in controlling path of creek bed.

May 1973 — Filled pond and drained marsh.

In the early 1970s, as many as 1400 head of cattle roamed freely through the streams of the Cove. A 1968 Bureau of Sport Fisheries and Wildlife Report included these comments:

… over 2000 acres leased to several entrepenuers are dedicated almost entirely to grass-clover for pasture, and carrying a tremendous number of animal units…woodlots are heavily grazed. The pastures were generally so heavily used at the time of our visit that grasses were cropped to the roots and streams were milky.

Fertilizer and lime were routinely applied to the pasture fields through the 1970s. Groundhog colonies were poisoned on a selected basis. In October 1974, Joseph Congleton of Trout Unlimited complained in a letter to Nathanial Reed, then Assistant Secretary of the Interior, that Abram's Creek through the Cove was frequently silted up following rainfall:

The cattle have turned most of these feeder streams into mud holes with the result that the silt they suspend in the stream now covers the entire ten to fifteen miles of the stream's length (Abrams) in the Park.

Clearly, the land use policy practiced in Cades Cove at that time favored agriculture over natural resources.

MANAGING AN HISTORIC LANDSCAPE IN CADES COVE

When the national park was formed in 1936, there was some discussion of damming up the valley to create a recreational lake or letting the land revert to forest, but the NPS decided to maintain Cades Cove as a "historical exhibit" instead. All buildings were removed except for those associated with the agricultural activity needed to keep the Cove lands open and for several pioneer era homesteads representing architectural styles of the early European settlement period. Structures date from 1818 through 1898.

A one-way loop road was constructed around the Cove, linking the historic structures, trails, and recreation facilities. The recreational facilities are distinctly oriented to an experience centered around the private automobile. The Cove has been meticulously manicured. Most fence rows have been removed, creating large expanses of pasture to be viewed by visitors from the perspective of the periphery. "Weeds" have been aggressively eliminated from the pastures.

The result is an English-style pastoral scene that delights tourists. Indeed, the Cove is the second most popular destination in the park. Favorite activities there include driving for pleasure, sightseeing, and viewing wildlife (Peine and Renfro 1988). Cades Cove is widely recognized as one of the most scenic spots in North America.

From a historical perspective, however, the scene created is very misleading, since for most of the occupation during the European settler period the Cove was divided into much smaller lots and the livestock were not located in the valuable bottomlands — which were instead largely dedicated to orchards and grain crops. In reality, the landscape of Cades Cove is managed by the National Park Service primarily for its scenic quality, not its historic authenticity. Again, the dominant purpose is to provide for the windshield tourist.

NATURAL AND CULTURAL RESOURCES PROTECTION AND MANAGEMENT (1976 TO 1987)

The next stage in the development of the Park's land use ethic, though still utilitarian and anthropocentric, is marked by a widened and deepened perspective. By the 1960s it was becoming apparent that the long-term preservation of the Park demanded planned responses to emerging environmental threats and an increased emphasis on resource protection. The passage of the National Environmental Policy Act (NEPA) in 1969 (PL 91-190), established a legal process to assess the environmental impacts from the actions of various levels of government. Environmental advocacy groups immediately began applying this tool to question the actions of public agencies. In the Smokies, NEPA had an indirect impact on decisions concerning tentative plans for extensive road construction. Examples of planned road development that were stopped include a transmountain road from Bryson City to Townsend, a connector road from Cataloochee valley to Interstate 40, a road along the north shore of Fontana Lake connecting Bryson City to U.S. Highway 129, and a little-publicized plan to construct as many as 10 additional motor nature trails in the Park. An access road to the north shore road was in fact constructed from Bryson City 6 miles toward Fontana Lake, but it ends just beyond a tunnel. Known locally as "the road to nowhere," it is a dramatic monument to the changing environmental ethic manifest in NEPA.

The transmountain road sparked the greatest controversy. The Sierra Club and other environmental groups protested the proposal at rallies in the summer of 1968. Their argument was that since the park was under consideration for wilderness designation by Congress, no plans for road construction could go forward. The issue remains sensitive today. Senator Jesse Helms of North Carolina filibustered on the floor of the U.S. Senate Chamber during the 91st Congress to block legislation designating Great Smoky Mountains National Park as wilderness. Local special interest groups still want the north shore road built to facilitate access to cemeteries and, more importantly, to increase tourist traffic through Bryson City.

This was the first case in which emerging natural resource values clearly took precedence over the utilitarian values that provided the primary ethical foundation for the establishment of the park.

In the mid 1970s, at the request of Congress, the National Park Service compiled a list of threats to the natural and cultural resources of the national park system. This effort brought a new era of awareness of the resource stewardship mission of the National Park Service. It marks another extraordinary shift in land use ethic away from an orientation to consumptive human values and toward acknowledgment of the long-term responsibility to sustain natural and cultural resources.

This shift was part of the growing environmental awareness in American culture. Many forces were at work here, but perhaps most significant was the change of perspective achieved by the emerging science of ecology. Ecology enlarged the scale of our understanding of the effects of human activities both in space and in time, and it taught us to think holistically. Rachel Carson's *Silent Spring* made the new perspective vivid to a large popular audience by invoking the specter of a world without birds. If we now had the power to create that kind of alteration, many reasoned, then we needed an enlarged moral vision to match it. The old, relatively short-term anthropocentric utilitarianism would no longer do.

What emerged in its place was an expanded and enlightened, but still anthropocentric, utilitarianism. Henceforth we would have to be more careful about the long-term effects of our actions and think more holistically, but human happiness (expanded now to more explicitly include the happiness of future generations) remained the central value. The idea was to preserve dwindling or threatened resources, but implicit in the very concept of a resource is assumption that it is eventually to be used for the enjoyment of human beings.

In the Smokies, the visionary Park superintendent Boyd Evison established the Uplands Field Research Laboratory and Division of Resources Management, which intensified management of the Park's natural and cultural resources. These organizations initiated many programs to deal with exotic and rare and/or endangered species.

By the late 1970s, the Park had acquired scientific and resource management expertise which greatly contributed to the in-house awareness of these problems and had achieved several refinements in management practices.

Black Bear Management

Great Smoky Mountains National Park is the premiere refugium for black bears (*Ursus americanus*) in the southeastern portion of North America. As described by Clark and Pelton in Chapter 10 in this volume, management of this species illustrates a changing orientation to land use ethics. An estimated 600 animals roam freely throughout the Park. Their numbers and condition are largely dependent on the fall nut (mast) crop (*Quercus* spp., *Carya* spp.). During times of food shortages and/or abundance of young males, individuals range widely beyond the boundaries of the Park.

The building of Interstate 40 through the mountains along the Pigeon River in the 1960s created a formidable barrier to their natural movement patterns along the southern Appalachian mountain range. Bears use the only two land bridges over the highway extensively, but neither the national forest nor national park bordering either side of the highway has recognized the significance of these corridors and altered their management plans accordingly. In fact, authorities in Swain County, NC have authorized the creation of a refuge dump likely to attract bears to be located near one of these two movement corridors. The significance of these movement corridors will likely grow in importance in the future when gypsy moth defoliation depletes the mast crop and the animals must extend their range to forage for the critical fall food sources. This phenomenon has already occurred in Shenandoah National Park, located 470 miles up the Blue Ridge Parkway to the northeast in the Appalachian mountain range (Kasbom et al. 1996).

Until the mid 1970s, black bear routinely begged for food along the roadsides, and at scenic overlooks, picnic grounds, and campgrounds of the Park. Their behavior served the human desire to see and interact with bears. With the change in land use ethic occurring in the mid 1970s, attempts were made to "naturalize" black bear behavior, breaking their dependence on human-related food sources. Trash cans in the park were fitted with bear-proof tops, an education program informed visitors of regulations against feeding the animals, and rangers more aggressively enforced the regulations.

In the mid 1980s, special agents of the U.S. Fish and Wildlife Service, with support from other federal and state officials, infiltrated an extensive network that involved the illegal commercialization of black bears whose home range was based in the Park. Operation Smoky* included investigation of the illegal hunters, their contacts for direct sale of bear parts, and those engaged in

---

* All information concerning Operation Smoky came from a personal communication with Bill Cook, at the National Park Service, Shenandoah National Park.

illegal exporting of the parts primarily to oriental markets. Ultimately, 84 people from 4 states were arrested for participating in the systematic killing, removal, and illegal sale of parts of 368 Park-based bears over a 3+-year period from 1987 through 1990. Assuming a total population of 600 bear in the Park, the investigation revealed that at least 20 to 40% of the bear population in the Park and nearby environs was being removed every year. Most of these bears were taken from the Cataloochee and Fontana Lake portions of the Park, but there were illegal hunters working systematically throughout the Park. Bear parts were being illegally exported at an alarming rate to satisfy the massive oriental market based on their medicinal and mystical values. The illegal activity was clearly a commercial operation.

Ethical conflicts were pervasive throughout the investigation and prosecution of Operation Smoky. For instance, during the trial a defense lawyer adamantly expressed his clients' "God given right to kill bears any time they want." There were conflicts among the hunters. Some were concerned about limiting the harvest to sustain the population while others were intent on maximizing economic gain. Another source of considerable conflict was with those officials who were not informed of the operation but thought that they should have been a part of the action. And finally, a third conflict was with those who felt that the issue would have been better handled via education and community outreach.

By the mid 1990s, Park officials began working actively with adjacent communities to address the problem of bears wandering outside the park boundary. Problems result from inadequately secured solid waste, tourists feeding the bears from condominium balconies, and so on. To date, that problem has not been resolved.

Management of black bears serves as an eloquent example of evolving environmental ethics. From a key source of food and clothing for native Americans to the unrelenting over harvest and loss of habitat and primary food source (the American chestnut), we have progressed to establishing a no-hunting sanctuary and established management practices to minimize human interference in the form of tourist interaction, poaching, and waste disposal. Today, the black bear serves as an inspiration and universal symbol of nature and wilderness values in the southern Appalachians.

WILD HOG CONTROL

The land use policy decisions associated with the European wild hog (*Sus scrofa*) are complex as well. This introduced animal reflects the utilitarian goal of adding variety to the consumptive activity of sport hunting. As Peine and Lancia note in Chapter 13 of this volume, the populations in western North Carolina and eastern Tennessee originated from an accidental escape of animals from a hunting enclosure near Hoopers Bald, NC (Stegman 1938).

The hogs currently occur throughout the park. In the spring, they move from lower elevations to root for herbaceous plants, such as the spring beauty (*Claytonia virginica*) corms on high elevation slopes and ridge tops. In late summer, hard mast (*Quercus* spp., *Carya* spp.) comprises 60 to 85% by volume of their diet (Henry and Conley 1972). Left uncontrolled, these animals would seriously disturb the forest floor, producing in extreme cases a rototilled appearance.

As the land use ethic changed in the mid 1970s, the National Park Service began operating a control program for this exotic species. Many techniques have been attempted, the most effective being trapping and shooting, particularly in the high elevations. Such activity has been curtailed on the North Carolina side of the park, due to opposition from hunting interests who see the hogs as a resource to restock hunting reserves outside the Park — the land use principle that led to the introduction of the exotic species in the first place. For many years, local citizens were allowed to trap hogs in the Park and transport them to hunting reserves not far away. These conflicting social values have long prevented the Park from operating a cost effective control program. Even today, captured wild hogs are occasionally transported to state game areas, perpetuating the double standard of conflicting environmental values.

The Deer Herd in Cades Cove

Another persistent wildlife issue with ethical overtones concerns the density of the deer herd in Cades Cove. The open meadows and substantial forest edges of the Cove sustain a large deer herd, which produces a distinctive browse line all around its periphery. As noted earlier, evidence exists that native Americans may have maintained these open lands for thousands of years, in part to enhance deer habitat. In 1982, 80% of the herd died from the hemorrhagic disease, blue tongue. The density of the herd may have been a contributing cause. The herd quickly replenished itself, however, and remains today at a relatively stable 400 animals (Wathen and New 1989).

There has been considerable debate over the years as to whether to artificially reduce the size of the herd in order to reduce impacts on native vegetation in the Cove. Should humans intervene with the balance of nature associated with this artificially maintained scenic landscape? On what scale can we weigh the values of vegetative and wildlife attributes of Cades Cove?

Fire Management

As previously mentioned and as described by Buckner and Turrill in Chapter 16 of this volume, fire has been a major part of the disturbance history impacting the ecosystem processes since prehistoric times. It was used routinely by Native Americans and early European settlers to create grasslands for wildlife and later for domestic cattle. Very abruptly when the park was established, fires were completely suppressed, first by the CCCs and then by National Park Service personnel. This total suppression policy began to be challenged as there became a growing awareness of the role of fire in sustaining biodiversity in the southern Appalachian landscape. Scientists believe that many plant communities, such as the extensive pine–oak and chestnut oak forests, originated from fires likely to have been primarily set by humans (Delcourt and Delcourt 1996).

Recently, however, the Park has adopted a fire management plan that recognizes the prehistoric and historic role of fire on the landscape and calls for prescribed natural burns to reintroduce this disturbance on the landscape. The fire issue provides yet another intriguing land use dilemma. Should prescribed burns be set to simulate the prehistoric and historic fire regimen?

SUSTAINABLE DEVELOPMENT STRATEGIES FOR THE SOUTHERN APPALACHIANS (1988 TO PRESENT)

The movement toward ecosystem management for sustainability represents another major shift in land use ethics. For the first time, a consensus is growing among public land managing agencies to adhere to land use practices based, not merely on short- or long-term anthropocentric interests, but on the protection of ecosystems and preservation of biodiversity for their own sake. Society is beginning to recognize the need to balance human consumptive interests in natural resources with the value of biodiversity and the naturally occurring ecosystem processes. A key concept is to begin to integrate resource management practices at a landscape scale. This shift in ethics is in part the result of finally looking back to assess the cumulative human impact on the environment throughout recorded history.

The Southern Appalachian Man and the Biosphere Cooperative (SAMAB)

This policy shift became more focused with the formation of SAMAB in 1988. As described by Hinote in Chapter 5 of this volume, SAMAB has provided a framework for the integration of a wide range of resource management initiatives. This Cooperative is the first regional biosphere program established in the U.S. The program's basic objective is to provide solutions to resources management and economic development problems in the region through cooperation and collaboration among federal agencies, state governments, and nongovernmental organizations. The SAMAB membership meets regularly to discuss opportunities for collaborative projects that minimize overlap and maximize use of scarce resources while focusing on research needs of the region and ways to achieve truly sustainable development. Because SAMAB membership consists of regional land managers and decision makers, along with scientists and other concerned parties, SAMAB provides a unique mechanism to implement an integrated approach to ecosystem management in a region of significant ecological resources. A central value of SAMAB is that it serves as an active forum to debate land use ethics in the region.

The Southern Appalachian Mountain Initiative

As described by Peine et al. in Chapter 15 of this volume, the Southern Appalachian Mountain Initiative has been formed by affected state and federal agencies to devise and implement an overall framework for the management of air resources of the region. Special attention is given to mitigating the deterioration of air quality that has occurred in Class I areas such as national parks and federally designated wilderness areas. The emphasis is on cooperation rather than regulation. Elements of the program include research on the effects of air pollution, monitoring of various air pollutants, modeling the long-range transport of air pollutants, and policy formulation to design regulatory measures to alleviate the problems. This group deals with the conflict between clean air-related values and economic development interests. This debate exemplifies a fundamental value shift, since the proponents of clean air base their arguments not only on its utilitarian benefits, but also on its contribution to biodiversity and ecosystem health. It is therefore not simply a clashing of interests within the old utilitarian paradigm, but a conflict of paradigms.

Pittman Center, a Future Gateway Community

There is a growing perception that the tourist industry is degrading the natural resources and cultural character of communities surrounding the Park (Jackus and Seigel 1993). The first incidence of this concern was expressed by then Park Superintendent Boyd Evison, who openly opposed the construction of the 15 story Park Vista hotel in Gatlinburg on a site adjacent to the Park. The issue concerned the aesthetic intrusion on views from the Park vs. economic benefits to the tourist industry. The hotel was constructed in spite of the objections and can be seen from vista points in the park as a prominent visual intrusion on an otherwise natural landscape.

As described by Peine in Chapter 17 of this volume, the community of Pittman Center, TN adjacent to the Park gateway community of Gatlinburg is developing an aggressive program of growth management. With the guidance of a planning commission, the community has completed an extensive planning process to define a vision for the future and develop growth management strategies utilizing a combination of regulations, public education, and design consultations with developers. They have traveled a most difficult road to protect the value expressed by their vision statement: "a community dedicated to preserving our mountain heritage."

Red Wolf Reintroduction

Wolves have always been associated with wilderness values. As described by Lucash et al. in Chapter 11 of this volume, red wolves ranged throughout the southeastern U.S. before European settlement of the region but have been considered extinct in the wild since 1980. Persecution of wolves came to America with the European culture. U.S. President Andrew Jackson reportedly likened Indians to wolves (Josephy, Jr. 1993, p. 18). Yet Native Americans revered canids. One of the major clans of the Cherokees was known as "The Wolves."

As described in Chapter 11 of this volume, the recovery process for the red wolf has been a long, arduous undertaking. In 1991, two wolf families were experimentally released into Cades Cove. This action was preceded by an extraordinary effort to educate the general public, park visitors, and special interest groups. As a result, there has been strong support for the reintroduction program nationally, regionally, and in communities adjacent to the Park. This represents an extraordinary benchmark in the evolution of environmental ethics, suggesting a growing acceptance of the value of nature for nature's sake and the acceptance of restoring ecosystems.

THE AUTOMOBILE AND THE PARK

Balancing human and ecological values will require rethinking the context of the Park experience. Eventually it will no longer will be practical to orient the Park experience from "behind the windshield" of a private automobile. Various schemes have been discussed for limiting vehicular traffic in the Park. Shuttle buses and bicycles might help, but the latter pose further problems so long as private automobile access continues. Park officials, while guardedly sympathetic to the idea of bike paths or greenways in or near the Park, have expressed concern about the potential increase of bicycle traffic on the Park's narrow roads, which could create significant hazards for the cyclists. Thus despite the evident need to reduce automobile traffic, there is considerable ambivalence about one of the major alternatives to it.

FUTURE CHALLENGES: HOW DEEP THE SPIRAL? (2000 AND BEYOND)

How will environmental ethics evolve in the future? Perhaps the 1990s will be known as the decade of enlightenment as to the connectedness of environmental processes, and of attempts to apply the amorphous concepts of ecosystem management and sustainable development to problem solving. The turn of the century may bring an era of practicality to the principles espoused in the 1990s.

A tentative look into the next century in the southern Appalachians suggests an era of unprecedented challenges. Gypsy moth defoliation will likely become widespread. Specific pests and pathogens like the hemlock adelgid and dogwood anthracnose will probably cause widespread mortality. Air quality is likely to continue to deteriorate. The Park will most certainly become surrounded by a wall of urban development, and gridlock from tourist traffic will likely become the norm, no matter how much land is dedicated to building more roads; and potentially the most devastating adverse trend of all is global warming.

Managers with limited resources will have to make difficult choices concerning the allocation of finite resources to rescue species and/or ecosystems in jeopardy. For instance, will there be enough remnants of the spruce–fir ecosystem to preserve and/or restore? And what is more important, managing what you know you can save, such as cove hardwood forests, or gambling on those elements at greatest risk? If global warming creates the drastic changes predicted, will managers retain any certainty of ecosystem processes? Will our society go beyond the rescue of the red wolf to the reintroduction of elk and/or cougar or the sustainability of pine/oak forests?

In a way, things seem to be coming full circle. Among environmentalists, at least, there is a growing conviction that ecosystem management should be directed, not primarily toward long-term human happiness, but toward the health of the whole, even if this sometimes contravenes human happiness. The so-called "deep ecologists," for example, hold that even nonsentient life forms such as plants, and nonliving systems such as rivers, have intrinsic value — that is, should be respected in their own right and not simply for their utility to human beings. Their views are reminiscent of the respect for and sense of kinship with nature that pervaded the thinking of the Cherokees.

But it would be a mistake to think of the Cherokees as deep ecologists, in the sense in which that term is used today. Deep ecology is an ethical stance shaped by elements of contemporary thought, including the science of ecology itself. Central to its understanding of the world are such concepts as "ecosystem," "biodiversity," "food chain," and "sustainability." The Cherokees had no equivalent concepts. Their world was inhabited by living, personal spirits, who became manifest in natural objects. The spirituality associated with deep ecology is typically a vague, thin pantheism. Moreover, the Cherokee's mythology helped them to live sustainably, in and with the land. For most contemporary advocates of deep ecology, that remains a distant dream.

Not a circle, then, but perhaps a spiral.

REFERENCES

Armstrong, D. 1851. Text of lease agreement on 25 acres of property in Sevier County TN to David Howser. Great Smoky Mountains National Park Archives. Gatlinburg, TN.

Ayres, H.B. and N. Ashe. 1905. *The Southern Appalachian Forest*. Great Smoky Mountains National Park Archives. Gatlinburg, TN.

Barrett, R. 1978. The feral hogs on the Dye Creek Ranch, California. *Hilgardia* 46:283–355.

Calvin, J. 1928. *Institutes of the Christian Religion*. Philadelphia. III, ix, 4.

Campbell, C. C. 1960. *Birth of a National Park in the Great Smoky Mountains*. University of Tennessee Press: Knoxville. 154 pp.

Carson, R. 1962. *Silent Spring*. Boston: Houghton Mifflin.

Cotterill, R. S. 1954. The Southern Indians: *The Story of the Civilized Tribes before Removal,* Norman: University of Oklahoma Press.

Cronon, W. 1985. *Changes in the Land: Indians, Colonists and the Ecology of New England*. New York: Hill and Wang.

Delcourt P.A. and H.R. Delcourt. 1996. Holocene Vegetation History of the Northern Chattooga Basin, North Carolina. Tennessee Valley Authority Project, Personal Services Contract No. TV-95990V. Knoxville, TN.

Dickens, R.S., Jr. 1976. Cherokee Prehistory: *The Pisgah Phase in the Appalachian Summit Region*. University of Tennessee Press: Knoxville.

Dorgan, H. 1987. *Giving Glory to God in Appalachia: Worship Practices of Six Baptist Subdenominations*. University of Tennessee Press: Knoxville.

Dunn, D. 1988. *Cades Cove: The Life and Death of a Southern Appalachian Community*. University of Tennessee Press: Knoxville. 335 pp.

Dykeman, W., and J. Stokely. 1978. *Highland Homeland: The People of the Great Smokies*. National Park Service, Dept. of the Interior. Washington, D.C. 189 pp.

Etnier, D.A. and W.C. Starnes. 1993. *The Fishes of Tennessee*. University of Tennessee Press: Knoxville. 681 pp.

Frome, M. 1980. *Strangers in High Places*. University of Tennessee Press: Knoxville. 391 pp.

Hudson, C. 1970. The Cherokee concept of natural balance. *Indian Historian*. Vol. 3, Fall.

Hudson, C. 1976. *The Southeastern Indians*. University of Tennessee Press: Knoxville.

Jackus, P.M. and P.B.Seigel. 1993. Attitudes toward Tourism Development in the Appalachian Highlands of Tennessee and North Carolina. Department of Agricultural Economics and Rural Sociology, University of Tennessee. Knoxville, TN. 144 pp.

Josephy, Jr. and Alvin M. 1993. Introduction, *The Native Americans, an Illustrated History*. Turner Publishing: Atlanta, Ga. 479 pp.

Kasbohm, J.W., M.R. Vaughan, and J.G. Krause. 1996. Effect of gypsy moth infestation on black bear reproduction and survival. *J. Wildl. Manage.,* 60: 408-416.

Kephart, H. *Our Southern Highlanders*. McMillan: New York. 522 pp.

King, D.H., 1979. *The Cherokee Indian Nation: A Troubled History*. University of Tennessee Press: Knoxville.

Lambert, R.S. 1958. Logging in the Great Smoky Mountains National Park: A Report to the Superintendent. Great Smoky Mountains National Park Archives. Gatlinburg, TN.

Leopold, A., 1966. The Land Ethic, in A Sand County Almanac with Essays on Conservation from Round River. New York: Ballantine.

Lewis, H.M., L. Johnson, and D. Askins. 1978. *Colonization in Modern America: The Appalachian Case*. Boone, NC.: Appalachian Consortium Press. 370 pp.

Licks, H.W. 1950. A Brief History of Great Smoky Mountains National Park. Great Smoky Mountains National Park Archives. Gatlinburg, TN.

Locke, J., 1980. *Second Treatise of Government*. Peter Laslett, New York and Toronto, New American Library, Second Treatise, Chapter V, "Of Property."

Martin, C. 1978. *Keepers of the Game: Indian-Animal Relationships and the Fur Trade*. Berkeley: University of California Press.

Mooney, J. 1900. The Cherokee River Cult, *J. Am. Folklore* 13: 1–10.

Mooney, J. 1972. *Myths of the Cherokee and Sacred Formulas of the Cherokees*. C. Elder Booksellers: Nashville, TN.

Peacock, J.L. and R.W. Tyson, Jr. 1989. *Pilgrims of Paradox: Calvinism and Experience among the Primitive Baptists of the Blue Ridge.* Washington: Smithsonian Institution Press.

Peine, J.D. and J. Renfro. 1988. Visitor Use at Great Smoky Mountains National Park. Res./Resources Manage. Ser. SER-90. National Park Service. Atlanta, GA. 93 pp.

Philips, M.K., R.Smith, V. Henry, and Chris Lucash. 1993. Red Wolf Reintroduction Program. Unpublished paper. U.S.Fish and Wildlife Service. Ashville, NC.

Shields, R.A. 1977. *The Cades Cove Story.* Gatlinburg: Great Smoky Mountains Natural History Association. 116 pp.

Soil Conservation Service. 1968–1973. Soil and Water Conservation Plan for Cades Cove. Great Smoky Mountains National Park Archives. Gatlinburg, TN.

Springer, M.D. 1977. Ecological and economic aspects of wild hogs in Texas. Pages 37–46 in G.W. Wood, Ed. *Research and Management of Wild Hog Populations.* Belle W. Baruch Forest Science Institute, Georgetown, SC.

Smith, A., 1930. *The Wealth of Nations.* 5th ed. London: Methuen & Co. (originally published in 1776).

Stegman, L.C. 1938. The European wild boar in the Cherokee National Forest. *TN J. Mammal.* 19:279–290.

Thomas, D.H. 1993. Part One: The World As It Was. In *The Native Americans: An Illustrated History.* Atlanta, GA: Turner Publishing. 479 pp.

Wathen W.G. and J.C. New. 1989. The White Tail Deer of Cades Cove: Population Status, Movement and Survey of Infectious Disease. Res./Resource Manage. Rep. Ser. SER 89/01. National Park Service, Atlanta, GA. 139 pp.

Wenz, P.S. 1988. *Environmental Justice.* Albany, NY: State University of New York Press.

White, L., Jr. 1967. The historic roots of our ecological crisis, *Science* 155: 1203–1207.

White, M. n.d. Contemporary Usage of Native Plant Foods by the Eastern Cherokees. *Appalachian Journal.*

White, M., 1978. *The Philosophy of the American Revolution.* New York: Oxford University Press.

Whitney, E. and L. White. 1993. Ecotheology and History. *Environ. Ethics* 15, 2, Summer 1993, 151–169.

Wilms, D.C. 1973. Cherokee Indian Land Use in Georgia, 1800–1838, Unpublished Ph.D. dissertation. Geography Department, University of Georgia. Athens.

Woodward, G.S. 1963. *The Cherokees.* Norman: University of Oklahoma Press. 375 pp.

Section II

The Case Study Setting and Institutional Framework

4 The Southern Appalachians: The Setting for the Case Study

J. C. Randolph, Michael L. McKinney, Caffilene Allen, and John D. Peine

CONTENTS

Introduction 63
Geological Features..65
    Piedmont...66
    Blue Ridge..66
    Valley and Ridge 66
    Appalachian Plateaus 66
Geological History 68
Climate 69
Soils...71
Vegetation...72
Wildlife..74
Aquatic Resources..76
Early Inhabitants 76
Early European Settlers 77
The Economy 77
Culture...78
Relevance to Ecosystem Management 79
References 79

INTRODUCTION

The objective of this chapter is to provide an overview of the geological, ecological, socioeconomic, and cultural factors that have shaped current patterns and distributions in the southern Appalachian region. This overview will provide the context in which natural resources management issues arise and decisions are made. With its rich biological and cultural diversity, the southern Appalachian region provides an excellent opportunity to illustrate the principles and goals of ecosystem management.

This relatively narrow, mostly mountainous corridor, which extends from northern Virginia to eastern Alabama (Figure 1), is well known for one of the most richly varied temperate forests in North America. Although modest in elevation when compared to many other mountain ranges, the physiographic complexity of this region combined with its relatively warm, moist climate results in exceptional natural diversity. This diversity traces back to the geologic origin of the region. Over 230 million years ago, the land was subjected to massive and severe compression in a collision with Africa which produced mountains, valleys, plateaus, and other features. In adapting to this physical heterogeneity, life evolved in a similarly heterogeneous manner, resulting in many relatively unique

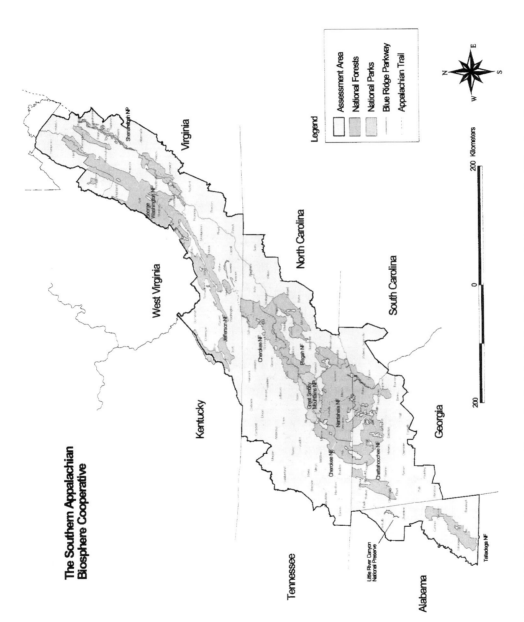

FIGURE 1   The southern Appalachian region included in the Southern Appalachian Man And Biosphere Reserve.

species and ecosystems. The same is true of cultural diversity: the rugged terrain often isolated settlers from each other and from other regions so that the Appalachian culture is not only distinct from other regions of North America, but also exceptionally diverse.

Although a comprehensive description of the natural and cultural resources of the southern Appalachian region is beyond the scope of this chapter, a recognition of their unique nature is essential as background for the examples of ecosystem management discussed throughout this book. The Southern Appalachian Assessment (SAMAB 1996), a five-volume report produced by a team of natural resource specialists representing 14 state and federal agencies coordinated through the Southern Appalachian Man and the Biosphere Cooperative, provides many details about this region. The Assessment as described in Chapter 7 of this volume provides complimentary information to that presented here.

GEOLOGICAL FEATURES

Geologically, the Appalachian region is dominated by a very old mountain range which begins in Alabama and extends through the northeastern U.S. into Canada. The highly deformed and folded sedimentary rocks that form much of the rugged terrain were deposited in the early Paleozoic era about 400 million years ago in an ocean that covered eastern North America. The greatest deformation and folding of these sedimentary rocks occurred when the continents of Africa and Europe collided with eastern North America in the late Paleozoic era, about 230 million years ago (Stearn 1979). Since that time, the Appalachian mountains have been gradually eroding. As a result, the landscape is more "gently rolling" and the elevations are lower than the much younger Rocky Mountains of the western U.S. and Canada.

Topography is the general configuration of the earth's surface, including its relief and location of waterways. While the topography of the southern Appalachian region is extremely varied, it is not chaotic. The Region can be divided into four distinct physiographic provinces: Piedmont, Blue Ridge, Valley and Ridge, and Appalachian Plateaus (Figure 2). We will now briefly review each of these provinces. Further detailed discussions of these are found in Pirkle (1982), Moore (1988) and Hatcher (1989).

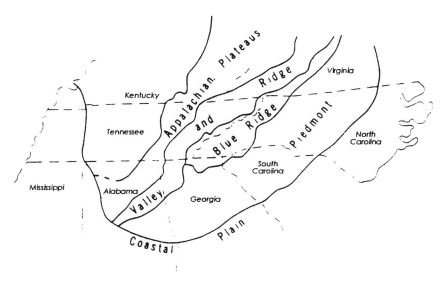

FIGURE 2 Physiographic provinces of the southern Appalachian region. (Modified from Pirkle, E.C., *Natural Landscapes of the U.S.*, Kendall/Hunt, Charlotte, NC, 1982.)

PIEDMONT

This "apron" of crystalline rocks is a transitional zone between the very old southern Appalachian mountainous "core" and the more geologically recent Coastal Plain sediments. The Piedmont has a clear boundary with the Coastal Plain, called the "fall line," or "fall belt" (Figure 3). The Piedmont province extends from Alabama to southern New York, a distance of about 1000 miles (1609 km). Although often referred to as a "plateau," it is not flat because differential erosion over geological time produced a rolling terrain ranging from 1800 feet (549 m) above sea level to about 400 feet (122 m) above sea level near the fall line. In some areas, "monadnocks" of durable rocks such as granite and quartzite stand out as isolated elevated areas such as the Stone, Pine, and Kennesaw Mountains. Historically, granite, marble and other minerals have been mined from the Piedmont.

BLUE RIDGE

These mountains extending from northern Georgia to southern Pennsylvania were formed when old crystalline rocks were thrust over the folded younger sedimentary rocks that form the Valley and Ridge province (Figure 3). The geological structures in this province are very complex because of these intense forces. The ancient crystalline rocks are Precambrian (over 600 million years old), consisting mostly of metamorphic rocks such as gneisses, slates, and quartzites. The southern part of this province is the most mountainous of the region, including 46 peaks over 6000 feet (1829 m) above sea level and 288 other peaks over 5000 feet (1524 m).

The mountain range that contains many of these high peaks is the Great Smoky Mountains in western North Carolina and eastern Tennessee (Figure 1). The Great Smoky Mountains National Park (GSMNP) contains many of the highest and most massive mountains in the eastern U.S. The highest peak in the Park is Clingman's Dome at 6643 feet (2025 m). The most abundant rocks in the Park are late Precambrian schists and other metamorphic rocks that were thrust over younger Paleozoic sedimentary rocks (Moore 1988) that are exposed in some locations such as Cades Cove (Figure 4).

VALLEY AND RIDGE

This province has the greatest length of all provinces in the southern Appalachian region, extending from central Alabama into New York. The Valley and Ridge province is formed from the erosion of tilted and folded Paleozoic sedimentary layers. The valleys formed because limestone and shale layers are relatively easily eroded in the warm, humid climate. The ridges occur where the more durable rock layers, such as sandstones, have resisted erosion.

Along the eastern side of the province is a large continuous lowland, usually called the "Great Valley" in eastern Tennessee (Figure 3) and the Shenandoah Valley in Virginia. Because of the many valleys in this province, many streams and rivers flow through them. Many are tributaries of the Tennessee River, which meanders through the Great Valley. Many of these streams and rivers are very picturesque, sometimes forming "water gaps," or narrows, and containing a wide variety of aquatic organisms. The isolation of these smaller water bodies has led to the evolution of many unique species of mussels, fish, and other aquatic organisms found nowhere else in the world.

APPALACHIAN PLATEAUS

This province is characterized by nearly horizontal sedimentary rocks of the Paleozoic age. It is called "plateaus" because the elevation of this relatively flat landscape is fairly high, about 1000 feet (305 m) above sea level, and consists of a number of plateaus. The Cumberland Plateau is the southernmost plateau. The plateaus were formed because the underlying horizontal rock layers consist of sandstones and conglomerates that are resistant to weathering and erosion. There are also vast areas of bituminous Paleozoic coal beds near the surface, which form one of the largest

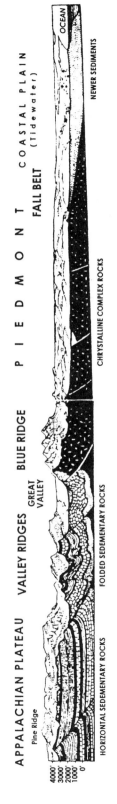

FIGURE 3 Geologic cross-section of physiographic provinces of the southern Appalachian region. (Modified from Pirkle, E.C., *Natural Landscapes of the U.S.*, Kendall/Hunt, Charlotte, NC, 1982.)

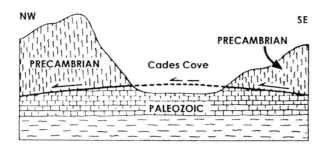

FIGURE 4 Geologic cross-section of Cades Cove.

continuous coal deposits in the world. Unfortunately, mining of these coal deposits has caused widespread landscape degradation and water pollution.

In parts of the Cumberland Plateau, the land surface has been deeply eroded by streams, producing deep valleys and leaving the remaining undissect surfaces as mountains, such as the Cumberland Mountains of eastern Kentucky and Tennessee. These "residual" mountains were formed from the remnants of late Paleozoic sandstones which resisted erosion while the surrounding shale and other softer sedimentary rocks eroded away. Along the eastern side of the Cumberland Plateau, a high escarpment rises from 500 to 1000 feet (152 to 305 m) above the lowlands of the Valley and Ridge province (Figure 3). This Cumberland Escarpment, like the Cumberland Mountains, was formed by the resistance to erosion of surficially exposed sandstones, such as the Clinch Sandstone formation.

GEOLOGICAL HISTORY

The Appalachian mountain region has a long and complex geological history (review in Stearn 1979). Igneous, metamorphic, and especially sedimentary rocks are common throughout the Appalachian region. Most rocks date to the early Paleozoic era (400 to 600 million years ago) and some are as old as the Precambrian era (over 600 million years ago).

The Appalachian mountains were formed when "continental drift" caused Africa to collide with North America at the end of the Paleozoic era about 230 million years ago. Because all continents float on a hot plastic layer of magma, they move slowly over large distances given geological time scales. Thus, the collision with Africa began over 400 million years ago when small land masses between Africa and North America made contact. By 230 million years ago, such motion had united all of Earth's major continents into a single super continent called "Pangaea."

The collision of North America with the western edge of Africa (Figure 5) formed the Appalachian chain and led to the provinces discussed above (Hatcher 1978, 1979). These horizontal forces produced folding and compression of rock layers. Many of these rock layers were thrust over one another; for example, the Blue Ridge province is thrust over the Great Valley. This thrusting also produced "faults," or planes of breakage, that now form many of the escarpments visible in the southern Appalachians.

The original Appalachian mountains, formed by the African collision, were highly eroded after about 10 million years. The present Appalachian mountains represent the remains of these much older mountains. A rejuvenation process has occurred which involves an uplifting of the region due to geological forces such as buoyancy on the mantle and the dissection of the uplifted rocks by nearly continuous erosion over the last 200 million years.

The earliest fossils known from the southern Appalachian region are marine organisms found in the shales and other marine rocks deposited in the early and middle Paleozoic (Stearn 1979). Common organisms at this time include trilobites, corals, and crinoids. Life on land becomes evident by the middle Paleozoic, when rocks of this region contain evidence of early plants. By

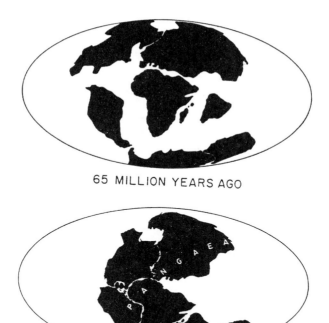

65 MILLION YEARS AGO

200 MILLION YEARS AGO

FIGURE 5 Unification of continents into the supercontinent Pangaea and its breakup 65 million years ago. Note location of Africa in relation to southeastern North America.

the late Paleozoic, huge forests of ferns and other primitive plants dominated the region and later formed coal deposits.

Fossils of dinosaurs are virtually unknown in the region because dinosaurs lived during the Mesozoic era, long after the Appalachians had been uplifted. However, there are fossils of numerous Pleistocene organisms such as mastodons, woolly mammoths, ground sloths, camels, and saber tooth cats that lived in the region about 2 million years ago.

CLIMATE

Just as the region's geological history has resulted in highly varied topography, that topographic variation results in many different climatic regimes. Generally the climate of the southern Appalachian region is described as " warm, humid" (Thornthwaite and Mather 1955). Average annual temperature of the region ranges from 50°F (10°C) to 60°F (16°C), with a pronounced southerly dip in the thermocline down the Appalachian Mountain range (Figure 6). The wide range of altitudes results in considerable differentiation of temperature patterns and strongly influences patterns of precipitation. During autumn through spring, prevailing westerly winds result in a succession of high and low pressure systems. However, during summer, high pressure over the Atlantic Ocean blocks movements of low pressure cells, which allows movement of warm, moist air from the south. These conditions often result in air stagnation.

The southern Appalachian region averages 45 inches (1143 mm) of precipitation annually, ranging from 40 in. (1016 mm) to 80 in. (2032 mm). Generally the precipitation in this area is produced by the interaction of warm, moist air moving northward from the Gulf of Mexico and interacting with cooler, drier air from the interior. Prevailing westerly winds also bring moisture-laden air, which is

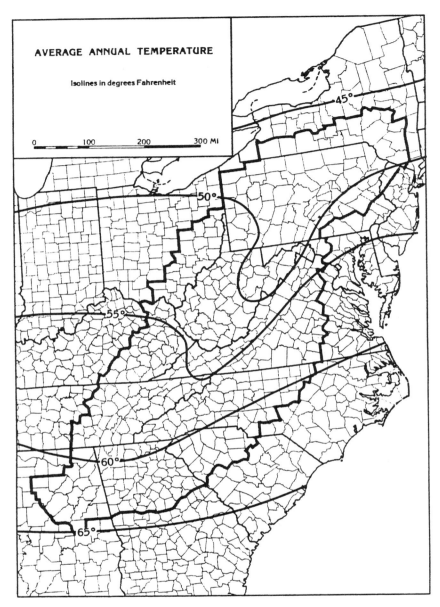

FIGURE 6   Average annual temperature in the Appalachian region. (From Baldwin, J.L., *Weather Atlas of the U.S.*, U.S. Department of Commerce, Environmental Science Services Administration, Washington, D.C., 1968.)

forced up the mountain ranges. As this air rises, it cools, producing condensation and precipitation. The windward (west-facing) slopes typically have higher precipitation than the leeward (east-facing) slopes. This pattern is particularly noticeable in the mountains of the southern Blue Ridge, where at higher elevations the greatest amount (80 in.) of precipitation occurs, in the southern Appalachian region (Figure 7). Tropical storms originating over the Gulf of Mexico may bring heavy rain to the southern Appalachian region, although storms originating over the Atlantic Ocean rarely pass over the Blue Ridge. Snowfall occurs regularly only in the higher elevations of eastern Tennessee and western North Carolina.

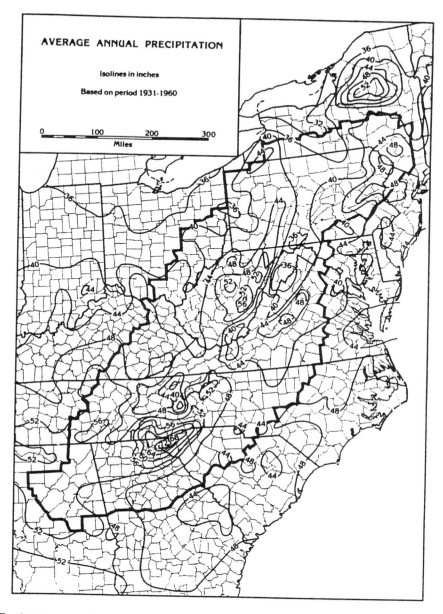

FIGURE 7 Average annual precipitation in the Appalachian region (From Baldwin, J.L., *Weather Atlas of the U.S.*, U.S. Department of Commerce, Environmental Science Services Administration, Washington, D.C., 1987.

SOILS

Geology, topography, and climate interact to produce two dominant and distinct soil orders (Figure 8). Inceptisols are young soils from mechanical weathering of lithic materials. Ultisols are the major soil order in the region. Ultisols are geologically old soils, which are produced by long periods of chemical and physical weathering. Ultisols typically are low in organic matter content and high in iron and aluminum oxides. Neither of these soil orders is well suited for agriculture. Accordingly, patterns of forest clearing for agriculture and abandoned agriculture reverting to successional forests are common.

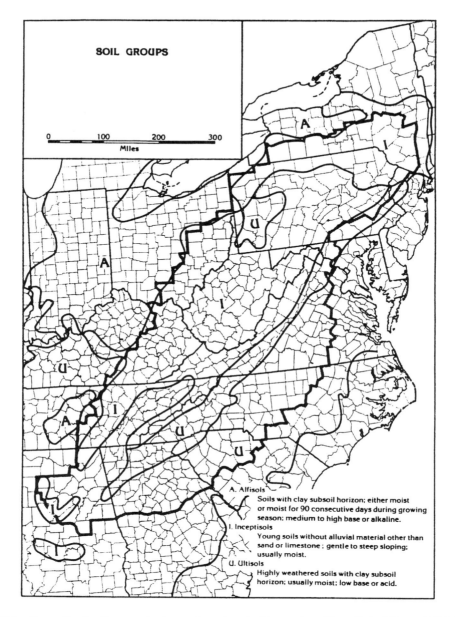

FIGURE 8 Generalized soil groups in the southern Appalachian region. (From Foth, H.D. and J.W. Schafer, *Soil Geography and Land Use,* copyright ©1980. Reprinted by permission of John Wiley & Sons, Inc.)

VEGETATION

In 1996, there are about 26 million acres of forests in the southern Appalachian region; 67% is deciduous forests, 17% is evergreen forests, and about 16% is mixed evergreen and deciduous forests. About 77% of these forests are in private ownership, 20% are in national forests and parks, and 3% are under state or other federal ownership (SAMAB 1996).

The vegetation of the southern Appalachian region reflects and is influenced by the diversity of bedrock and soil types, topography, and microclimate. The original forest vegetation of the southern Appalachian region can be classified into two major forest associations: (1) mixed mesophytic and

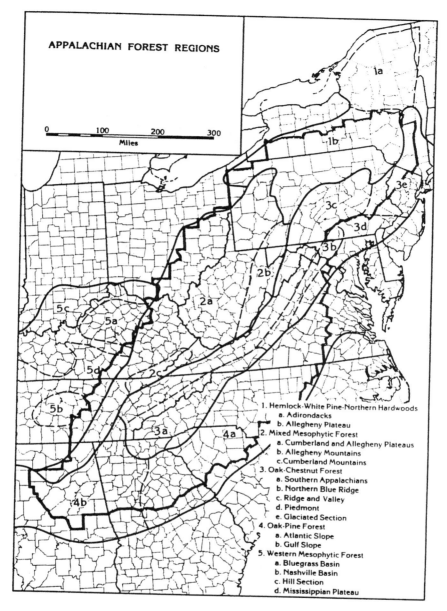

APPALACHIAN FOREST REGIONS

0   100   200   300
Miles

3e
3c
3d
1a
1b
3b
2b
2a
5c
5a
5d
2c
5b
3a
4a
4b

1. Hemlock-White Pine-Northern Hardwoods
   a. Adirondacks
   b. Allegheny Plateau
2. Mixed Mesophytic Forest
   a. Cumberland and Allegheny Plateaus
   b. Allegheny Mountains
   c. Cumberland Mountains
3. Oak-Chestnut Forest
   a. Southern Appalachians
   b. Northern Blue Ridge
   c. Ridge and Valley
   d. Piedmont
   e. Glaciated Section
4. Oak-Pine Forest
   a. Atlantic Slope
   b. Gulf Slope
5. Western Mesophytic Forest
   a. Bluegrass Basin
   b. Nashville Basin
   c. Hill Section
   d. Mississippian Plateau

FIGURE 9 Appalachian forest regions. (From Brawn, E.L., *Deciduous Forests of Eastern North America*, Hafner Press, New York, 1950. Reprinted 1974.)

(2) oak–chestnut (Figure 9). The vegetation distribution shown in Figure 9 is a very generalized map developed for a large region. In reality, the vegetation of the southern Appalachian region is highly variable and changes quickly over relatively short distances. Consequently, no regional map of either the presettlement or current vegetation has ever been produced that reflects the actual heterogeneity of the landscape. In addition to the topographic influences discussed earlier, these forests also are influenced by other natural forces such as fire, drought, and flooding, each adding to the heterogeneous pattern.

Additionally, these forests have experienced a long history of human modifications, some of which are discussed in the following sections. The natural vegetation of the southern Appalachian region also has been subjected to a wide variety of pests, pathogens, and anthropogenic pollutants. Except for some small areas of original forests protected by establishment of parks and preserves, most of these forests have been harvested for timber and timber products since early settlement times.

The mixed mesophytic forest association occurs over the unglaciated Allegheny Plateau, the Cumberland Mountains in eastern Kentucky and western Virginia, and all but the southern end of the Cumberland Plateau. The western boundary is along the western edge of the Cumberland and Allegheny Plateaus and the eastern boundary is along the Appalachian Front. In the more northerly part of this range, American beech (*Fagus grandifolia*), sugar maple (*Acer saccharum*), and hemlock (*Tsuga canadensis*) are dominant tree species. In the more southerly area of this range, white oak (*Quercus alba*), red oak (*Quercus rubra*), tulip poplar (*Liriodendron tulipifera*), and American basswood (*Tilia americana*) are the dominants. Vast tracts of these forests were cleared and burned to clear land for agriculture, most of which, even after repeated attempts, has proven unsuccessful and ill-suited for these soils and topography. Fortunately, several areas in this region have been set aside in publicly held forests and parks, as well as some protected private ownership.

The second major forest association of the southern Appalachian region usually is described as "oak–chestnut," even though a blight eradicated the chestnut (*Castanea dentata*) in the early part of this century. This forest once covered almost all of the Valley and Ridge and the Blue Ridge physiographic provinces. The dominant tree species at elevations lower than about 4500 feet (1318 m) are red oak, chestnut oak (*Quercus prinus*), white oak, and tulip poplar. At elevations above about 4500 feet (1318 m), the oak–chestnut forest begins a transition towards a more northern hardwood forest type; above about 5500 feet (1672 m), red spruce (*Picea rubens*) and Frasier fir (*Abies fraseri*) dominate. In eastern Tennessee and western North Carolina, the interactions of topography, climate, and soils result in one of the most biologically diverse areas in North America.

The Southern Appalachian Assessment (SAMAB 1996) identified 16 broad vegetation classes and 31 rare community types in the southern Appalachian region. The eight forest types, including early to late successional stages, are summarized in Table 1. However, even the SAMAB classification of vegetation is broad. The Nature Conservancy recognizes about 200 community types in the region (P. White, pers. commun.) and is working towards a more comprehensive, detailed assessment. Whittaker (1956) described the vegetation of the Great Smoky Mountains as the most complex pattern of vegetation in North America. Vegetation varies with elevation and soil moisture, as influenced by insulation, slope aspect, slope steepness, and slope position. Hemlock and mesic hardwood species occur in coves; oaks and mixed hardwood species on open slopes; beech and northern hardwood species on moist, mid-elevation sites; spruce and fir at cool, wet, high elevation sites; and pines on low to mid-elevation dry sites. The establishment of the GSMNP in 1931 is perhaps the best example of the recognition of the value of these diverse and often unique environments.

As occurred in most other forests of the eastern U.S., valley bottoms and flatter, accessible ridges were cleared for agriculture. In more rugged terrain, forests were somewhat protected from land clearing for agriculture and settlements, but as demand for timber products increased, the more commercially valuable species were logged. During the period from about 1880 through the late 1920s, very destructive timber harvest practices were used, often followed by drastic fires and extensive soil erosion, which damaged large areas of forests in the southern Appalachian region. Second-growth forests now have become well established and increasingly are given varying degrees of protection under both public and private ownership.

WILDLIFE

The southern Appalachian region long has been recognized for an abundance and diversity of wildlife. More than 25,000 species of plants and animals are thought to occur in the southern

TABLE 1

The Current Acres by Successional Class and Forest Type Group for All Ownerships in the Southern Appalachian Assessment Area Based upon FIA, CISC and LANDSAT Data

FIA Forest Type Group	Current Timberland Acreage and Percents					
	Grass/Seedling/Shrub Stage		Sapling/Pole Stage		Mid Successional Stage	
	Acres	%	Acres	%	Acres	%
Maple–beech–birch forests	7,445	0.4	95,671	1.8	356,503	2.8
Oak–hickory forests	814,009	43.0	3,187,729	58.6	8,395,027	66.1
Elm–ash–cottonwood forests	15,937	0.8	22,529	0.4	129,023	1.0
White pine–hemlock forests	67,107	3.5	245,249	4.5	280,491	2.2
Spruce–fir forests	0	0	896	0	11,481	0.1
Southern yellow pine forests	572,418	30.3	602,435	11.1	1,879,563	14.8
Longleaf pine forests	7,725	0.4	1,060	0	19,385	0.2
Oak–pine forests	406,623	21.5	1,281,636	23.6	1,635,265	12.9
Totals	1,891,264	8	5,437,205	22	12,706,738	52

FIA Forest Type Group	Late Successional Stage		All Stages Totals	
	Acres	%	Acres	%
Maple–beech–birch forests	66,154	1.5	525,773	2
Oak–hickory forests	3,174,064	70.9	15,570,829	64
Elm–ash–cottonwood forests	19,579	0.4	187,068	1
White pine–hemlock forests	24,840	0.6	617,687	2.5
Spruce–fir forests	67,208	1.5	79,585	0.3
Southern yellow pine forests	387,507	8.7	3,441,023	14.0
Longleaf pine forests	27,485	0.6	55,655	0.2
Oak–pine forests	711,309	15.9	4,034,833	16.5
Totals	4,478,146	18	24,513,353	

From SAMAB (1996).

Appalachian region. Of these, there are 2250 species of vascular plants, 80 species of amphibians and reptiles, 175 species of birds, and 65 species of mammals, the remainder being invertebrates and lower plants. Many larger mammalian species (e.g., cougar (*Felis concolor*); red wolf (*Canis rufus*)) have been extirpated, while other species (e.g., black bear (*Ursus americanus*)) occur primarily in protected areas such as the GSMNP and federal forests. Other mammals such as white-tailed deer (*Odocoileus virginianus*), eastern fox squirrel (*Sciurus niger*), and raccoon (*Procyon lotor*) have thrived.

The diversity of suitable habitats also permits avian species to occur in great diversity in the southern Appalachian region. Popular game birds such as eastern wild turkey (*Meleagris gallopavo*) have expanded their range and density in parts of the southern Appalachian region during the past 25 years, whereas other game birds such as ruffed grouse (*Bonasa umbellus*) and bobwhite quail (*Colinus virginianus*) have declined. Avian species requiring unfragmented, closed-canopy, deciduous forests, as well as species with habitat requirements for rare and unusual communities (e.g., heaths and grassy balds, seeps, bogs, etc.), are becoming increasingly threatened by human activities.

The southern Appalachian region also is well known for a diversity of reptiles and amphibians, notably several unusual species of salamander. Typically, reptiles and amphibians have relatively limited mobility and frequently have requirements for specialized habitats. Ten species (nine species of salamander and one species of turtle) are listed in the Southern Appalachian Assessment (SAMAB 1996) as having potential problems with loss of habitat and future viability.

AQUATIC RESOURCES

Water is an important part of the southern Appalachian region. The region contains more than 556,000 acres of river and lake surface (SAMAB 1996). The southern Appalachian region contains parts of 73 major watersheds. Nine major rivers have headwaters in the southern Appalachians. Although water quality has improved since the adoption of the Clean Water Act in 1972, the trophic status of lakes in the region varies widely. Of lakes greater than 500 surface acres, 38% are classified as eutrophic. There are 15 watersheds where greater than 20% of their stream miles have impaired water quality (SAMAB 1996).

As is true for vegetation and terrestrial animals, the unique intersection of geological and climatological features results in a high diversity of aquatic species and communities. Unfortunately, human activities such as effluent discharges, land development, construction, mining, agriculture, and forestry operations influence water quality and aquatic life. Also, in the southern Appalachian region, 54% of stream miles have high sensitivity to acid deposition. The Southern Appalachian Assessment concluded that 260 aquatic species, including 97 species of fish, are endangered, threatened, or at risk in the southern Appalachian region.

EARLY INHABITANTS

The earliest inhabitants of what is now known as Appalachia were Paleo-Indians, who were predominantly hunters. By 2500 B.C., however, most of the groups in the area concentrated more on survival through agricultural means. About 500 years before the Europeans arrived in the Appalachian region, dramatic cultural changes occurred among the southeastern Indians, a change that signaled the beginning of the Mississippian period (Hudson and Tesser 1994).

When the first Europeans ventured into the Appalachians, they encountered the Mississippian Indian culture. The first European influence in the Appalachians was Hispanic. Spanish soldiers explored the area in the mid-1500s, in hopes of finding treasure. Among the Spanish soldiers conducting expeditions in the Southeast were Lucas Vasquez de Ayllon (1526), Panfilo de Narvaez (1528), Tristan de Luna (1559 to 1561), and Juan Pardo (1566 to 1568). However, the Spanish leader most remembered for his expeditions through the Southern Appalachians is Hernando DeSoto (1539 to 1543). DeSoto conducted an expedition through the Appalachians in the mid-1500s, after hearing stories from the Appalachee Indians in Florida of the great stores of gold and silver to be found in the mountains. Although he searched extensively, DeSoto's efforts proved fruitless, as no treasures were discovered. The first of the myriad travel narratives written about the Appalachians was by a member of DeSoto's expedition — a man known today as the Gentleman of Elvas. Some accounts of history state that the Appalachians received their name from DeSoto, who may have named them after the Appalachee Indians.

The Native American population in the southern Appalachians was about one million when the first Europeans began to arrive. Among the various tribes were the Powhatan, Shawnee, Cherokee, Catawba, Choctaw, Tuscarora, Seminole, Tunica, Yuchi, Natchitoches, and Chitmacha. As European settlements were established, the Indians were forced to relocate, and mass expulsions to the Indian Territory occurred between 1820 and 1840. At least 50,000 Cherokees, Chickasaw, Choctaw, Creek, and Seminole were driven from their home areas in several southern states. The mass exodus, resulted in high mortality rate as the Indians journeyed from northern Georgia through Tennessee, western Kentucky, southern Illinois, southern Missouri and into Oklahoma.

The African–American presence in the southern Appalachian region came well before the arrival of European settlers. The progeny of black explorers and runaway slaves became the first and most assimilated: the rural blacks of Appalachia who were landowners at the time of Emancipation. A second wave were the blacks who came from the lowlands of the South to be laborers in the salt mines and workers on the rail lines that traversed Appalachia by the turn of the century. The largest number of blacks in the region migrated between 1900 and 1930, most notably from Alabama, to southeastern Kentucky, West Virginia, and southeastern Virginia. Another generation of blacks was born in the region after 1925, the time of the precipitous decline of the black population in the coal mining sections, and a fifth generation of Black Appalachians was born after World War II (Turner and Cabbell 1985).

EARLY EUROPEAN SETTLERS

The early European settlers in the southern Appalachian region were generally of three ethnic origins: Scotch-Irish, English, and German. Large numbers of Ulster Scots left the British Isles and came to America during the early part of the 18th century. They originally came to Maryland and Pennsylvania, but soon found that the lands along the Delaware and the Chesapeake had been occupied by the arrivals from England. Therefore, they moved towards the west, and following the great Appalachian Valley, began their journey southward into the Piedmont and mountains of North Carolina and Tennessee.

Another group of settlers, equally large in number, was of English origin. Many of them made the migration down into the southern Highlands as they found that most of the good land had already been taken by earlier settlers. They migrated down from Virginia and southward in order to escape discrimination, persecution, and taxes levied to support the Anglican Church.

A third and smaller group of settlers, the Germans, joined the movement of the Scotch–Irish from Maryland and Pennsylvania during the second quarter of the 18th century. They had fled a war-ravaged homeland and religious persecution to find land and a better life in America.

By the time large numbers of white settlers appeared in the Appalachian region in the mid-18th century, the Mississippian Indian culture had been replaced by the Cherokee. War between the European settlers and the Cherokees was common, but cooperative efforts, especially in education, also occurred.

THE ECONOMY

The family farm was central to the preindustrial Appalachian economy. Each mountain homestead functioned as a nearly self-contained economic unit, depending upon the land and the energy of a single family to provide food, clothing, shelter, and other necessities of life. Whereas farms in the Midwest and nonmountainous South moved steadily towards a single cash crop, mountain family farms remained diversified and independent, producing primarily for the needs of the family. By 1880, Appalachia contained a greater concentration of noncommercial farms than any other area of the nation (Raitz and Ulack 1984).

The typical mountain farm of the preindustrial period consisted of a mixture of bottomland and rugged mountainside. Corn was the staple crop, occupying about 50% of the acreage under cultivation, but oats and wheat were also harvested, as well as hay, sorghum, rye, potatoes, buckwheat, and other crops. By the late 19th century, large portions of the mountain hillsides had been cleared (usually by burning or girdling trees) for raising cattle, sheep, mules, and fowl. However, the greatest proportion of the farm acreage remained in woodland where the family hogs grazed throughout much of the year.

After 1900, extractive industries such as logging and coal mining competed with mountain farmers for the use of the woodlands. During the first 3 decades of this century, private companies acquired large tracts of mountain woodland. Entire valleys were given over to railroads, coal mines, and coal towns, while forested slopes were denuded to provide timber for underground mines and lumber for coal towns. By 1930, only 60% of the land in Appalachia was still owned by farm families (Eller 1978).

As mountain families abandoned the farms after World War II, coal companies expanded their land ownership and introduced the new technique of strip mining. Companies found that bulldozers and power shovels removed the overburden covering coal seams at a fraction of the cost of underground mining. Strip mining, nevertheless, removed soils and vegetation, as well as overburden, transforming mountain lands into barren slopes (Caudill 1963).

Although strip mines claimed much of the abandoned farm land, some was converted to federal forests. The national forests traced their origins to the Clarke-McNary Act of 1924, which permitted the federal government to acquire "cut over" lands for timber management purposes (Allen and Sharpe 1960).

The introduction of strip mining, the expansion of federal forests, and the migration of marginal farmers contributed to the decline of agriculture in the Appalachians. Today, agriculture is confined to larger valleys, where level terrain permits intensive commercial operations.

The Appalachian Redevelopment Act was passed in March of 1965. Under the Act, Congress established the Appalachian Regional Commission (ARC). The goal of ARC is to provide a cooperative federal-state framework for planning a coordinated social and economic development program for the Region. The Act was passed in response to the severe economic and social conditions which existed in large sections of Appalachia in the late 1950s and early 1960s.

Appalachia historically has had a highly specialized economy, heavily dependent upon exploitation of the region's natural resources. Still largely rural, the region is deficient in service and light manufacturing employment which could help provide new opportunities when workers are displaced from jobs in traditional heavy manufacturing and extractive industries.

CULTURE

The early settlers shared many common characteristics, which are important in understanding their way of life. Many of these traits can be found in those modern-day residents of the region who trace their ancestry back to these early settlers. These were a proud people, proud of their cultural heritage and how they had overcome the many obstacles to their survival. Religion was an integral part of their lives and most were strongly individualistic and self-reliant. Being conservative, they moved cautiously towards change and were sensitive to attitudes of "outsiders."

During the days of the pioneers, the family was the only basic economic unit within a self-sufficient agricultural setting, but kinship also set the context within which politics and government, as well as organizations for religion, education, and other social relationships developed. The influence of the family and kinship group was felt in almost every aspect of mountain life.

Until well into the 20th century, life for residents of the southern Appalachian region was tied to the land and natural resources. Perhaps more than in other rural areas, physiography shaped cultural and social patterns in the mountains. Each community occupied a distinct cove, hollow, or valley and was separated from its neighbors by a rim of mountains or ridges. Land ownership patterns usually terminated at the ridgetop, reinforcing the community's identity and independence. However, the hillsides themselves were generally considered to be "public land" open to the use of all members of the community. Economic and social activities were largely self-contained within these geographic "bowls," with individual households relying upon themselves or their neighbors for both the necessities and pleasures of life. The land was such a dominant factor in the mountain culture that neighborhoods often drew their names from the creeks near the settlement (e.g., Spring Creek community, Walker's Branch community).

Even today, this strong attachment to the land exists. Although fewer modern-day residents make their living directly from the land, they continue to share the attitude that land is to be used. Within the Region, there is the widely held belief that private ownership of land provides a legal right to do with it as one pleases. Obviously, understanding of such attitudes and beliefs is important to effective environmental management.

RELEVANCE TO ECOSYSTEM MANAGEMENT

The southern Appalachian region is ideally suited to the application of the principles of ecosystem management, and the experiences documented in this volume provide graphic illustration of the diversity and complexity of issues challenging the ecosystem manager, particularly on a landscape scale. Some of the key factors contributing to the strength of the region as an ideal setting to demonstrate these principles are as follows:

- The diversity and abundance of natural and cultural resources residing within the complex topography;
- The degree to which the native environment is intact and/or under recovery;
- The size and configuration of protected lands being managed within the principles of sustainability;
- The presence of regional interagency organizations dedicated to integrated ecosystem management;
- The presence of robust programs to inventory, monitor, assess, and research the natural resources and human dimensions of the biogeographic region;
- The variety and severity of threats to natural resources confronting the ecosystem manager occurring at the species, community, and ecosystem levels;
- The abundance of creative management and education actions being pursued in the region to mitigate these perturbations; and
- Linkage to the international system of biosphere reserves where ecosystem management in the southern Appalachians has received recognition for it's leadership role.

The lessons learned in the southern Appalachians in the struggle to adopt the principles in ecosystem management are generally applicable throughout the international system of biosphere reserves.

REFERENCES

Allen, S. W. and G. W. Sharpe. 1960. *An Introduction to American Forestry*, 3rd ed. McGraw-Hill: New York.
Baldwin, J.L., 1968. *Weather Atlas of the United States*. U.S. Department of Commerce, Environmental Science Services Administration. Washington, D.C. Reprinted by Gale Research: Detroit, MI. 1975.
Braun, E.L. 1950. *Deciduous Forests of Eastern North America*. Hafner Press: New York, reprinted 1974.
Caudill, H. M. 1963. *Night Comes to the Cumberland: A Bibliography of a Depressed Area*. Little, Brown: Boston.
Eller, R. D. 1982. *Miners, Millhands, and Mountaineers: Industrialization of the Appalachian South, 1880–1930*. University of Tennessee: Knoxville.
Foth, H.D. and J.W. Schafer. 1980. *Soil Geography and Land Use*. John Wiley & Sons: New York. p. 38.
Hatcher, R.D., Jr. 1978. Tectonics of the Piedmont and Blue Ridge, Southern Appalachians: Review. *Am. J. Sc.*, 278: 276–304.
Hatcher, R. D., Jr. 1989. Tectonics synthesis of the U.S. Appalachians. In Hatcher, R.D., Thomas, W.A., and Viele, G., Eds. *The Appalachian-Ouachita Orogen in the U.S.* Geological Society of America: Boulder.
Hudson, C. and C. C. Tesser. 1994. *The Forgotten Centuries: Indians and Europeans in the American South*, 1521–1704. University of Georgia Press: Athens.
Moore, H. 1988. *Roadside Guide to the Geology of the Great Smoky Mountains*. University of Tennessee Press: Knoxville.
Pirkle, E.C. 1982. *Natural Landscapes of the U.S.* Kendall/Hunt: Charlotte, NC.
Raitz, K.B. and R. Ulack. 1984. *Appalachia: A Regional Geography*. Westview Press: Boulder, CO.
SAMAB (Southern Appalachian Man And Biosphere Cooperative). 1996. The Southern Appalachian Assessment: Summary Report, Report 1 of 5. USDA Forest Service, Atlanta, GA. 118 pp.
Stearn, C.W. 1979. *Geological Evolution of North America*. Wiley: New York.

Thornthwaite, C.W. and J.R. Mather. 1955. *The Water Balance*. Vol. 8. Publications in Climatology. Drexel Institute of Technology: Centerton, NJ.

Turner, W. H. and E. J. Cabbell, Ed. 1985. *Blacks in Appalachia*. University of Kentucky Press: Lexington.

White, P.S. 1996. University of North Carolina at Chapel Hill. Personal communication.

Whittaker, R.H. 1956. *The Vegetation of the Great Smoky Mountains*. *Ecol. Monogr.* 26:1–80.

5 Framework for Integrated Ecosystem Management: The Southern Appalachian Man and the Biosphere Cooperative

Hubert Hinote

CONTENTS

Evolution of the Concept of Biosphere Reserves 81
   The First Biosphere Reserves, 1974 to 1976 83
The Multiple Functions of Biosphere Reserves 84
Laying the Foundation for the First Regional Biosphere Reserve 85
The Southern Appalachian Biosphere Reserve 88
Formation of the SAMAB Program 89
The SAMAB Organization 91
The SAMAB Program 93
Funding the SAMAB Program 95
Future of the SAMAB Program 96
Transferring the SAMAB Experience 96
Summary 97
References and Notes..98

EVOLUTION OF THE CONCEPT OF BIOSPHERE RESERVES[1]

The concept of biosphere reserves has emerged from UNESCO's Man and the Biosphere (MAB) Program. The MAB Program is one of the better known of the science programs of the United Nations Educational Scientific and Cultural Organization (UNESCO). MAB's ambitious purpose is to enjoin the cooperation of science and society in establishing harmonious relationships between people and the environment.

MAB itself originated from a 1968 conference on the "Rational Use and Conservation of the Resources of the Biosphere." One of the recommendations of that conference dealt with the "utilization and preservation of genetic resources" and proposed to make specific efforts to conserve representative samples of significant ecosystems, original habitats of domesticated plants and animals, and remnant populations of rare and endangered species. Another recommendation dealt with the "preservation of natural areas and endangered species."[2]

In 1969, when scientific consultations were being held to formulate the elements of the MAB Program, the idea emerged of "a coordinated worldwide network of national parks, biological reserves, and other protected areas," serving conservation as well as research and education needs. Because these multifunctional biological reserves were to be set up within the framework of the

MAB Program, they occasionally were referred to as "biosphere reserves," but without a precise definition of what that meant.

In November 1970, the UNESCO General Conference initiated a "long-term intergovernmental and interdisciplinary program on Man and the Biosphere focusing on the general study of the structure and functioning of the biosphere and its ecological regions, on the systematic observation of the changes brought about by man in the biosphere and its resources, on the study of the overall effects of these changes on the human species itself, and on the education and information to be provided on these subjects." The conference invited UNESCO's member states to set up national committees for participation in the program and established an International Coordinating Council to set policy and priorities.

When the International Coordinating Council met at its first session in November 1971, the MAB Program was focused around 13 research themes (a 14th theme was added later), with theme number 8 being identified as "Conservation of natural areas and the genetic material they contain" and spelling out the idea of a worldwide network of protected areas.[3] Biosphere reserves were mentioned under this theme (theme number 8, and under this theme only) and were at the same time proposed "as basic logistic resources for research where experiments can be repeated in the same places over periods of time, as areas for education and training, and as essential components for the study of many projects under the program". Thus the idea and the term "biosphere reserves" were officially launched, but in a hazy manner, without much clarity about their role and nature.

One of the key recommendations at the 1972 United Nations Conference on the Human Environment at Stockholm was the establishment of a global network of protected areas that would conserve representative examples of the world's ecosystems. These areas would serve as benchmarks of environmental quality, help preserve gene pools, and provide a framework for international scientific cooperation. This recommendation gave a major boost to the development of biosphere reserves.

From 1973 through 1975, UNESCO, in cooperation with others, convened panels of experts to clarify the scientific mission of biosphere reserves and to set criteria for their selection. The association of the biosphere reserve concept with MAB theme 8 seemed to blur everything but a conservation/protection role. In 1974 this bias was partly mitigated by a special task force convened by UNESCO and the United Nations Environment Program (UNEP) which drew up a set of objectives and a set of characteristics for biosphere reserves.[4] The multiple functions of biosphere reserves covering three basic needs were indicated, namely:

1. The need for reinforcing the conservation of genetic resources and ecosystems and the maintenance of biological diversity (conservation role)
2. The need for setting up a well-identified international network of areas directly related to MAB research and monitoring activities, including accompanying training and information exchange (logistic role)
3. The need to associate environmental protection and land resource development as a governing principle for research and education activities (development role).

It is obvious that these needs can be interpreted in different ways because they cover a wide variety of situations. Moreover, they provide no framework or priorities for selection. A generalized zoning pattern, however, was proposed which was intended to combine the different interests or needs of biosphere reserves. This zonation is shown, in an updated form, in Figure 1.

Also, in 1974, UNESCO proposed an ambitious 6-year program to develop an international network. It soon became apparent that the United Nations organizations would be unable to fund this ambitious program; thus a bold measure was needed to maintain enthusiasm. This came in the form of a provision in the Nixon–Brezhnev Summit Communique of July 1974.[5] In this communique, both sides agreed to support the implementation of MAB and to "designate in the territories of their respective countries certain natural areas as biosphere reserves for protecting valuable plant and animal genetic strains and ecosystems, and for conducting scientific research needed for more effective actions concerned with global protection." Following this communique, U.S. scientists

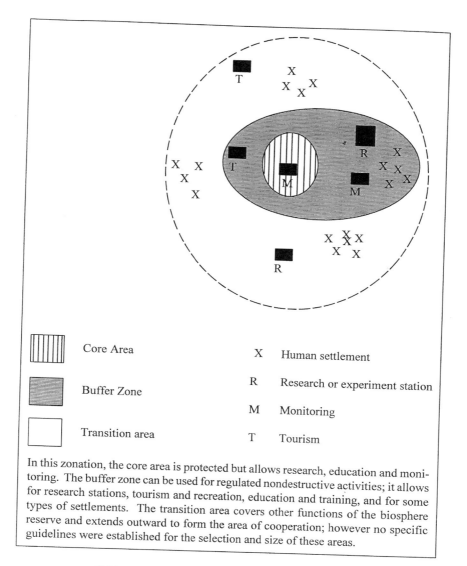

In this zonation, the core area is protected but allows research, education and monitoring. The buffer zone can be used for regulated nondestructive activities; it allows for research stations, tourism and recreation, education and training, and for some types of settlements. The transition area covers other functions of the biosphere reserve and extends outward to form the area of cooperation; however no specific guidelines were established for the selection and size of these areas.

FIGURE 1   Concept zonation of a biosphere reserve.

worked to evaluate and name sites in time for the International Coordinating Council meeting in November 1974. At that time, the U.S. scientists' choice favored protected sites with a long history of experimental, ecological research under the Department of Agriculture, and large conservation areas, mostly managed by the National Park Service.

The First Biosphere Reserves, 1974–1976

The U.S. unilaterally established its first biosphere reserves in 1974. The national goal was to establish a representative site in as many biogeographical provinces as possible. In the Eastern Forest, the Great Smoky Mountains National Park (GSMNP) offered an outstanding area for conservation, public education, and baseline studies. However, its management policies limited possibilities for manipulative research and demonstration of sustainable human uses. It was, therefore, decided to pair the park with the Coweeta Hydrological Laboratory — about 40 km from the southern boundary of the park (Figure 2), where long-term experiments focused on the effects of

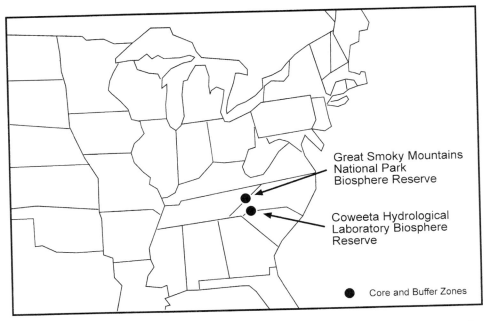

FIGURE 2   Biosphere reserves in the eastern forest biogeographical province, 1974 to 1976.

forest management practices in small watersheds. Coweeta had previously participated in the International Biological Program. Both sites contained suitable core areas and buffer zones, although in very different proportions. At the time, it was hoped that designation of the two biosphere reserves would strengthen cooperation between their federal managers in solving complex problems of conservation and the use of forest resources.

In 1976, when UNESCO officially designated biosphere reserves — 59 in 8 counties — paired biosphere reserves in the Eastern Forest were included. Thus, the GSMNP, managed by the National Park Service, and the Coweeta Hydrological Laboratory, managed by the U.S. Forest Service — both in the southern Appalachian region of the U.S. — were among the world's first UNESCO designated biosphere reserves.[6]

THE MULTIPLE FUNCTIONS OF BIOSPHERE RESERVES

From the early beginnings, biosphere reserves were to serve multiple functions covering the three basic needs or concerns (Figure 3).

By and large, the initial list of internationally designated biosphere reserves did not effectively convey the innovative multifunctional approach embodied in the concept. In the early years, the main criteria used for selection appeared to be the conservation and research roles. Almost all designated biosphere reserves were already protected areas, such as national parks or nature reserves,* and, in most cases, the designation was not adding new functions. Moreover, research work conducted in these areas was generally of an academic nature and was not clearly related to ecosystem and resource management, nor did it address the relationship between environment and development. An appropriate balance between the three needs, shown in Figure 3, was not reached by the designation process.†

---

* A national park was normally considered to correspond to a core area and a buffer zone, as shown in Figure 1; therefore in this scheme it was natural for them to be designated a biosphere reserve. However, all national parks were not designated as biosphere reserves.

† There were exceptions to this generalized statement but the exceptions were rare.

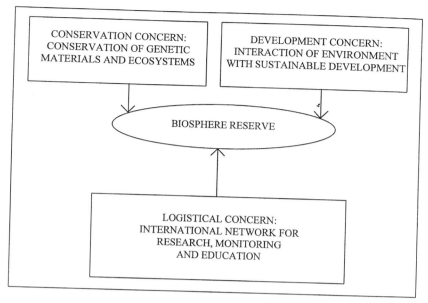

FIGURE 3 Conceptual combination of the different concerns of biosphere reserves.

As early as 1977, the idea of "clustering" was endorsed by the MAB International Coordinating Council. A "cluster" reserve is shown in Figures 4 and 5. In the late 1970s, the idea of "clustering" was aimed towards accommodating the many situations where all the functions of biosphere reserves cannot be performed in contiguous areas (e.g., national parks) and where a regrouping and coordination of activities between several discreet areas is required.[*] In the early 1980s, the U.S. began to establish multiple site biosphere reserves. The intent was to build, through voluntary linkages, large, ecologically delineated conservation units and thereby to encourage cooperation among the administrators of complementary and often contiguous protected areas.[†]

LAYING THE FOUNDATION FOR THE FIRST REGIONAL BIOSPHERE RESERVE

In most geographic provinces of the U.S., partnerships among paired, cluster, and multisite biosphere reserves did not materialize. Managers tended to view designation as an honor or award,[‡] rather than a call to develop cooperative programs to implement the multiple functions of biosphere reserves. Although cooperation involving sites managed for different purposes under various management and/or ownership was clearly needed, the separate designations of parks, protected natural areas, and research sites appeared to be increasing the number of biosphere reserves without a corresponding improvement in their functions. To address this problem, the U.S. MAB (USMAB)

---

[*] By the late 1970s the two designated biosphere reserves in the Southern Appalachians had expanded their joint research on important regional environmental issues, such as air pollution and acid deposition. They were joined in these efforts by the Oak Ridge National Environmental Research Park, an important federal site for basic environmental and ecosystem research, including studies of radionuclides. Because of national security concerns at the nearby Oak Ridge National Laboratory, the site could not be nominated as a biosphere reserve at that time. It nevertheless served as an affiliated site in the expanding research partnership. The multisite partnership came to be known as the cluster concept of biosphere reserves.[7]

[†] Multisite Biosphere Reserves are groups of two or more administrative units designated together as a single biosphere reserve. In the U.S., multisite biosphere reserves were designated primarily during the period 1980–1986. For example, the California Coast Ranges Biosphere Reserve was established in 1983 with eight sites.

[‡] Like the motion picture industry would view an actor/actress receiving an Oscar or as the sports world would view a college football player receiving the Heisman trophy.

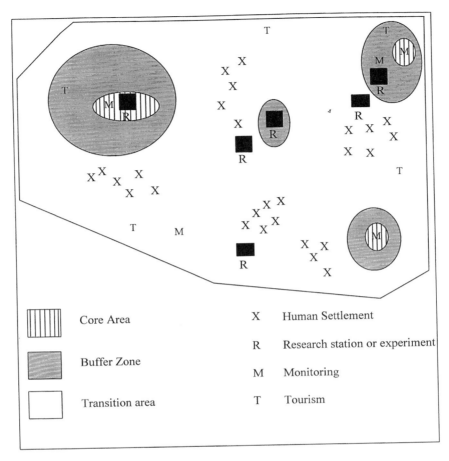

FIGURE 4   Cluster biosphere reserve concept.

program began to sponsor studies to help support local efforts to organize cooperative biosphere reserve programs.

In 1985, a USMAB Biosphere Reserve Selection Panel on biosphere reserves in the eastern forests recommended expansion of the biosphere reserve network.[*] The Panel identified several subprovinces and proposed possible biosphere reserve configurations, among them a new Southern Appalachian Biosphere Reserve in the southern part of the Blue Ridge subprovince (Figure 6). The proposal recommended designating the existing federal biosphere reserve cluster and several state and private sites as initial units of an integrated regional biosphere reserve. In 1986, a subsequent MAB-sponsored study documented considerable interest among government agencies and local groups in the concept.[†] Subsequently, in 1986, the USMAB National Committee endorsed the nomination of the Southern Appalachian Biosphere Reserve and initiated planning of a model Biosphere Reserve Regional Project.

---

[*] Also in 1985, the U.S. Strategy on the Conservation of Biological Diversity, an Interagency Task Force Report to Congress, recommended that support might be directed toward "...the potential role of Biosphere Reserves as centers for developing the information and skills needed for sustainable conservation of regional ecosystems and for continuing assessment and improvement of resource management through research."

[†] The study was conducted by a native son of the Southern Appalachians, who was also one of the early architects of the U.S. Biosphere Reserve Program.

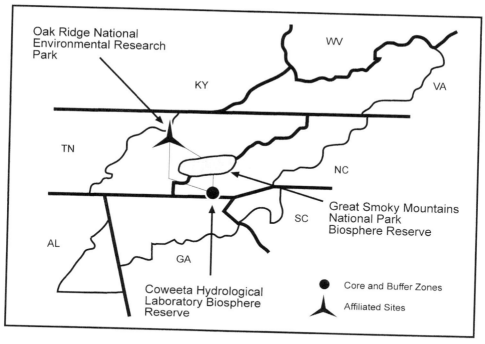

FIGURE 5 Cluster biosphere reserve concept 1978.

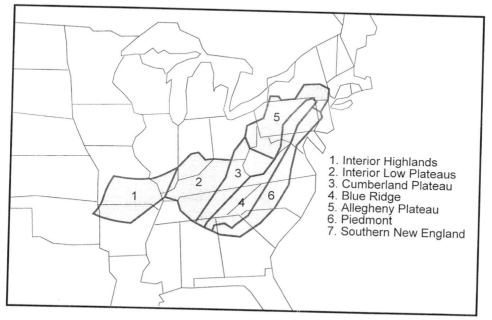

FIGURE 6 Eastern forest biogeographical province: subprovinces for biosphere reserve selection.

The timing of the study and subsequent endorsement by the National Committee was propitious. UNESCO had recently adopted an Action Plan for Biosphere Reserves; this plan gave encouragement to biosphere reserve managers interested in working with others in their region to find solutions to complex conservation and development problems.* Moreover, in the southeastern U.S., the National Park Service was launching a regional program to find ways to help national parks cooperate more effectively with other agencies, private organizations, and local people in their region. The then Southeast Regional Director of the National Park Service proposed pilot projects in cooperative system planning, and he agreed that the MAB approach could be used in the southern Appalachians. The Regional Director described the critical problems facing the region as increasing urbanization, pollution, competition for consumptive resources, and the shrinking of personnel and fiscal resources. He stressed the need to begin a process of identifying regional issues and developing objectives and strategies to address them on a scale reaching beyond park boundaries, indicating that "…these efforts should draw their strength from interagency cooperation aimed at achieving common goals — an ecosystem approach which should be discussed with leaders in the area".†

THE SOUTHERN APPALACHIAN BIOSPHERE RESERVE

As mentioned previously, two of the original 59 biosphere reserves designated in 1976 are in Southern Appalachia. Over the next decade (1976 to 1986), a number of MAB-related activities occurred in Southern Appalachia; for example:

- In 1976 the first bioregional MAB workshop was held at GSMNP.
- In 1977 the first pilot study sites to develop the criteria and methodology for pollutant monitoring in biosphere reserves were selected in the GSMNP
- In 1978 the Southern Appalachian Research and Resource Management Cooperative was formed — a cooperative of six major state universities and three federal agencies. This cooperative was based on MAB principles
- In 1978 an international workshop (sponsored by USMAB, UNESCO, and UNEP) was held in the southern Appalachians to develop recommendations for long-term ecological monitoring in biosphere reserves around the world
- In 1980 GSMNP was selected as a prototype for a USMAB report series on the history of scientific activities in biosphere reserves.[8]
- In 1984 a conference on the Management of Biosphere Reserves was convened as a major event in the 50th anniversary celebration of establishment of GSMNP as a national park.[9]
- In 1986, UNESCO recognized the southern Appalachians as one of two areas in the U.S. which best exemplified biosphere reserve concepts
- Also in 1986, the USMAB National Committee endorsed the nomination of the Southern Appalachian Biosphere Reserve and initiated planning of a model Biosphere Reserve Regional Project.

In 1988 Southern Appalachia was officially designated a multiunit biosphere reserve. Three management units were designated:

1. Great Smoky Mountains National Park, administered by the National Park Service;
2. The Coweeta Hydrological Laboratory, administered by the USDA Forest Service; and
3. The Oak Ridge National Environmental Research Park, administered by a private contractor for the U.S. Department of Energy.

---

* In 1983, UNESCO and UNEP jointly convened the first International Biosphere Reserve Congress in Minsk (Belarus) in cooperation with FAO and IUCN. The Congress activities gave rise to an "Action Plan for Biosphere Reserves." The Minsk Action Plan prescribed nine goals and 35 recommended action for the biosphere reserve network to achieve in the next 5 years. Some of the actions were quite ambitious and, for a variety of reasons, only a few of them were ever fully implemented.
† In this same time frame, the U.S. Forest Service was beginning to emphasize biodiversity conservation and ecosystem sustainability in managing multiple uses on national forests; and other agencies and institutions were interested in forming partnerships on specific issues.

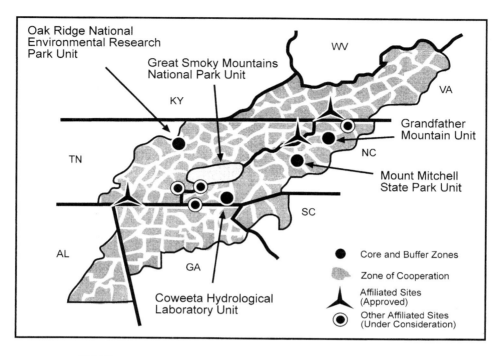

FIGURE 7   Southern Appalachian Biosphere Reserve, 1995 (5 units).

Since 1988 two additional management units have been added: Mount Mitchell State Park, administered by the state of North Carolina; and Grandfather Mountain Corporation, with guidance from the Nature Conservancy. Figure 7 shows the Zone of Cooperation and the location of the current management units of the Southern Appalachian Biosphere Reserve. Figure 8 is a schematic of the functions of the Southern Appalachian Biosphere Reserve. Note in Figure 8 that the terminology adopted for the Southern Appalachian Biosphere Reserve is somewhat different from that shown in Figure 1, but the concepts generally remain the same.

FORMATION OF THE SAMAB PROGRAM

As stated earlier in "Laying the Foundation..." in 1986 USMAB endorsed the nomination of the Southern Appalachian Biosphere Reserve and initiated planning of a model biosphere reserve regional project. This planning was in the form of a feasibility study based on the concept of a biogeocultural region. In August 1987, site managers and administrators from federal and state agencies met to explore the development of a cooperative organization based on the MAB framework and the aforementioned biogeocultural regional feasibility study. Those present agreed that in order for any activity or organization to be successful it would:

- Recognize the need to accommodate compatible economic development with the appropriate enhancement, conservation, and protection of natural and cultural resources;
- Have the capability to provide economic, natural and cultural resource data relevant to specific regional issues;
- Be recognized by interested parties as an organization of reason, influence, and credibility; and
- Be a source of information for decision makers and opinion leaders interested in and responsible for decisions which could affect natural and cultural resources.

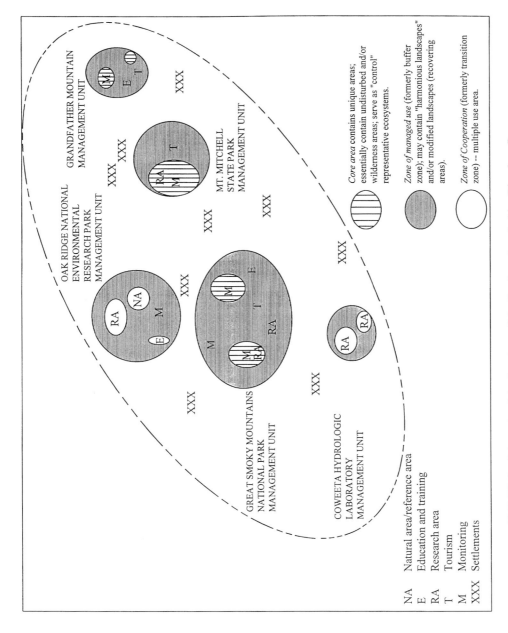

FIGURE 8 Schematic of the functions of the Southern Appalachian Biosphere Reserve.

NA   Natural area/reference area
E   Education and training
RA   Research area
T   Tourism
M   Monitoring
XXX   Settlements

These stipulations, utilizing the MAB framework, ultimately became the operational guidelines for the Southern Appalachian Man and Biosphere (SAMAB) program.

In August 1988, six federal agencies and bureaus signed an "Interagency and Cooperative Agreement for the Establishment and Operation of the Southern Appalachian Man and the Biosphere Cooperative." Those signing the agreement were the Southeast Region, National Park Service; Southeast Region, U.S. Fish and Wildlife Service; Southern Region, USDA Forest Service and Southeastern Forest Experiment Station; Atlanta regional office, Economic Development Administration; Tennessee Valley Authority; and the Ecological Research Division, U.S. Department of Energy. As of February 1996, the agreement had been signed by six other federal agencies: U.S. Environmental Protection Agency, U.S. Geological Survey, National Biological Service, U.S. Army Corps of Engineers, Appalachian Regional Commission, and the USDA's Natural Resource Conservation Service. The Departments of Natural Resources, Conservation, and Environment from three states — Tennessee, North Carolina, and Georgia — have also signed the Agreement. Other federal (e.g., the Federal Highway Administration) and other state agencies are currently considering membership in the SAMAB program.

With this extensive level of membership and diversity of interests, the SAMAB program easily has expanded and will expand on the expertise to thoroughly comply with its stated mission for the Southern Appalachian Biosphere Reserve. The mission is to:

> ...promote the achievement of a sustainable balance between the conservation of biological diversity, compatible economic uses and cultural values across the southern Appalachians. This balance will be achieved by collaborating with stakeholders through information gathering and sharing, integrated assessments, and demonstration projects directed toward the solution of critical regional issues.

To accomplish this mission, SAMAB is promoting environmentally sound, sustainable resource management and economic development through research, management, and educational activities. These involve participation by all levels of government and private interest groups in the southern Appalachians.

THE SAMAB ORGANIZATION

SAMAB partners have set up an organization that is intended to facilitate the broadest possible participation of all levels of government and the private sector. Figure 1 in Chapter 4 illustrates the organizational structure of the SAMAB program (Figure 9). It presently consists of two organizational entities

- The SAMAB; and
- The SAMAB Foundation.

The SAMAB Cooperative consists of federal and state agencies that have voluntarily signed the "Interagency and Cooperative Agreement." The policy board of the Cooperative are senior federal and state officials who signed the agreement; and it is managed by an executive committee of managers which oversees and arranges support for the Cooperative's activities and has overall responsibility. The Executive committee members are appointed by the senior officials who signed the agreement. The coordinating office, jointly funded by the SAMAB Cooperative member agencies, is charged with day-to-day operations and representing and coordinating the regional biosphere reserve program. Six permanent committees, made up of representatives from both the public and private sectors, define issues, develop a plan of work, and implement activities in SAMAB's major program areas: research and monitoring, resource management, sustainable development, cultural and historical resources, environmental education, and training and public affairs.

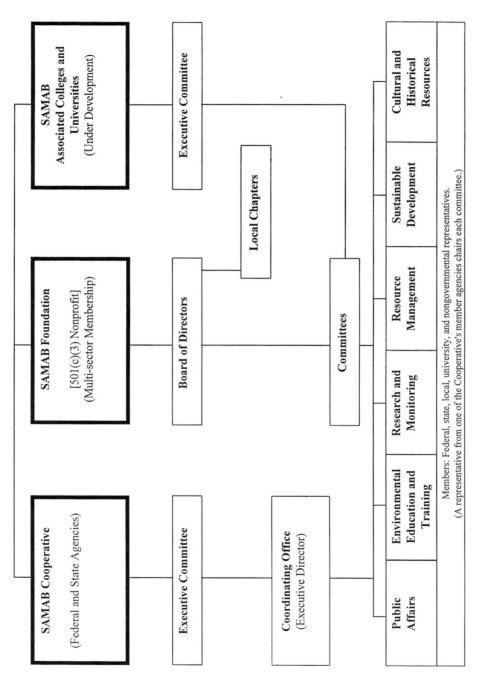

FIGURE 9 The Southern Appalachian Man and the Biosphere (SAMAB) Program.

In 1990, the nonprofit SAMAB Foundation was formed to complement the Cooperative; to involve other interest groups such as private industry, other nonprofit organizations, and special interest groups; and to help find means to support the program. The Foundation has established its own board of directors; members of the board consist of (1) private industry (e.g., Duke Power, Georgia Power, The Chevron Companies, WBIR-TV); (2) nongovernmental organizations (e.g., National Parks and Conservation Association, Environmental Defense Fund, Sierra Club); (3) universities/colleges (e.g., University of Tennessee-Knoxville, Carson-Newman College, Appalachian State University); and (4) local communities (e.g., Pittman Center, Tennessee). The Foundation was expected to become a significant fund-raiser.

The Foundation and the Cooperative work together to identify important natural resource and economic development issues. Independently and together, they develop means for addressing these issues.

A third entity is presently being developed, the SAMAB Consortium. The Consortium will consist of colleges and universities that affiliate with the SAMAB program.

THE SAMAB PROGRAM

In its formative years, the SAMAB partners coordinated significant region-wide programs in some aspects of the following subject areas/issues:[*]

Environmental Monitoring and Assessment

- Forest health monitoring — threats to forest health in the southern Appalachians
  - Three workshops across the region describing exotic insects and diseases affecting the forests
  - Approximately 100 plots already providing data. There are 100 additional plots expected in the next 1 to 2 years
- Landscape ecology/landscape monitoring
  - Held two workshops on Integrated/Ecological Assessments
  - In cooperation with EPA's EMAP program, significant research is being funded on developing landscape scale modeling and analysis.

Sustainable Development/Sustainable Technologies

- Two regional workshops for better understanding and implementation strategies
- Community strategic planning/tourism
  - Assistance in developing a strategic plan led to additional grants to the community for implementation
  - Outreach program to other communities underway
- Geographic information systems
  - Regional geographic information system underway
- Workshops on forestry best management practices

Conservation Biology

- Wetlands
  - Regional conference led to publication of Book entitled "Wetlands of the Southeastern United States"

[*] More detailed discussion on various aspects of these issues is given in other chapters.

- Economic use(s)/protection of native plants
  · First regional workshop clarified local interest; data needs; and opportunities for achieving sustainable economic development of biological resources
- Range of native brook trout
  · Workshop led to additional funded research
- Neotropical migratory birds
  · Cooperative support led to additional monitoring and education programs

Ecosystem Management

- Testimony to the U.S. Senate subcommittee on Agricultural Research, Conservation, Forestry and General Legislation
  · Recognized by White House's Interagency Task Force on Ecosystem Management as a demonstration area for ecological assessment and ecosystem management
- Air quality management: threats to Class I airsheds
  · Brochure on "Understanding Air Pollution in the Southern Appalachians"
  · Workshop led to the creation of the Southern Appalachians Mountain Initiative (SAMI) — an eight-state consortium of public and private groups to address air quality impacting Class I areas in the region
  · Assisting in developing a framework for preparing air quality management plans on public lands in the southern Appalachians
- Partner with U.S. Forest Service on the Chattooga Ecosystem Demonstration Project
- Regional demonstration — Southern Appalachian Assessment

Environmental Education and Training

- Directory of environment education and training (member organizations)[*]
- Videos, posters, and teacher guides on:[12]
  · Reintroduction of the red wolf into the GSMNP in cooperation with WBIR-TV (NBC affiliate), Knoxville, Tennessee (video won an Emmy Award and the educational poster was selected by Urban America as one of top 20 posters in America in 1992)
  · "Water: From the Mountains to the Coast", also in Cooperation with WBIR-TV, Knoxville
  · Dogwood anthracnose, in cooperation with several nongovernment organizations

Cultural and Historical Resources

- Workshop(s) led to ongoing development of a cooperative program to preserve and promote regional cultural resources
- Developing databases on regional cultural resources

Public Information and Education

- Newsletter (s) — four/five annually
- Spring Planning Workshop and Fall Annual Conference
- General information about SAMAB and the Southern Appalachian Biosphere Reserve.
- SAMAB Home Page on the Internet

---

[*] These were distributed to all schools and public libraries in the zone of cooperation.

In 1994, SAMAB initiated work on one of the most significant projects to date, the Southern Appalachian Assessment (SAA). This integrated assessment assembled existing data and evaluated past trends, current conditions, and future risks to the economic, ecological, and cultural resources of the region. The SAA was completed in July 1996 with publication of four resource-specific technical reports (Atmospheric, Aquatic, Terrestrial, Social/Economic/Cultural) and an integrated summary. Other products of the SAA include: (1) a comprehensive database made available to interested parties through a variety of media, including a SAMAB homepage on the Internet; and (2) identification of gaps in both available data and understanding of system function that will guide future research and monitoring activities. It is hoped that the results of the SAA will enable SAMAB partners to work together to protect the unique resources of Southern Appalachia, while promoting economic development that is sustainable. The SAA has been recognized as one of three prototypes by the National Assessment Program under the Office of Environmental Policy and is expected to assist in setting the standards for anticipated integrated assessments conducted across the country.

FUNDING THE SAMAB PROGRAM

Funding support for the SAMAB Cooperative may be categorized as follows:

- Direct support for the Coordinating Office (salaries, benefits, travel and related costs directly associated with staffing the coordinating office)
- Pooling funds to implement specific programs/projects (e.g., Red Wolf Education program)
- In-kind services
  · Staff serves on executive committee, chairs the various committees, and works on specific programs/projects
  · Administrative/clerical services (office, telephone, postage, printing services, etc.)

The SAMAB Cooperative Executive Committee is charged with generating funds and in-kind services to plan and implement the program. These funds and in-kind support generally come from the local managers and/or their regional offices. Since many of the Executive Committee members are local managers, they have willingly supported the program/projects because they have recognized the value of cooperation, coordination, and integration. It should be noted that no new or additional funds have been provided to the local managers because of their participation; rather, they see the SAMAB program as an effective way to identify and address local and regional issues that reach beyond the mandate and scales of their respective agencies. SAMAB cooperation allows them to take an ecosystem and adaptive management approach to identifying and addressing problems facing the region.

Support, both financial and in-kind, has grown as the program has matured; but local managers have limited ability to redirect funds and personnel. Gaining the support of regional and national agency administrators to commit time, attention, and money to support SAMAB efforts has limited its effectiveness in addressing a larger variety of regional issues. However, as the program has grown in stature and recognition, regional administrators and other agencies have increasingly provided support.

SAMAB still lacks a reliable financial base on a long-term basis. This limits its ability to undertake activities that reach beyond the current budget cycle of local managers. Nevertheless, it has found ways to undertake a number of research and education projects that better inform the public and encourage better management practices.

The SAMAB Foundation may help raise funds, but to date it has not been successful in raising enough funds to significantly support regional projects, needed staff, and administrative expenses. The funds that the Foundation has raised have been used to support programs/projects; but much more is needed. The Foundation is working to attract more private sector partners and to involve local people more directly in SAMAB activities.

FUTURE OF THE SAMAB PROGRAM

The future of the SAMAB program looks encouraging. Recognition and support for SAMAB has grown steadily. This is a reflection of the fact that the concept and principles on which it was established are sound and future oriented. It is also a strong reflection of individual leadership within each of the member organizations. More specifically:

1. The concepts and principles on which SAMAB was established are more valid today than ever before, especially the need to identify and address issues that reach beyond the scale and mandate(s) of any single agency, and the need for developing coordinated and consistent databases for multiple uses in both the public and private sectors.
2. The structure of the program places major emphasis on the development of strong public–private partnerships for both funding and decision-making (this is extremely important in Southern Appalachia, where about one third of the land base is publicly owned and managed).
3. The products and services of the program include environmental and economic problem solving options.
4. The federal government's emphasis on Performance Reviews, Ecosystem Management, Adaptive Management and other effectiveness and efficiency measures fits with the vision and action plan of SAMAB.
5. A wide range of stakeholders — local citizens, environmental groups, private industry, and government officials (local, state, national, and international) — have recognized the SAMAB program as a model for developing cooperative and coordinated research, resource management, and educational programs/projects.

In 1994, SAMAB developed an Action Plan for the 1994 to 1996 period and is in the process of implementing that plan.[10] SAMAB has done much with good faith and limited funds to fully achieve its goals. Participating agencies and the MAB National Committee need to consider ways to support biosphere reserves through national programs that provide adequate funding to enable regional MAB Cooperatives to contribute effectively and efficiently to ecosystem management.[11]

TRANSFERRING THE SAMAB EXPERIENCE

It should be clear that the process for designating biosphere reserves is distinct from that of putting an institutional structure in place to make them serve the multifunctional roles embodied in the concept.

The designation process — "a hard technology" — transfers rather easily. National committees and UNESCO have a standard procedure that must work its way through the process — initiated locally, approved by the respective MAB National Committee, and approved by the MAB International Coordinating Committee. However, when considering a regional biosphere reserve based on the utilization of ecosystem principles and sustainability (as discussed in Chapter 1 by Dr. Frank McCormick), adaptations may need to be made. One country or region cannot simply duplicate the exact combination of strategies used elsewhere, because each region or ecosystem presents a unique pattern of resources, economic problems, culture and administrative or political boundaries; it also presents a unique pattern of opportunities. But when adaptations are required, they are generally a matter of course for trained scientists and technologists.

There is much more to emulating the SAMAB experience, however, than just providing a rationale for the designation of a regional biosphere reserve. Administrative style, planning techniques, methods of cooperation and coordination with other parts of the public sector, methods of relating to the public in general, financing methods and controls — "soft technologies" — play a major and significant role. Basically, this is the system that plans, coordinates, and integrates the various component parts to make the unified concept work in practice. These "soft technologies" of the SAMAB experience are not easily understood by outside observers, because they are difficult to codify in a manner that permits easy understanding, much less easily transfer to a significantly different sociopolitical environment.

Nevertheless, considering the fact that SAMAB is still growing in stature and support, it is a demonstration model that should be carefully studied for applicability to other regions of the world. Moreover, it is a dynamic model that is still evolving, e.g., more stable funding sources are being sought, there is a need to get the colleges and universities more intimately involved in the program through the development of the SAMAB Consortium, a subcommittee structure dealing with specific and specialized issues may be needed, and a way needs to be found to more adequately reward those who provide in-kind services.

SUMMARY

The MAB program is now about 25 years old. It can reasonably be assumed that from the beginning it has been concerned with the rational use and conservation of the resources of the biosphere, and that the best known of the MAB programs is the establishment and recognition of biosphere reserves. As of 1995, there were more than 325 biosphere reserves in 82 countries.

The concept, recognition, and implementation of biosphere reserves has had a somewhat hazy and ill-defined history. However, from this beginning, the concept has evolved and blossomed over the past decade. Biosphere reserves are designed to conserve, gather, analyze, communicate, and employ information for the purpose of sustaining natural and managed ecosystems. While no model of a biosphere reserve applies universally in practice, any biosphere reserve — no matter where it is located in the world nor its particular zonation as long as it contains a core area — must to some extent address itself to the multifunctional roles of conservation, development, and a network for research, monitoring, and education. The relative combined effect sets biosphere reserves apart from protected areas, makes them unique, and offers the greatest opportunity for maintaining landscape diversity.

The Southern Appalachian Biosphere reserve, established in 1988, is a multiunit, multifunctional biosphere reserve. Steps have been taken by the administrative authorities in Southern Appalachia to implement a fully functional biosphere reserve. The first step related to the recognition of Southern Appalachia as an ecological, biogeocultural, and/or ethnological region has been taken. The second step consisted of putting a coordinating group in place. While not yet at the final stage, a fully operational and coordinated program for identifying, integrating, and addressing issues affecting the entire zone of cooperation of the Southern Appalachian Biosphere Reserve, progress is being made to enlist the continuing support and cooperation of the local people (through a series of programs, projects, and activities) without which long-term sustainable conservation of species or ecosystems is seriously open to question.* Bioregional or ecosystem-based institutions focused

---

* "The Southern Appalachian Man and the Biosphere Cooperative, although containing Federal agency partners, has developed an identity separate from the agencies. This gives the Cooperative a unique ability to forge cooperation in all aspects of science and information dissemination. Many interviewees viewed the Cooperative as a resource and facilitator. Individual Management agencies might, by contrast, be perceived as a threat. The Southern Appalachian Man and the Biosphere Cooperative has become accepted as a translator of technology. It facilitates science by increasing awareness among agencies of other agencies' missions and functions. It also helps eliminate duplication of effort in research activities and it encourages software compatibility for data sharing." *The Ecosystem Approach: Healthy Ecosystems and Sustainable Economies*. Report of the Office of Environmental Policy's Interagency Ecosystem Management Task Force, vol. 1, June 1995, pp 43–44.

around an "ideal biosphere reserve" offer a sound basis for maintaining landscape diversity and sustaining the natural and human resources of an ecosystem.

REFERENCES AND NOTES

1. For a more detailed explanation, see "Developing and Focusing the Biosphere Reserve Concept" by Michel Batiste, a reprint from vol. XXII (3), July–September 1986, UNESCO, *Nature and Resources*. Much of this discussion is taken from that report.

2. MAB's immediate predecessor was the International Biological Program (IBP), which was launched in the 1960s through the joint efforts of UNESCO, nongovernmental scientific organizations, and participating governments in the developed countries. IBP's interdisciplinary projects amassed huge banks of basic information on deciduous forests, grasslands, tropical forests and other natural regions of the world. In view of the decade's end and potential termination of the program, many scientists saw the need to maintain the program's momentum while expanding and redirecting its efforts.

3. Projects (themes) 1 through 7 focused on major natural regions or biomes (for example, tropical forests, temperate forests and lands, islands and so on). Projects (themes) 9 through 14 were concerned with types of interactions between man and the environment (for example, cultivation systems, urban ecosystems, perception of environmental quality, and so on) with the objective of improving these relationships though scientific understanding.

4. "Task Force on Criteria and Guidelines for the Choice and Establishment of Biosphere Reserves", UNESCO, 1974, 61 pp (MAB Report Series 2). For susequent criteria in the U.S., see "Guidelines for Selection of Biosphere Reserves," an Interim Report. USMAB Publication No. 1. National Park Service, Department of Interior. Washington, D.C. 1979.

5. Nixon-Brezhnev, Joint US-Soviet Communique. Dated July 3, 1994, between U.S. President Richard M. Nixon and General Secretary L.I. Brezhnev of the Central Committee of the Communist Party of the Soviet Union.

6. For a more detailed history, see "Biosphere Reserves, Their History and Their Promise" by William P. Gregg, Jr. and Betsy Ann McGean. *Orion Nature Quarterly*, vol. 4(3), Summer 1985, pp 41–51.

7. Johnson, W.C., J.S. Olson, and D.E. Reichle. 1977. "Management of experimental reserves and their relation to conservation reserves: the reserve cluster." *Nat. Resour.* 13 (1): 8-14.

8. Great Smoky Mountains Biosphere Reserve: History of Scientific Study. USMAB Report 5, March 1982. USDI National Park Service, Southeast Regional Office, Atlanta, GA. 276 pp.

9. Peine, J.D., Ed. 1985. *Proceedings on the Management of Biosphere Reserves*. Great Smoky Mountains National Park Biosphere Reserve, Gatlinburg, TN, November 27 to 29, 1984. USDI National Park Service, Southeast Regional Office, Atlanta, GA.

10. The SAMAB Action Plan closely follows the guidelines/goals/objectives of the Strategic Plan for the U.S. Biosphere Reserve Program. Department of State Publication 10186, Bureau of Oceans and International Environmental and Scientific Affairs. Washington, D.C. December 1994. See also "The Seville Strategy and the Statutory Framework of the World Network of Biosphere Reserves" UNESCO, Paris, France. 1996.

11. "The Constable Commission" report, among other things, recognized the problem of inadequate funding of the MAB program and made recommendations for dealing with it. Department of State, USMAB program. Washington, D.C. May 1995.

6 A Comparison in the Management of Two Biosphere Reserves: Prioksko-Terrasny Reserve, Russia and Great Smoky Mountains National Park, U.S.

John D. Peine, Vladimir A. Koshevoi, Karen P. Wade, Igor A. Zhigarev, and Margarita A. Davydova

CONTENTS

Introduction 100
Overview of the Biosphere Reserves 100
   Overview of the National Systems of Biosphere Reserves 100
      Russia 100
      United States..101
   Prioksko-Terrasny Biosphere Reserve 102
   Great Smoky Mountains National Park Biosphere Reserve 103
Changes in National Governments 104
   Russia 104
   United States 104
Changing Management of Biosphere Reserves..105
   Mission 106
   Organization 106
   Budget..107
   Walking the Tight Rope: Setting Management Priorities 108
   Programs and Partnerships in Ecosystem Management..108
      Species Reintroduction..109
      Research and Monitoring 110
      Air Quality..111
      Native Brook Trout Management 112
      Black Bear Management 112
      Strategic Planning for Wilderness Management..112
      Environmental Education: Linkages between Biospheres.......................................113
References 114

0-57444-053-5/99/$0.00+$.50
© 1999 by CRC Press LLC

INTRODUCTION

The primary intent of this chapter is to provide a forum for managers of internationally significant ecosystems to discuss their challenges and methods to achieve a balance in the ever changing world of the management of protected natural areas. Although the two international biosphere reserves represent somewhat divergent primary missions and very different cultural contexts, the challenges to the managers are similar in nature. The day to day struggle is to protect a sanctuary of biological diversity within the context of a regional landscape highly manipulated by human activity. Managers are required to pay attention to an ever expanding venue of concerns and responsibilities during a period of drastically reduced funding and volatile political activity.

OVERVIEW OF THE BIOSPHERE RESERVES

The Man and the Biosphere Program (MAB) was established at the 1970 General Conference of the United Nations Educational, Scientific and Cultural Organization (UNESCO). At this conference, the International Coordinating Council for MAB was chartered. MAB provided the first formal mechanism for bringing together and coordinating diffuse national and international research, conservation, and training activities through an international network of what has now become 329 multifunctional areas designated as biosphere reserves in 83 countries, nominated by MAB national programs (Gregg, Chapter 2 in this volume). The criteria for selection are legal protection for conservation of core areas, active science infrastructure, involvement of regional stakeholders, and the potential to demonstrate sustainable human use of the ecosystem (U.S. Department of State 1995).

OVERVIEW OF THE NATIONAL SYSTEMS OF BIOSPHERE RESERVES

Russia

According to the laws of environmental protection in Russia, biosphere reserves are listed as Nature Reserves. The primary objective of the reserve system is to preserve and research natural ecosystem processes and plant and animal communities of typical and unique ecosystems. Maintaining biodiversity in the country is the central goal. The designated biosphere reserves in Russia are part of the global network of protected natural areas dedicated to the protection of natural heritage under the aegis of the UNESCO MAB Program.

As of 1994, 84 state natural reserves exist in Russia, covering over 28 million hectares. Continental and island areas encompass 23.5 million hectares equaling 1.45% of the entire Russian territory. Among these reserves there are 16 designated biosphere reserves as listed in Table 1. Five additional nature reserves have been nominated to UNESCO, including Teberdinsky, Komandorsky, and Bolshoy Arktichesky as additional units of the international network of biosphere reserves.

Some of the newly proposed biosphere reserves will be zoned. Always before, the Russian reserves consisted entirely of core protection zones. The proposed Komandorsky Biosphere Reserve, for instance, includes zonation for natural undisturbed ecosystems, regeneration to a natural state, agricultural activity, and a buffer from human activity in the surrounding landscape. In this case, buffer zones had existed earlier, but they were not within state natural reserve boarders. In general, buffer zones for biosphere reserves range in size from 0.5 to 2 kilometers beyond the boundary of the core protected areas. Human activity in the buffer zones are limited. Buildings, agricultural activity, hunting and fishing are generally prohibited. Any deviations from this policy are agreed to with the local government. The boundaries of the protected buffer zone are constant and the area encompassed tend to exceed that of the core zone of protection by 20 times. These buffer zones are usually controlled by the ecosystem managers of the biosphere reserve.

TABLE 1
Biosphere Reserves in Russia

Name	Hectares	Date founded	Location
Astrakhansky	6,816	1919	The Volga Delta
Bilhalsky	165,724	1969	Lakeshore of Bikal Lake
Bargugnsky	374,322	1916	Lakeshore and aquatory of Bikal Lake
Voroneghsky	31,053	1927	Center of Russian Plain
Karkazsky	288,277	1929	Western Big Cancas Mountains
Kaonotskiy	1,142,000	1934	Eastern coast
Laplandsky	278,436	1930	Seaside of Barentz Lake, Kolsky Peninsula
Oksky	55,722	1935	Center of Russian Plain
Pechozo-Llyichisky	721,322	1930	Western North Uzal Mountains
Prioksko-Terrasny	4,945	1948	Moscow region
Sayano-Shusheneksy	390,368	1976	Sayani Mountains
Sihote-Alsnky	351,950	1973	Coast and aquatory of Japanese Sea
Sohondinsky	210,985	1973	Eastern part of Bikal region
Tsentralno-Lesnoy	21,380	1931	Central part of Russian Plain
Tsentraino-Salinsky	972,017	1985	Region of contact between Western and Middle Siberia, mid enssey
Tsentraino-Chernozensky	4,874	1935	South of Russian Plain

United States

The USMAB (USMAB) Program consists of programs in high latitude ecosystems, marine and coastal ecosystems, human dominated ecosystems, temperate ecosystems, tropical ecosystems, and biosphere reserves. There are 47 existing U.S. International Biosphere Reserves dedicated to conservation of biological diversity, including some with demonstration projects concerning the sustainable use of natural resources. Like the biosphere reserves in Russia, the vast majority of biosphere reserves in the U.S. are strictly reserves, most of which are, in fact, national parks with minimal concern with the demonstration of sustainable living. Unlike Russia, there are no buffer zones in U.S. biosphere reserves restricting activity on lands adjacent to core protected areas.

A strategic plan for U.S. Biosphere Reserves developed in 1994 is intended to nudge these reserves more in the direction of demonstrations of sustainable practices. Key goals include (U.S. Department of State 1995):

- Create a network of biosphere reserves that represents the biogeographical diversity of the U.S. and fulfills the internationally established roles and functions of biosphere reserves
- Foster cooperative partnerships among all stakeholders in biosphere reserves
- Acquire and integrate knowledge for sustaining biodiversity, cultural values, and viable economics within an ecosystem/landscape context
- Promote public awareness and education that strengthens the commitment of stakeholders to MAB concepts
- Establish mechanisms for sharing and disseminating data and information among U.S. Biosphere Reserves and others

As emphasis on the biosphere reserves expands in the U.S., expectations are raised that the system of biosphere reserves will steadily increase the number of demonstration projects which truly emulate the full intent of the UNESCO MAB Program.

Prioksko-Terrasny Biosphere Reserve

The Reserve was founded in 1948 and designated an international biosphere reserve in 1979. Located 100 km to the south of Moscow, the capital of Russia, the Reserve is situated on the left bank flat terrace slopes of the valley of the Oka River in the vicinity of the city of Serpukov. The largely forested reserve that occupies a territory of 4945 ha lies within the borders of the Moscow botany geographical region of the East-European Province (Razumovsky 1981). On all sides, the Reserve is surrounded by the recreation forests of the Experimental — Manufacturing's (or industrial) Forestry Association (Lesnichestvo) "Russian Forest" (*Russkiy Les*). These forest lands form the buffer zone around the core zone of protection so human influence is minimal. This buffer zone is particularly important since the Reserve is located in such a densely populated region. Numerous research projects are carried out in the buffer zone. Their focus is largely to measure the human influence on forest resources from recreation and other activities. This research sets a standard for this urban territory in Russia. There is interest in expanding the boundaries of the reserve to include the buffer zone at some time in the future when the economy recovers and funds become available.

Within a distance of some 10 km around the Oka River Reserve, wide areas are occupied by coniferous, broadleaf, steppe, and mixed forests; swamps; flood plain meadows and steppes. Ancient pine woods grow on deep sediments of sand occurring on the flood plain terraces of the steppe territories, which are separated from the main northern border of steppe territory in Russia 200 to 600 km to the north. The relatively small reserve encompasses the entirety of this complex mosaic of floral communities.

Botanists worldwide are astounded by the rich variety of plant species in the Reserve. There are 892 species of flora on the reserve; 29 of them are rare and endangered, and the populations of 18 species are being slowly depleted toward extinction. The greatest number of rare plants are associated with the free flood meadow plains along the Oka River. The natural habitats of feather grass (*Stipa joannis*) and the Russian hazel (*Fritillaria ruthenica*) found in the Reserve are situated hundreds of kilometers to the south of the Oka River valley on the steppes of the East European Plain and Kazackstan.

Scientists throughout the world describe the local flora as the Oka flora. The secret as to its origin is not known. Some speculate that the seeds were carried down river by the water from distant parts of the watershed; others contend that the plant assemblage is an artifact of plant migration as a result of previous ice ages; and still others speculate that the seeds were carried in the hooves of horses and carts of nomads, who in the days of old, moved from south to north attacking various areas of Russia (Davydova and Koshevoi 1989). Who knows, maybe all of these hypotheses are partially correct!

In the spring, flood waters cover vast flood plains carrying fertile black silt from the upper reaches. They sometimes cover hollows in the upper more ancient levels of the flood plain as part of the karst relief formations common for the central Russian upland. In the northern portion of the Reserve, the dry flat bottom valleys closely border the dry lichen pine forests, the kind found in Western Siberia and the upper reaches of Berezina. The further from the river, the more oak forests mix with coniferous species.

There are more than 200 species of vertebrates found in the Reserve, including 130 species of birds. There are 53 species of mammals. Large mammals include, among others, elk (*Alkes alkes*), European bison (*Bonasus capreolus*), Siberian roe deer (*Capreolus capreolus*), otter (*Lutra lutra*), beaver (*Castor fiber*), wild boar (*Sus scrofa*), squirrel (*Sciurus vulgaru*), European brown and polar hares (*Lepus timidus*), fox (*Vulpes vulpes*), European ermine (*Mustela erminea*), weasel (*Mustele putoris*), pine martin (*Martes martes*), and the European pole cat (*Mustele putoris*) (Davydova and Koshevoi 1989). The European bison are maintained in a nursery to provide animals for reintroduction.

The Reserve is situated in a densely populated region. There is a long-standing tradition in Russia to gather fresh berries and pick mushrooms. Sometimes there is a problem in that too many people are pursuing such activity on the Reserve and interfering with the protected ecosystem which is primarily devoted to research. There is a 2-m high fence around the core protected area to discourage casual trespass, but enforcement of such regulations is problematic.

GREAT SMOKY MOUNTAINS NATIONAL PARK BIOSPHERE RESERVE

The Park was established in 1934 by an act of the U.S. Congress. Other major eastern national parks established in the region during the Great Depression include Mammoth Cave National Park, Shenandoah National Park, and the Blue Ridge Parkway. Great Smokies, Mammoth Cave, and Shenandoah share the distinction of being three of the few national parks established primarily from privately held lands. Of over 360 units of the national park system, only 54 are designated as national parks.

The centers of interest for the establishment of the Park were Knoxville, Tennessee and Asheville, North Carolina. The primary lobbyist was the American Automobile Association, which at the time was promoting the development of tourist oriented destinations for automobiles. Much of the work to develop the Park was accomplished by the Civilian Conservation Corps (CCC), an agency formed during the Great Depression to provide work for the unemployed young male adults. At its peak in the late 1930s, over 4300 men were at work building roads, trails (800 miles), scenic overlooks, campgrounds, and picnic grounds in the Park. The craftsmanship of their labors has left a wonderful legacy that for the most part remains intact to this day. The Park receives the most daily visits of any of the national parks in the system, although it is third to the visitation experienced on the Blue Ridge Parkway and in Golden Gate National Recreation Area (also units of the national park system). The growth of the tourism industry surrounding the Park has accelerated in the last decade. Visitation at tourist attractions outside the Park, such as theme parks, music theaters, and outlet shopping malls has recently been reported to exceed that of the national park. Sevier County, the center of the tourism industry adjacent to the Park, is one of the fastest growing counties in the entire state of Tennessee (Peine, Chapter 17 in this volume).

The Park was designated as an International Biosphere Reserve in 1976, and a World Heritage Site in 1984, primarily for its rich biological diversity within the temperate forest biome that occurs on the complex mountain topography within the Blue Ridge Province of the southern Appalachian mountains. It is currently one of five core areas within the region covered by the Southern Appalachian Biosphere Reserve, which was created in 1988. The Reserve's Southern Appalachian Man and the Biosphere Cooperative (SAMAB) is composed of 14 federal and state member agencies (Hinote, Chapter 5 in this volume). Others are expected to join. The Park landscape consists of numerous sharp-crested, steep-sided ridges forming mostly narrow valleys. Elevations vary from 270 to 2025 m above sea level. Forty-five major watersheds and 3500 km of streams are contained in the 210,600-ha reserve constituting the Park (Parker and Pipes 1990). There are over 1600 species of vascular plants recorded in the Park, including over 130 tree species, distributed in such a way as to form a virtual collage of forest associations on the landscape. The forests of the Smokies are known as one of the most biologically complex in North America and have been the focus of hundreds of research projects over the last three quarters of a century. About 20% of the forests have never been harvested; therefore, the Park contains one of the largest uncut forests left in the eastern U.S. The Park is also a primary viewing area for wildlife found in temperate forests and harbors the nation's most diverse collection of European pioneer era homestead buildings (circa 1818 to 1904).

Primary threats to natural resources and visitor experience include air pollution, the invasion of various alien species (including European wild boar, invasive plants, and forest pathogens), suppression of the prehistoric fire regime, and significant adjacent land use conversion. Today's major struggle, however, is in meeting the mandate of the National Park Service; i.e., to balance the press of human impacts with the needs of the ecosystem so that future generations can continue to enjoy the wonders of the Smokies. The difficulty in doing this is compounded by ever increasing operating requirements which drive operating costs up at the same time that revenues decline. Research activity has slowed significantly because of a lack of funding and the situation is worsening as a result of Congressional activities which tend to target scientific research for major reductions in appropriations.

CHANGES IN NATIONAL GOVERNMENTS

RUSSIA

At the time of the break up of the former Union of Soviet Socialist Republics (U.S.S.R.) in 1991, Russia had 75 reserves encompassing an area of approximately 6.5 million square kilometers, equalling encompassing 29% of the national area. Despite the political, social, and economic upheaval following the relatively bloodless coup in August of 1991, the nature reserve movement continued and even prospered in the sense that the number of designated protected areas grew dramatically. This phenomenon may have, in part, been in response to speculation that as a result of the process of democratization and privatization, some natural landscapes would become vulnerable to development. In 1992 to 1993, eight new nature reserves were established, adding 8.2 million hectares to the reserve system. The primary contribution to reserve lands was the addition of the Bolshoi Arktiichesky Reserve, situated on the Timir peninsula, incorporating 4.1 million hectares, and the Komandorsky Terra-Maranl Reserve (3.6 million hectares) on the base of the Komandorsky archipelago near the eastern coast of the Kamchatka peninsula. During that time period, no other country in the world added a greater amount of land area to its system of protected areas.

There has also been a fundamental shift concerning the rationale for establishing nature reserves. New considerations include social and economic development interests expressed at the local and/or regional level. There is a growing interest in preserving indigenous cultural activities and life styles. Possibly the most fundamental shift in Russian philosophy in the role and function of biosphere reserves is consideration of the sustainable use of natural resources, rather than being just strictly concerned with the maintenance of protected areas where all but scientific activities are excluded.

Since the early 1980s, there has been a movement in Russia to create a system of national parks (Knystautas 1987; Soloviev 1993). Since then, 24 parks have been created, mostly near urban areas or where outdoor recreation activity has been well established. The design and management of these areas seems to have been largely based on European models of parks, with emphasis on outdoor recreation. Many parks contain small villages and permit some forms of consumptive use, such as subsistence hunting, grazing, selective timber harvest, berry picking and mushroom hunting, and small-scale commercial fishing. All of these activities reflect the principles of sustainable living and lend themselves extremely well to the primary intent of the MAB program.

Several exchanges of officials from the U.S. and Russia have been made to refine the national park and preserve management concepts in Russia. The U.S. National Park Service has been actively involved in this exchange (Haskell 1995). Frankly, it is suggested that the Russian government officials proceed with caution in adopting a U.S. formula for national park policy which, in many instances, is highly exclusionary and restrictive as to the types of sustainable resource utilization allowed; and is therefore not a particularly appropriate model for biosphere reserves within the cultural context of the landscapes of central Europe.

The Russian government has taken away management authority for nature reserves from institutes dealing with the consumptive uses of natural resources, such as the Department of Agriculture and the Department of Forestry and Wildlife and Fisheries and given it to such institutions as the Department of Natural Resources, Academy of Science, and to the Department of Higher Education of Russia. On the other hand, local authorities have sometimes seized territory within the boundaries of nationally designated nature reserves.

In March 1995, a plan was adopted to save biodiversity in Russia through a system of protected areas. The strategy centered around further development of the existing system of national parks and nature reserves and their staffing by professional managers to insure adequate protection and appropriate utilization by the people. In August 1995, the Department of Protected Areas was established within the Ministry of Nature of the Russian Federation. The plan calls for the development of infrastructure for these protected areas and the establishment of zones for collaborative management. In addition, the role and function of the designated areas have been expanded. For instance, national parks must conduct programs in environmental education and scientific research.

Under this new strategy, the classification of protected areas will include natural areas, some of which are designated to have global significance; national parks; zakazniki (no equivalent English translation), which are areas of specialized biological interest in which some limited human activity is allowed; and natural monuments which have regional status. Role and function statements have been prepared for each of these classifications along with goals for funding, staffing, and infrastructure.

The largest problem facing reserve managers in Russia is a lack of adequate funding. The fiscal chaos is not unique to the management of natural resource areas and is replete throughout the public sector in Russia. It has been reported that administrators and staff members in reserves sometimes do not receive their salaries for several months at a time. The extraordinary rate of inflation of the ruble has drastically reduced the purchasing power of wages. There are generally no funds available for equipment purchases or the maintenance or operation of vehicles and motor boats.

Research activity has been particularly hard hit by a lack of adequate funding. Some scientists working in remote natural reserves have been reported to be living in near survival conditions. There is great financial difficulty in publishing scientific works and holding conferences, professional meetings, and symposiums. The limited funds available for research tend to be allocated to maintaining at a minimum level long standing field research. These stark circumstances have resulted in a reduction of interest among talented young people to enter professions related to natural science and natural resource management.

UNITED STATES

Changes in national politics, although not as extreme as those in Russia, have dramatically shifted in the last 3½ years. When President Bill Clinton's administration came into power in January of 1993, the Democratic Party controlled both houses of the U.S. Congress. There was a dramatic shift in policy concerning the management of public lands compared to that of the previous Republican president, George Bush. Principles of sustainable forestry were adopted by the U.S. Forest Service. A presidential commission on sustainable development was formed. The National Park Service radically reorganized, primarily to align management with biogeographic regions. The intent was to become more oriented to ecosystem management and efficient in providing technical and financial assistance to managers with similar natural resources. National parks in the southern Appalachian region, for instance, formed a "cluster" mini-region in order to share expertise and formulate collaborative activity in ecosystem management.

The political climate changed drastically again on November 4, 1994 when the Republican party won a majority of seats in both houses of the U.S. Congress. Natural resources programs experienced major budget cuts under the guise of balancing the federal budget. Projects related to ecosystem management and ecosystem assessments have been targeted by the conservative right wing of the Republican party for funding elimination. The future of these initiatives is uncertain as of this writing. In any case, there has been a long-term reduction of the buying power of all federal land managing agencies. This situation has been exacerbated for the National Park Service by having to absorb pay increases for park rangers and benefit increases for all employees, leaving little financial opportunity for anything more than trying to sustain core programs in resource management. Attempts are being made by conservative members of the new U.S. Congress to systematically dismantle regulations of the federal government concerning environmental protection. This current political atmosphere does not lend itself to experimentation to devise new and better ecosystem management activities.

CHANGING MANAGEMENT OF BIOSPHERE RESERVES

MISSION

Prioksko-Terrasny Biosphere Reserve

The primary mission of the reserve is conservation of the biological diversity and the pursuit of scientific research concerning ecosystem processes and species and community dynamics in an area that incorporates an extraordinary number of plant communities representing a divergent assemblage of biogeographic regions. The reserve serves as a control from which to measure the influence of humans on the surrounding Russian forests. The reserve serves as the key site in all of Russia for the propagation of the European bison. Finally, a function of growing importance is the use of the area for environmental education. The reserve is generally not used for recreation which is actively pursued in the surrounding forest lands and therefore serves as a baseline from which to assess the impacts of such activity.

Great Smoky Mountains National Park Biosphere Reserve

The Park plays a primary role within the context of the Southern Appalachian Biosphere Reserve as one of the measuring sticks for what is ecologically "natural" and "sustainable" in the southern Appalachians. In that role, it is increasingly apparent that the Park is conducting pioneering work. Although natural resource monitoring has been done in a variety of forms for decades in the Park, long-term ecological monitoring specifically designed to define trends in ecosystems and ecological communities has only recently begun. As a new program, the science of long-term ecological monitoring is still evolving. This occurs just at a time when the National Park Service and its cooperators are subject to constantly shifting national priorities set by politicians largely ignorant of scientific processes and methods, dramatically shifting budget variations not amenable to good management, and chronic staffing constraints — all of this in an organization where, traditionally, science budgets have not received high priority nor has the need for good science been articulated effectively enough to get consistent departmental and congressional support. The lack of effectiveness by the National Park Service in securing funds and positions for research may be due to the relatively small size of the agency and/or to the fact that the primary mission is perceived to be less as stewards of core protected lands and more as servants of the visitor experience.

The Smokies was selected by the National Park Service as one of the four national parks to be funded for prototype long-term ecological monitoring at $500,000 per year (Smith et al., Chapter 8 in this volume). As a result, the Park is in a position to add to a sound scientific information base previously built by the work of many university scientists and the Uplands Field Research Laboratory, a biological research facility located in the Park from 1975 through 1992, and largely funded by the Southeastern Regional Office of the National Park Service. Gaps in scientific information have been identified, and such work is well underway to build valid indicators of the status and trends of the Park's health within which to form the necessary scientific based judgments for resource management decisions. It is fairly obvious to any observant visitor to the Park that resources are presently being degraded or are at risk from overcrowding and inadequate operational support to manage it. Utilizing scientific information on which to base decision-making, when coupled with funding to do necessary planning and environmental compliance, will help make the hard decisions to deal with overcrowding and cope with conflicting requests for incompatible resource development. Such information will also make it easier to explain to the public what kinds of practices are sustainable in support of the economy, environment and society within the SAMAB zone of cooperation.

ORGANIZATION

Prioksko-Terrasny Biosphere Reserve

The organization which manages all the reserved territories of the Russian Federation is called The Direction of the Reserve Territories in the Ministry of Conservation of Environmental and Natural Resources. Decisions concerning the preservation of the territory, exploitation of the natural resources, assignments for the staff and scientists, main direction and development of scientific plans, education, etc. are at the disposal of the Directory of the Scientific (Soviet) Consulting Group of the Reserve. The Scientific Consulting Group (Soviet-Advise) solves tasks of the Reserve development, and adopts different plans and ideas. Reports from all departments are made at the meetings of the Group. The members of the Group are elected among the scientific specialists of the Reserve for 5 years, plus the Director of the Reserves and the Assistant Directors of the Departments who all take part in management of the Reserve.

The management structure of the Reserve includes: the Director of the Reserve; Scientific Department (20 specialists); Forestry Department including one territory protection system which consists of different full-time working groups of forester rangers which take care of sections of territory, and a system of forestry management (12 people); an administrative and maintenance department, the main task of which is solving different economical and maintenance problems; and a bison nursery (10 people). The total number of staff members is 60 people.

Great Smoky Mountains National Park Biosphere Reserve

The Department of Interior and its agency, the National Park Service, manage the Park and comply with the World Heritage Convention and the Biosphere Reserve designations under UNESCO.

Decisions concerning the preservation and use of the Park are driven largely by specific enabling legislation, policy of the Department and agency, and the guiding document for management, the Park's General Management Plan, which is developed with public participation (national in scope) directing the Park for a 10- to 15-year time frame.

The Park Superintendent and staff are responsible to one of the seven Field Directors to manage park operations. Operational units (called divisions) are maintenance, ranger activities, interpretation, administration, and resource management and science. The management structure of the resource management division consists of a division chief, a natural resources management section with specialties in the management of vegetation, wildlife, fisheries, and air quality; a fire management section; a cultural resources section; and the ecological inventory and monitoring section. The division's resource management and science program is guided by a resource management plan.

Park Superintendents generally rotate every 5 to 10 years, with similar turnover of division chiefs. The total permanent staff of the Park is approximately 200, plus 100 to 150 seasonal workers.

BUDGET

Prioksko-Terrasny Biosphere Reserve

Generally, there is no difference in the budget of a nature reserve that has been designated an international biosphere reserve vs. those that are not so designated. Prior to 1990, a significant portion of the Reserve budget was designated for scientific research, resource monitoring, and environmental education. Now, the budget is distributed as follows: 20% staff salaries, 20% equipment and maintenance, 20% for the operation of the bison nursery, and 40% designated as "current payment" or operating expenses, such as utilities and office expenses. The Reserve does not get many funds from the federal budget. Federal funding for the Reserve has been cut 10 times over the last 6 years. Money for scientific research has been almost totally shut down. The amount spent is "just a drop in the big sea." Only about 7 to 8% of the revenue to operate the Reserve comes from various outside grants and funds.

Great Smoky Mountains National Park Biosphere Reserve

The Park's approximate $10 million per year operating budget is appropriated by Congress. Short-falls in natural resources and science funding, as well as shortfalls in other Park divisions, can sometimes be made up by competing with other parks for moneys held in the agency's Washington, D.C. office or Southeast Field Area office. These funding sources are for short-term use only, are often erratic in nature, and require yet additional administrative time from the professional staff for proposals and special reports. The Park generally is successful in competing for a total of approximately $2 million per year of these funds for all Park divisions. No specified funds have been made available to the Park for biosphere reserve activities.

WALKING THE TIGHT ROPE: SETTING MANAGEMENT PRIORITIES

PRIOKSKO-TERRASNY BIOSPHERE RESERVE

Managing the Reserve during this current transition of the federal government has been an enormous challenge. The excellent scientific research carried along over many years has been disrupted. This activity has been a traditional core mission of the Reserve. As the emphasis changes to include environmental education and public service at the Reserve, the meager resources available for operation become further stretched. During these difficult times, the priorities are to protect the natural resources in the reserve and to maintain the Central European Bison Nursery at a minimal operative level. Like the efforts in the Smokies, partnerships are constantly being sought to provide environmental education and volunteer guides and other services for visitors. New ideas are always being sought to improve and enhance environmental education activities. Of greatest concern is to find a means to provide a minimum level of support to scientists for their continued data gathering related to long-standing research programs.

GREAT SMOKY MOUNTAINS NATIONAL PARK BIOSPHERE RESERVE

Difficult budget times create an opportunity that should not be missed. It forces Park staff to look beyond their normal sources of funds and working relationships to get critical jobs done. This fosters new partnerships, which in turn infuses new perspectives and energy, which is invaluable in tackling the complex issues confronting ecosystem managers. The sources of virtually all of the primary threats to the Park resources come from beyond the borders of that core protected area. Strengthening partnerships with other stakeholders concerning these key issues is of critical importance if anything positive is to occur. Priorities during these difficult budget times are similar to those of the Prioksko-Terrasny Biosphere Reserve: to protect the natural and cultural resources, provide a minimum level of quality visitor services, and find ways to leverage outside scientists to continue their research in the Park.

PROGRAMS AND PARTNERSHIPS IN ECOSYSTEM MANAGEMENT

Partnerships have become the watchword of today's public sector environment in both countries. Administrators of Russian biosphere reserves are particularly looking forward to establishing international sponsors to get new ideas and resources to deal with the milieu of problems facing management of natural areas. This section is intended to highlight examples of how ecosystem managers have leveraged resources to deal with specific issues related to ecosystem management.

SPECIES REINTRODUCTION

Priokso-Terrasny Biosphere Reserve

The Central European Bison Nursery was established in the reserve in 1948 for the following purposes:

- To protect the existing population
- To study the genetic diversity according to morphologic, physiologic, immunologic, cytologic and molecular perspectives
- Cross-breeding to expand the genetic diversity of the herd
- To study behavior at various life stages
- To reintroduce bison on landscapes beyond protected areas to establish viable wild herds
- To secure funds for all the above

This species of European bison is a large-hoofed mammal which has long since been extirpated from the wild. At the beginning of the current millennium, the European bison inhabited all of Europe except for the British Isles and the Pyrrenean and Apennine Peninsula. In the summer, the bison feed on grass and leaves of deciduous trees and shrubs. In the fall, acorns are their favorite food, and in winter they peel off bark and dig through snow for evergreen boughs. The European bison were actively hunted throughout history by kings, czars, and their favored colleagues on great hunts in Germany, Poland, Russia, and other countries.

By 1927, there were only 48 European bison living in zoos and menageries. In 1948, two pair, of pure-bred European bison were brought to the reserve. Today, the reserve maintains two small pure-bred herds of European and American bison. They also maintain a herd of cross bred bison to "improve" the herd and diversify the gene pool. Unlike in the U.S., there is little concern for producing a pure strain of the original species before returning it to the wild.

Every year, some of the bison are taken to different places in the former Soviet Union. As of this date, 250 individuals have been released to the woods of Lithuania, the Caucasus Mountains, and the Carpathian mountains in order to restore the population in the wild. There are similar bison nurseries in other reserves in the former Soviet Union, but the nursery of the Reserve ranks first among them. Currently there are about 2000 bison living on the territories of the former U.S.S.R.

Great Smoky Mountains National Park Biosphere Reserve

Several species have been reintroduced to the reserve, including the river otter (*Lutra canadensis*), Peregrine falcon (*Falco peregrinus*), Smoky Mountain mad tom (*Noturus baileyi*), and red wolf (*Canis rufus*). In every case, the project involved a large number of collaborators and required numerous releases of individuals that were closely monitored as to their ability to survive. Education of visitors to the reserve as well as people living around the region surrounding the reserve has always been an important part of these efforts. There never seems to be enough funds available to conduct these reintroduction projects at an ideal level of intensity, but there is always considerable public and agency support for the efforts so that adequate resources are usually assembled for the operation. Invariably there are many individuals who are willing to dedicate their time to the effort in order to insure its success.

Although each of the activities described above involve partnerships, the reintroduction of red wolves into the Park as described by Lucash et al. in Chapter 11 is a particularly good example. The U.S. Fish and Wildlife Service, a sister agency of the National Park Service, has overall responsibility for the protection and enhancement of threatened or endangered species. Since the Park is located within the wolf's former range, the U.S. Fish and Wildlife Service initiated a request to assess reintroduction feasibility, which is totally compatible with National Park Service policy. However, unlike other reintroductions, this time a predator was being considered for reintroduction and public sentiment for such a proposal was unknown.

Before the first wolf was released, U.S. Fish and Wildlife Service and Park staff collaborated on the development of a brochure, fact sheet, educational materials, poster, and slide program intended to increase public awareness about the plight of this endangered species and the need for reintroduction. Public meetings were jointly conducted specifically to address issues and concerns raised by Park neighbors and to garner support for the proposal. Although the initial proposal was ultimately modified as a result of these meetings, there was almost unanimous support for an experimental release of radio collared wolves into the Park.

The first release occurred in 1991 as an experiment. Subsequently, additional adult pairs and family groups have been released and, although the project has experienced only limited success, U.S. Fish and Wildlife Service and National Park Service biologists continue to closely coordinate efforts directed at the reintroduction of this endangered animal.

RESEARCH AND MONITORING

Prioksko-Terrasny Biosphere Reserve

Monitoring of the state of the environment has been conducted since the foundation of the Reserve in 1945. Stations for the monitoring of air, water, and soil were established on the territory of the Reserve in 1979. Analysis of data includes seasonal and daily averages. Pollution parameters measured include chlorine organics, heavy metals, acidity, dust, radioactivity and anion–cation analysis. Samples are taken from the air, water, river bed sediments, and tissue samples of selected plants and animals. All processing of samples and data analysis is made by the specialists from the Institute of Soil Studying and Photosynthesis of the Russian Academy of Science (RAS). Scientists also monitor forests and forest pathology using maps of the state of the forest plantations with a 5 year remeasurement cycle.

Biomonitoring includes shadowing (observing) the state of the population and dynamics of numbers of soil invertebrate organisms, birds, small mammals, and ungulates and has been ongoing since the reserve was established. More than a dozen scientists from the Academic Institutes, Moscow State and Moscow Pedagogical (Teacher and Training) Universities are taking part in this work. A separate major project is to monitor the populations of 50 rare species of plants and animals, some of which are in danger of going extinct.

The reserve has served as a base line of ecosystem analysis from which to compare human impact on surrounding forests. During the last 10 years, specialists from the zoological and ecological departments of the Moscow Pedagogical University have been engaged in a complex scientific program to evaluate the changes in ecological systems under the influence of recreational activity on the "Russian Forest." Study includes changes in soil characteristics, vegetation, and invertebrate animals and small mammals. Much attention is paid to study of the stability and organization of the ecosystem as a whole, as well as its components.

Great Smoky Mountains National Park Biosphere Reserve

The Park has long attracted researchers, due to its relatively large unmanipulated landscape, amount of original uncut forest communities, and unhunted wildlife populations. Also, the scientifically intriguing circumstance of having one of the highest visitations for a national park overlaid onto one of the most biologically diverse temperate forest areas has attracted biologists, ecologists, and sociologists alike. Hundreds of research reports and published articles have been produced. During a recent 15-year period, almost 350 projects were carried out in the area of botany/plant ecology alone (White 1987).

Each year, about 100 permits are approved for a wide variety of research, monitoring, and inventory activities. Most of the permittees are associated with universities. In 1993, almost all research grade scientists were removed from the Department of Interior's land management agencies (including the National Park Service) and were placed in a separate new agency known as the National Biological Service. This new agency has responsibility for providing professional research support to the National Park Service, but there have been many reorganization problems, and the Congress substantially cut funding for the National Biological Service and submerged it into yet another agency, the U.S. Geological Survey. At this writing, the future of ecological research by and for the National Park Service is unclear. This makes partnerships with other agencies and organizations, including biosphere reserves in other countries, increasingly important.

AIR QUALITY

Priokso-Terrasny Biosphere Reserve

As mentioned above, the Reserve has a long history of sophisticated monitoring of airborne pollutants and research to track where the trace elements precipitating out of the atmosphere collect in the environment; i.e., ground and surface water and soil chemistry and plant and animal tissue.

Great Smoky Mountains National Park Biosphere Reserve

As describe by Peine et al. in Chapter 15 of this volume, air pollution is one of the most significant problems facing the Park. The burning of fossil fuels (coal, oil and gasoline by industry, utilities, and automobiles) produces oxides of sulfur and nitrogen. These emissions convert to ozone, sulfates, and nitrates and are deposited in the Park at very high levels. Because air quality problems in the southeastern U.S. are regional, solutions cannot come without regional cooperation.

The Park, the largest Class I area (designated as such by the Clean Air Act of 1977 to require the nation's highest standard of air quality) in the temperate forests of the eastern U.S., and the Tennessee Valley Authority, the largest electric utility in the U.S., have joined forces through an interagency agreement. The purpose of the partnership is to fill crucial gaps in knowledge of tropospheric ozone and regional haze issues. Specific ongoing, collaborative projects include: (1) enhanced monitoring of a variety of air pollutants including ozone, sulfur dioxide, carbon monoxide, nitrogen oxides, and hydrocarbons; (2) determining the primary mechanisms for ozone formation, photochemistry, and transport mechanisms; (3) assessing contributions of sources of pollutants to ozone levels; (4) determining tree physiology and ozone dose–response relationships to mature tree species; and (5) determining factors controlling visibility impairment from regional haze. Knowing the concentrations of pollutants in ambient air, determining the effects of those pollutants, and determining the sources of pollutants are necessary if any action is to be taken to protect Park resources from air pollution. Without a clear demonstration that Park resources are affected by air pollutants, the U.S. Environmental Protection Agency and state environmental agencies are unlikely to take action to protect them. The partnership between the Park and Tennessee Valley Authority is an excellent example of two very different agencies working on a common regional problem and generating useful, scientifically credible information for policy decision-makers. The National Park Service and Tennessee Valley Authority together are also working directly with other regional air quality initiatives and assessments including the Southern Appalachian Mountains Initiative (SAMI), the Southern Oxidants Study (SOS), and the Southern Appalachian Man and the Biosphere Cooperative (SAMAB).

Native Brook Trout Management

Great Smoky Mountains National Park Biosphere Reserve

Cooperation between Park fisheries staff and Trout Unlimited Inc. is based upon open communication that identifies fisheries and water quality projects of importance to the Park which need
volunteer support for implementation. Trout Unlimited volunteers assist with large stream monitoring projects on three streams, including Abrams Creek, Little River, and Cataloochee Creek.
They provide assistance with native brook trout distribution surveys and restoration efforts. They
are currently involved in a cooperative project to assess stream acidification over the next 5 years
in selected stream reaches. Trout Unlimited has also raised funds for stream bank stabilization work
in Cades Cove, for the purchase of needed scientific equipment, and for support of seasonal staff.
During fiscal year 1994, their involvement and financial support exceeded $36,000 in value to the
Park. See Guffy et al., Chapter 12 of this volume, for more information on the trout fisheries
program.

Black Bear Management

Great Smoky Mountains National Park Biosphere Reserve

A public education strategy to deal with the availability of human food and garbage to black bears
in and around Great Smoky Mountains National Park is being developed by the Park in conjunction
with state wildlife agencies, city officials, and Park partners. A television segment for cable TV, a
brochure, bumper sticker, lesson plans, and a poster series are just a few of the materials being
developed. Great Smoky Mountains Natural History Association, Friends of Great Smoky Mountains National Park, and the Great Smoky Mountains Conservation Association are all providing
funds for the various educational materials. See Clark and Pelton, Chapter 10 of this volume, for
more information on this program.

Strategic Planning for Wilderness Management

Great Smoky Mountains National Park Biosphere Reserve

The Park includes 191,068 ha recommended for inclusion in the National Wilderness Preservation
System. With a difference of nearly 1770 m in relief, over 1287 km of horse and hiking trails and
106 designated camping areas, shelters, and vehicle-access horse camps provide access into the
Park's backcountry for the 500,000 to 700,000 visitors annually who come to the Park seeking a
backcountry wilderness experience.

Visitation figures suggest that backcountry camping, private horse riding, total backcountry
visitation, and day-hiking continue to increase annually. At the same time, however, the Park's
ability to maintain backcountry trails and facilities declined, resulting in an increase in trail
deterioration and visitor complaints about poor trail conditions. The decline in attention to the
backcountry coincided with an increase from 8.0 to 9.3 million Park visits and static purchasing
power.

In an attempt to address increasing staff and visitor concerns about resource impacts, the Park
initiated an assessment of trail conditions under a cooperative agreement with Virginia Polytechnic
Institute (Marion 1994). The assessment formed the cornerstone for the development of the strategic
plan for managing backcountry recreation (GSMNP 1995), again under cooperative agreement with
Virginia Polytechnic Institute.

The development of the strategic plan did not follow normal protocols, however, since National
Park Service planning documents usually are developed in draft form for eventual public review.
In this case, the stakeholders were involved from the beginning. Four important points are thus
relevant:

1. All of the discussions about reasons for the National Park Service procedures (legislation, policies, financial capability, etc.) were addressed in the beginning of the exercise so that participants understood the Park's constraints and limitations
2. The opportunity was available to involve participants, with oftentimes divergent viewpoints, in discussions that focused on problem solving
3. Consensus building was not a goal, particularly because of divergent viewpoints, but rather an understanding of what the real issues were and what needed to be done to correct deteriorated trail conditions
4. Stakeholders needed to become involved in solving the problems since the Park no longer has the resources to effect positive change

This year-long planning process spawned a new beginning in how the Park will manage its wilderness resources. The intensive communications process brought together, for the first time, individuals and groups with differing points of view to focus on the common theme of trail conditions. As implemented, the strategic plan focuses on a dramatic and proactive increase in trail maintenance via volunteerism as being the cornerstone for program success. The repairing and maintenance of trails, shelters, and campsites will increase as will visitor enjoyment.

One result of this effort was the development of the Appalachian Trail Task Force, composed of organized hiking and horse riding groups and the Appalachian Trail Conference, who entered into a memorandum of agreement for joint management of the Appalachian Trail where horse riding is permitted. The Task Force also developed and printed a publication entitled "Gentle on the Land," a hiker and horseback rider code of conduct which is similar to the "Leave No Trace" materials popularized by the National Outdoor Leadership School. Under a challenge-cost-share with the leadership school, the Park also published a brochure entitled "Leaving No Trace in Great Smoky Mountains National Park." Although the Smoky Mountain Hiking Club and the Appalachian Trail Council have been partners in maintaining the Appalachian Trail for quite some time, organized horse riding groups have recently stepped forward to adopt trail maintenance on certain horse trails. In addition, the Park has reorganized to accommodate this new thrust by establishing a full-time position dedicated to coordinating volunteerism and partnership efforts, and an additional full-time position dedicated to coordinating backcountry management functions across division boundaries.

Through this strategic planning process, the Park is now intimately involved with those who benefit most in a collaborative effort to manage the Park's backcountry and protect wilderness values.

ENVIRONMENTAL EDUCATION: LINKAGES BETWEEN BIOSPHERES

As the Russian government privatizes land ownership, there is a growing need to educate the citizenry as to the importance of protecting unique natural areas. Thus the role of environmental education is recognized to be of growing importance in Russian society.

An active collaborative program between colleagues in the southern Appalachians and Russia has been ongoing since 1991 in the field of environmental education. The exchange has been organized principally by environmental educators from Moscow State Pedagogical University (MSPU) and individuals associated with the Great Smoky Mountains National Park Biosphere Reserve.

To date, more than 400 people have participated in the exchange. Each year the exchange is somewhat different, visiting various national parks and/or biosphere reserves and experiencing various manifestations of environmental education. A total of four Russian students have worked as staff members at the Great Smoky Mountains Institute at Tremont for a year long period and taken back to Russia new knowledge and skills. Many of these ideas may eventually be applied in Russian biosphere reserves, particularly when environmental education centers for these areas become established. In turn, staff members from Tremont visited Russia in the summer of 1994 to learn Russian techniques in environmental education and to share approaches to educating people concerning the natural environment.

In Russia, environmental education has been incorporated into the school curriculum in a fragmented fashion. It has never been considered as an independent subject for study. The national parks and biosphere reserves in Russia are just now beginning to be used to stage environmental education activities to demonstrate the interdisciplinary nature of environmental education. Faculty at Moscow State Pedagogical University have recently been asked by the Russian government to develop curriculum for training students to become teacher/naturalists working in national reserves and parks.

A new activity to begin in the summer of 1996 is the exchange of groups of students from a school in each country. The schools and students have been selected and are exchanging information on the fauna and flora of two national parks that they will be studying. Parents of the participating students have agreed to host an exchange student during the program.

Many ideas are flowing between the two groups now as everyone learns from each other. Such examples demonstrate that people in the two countries are greatly enriched by their cultural interaction.

REFERENCES

Davydova, M. and V. Koshevoi. 1989. *Nature Reserves in the USSR*. Progress Publishers: Moscow, Russia. 106 pp.

Great Smoky Mountains National Park (GSMNP). 1995. A Strategic Plan for Managing Backcountry Recreation in Great Smoky Mountains National Park Gatlinburg, TN: Great Smoky Mountains National Park.

Haskell, D. A. 1995. Forging the National Park Concept in the Russian Federation. *The George Wright Society Forum*. 12(2): 26–39.

Knystautas, A. 1987. *The Natural History of the USSR*. McGraw Hill: New York.

Marion, J.L. 1994. An Assessment of Trail Conditions in Great Smoky Mountains National Park Gatlinburg, TN: U.S. Dept. of Interior, National Park Service, SE Region, Great Smoky Mountains National Park. Final Research.

Parker, C. and D. Pipes. 1991. Watersheds of Great Smoky Mountains National Park: A Geographic Information System Analysis. Rep. No. 91/01. National Park Service. Atlanta, GA. 126 pp.

Razumovsky S.M. 1981. The laws of the dynamics of 4945 hectares; Moscow,: *Science*. Moscow, Russia. 232 pp.

Soloviev, V. 1993. National Parks in Russia. Unpublished report from the Director of National Park Management, Russian Federal Forest Service.

U.S. Department of State. The United States Man and the Biosphere Program. 1995. Publication 10187, Bureau of Oceans and International Environmental Affairs. 26 pp.

White, P.S. 1987. Terrestrial Plant Ecology in Great Smoky Mountains National Park Biosphere Reserve: A Fifteen-Year Review and a Program for Future Res. Research/Resources Manage. Rep. SER-84. National Park Service, Atlanta, GA. 70 pp plus appendices.

Section III

Components of
Ecosystem Management

7 Conducting Regional Environmental Assessments: The Southern Appalachian Experience

Cory W. Berish, B. Richard Durbrow, James E. Harrison, William A Jackson, and Kurt H. Riitters

CONTENTS

Ecosystem Management 117
Environmental Assessments..118
Why Focus on Southern Appalachia? 120
Southern Appalachian Environmental Assessment 121
Southern Appalachian Assessment Results 123
    Human Use Patterns at a Landscape Scale 124
    Terrestrial Team Results..127
    Aquatic Team Results 136
    Atmospheric Team Results 142
    Social, Cultural, Economic Team Results 154
Concluding Remarks..158
Acknowledgments..159
Appendix — Questions..160
References 162

ECOSYSTEM MANAGEMENT

Over the last 25 years states, working partnerships among numerous federal and state agencies and local organizations, have made tremendous strides in cleaning up our environment. The focus of these efforts has been on reducing many sources of terrestrial, water, and air pollution (EPA 1992). But, as declines in neotropical migratory bird populations (DeGraaf and Rappole 1995) and many aquatic species (Etnier and Starnes 1991) demonstrate, traditional environmental regulation strategies alone simply cannot assure the protection of ecological integrity. Large-scale disturbances on ecosystems often demonstrate fairly uniform responses which allow identification of distress to ecosystems (Rapport et al. 1985). Ecological integrity, in this respect, can be defined as the ability of a living system to recover by resembling the preexisting state after disturbance to the ecosystem (Regier 1993).

Increasingly, many state and federal agencies are turning to ecosystem management to accomplish their environmental protection goals. The primary goal of ecosystem management is to sustain the integrity of an ecosystem for future generations while equitably providing goods and services for today's population (see McCormick, Chapter 1 of this volume). This goal can be achieved by integrating sociological, ecological, technological, and economic information to identify optimum land uses and ensure ecosystem stability. The goals and mission of the Man And Biosphere program (USMAB 1989) and specifically the Southern Appalachian Man And Biosphere (SAMAB) program epitomize the foundation of a sustainable development approach to ecosystem management. See Hinote, Chapter 5 of this volume.

There is, however, no one definition of ecosystem management. Nor is there a consistent approach across federal and state agencies using ecosystem management to protect natural resources. The underlying concept is deceptively simple: ecosystem management begins with a desire to preserve the integrity of a place. One definition identifies integrity as "the interaction of the physical, chemical, and biological elements of an ecosystem in a manner that ensures the long-term health and sustainability of the ecosystem" (EPA 1994a).

Managing for ecological integrity is defined by Norton (1992) as protecting the total diversity of organisms, populations and systems and also the functional processes that maintain the diversity of a place. In a recent review, Grumbine (1994) cites five common ecosystem management goals:

1. Maintain viable populations of all native species *in situ*
2. Represent within protected areas, all native ecosystem types across their natural range of variation
3. Maintain evolutionary and ecological processes (i.e., disturbance regimes, hydrological processes, nutrient cycles, etc.)
4. Manage over periods of time long enough to maintain the evolutionary potential of species and ecosystems
5. Accommodate human use and occupancy within these constraints

Grumbine points out that four of the five common ecosystem management goals are ecological value statements related to the protection of environmental integrity of a place. The fifth goal incorporates the human dimension or anthropogenic influence on natural resources. The inclusion of human influence is critical to the concept of ecosystem management. This stems from the knowledge that no ecosystem is immune from human induced impacts. Thus, the fifth ecosystem management component must include local stakeholders in a commitment to problem solving that includes the previous four goals. Potential solutions to ecosystem problems affect a broad assortment of individuals and institutions. Involving those diverse interests may raise barriers to effective action, such as conflicting mandates, physical distance between collaborators, and the challenge of harmonizing landowners' wishes with the requirement of the natural system. Natural resource managers who use ecosystem management recognize those challenges yet believe collaboration yields more stable solutions than legal regulation alone.

ENVIRONMENTAL ASSESSMENTS

Ecosystem assessments consider the potential land uses, products, values, and services available in the community for the long-term sustainability of the region (Overbay 1992). Ecosystem assessments come in many forms depending upon the final product desired. For example, environmental impact assessments focus on the comparative effects on the environment from a given action, such as a new highway or dam. Resource management assessments focus more on the predictive impact from a specific natural resource management decision (Suter 1993). Resource management decisions, such as the amount of timber to harvest from a given area, will focus on site- or species-specific issues in relation to the management of the natural resource.

A broader approach can be used to gain a general understanding of both comparative and predictive impacts between the physical environment and humans. This broad approach is called an environmental assessment. The purpose of environmental assessments is to contribute to the long-term sustainability of an area through the credible evaluation of the ecological effects of human activities on natural systems (Suter and Barnthhouse 1993). The scope of environmental assessments can vary from a site-specific damage assessment of a facility such is associated with the Comprehensive Environmental Response, Compensation and Liability Act, to a very large-scale regional assessment such as an assessment of the Chesapeake Bay (Costanza et al. 1992).

The regional environmental assessment approach to management recognizes the interrelationship between the physical natural environment and humankind. This relationship balances on the environmental and economic attributes of a region by linking the goals of environmental protection and economic development (GLC 1994). The regional environmental assessment approach incorporates ecological, economic, social, and cultural values into the assessment process. At a regional scale, the more general assessment approach provides an understanding of potential issues and community concerns in the area. This provides an opportunity to identify community issues and target specific indicator stresses to restore or sustain ecosystem integrity and desired conditions.

A multiagency Regional Ecosystem Office (REO) recently put together a six-step analytical procedure to provide the general framework on how to assess environmental conditions at the watershed scale (REO 1995). This analysis process is analogous to EPA's Watershed Assessment Approach (EPA 1993a). The REO procedures include: (1) characterizing the watershed area to be studied; (2) defining the issues and key questions of concern; (3) establishing existing environmental conditions in the watershed; (4) referencing the identified baseline conditions; (5) interpreting the findings; and (6) developing recommendations for watershed protection. This multistep procedure is being used in watershed projects across the U.S. (REO 1995). The protocol first places the watershed in the larger context of where it is located within a geographic area and then defines baseline environmental conditions, such as natural resources, political boundaries, topography, soil, land cover, species assemblages, and other features. The key step in the process is, arguably, number (2), the development of important issues and concerns. Key concerns are best developed by working with the local stakeholders to determine what is valued in their geographic area. In many cases, a relative initial prioritization of local concerns is developed in this step. Step three is the determination of the integrity and health or current conditions of the physical, biological, and human components of the landscape. This differs from the first step in that ecosystem health is identified in relation to the concerns of the community. In step four, current conditions are compared to the historical condition of respective physical, biological and human components of the area under investigation. Step five is the interpretation phase, the phase where current conditions are described, trends analyzed, and descriptive maps developed. In the final phase, potential management activities are developed.

Throughout the process, key stakeholder involvement is required for success. This element is most evident in the second step of the process; when issues are addressed in relation to community identification of key questions. Having identified the questions and issues to be addressed by the assessment; endpoints can then be developed. End points are the characteristics of value to the community that are identified during the issue formulation and question development. Thus, assessment endpoints are the formal expressions of the actual environmental value that is to be protected which reflect the community's values (Suter 1990). The assessment endpoints should be quantifiable under some measure, although this is not mandatory. The general criteria for an endpoint include the following: (1) susceptibility to a specific stressor; (2) societal value; and (3) ecological relevance (EPA 1992).

Subsequent to the regional environmental assessment process, the community identified assessment endpoints can be managed to protect the integrity and health of the ecosystem. Integration of specific stressors in the region provides the basis for targeting specific areas at risk. Thus, the integration creates a foundation on which to build greater understanding of ecosystem stressors and human influences that may be contributing to the degradation of a specific site. In this respect, the assessment utilizes the information obtained at a broad scale to highlight potential stressors associated with impaired ecological integrity and health of the ecosystem. Risks can then be managed for ecosystem integrity and health.

Ecological risks of concern across the country include risks to migratory birds, wetlands, commercial fisheries, natural area preserves, and privately owned lands (Troyer and Brody 1994). EPA defines ecological risk as the likelihood that adverse ecological impacts are occurring as the result of exposure to one or more stressors (RAF 1992). To evaluate risk to the environment, EPA developed a flexible ecological risk assessment framework around three sequential phases: (1) problem formulation; (2) analysis; and (3) risk characterization (RAF 1992). In essence, problem formulation is the planning and public involvement phase that links management goals to the risk assessment process. The analysis phase is where data profiles are developed and characterized. In the risk characterization phase, the data are integrated and results are described in relation to defined exposure endpoints. Additional information on the development of baseline natural resource condition profiles are presented in EPA's EMAP Assessment Framework (EPA 1994b).

The environmental assessment process used in the Southern Appalachian Assessment (SAA) provides a significant foundation for future risk assessment modeling. The risk assessment process relies on a number of technical methods to integrate known environmental conditions into a model that can be used to identify potential future impacts on ecosystem health and integrity. The less vigorous regional environmental assessment defines existing conditions at a given time and in relation to historical conditions. The SAA was not designed to define risk in the pure sense of the term. Instead, the SAA provides a broad understanding of ecological conditions and identifies relationships that help to focus on potential risks to the environment.

WHY FOCUS ON SOUTHERN APPALACHIA?

The southern Appalachian area has long been known as a center of temperate forest diversity (Whittaker 1972). This area is exceptionally rich in plant and animal diversity, harboring nearly 3000 known species and estimates of 17,000 unknown species. The focus of species diversity conservation efforts typically addresses research attention on top predators, large mammals, or rare species in the ecosystem (McMinn 1991). Many large predators and grazing animals have already been extirpated from the southern Appalachians, including the mountain lion (*Felis concolor*), golden eagle (*Aquila chrysaëtos*), timber wolf (*Canis lupis*), bison (*Bison bison*), and elk (*Cercus canadensis*)(Boone and Aplet 1994). The future of many remaining species is questionable. However, efforts to reintroduce native species, such as the red wolf (*Canis rufus*) and golden eagle, are currently underway in the SAA region.

The National Biological Service (NBS) recently published a list of ecosystems that are critically endangered, endangered, or threatened (Noss et al. 1995). Some listings are generic in nature, such as old-growth and virgin forests, while others are geographically explicit. The southern Appalachians contain ecosystems that fall into each of the listing criteria established by Noss. The ecosystems specifically included in the listing are spruce–fir forests at greater than 98% decline, mountain bogs at 85 to 98% decline, and a more general listing of riparian forests at 70 to 84% decline (this figure is nationwide). The integrity of ecosystem health for these and ecosystems in general is often reduced by developmental pressure. The vast land area of the southern Appalachians contains numerous rare community types and associated species assemblages which are adversely effected by this disturbance to ecosystem integrity.

Diversity in southern Appalachia is also reflected in the existing landforms and aquatic systems of the region. The topographic features of the SAA region vary from ridged mountain tops to level valley floors. The headwaters of eight major rivers flow from the mountainous landscape, including the James, Alabama, Savannah, Cooper, Pee Dee, Roanoke, Tennessee, and Appalachicola. The mountains provide drinking water for much of the Southeast. Controlling soil erosion, protecting water quality, and allowing the forest to act as a natural purification system are only a few of the benefits provided by the continuity of the diverse natural forested landscape in the region.

The local economy also depends upon the southern Appalachian forests as the base for numerous industries in the region, such as agriculture, fisheries, mining, forestry, and manufacturing. The timber industry and associated wood product manufacturing are very important in many selected counties across the region. This trend may change in the future. The service industry is rapidly growing, based on the scenic beauty, wildlife, and recreational opportunities of the area. The service industry revolves around the recreational value of the forests rather than the marketability of cut timber as a commodity. Currently 8.5 jobs are being created in the region's service industry for every job added by resource extraction industries (Morton 1994).

The member agencies of SAMAB believe that the southern Appalachian area has the potential for future environmental degradation because of its unique setting, including the immense biological wealth, pleasant climate, and unique cultural resources. The consequence of disturbance is demonstrated by diminished forest health and integrity, reduced water quality, a loss of terrestrial and aquatic species diversity, an increase in exotic species, and continued loss of rare communities.

Consequently, some aspects of southern Appalachia's heritage are at risk. Before the region's character is irreversibly altered, the SAMAB partners hope to illuminate some of the areas with the worst environmental problems and to help implement measures that will reverse undesirable trends. Thus, in 1994 the SAMAB members decided to coordinate their efforts and conduct a multiagency environmental assessment of the southern Appalachian region. The goal of the regional environmental assessment was to facilitate sustainable development and implement the goals of ecosystem management by providing comprehensive ecological, social, and economic data as a foundation for natural resource management.

SOUTHERN APPALACHIAN ENVIRONMENTAL ASSESSMENT

The Southern Appalachian Man and the Biosphere (SAMAB) program is a consortium of federal and state agencies striving to promote sustainable development and the concepts of ecosystem management in southern Appalachia. Current membership includes: the USDA Forest Service, the U.S. Environmental Protection Agency, the Tennessee Valley Authority, the National Park Service, the U.S. Fish and Wildlife Service, the U.S. Geological Survey, the Department of Energy's Oak Ridge National Laboratory, the Army Corps of Engineers, the Appalachian Regional Commission, the Economic Development Administration, and the States of Georgia, North Carolina, and Tennessee. Current SAMAB projects, organizational structure, and zone of cooperation were described by Hinote in Chapter 5 of this volume.

In the fall of 1994, the cooperators of SAMAB decided to coordinate their efforts in conducting a single multiagency environmental assessment of the southern Appalachian region. The assessment effort entitled the "Southern Appalachian Assessment" (SAA) is to document the status, conditions, and trends of ecological resources in southern Appalachia. The environmental assessment is to be used to identify and prioritize areas in need of environmental management activities and to provide baseline information for National Forest Plan revisions. Although most of the information used in the assessment is not new, the aggregation of pertinent information across multiple subject areas provides short- and long-term insights into natural resource planning activities.

Environmental managers asked two simple, pragmatic questions to motivate the assessment process (Table 1). These questions set the stage and determined the approach that each assessment team, based on natural resources in the region, would adopt in answering more detailed questions about different components of the Appalachian ecosystem. There was an emphasis on "where" as well as "what" that led to a spatially explicit assessment framework. There was also attention given to the "relative" conditions of the ecosystem, which meant that a premium was placed on the capability to rank many quantities in general as opposed to a few quantities in detail. Furthermore, there was an implied integration of agency information to obtain an overall view of existing conditions. Assessments are envisioned in the near future for specific, potentially high-risk concerns

TABLE 1
Motivating Questions for Southern Appalachian Regional Ecosystem Assessment

Question 1	What is the overall environmental or ecological quality and where is it relatively high or low?
Question 2	What is the overall environmental or ecological risk and where is it relatively high or low?

that were identified or believed to be present, leading to more detailed questions for more sophisticated analysis and decision making.

The regional SAA approach stands in sharp contrast to issue- or media-specific regional assessments, which start with a particular question of demonstrated concern or with a law or regulation of interest. A recent example is the National Acid Precipitation Assessment Program (NAPAP) which dealt specifically with oxides of nitrogen and sulfur and their effects on a limited set of human and ecological resources, pursuant to regulations under the Clean Air Act. In cases like that, there is no need to place the assessment questions in the context of all other issues that could have been addressed. In contrast, the SAA's regional focus began with a very broad set of issues, and addressed each of them in much less detail, in order to identify the priorities for subsequent work.

The rationale for adopting a broad-scale perspective as the first step in regional assessments can be summarized as follows. Some of the most important changes in ecological condition can occur at the broad spatial scale of whole landscapes and involve interactions among many embedded resources (EPA 1994c). These phenomena, such as air pollution impacts, are not easy to study except at a broad spatial scale. The reason has nothing to do with our computing facilities or our understanding of fine-scale science. Even precise fine-scale knowledge cannot be translated into broad-scale information, if only because otherwise hidden phenomena come into play when a broader context is considered (Overton 1977; O'Neill et al. 1986; Allen et al. 1987; Allen and Hoekstra 1992). If the goal is to assess broad-scale phenomena, it is probably more efficient to simply use that scale to start with, as opposed to building up a regional picture from a series of finer-scale analyses. In other words, there is little to gain and possibly much to lose by modeling the system at an inappropriate scale.

A top-down approach has implications for the types of environmental indicators that should be studied in the SAA. Here, attention is focused on "triage" — describing regional conditions without necessarily explaining them. For this purpose, integrative measures of condition are preferred over specific and precise bioindicators (Materna 1984; Smith 1984; Waring 1984). This is in contrast to assessments based on "diagnoses" — interpreting particular cause–effect relationships for which specific bioindicators are needed (Treshow 1984; Ratsep 1990). The power of a regional environmental assessment is reduced when only diagnostic models are employed; some causes or effects will be overlooked because bioindicators are (by design) insensitive to many of them. An environmental assessment requires greater breadth of coverage and it is simply impractical to adopt anything other than a triage approach.

The environmental assessment process evaluated the likelihood that ecological conditions are currently being adversely effected or may soon be adversely affected. The assessment identified ecological exposure to one or more stressors in an attempt to quantitatively or qualitatively express the risk to the ecosystem (EPA 1992). This process laid the basic foundation on which the structure and principles of the environmental assessment could be built. Evaluation of risk focused on single species, natural communities, and/or whole ecosystems (EPA 1996). Within the SAA region, the integration of single and multiple chemical, biological, and physical stressors were used to evaluate the ecosystem. Community involvement identified valued resources which appeared to be degrading and played a significant role in determining which stressors to evaluate. This provided direction for the research and the framework in which the analysis took place.

TABLE 2

Examples of Southern Appalachian Assessment Questions Developed by Each Team with Local Community Involvement

Terrestrial

Question 1   Based on available information and referenced material, what plant or animal species occur within the SAA area, and what are their habitat locations?

Question 2   How is the health of the forest ecosystem being affected by native and exotic pests?

Aquatic

Question 1   What is known about the current status and apparent trends in water quality, aquatic habitat, and aquatic species within the southern Appalachian study area?

Atmospheric

Question 1   What are the major air pollutants which could impact the southern Appalachians, and what areas receive the greatest exposure?

Question 2   To what extent are aquatic resources in the Southern Appalachian Assessment area being affected by acid deposition?

Social–Cultural

Question 1   How has the social pattern (demographics, occupations, lifestyles, cultures, etc.) of southern Appalachian communities changed over the past two decades?

In order to carry out the goals and objectives of the assessment, the SAA was organized into a number of different teams. These included four basic resource teams: atmospheric, terrestrial; aquatic; and social, cultural, and economic. The Terrestrial Team was subdivided into Forest Health and Plant and Animal Subteams due to the diversity of material studied. Similarly, the Social, Cultural, and Economic Team was subdivided into the Human Dimension, the Recreation, the Timber Supply and Demand, and the Roadless and Wilderness Subteams. Each team was charged with developing, with appropriate stakeholder involvement, detailed assessment questions, assembling the data, characterizing the data and producing a final report. In addition, the SAA utilized three main workgroups to facilitate the process: the Public Involvement, the Technical Writing, and the Geographic Information System (GIS) workgroups. The multiagency SAA GIS Workgroup was critical to the success of the project. The GIS Workgroup helped with the information needs assessment, database production and quality control, managing the complex data set, and providing GIS analysis and GIS product support.

The agencies involved in the assessment worked very closely with numerous interested stakeholders throughout the assessment process. The stakeholders within the geographic area possess critical information on what is valued in their "backyard." The USDA Forest Service provided the resources in conjunction with numerous other agencies, to host a number of public hearings during the question formulation stage. The SAA staff encouraged citizens to participate in public meetings and to respond to information presented in periodic newsletters. Table 2 shows a subset of key assessment questions from each resource team. The complete list of questions is presented in Appendix A.

SOUTHERN APPALACHIAN ASSESSMENT RESULTS

The first SAA product was the development and compilation of issues and questions through community outreach. Solicitation of public and member agency input, through public meetings, phone calls and mail-outs, played a vital role in directing the SAA. The initial input placed into perspective the natural resources valued in Appalachia and also some of the major perceived environmental stresses. Those values and apparent stresses helped to form the structure of the assessment. Some of the major environmental stressors and issues identified for the region include:

population growth and urbanization; second home and recreational development; acid and air toxic deposition; mine runoff and leaching to surface waters; erosion and siltation from development, mining, logging, and logging related impacts; nonpoint source runoff from agriculture and urban areas; loss of riparian areas and old-growth forests; and the introduction of exotic species and pathogens.

The GIS Workgroup helped define the assessment area based on elevation, ecoregion, and public input. The SAA region includes the area extending southwest from the upper Potomac River in western Virginia, to northern Alabama and Georgia (Figure 1). The area includes approximately 37.4 million acres in 7 states and 135 counties and includes 7 national forests and 2 national parks. Both geographical and political criteria were used to define the SAA region, a logical unit for purposes of ecosystem management (GAO 1994). The SAA region is based mainly on ecological "Sections" (Bailey et al. 1994; Bailey 1995). However, some areas north and west of the region were excluded because of administrative constraints. This is most notable in the omission of southern Appalachian Forest Service lands in West Virginia. The eastern and southern boundaries were based on the SAMAB zone of cooperation, which generally follows ecological Section boundaries. The majority (~85%) of the area is privately owned; national forests and parks make up about 90% of the remaining public lands.

A variety of physiographic conditions are contained in the assessment region (Figure 2). The Blue Ridge Mountains ecological Section is the high-elevation "backbone" of the assessment region and makes up 28% of the total area (Table 3). The Ridge and Valley Sections (west of the Blue Ridge) and the Piedmont Section (east of the Blue Ridge) comprise 43% and 19% of the region, respectively. The remaining area is mostly in Sections of the Cumberland Mountains and Plateau. Land cover is unevenly distributed within ecological sections (see Table 3).

HUMAN USE PATTERNS AT A LANDSCAPE SCALE

One way to simply summarize intensive human use of the landscape is by aggregating land use/land cover classes which include developed areas and nonnatural vegetation. The proportion of an area that is used for development or agriculture is a measure of human use known as the U-index (EPA 1994). Here intensive human influence is defined to include the developed/barren, cropland, and pasture/herbaceous classes. (Small areas of rock outcrops and mountain top balds may be included in the barren and herbaceous classes, respectively). By this simple definition, intensive human influence on landscapes in the study area ranges from 0.0% to 74.6%.

A recent study (Wickham et al. 1997) of the northern part of the southern Appalachian region found that, on a watershed-by-watershed basis, the U-index was associated with many measures of land cover spatial patterns, including contagion (O'Neill et al. 1988), patch shape, and fractal dimension (Krummel et al. 1987). In turn, these spatial patterns are known to be associated with the conditions of embedded ecological processes (Turner 1989). An example is the condition of wildlife, for which metapopulation dynamics are strongly influenced by the amount, connectivity, and patch size of suitable habitat and land cover (Weins et al. 1993).

A map of the U-index (Figure 3) was created in the following fashion. A square, 180-ha "window" was placed on the land cover map and the value of the U-index was estimated from the proportions of land cover classified as urban or agriculture contained within the window. The window was then moved in steps over the entire assessment region, and the resulting index values were mapped at the locations of the window placements. The final result was a map of the U-index which shows the general patterns of intensive human uses (i.e., for development and agriculture) over the region. The "window" technique is similar to spatial filtering, which is a common technique in image processing (Gonzales and Woods 1992) and which has been used, for example, to map wildlife habitat potential over large areas (Riitters et al. 1997).

It is almost an adage that humans tend to simplify their environment. But at landscape scales the U-index map displays complicated patterns of use. The transition from simple to complicated

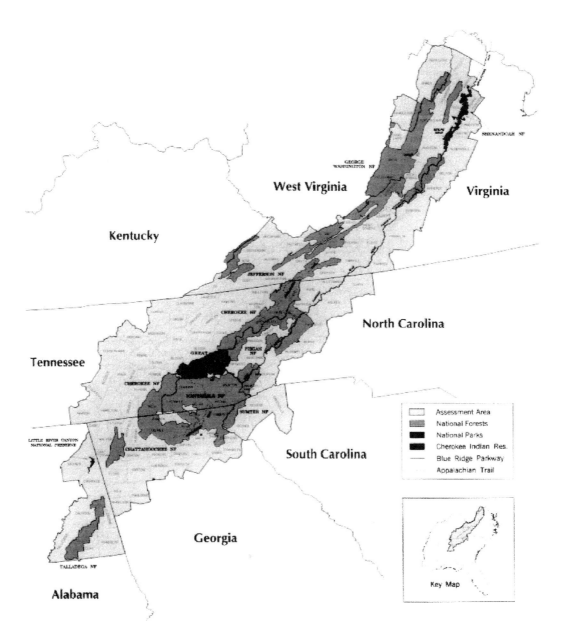

FIGURE 1   SAA region overview with state boundaries. View includes federal lands and reservations.

patterns might be a measure of the intensity to which humans have reconstructed the landscape, or conversely, how geophysical processes constrain human activity. For example, the land developed for agricultural purposes can be viewed as human impact on the landscape or the limitations placed on human activity by rugged terrain or soil type.

The utility of ecological Sections as a basis for ecosystem management might be a result of the association with human use patterns. A comparison of the maps of ecological Section boundaries (refer to Figure 2) and the U-index (refer to Figure 3) suggests a relationship of human settlement patterns to regional geophysical patterns. Average U-index values range from about 10% in mountainous ecological Sections to about 40% for the Central Ridge and Valley Section. Thus, the

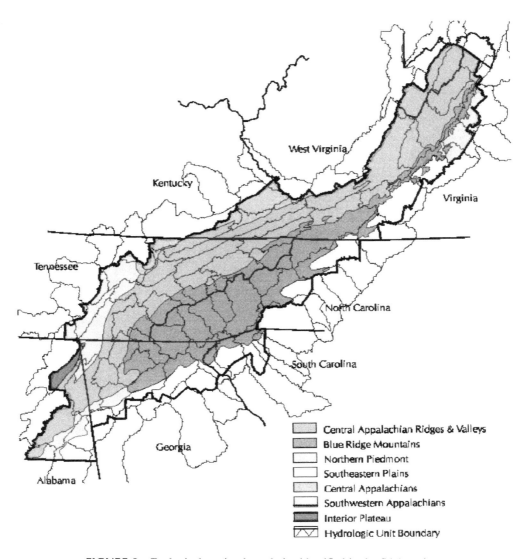

FIGURE 2 Ecological section boundaries identified in the SAA region.

distribution of human land use is uneven over the assessment region, and this information helps to understand why the conditions of terrestrial and aquatic resources also exhibit spatial variation in the region. For example, Figure 4 shows subdivisions of the SAA area that are defined by portions of hydrologic units within ecological Sections. Each area is classed according to its potential for aquatic resources integrity problems based on the relative level of intensive human influence across the landscape.

The diverse landforms in mountainous landscapes retain varied characteristics in terms of aspect, slope, and elevation, providing dramatic variability over time and space (Barry 1981). Although mountain peaks provide a great deal of isolation and undisturbed natural vegetation, the more suitable lowlands of human habitation and utilization continue to influence this secluded ecosystem and others. Air pollutants and changing climatic conditions are capable of reaching the highest, as well as the lowest, points in the southern Appalachians. These two stressors on the environment can act alone, in tandem, or with other stress, such as plant disease or insect infestation (see Peine and Berish, Chapter 19 of this volume). The terrestrial vegetation and water quality changes resulting

TABLE 3
Land Area by Ecological Sections in the Southern Appalachian Assessment Region

Ecological Section	Area (km²)	Area (%)[a]
Blue Ridge Mountains	43,003	28.4
Northern Ridge and Valley	33,159	21.9[b]
Central Ridge and Valley	18,292	12.1
Southern Ridge and Valley	14,180	9.4
Southern Appalachian Piedmont	27,977	18.5
Northern Cumberland Plateau	8,458	5.5
Southern Cumberland Mountains	2,158	1.4
Southern Cumberland Plateau	3,201	2.1
Allegheny Mountains	881	0.6

[a] Does not sum to 100% due to rounding error.
[b] Includes a small portion of the Northern Cumberland Mountains Section.

Source: Bailey et al., 1994.

from these stressors impart significant risks in altering current species distribution and potential habitat range in the SAA region, especially on higher elevation and isolated rare communities. Higher elevation forests are often subjected to numerous other stressors, such as ozone and heavy metals. Recent research (Gawel et al. 1996) indicates that heavy metals may be partially responsible for visible red spruce (*Picea rubens*) declines.

TERRESTRIAL TEAM RESULTS

The southern Appalachian ecosystem encompasses close to 37 million acres and is considered the single most biologically diverse land area in the U.S. About 77% of the land is managed by private land owners, while the remainder is retained in public trust. The USDA Forest Service is charged with maintaining the public portion according to the 'multiple use' philosophy; ensuring that as many resources as possible are used in a fashion compatible with other uses that are sustained by the land (Laitos and Tomain 1991). In all, public land accounts for nearly 4 million acres and represents the largest concentration of public lands east of the Mississippi River (Morton 1994). Private land uses are typically not subject to long-term sustainable management practices. The general exception is that sustainable private forest management can be required under the Endangered Species Act, if threatened and/or endangered animal species are present.

Recent estimates place the total potential number of plant (flora) and animal (fauna) species in the SAA region at about 20,000 (Boone and Aplet 1994). A total of 51 known species were identified in the region as federally listed threatened and endangered terrestrial species with the help of U.S. Fish and Wildlife Service and each state's Natural Heritage Program (SAMAB 1996e). Although the Terrestrial Team analysis did not focus exclusively on threatened and endangered species, the diversity of species and land cover types did play a strong role in the approach to address community stakeholder concerns. One question raised during the community valuation process required the compilation of various information databases to include species identification and location in the assessment region. Local community groups understood the relationship between species diversity and stresses to habitat. Providing this type of information allows resource managers and community organizations to work together to address a number of related questions concerning ecosystem integrity in the area. To address the question, the Terrestrial Team worked closely with the Nature

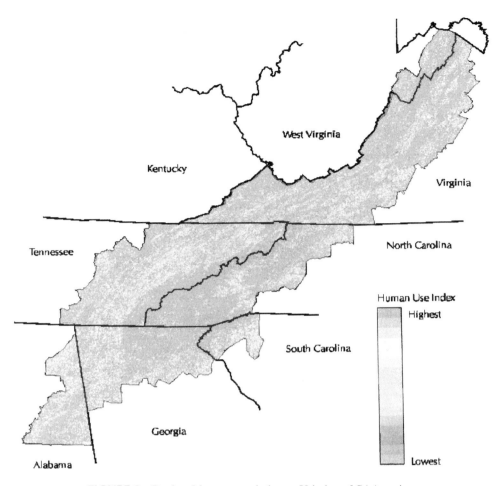

FIGURE 3   Regional human use index or U-index of SAA region.

Conservancy to obtain past and current area disbursement of known species. Integration of locational database information with GIS land cover types helped to provide an answer to this critical question.

The Terrestrial Team worked closely with the GIS Workgroup during the assessment of flora and fauna conditions in the SAA area. The coordinated effort helped to identify over 25 million acres of deciduous, evergreen, and mixed forest types. According to a map derived from Landsat Thematic Mapper images (Pacific Meridian Resources 1995), about 70% of the region is forested (Table 4).

The breakdown among deciduous, coniferous, and mixed forest types are 53%, 6%, and 10% respectively. Analysis of existing data indicates that about 2% of forest cover is lost each year. The trend is expected to continue at the same rate for the next 25 years (predominantly on private lands). Agriculture-pasture is the dominant nonforest cover type (17%), with lesser amounts of cropland (3%), developed (3%), water (1%), and miscellaneous (5%) cover types. Forest cover is unevenly distributed within the region, occurring in large tracts at higher elevations and along ridgelines, and in smaller tracts elsewhere. Agriculture land cover occurs primarily at low elevations, particularly in the valley parts of the Ridge and Valley ecological Sections (Figure 5). Using GIS technology in the assessment of land cover provided the opportunity to integrate multiple database information systems and delineate habitat ranges by predominate land cover characteristics.

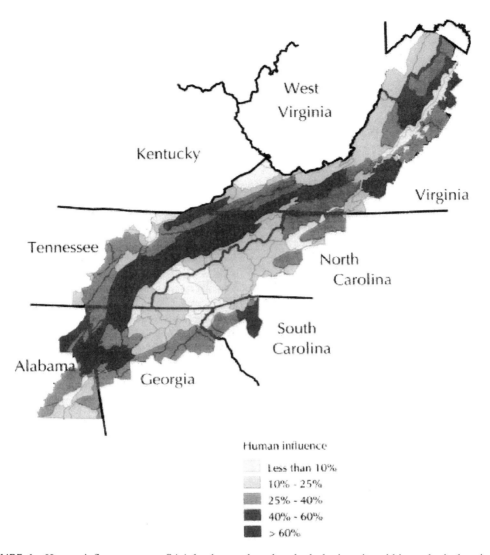

FIGURE 4   Human influence across SAA landscape, based on hydrologic units within ecological regions.

Potential black bear (*Euarctos americanus*) habitat areas were identified using GIS technology (Figure 6). A total of 21 million acres of black bear habitat is currently located within the SAA regional boundary. A county analysis of past and present black bear range is shown in Figure 7. Existing trends suggest that there will be little net loss of bear habitat on public lands in the near-term future. In fact, there has been a moderate expansion of suitable black bear habitat. Also, the northern and southern population centers have completed a link between the two populations (see Figure 7). Of concern, however, is the loss of habitat on private lands predicted to occur over the next decade. Currently, about 75% of suitable black bear habitat lie within private lands. The loss of black bear habitat is associated with the continued development trends in the region. Expected growth over the next decade of 35% in urban areas and over 50% in rural areas will account for the majority of the expected habitat loss over the next decade.

Black bear habitat loss will effect other species as well. Loss of privately owned forest land may create gaps in federal land continuity, creating isolated habitats that impact long-term survival

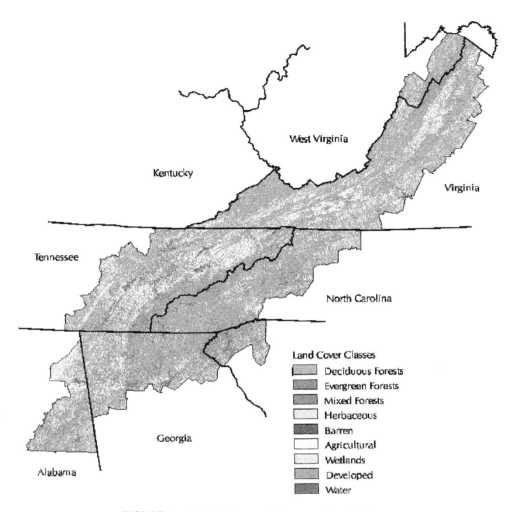

FIGURE 5   SAA study area land cover classifications.

**TABLE 4
Land Area by Land Cover Type in the
Southern Appalachian Region**

Cover type	Area (%)
Deciduous forest	53.0
Coniferous forest	6.4
Mixed forest	10.3
Agriculture — pasture	16.8
Agriculture — cropland	3.4
Developed	3.3
Water	1.5
Other	5.3

Source: Pacific Meridian Resources, Inc. 1995.

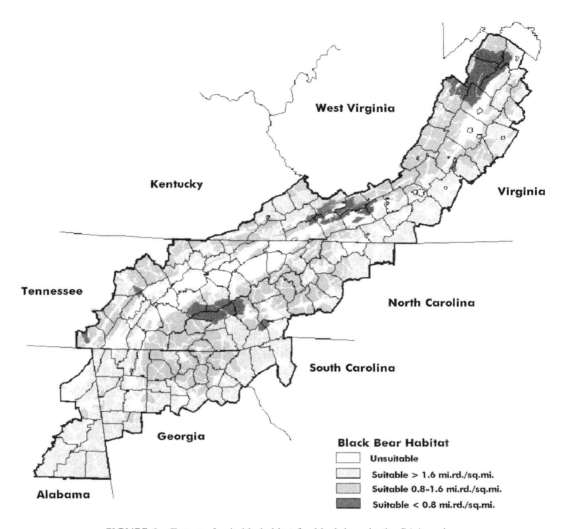

FIGURE 6   Extent of suitable habitat for black bear in the SAA region.

of species. The vast number of species in the SAA area prohibited the assessment from addressing each species individually. Identifying the integrity and health of every ecosystem for the potential 20,000 species in the region was not practical. For example, many species have not been identified as yet and the cost of such an assessment would be extreme both in time and dollars. The analysis did include 471 other species. Approximately 88% of the species identified in the assessment are considered threatened or endangered under criteria stipulated under the Endangered Species Act. Using species listed under the Act provided an economy of scales by integrating existing research into the environmental assessment. The remainder of the species fall under the category of game species. Game species maintain high public interest and management priority for economic and special habitat needs.

The SAA classified species habitat into 19 groups which allowed most of the 472 species to be categorized into a designated habitat. Potential habitat for many of the major game species encompassed the total SAA region. State Heritage Programs and U.S. Fish and Wildlife data in conjunction with GIS technology provided significant opportunities to visualize trends in population disbursement between 1970 and 1995. The information can be used as a powerful tool in analyzing the success or failure of species management practices.

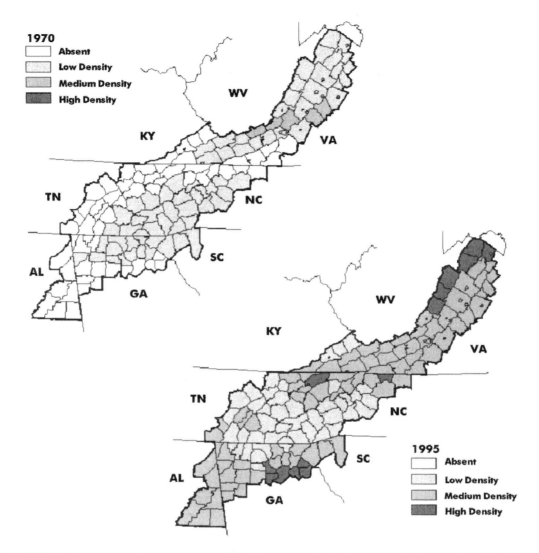

FIGURE 7   Past extent and current extent of black bear range indicating renewed linkage of two black bear populations.

During the past 25 years, management strategies for game species have focused on conservative harvest and restoration efforts, often with marked success. For example, game management has increased the total numbers and range for wild turkey (*Meleagris gallopavo*) and white-tailed deer (*Odocioleus virginianus*) (Figures 8 and 9). The range expansion and population increase trends are expected to level off during the next 15 years.

Well-known animal species play a critical element in developing public opinion and support for species management. A more difficult ecosystem management test approach concerns the buffer to vegetation and the general habitat requirement of all species. The importance of risk to vegetation and habitat is exemplified in the SAA analysis of the high-elevation spruce–fir–northern hardwood forest. A total of 23 fauna species make their home in this high-elevation habitat that covers some 184,000 acres in the SAA region (Figure 10). The Terrestrial Team and GIS Workgroup were able to merge database information over a 25-year period to gain insight to potential stress trends on this unique habitat.

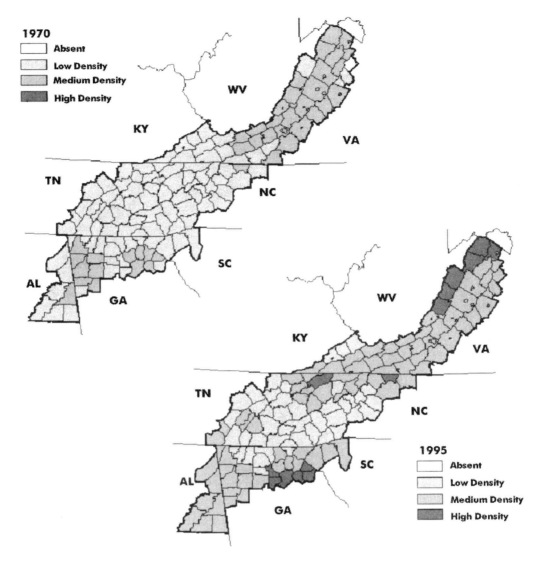

FIGURE 8   White-tailed deer habitat before and after implementation of conservation and harvest strategies.

The rugged terrain impeding human encroachment is not enough to stop the influence of urban and rural growth in the valleys. Population increases are creating additional air pollutants in the form of particulate matter and acid rain. These pollutants are intensifying stress on geographically unique ecosystems like the high-elevation spruce–fir forest. A recent study on the long-term effects of acid rain indicates that calcium-to-aluminum ratios for soil waters in high-elevation forests may have an adverse effect on growth or nutrition of forest vegetation (Likens et al. 1996). The slow recovery of acidified forests provides an opportunity for exotic species and pathogens to invade the weakened natural condition of the forest ecosystem; introducing new threats to species survival.

The Forest Health Subteam of the Terrestrial Team considered and prioritized damage and impacts resulting from the introduction of exotics and/or pathogens on terrestrial forest species. They found that a number of important woody shrub and tree species across the study area are being negatively impacted. For example, dogwood anthracnose (*Discula destructiva* Redlin) is a virulent pest of the flowering dogwood (*Cornus florida*), and the disease vector is found in every

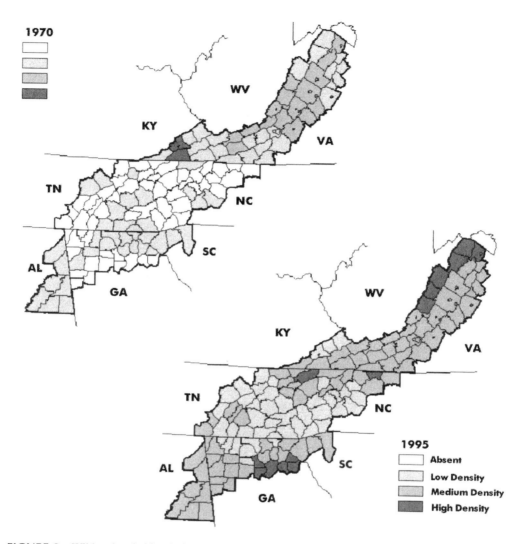

FIGURE 9   Wild turkey habitat before and after implementation of conservation and harvest strategies.

county of the southern Appalachians. Very little resistance to this pathogen has been found in the wild. The potential for infection increases with elevation and shade. The long-term projection for dogwood as a dominant understory flowering shrub is not good (Anderson et al. 1994). Additionally, the loss of dogwood as an understory dominant would forever change the "look" of southern deciduous forests, especially in early spring.

Similarly, the future of Carolina hemlock (*Tsuga caroliniana*) and Eastern hemlock (*Tsuga canadensis*) is threatened by the hemlock woolly adelgid (*Adelges tsugae*). The Fraser fir (*Abies fraseri*) has been severely impacted by the balsam woolly adelgid (*Adelges piceae*). Another tree species with an uncertain future is the butternut (*Juglans cinera*), as it is being infected by the butternut canker (*Sirococcus clavigignent-juglandacerum*) across the South. With time, the butternut canker fungus kills the host butternut tree (USDA FS 1994). The Forest Service has estimated that 90% of butternut trees have been extirpated across their southern range (USDA FS 1994).

Another very important and potentially devastating forest pest is the European gypsy moth (*Lymantria dispar* L.). Gypsy moths are known for their destructive defoliation of urban and forest tree species. As discussed in the terrestrial health report, the moth was introduced into the Northeast

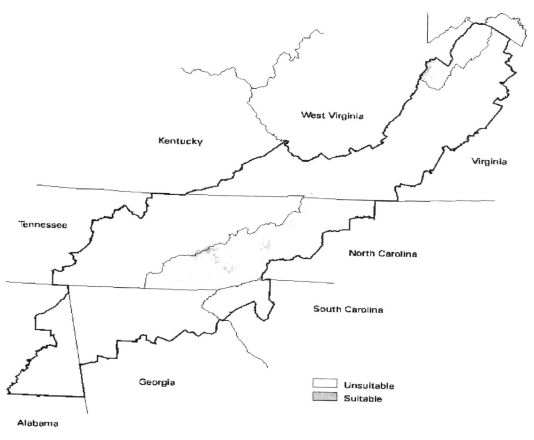

FIGURE 10 Limited spatial distribution of habitat suitability for high elevation spruce–fir species.

in the late 1860's and has spread west and south since that period. It is estimated to be established in all or part of 16 states, including Virginia and West Virginia of the southern Appalachian region. In the last 10 years alone, gypsy moths have defoliated more than 4 million acres in Virginia and 1 million acres in West Virginia (USDA FS 1994). Tree mortality intensifies with repeated defoliation and the interaction of drought and tree condition.

At the present time, only the northern portion of the SAA region is infested with the gypsy moth. Isolated outbreaks of the moth have been eradicated in a number of southern counties. The team modeled the spread of gypsy moth and estimate that 90% of the area could be infested with gypsy moths by the year 2010.

The repercussions to ecosystem health and integrity from a 90% destruction of a single species are great. In 1904, for instance, the first recorded case of chestnut blight (*Cryphonectria parasitica*) was reported in the U.S. American chestnut trees made up 25 to 50% of the hardwood stands in the southern Appalachians in 1929 and virtually all of the counties in the SAA region had been infected with this rapidly spreading fungus, typically transported through the air as microscopic spores. By 1940 the disease killed most of the chestnut trees in the region and severely affected the natural resources provided by this ecosystem. The total loss of the American chestnut from the southern Appalachian region may be averted, if management strategies can be implemented to protect the remaining genetic diversity of the species and silvicultural practices employed to enhance chestnut survival (1996e). For more information on pests and pathogens in the southern Appalachians, see Schlarbaum et al., Chapter 14 in this volume.

AQUATIC TEAM RESULTS

The Aquatic Team and community groups identified a number of concerns in relation to the current status and apparent trends in water resources in the southern Appalachians. Focusing on water quality, aquatic habitat, and aquatic species at a broad spatial scale helped to formulate specific questions for the assessment. One question, of particular interest to the community, concerned the sensitivity of streams to acid deposition.

In order to address this concern, the geology associated with various soil types was mapped, based on the ability of the soil to neutralize acidic deposition. Acidic deposition results from a number of sources in the southern Appalachians, including manufacturing processes, coal-fired generating plants, and vehicle emissions. The nitrogen and sulfur emissions from these sources react in the atmosphere with other chemicals to form nitric acid and sulfuric acid. These emissions fall back to the earth's surface with rain, fog, mist, frost, snow, and/or dust.

The acidic compounds can be assimilated or neutralized, depending upon the buffering capacity of the stream and adjacent watershed soils. In the southern Appalachians, the ability of the soil to neutralize acidic deposition is determined largely by the bedrock geology of the watershed. Assistance was provided by the U.S. Geological Survey (Reston, VA) to produce a map that identifies stream sensitivity to acid deposition (Figure 11) based on the geology of the watershed (Peper et al. 1995).

The methodology used in identifying the various geologic features in the area were developed at a broad spatial scale to gain a general idea of the acid deposition problem. The GIS team then overlaid a map of the streams in order to identify the number of stream miles that are susceptible to acidic impacts (SAMAB 1996e). A more detailed study of nitrogen and sulfur concentrations emitted in an area may provide useful insight to specific areas at risk, if used in conjunction with this assessment product.

The northern portions of the SAA area (contained in an area referred to as the Mid-Atlantic Highlands) has one of the highest rates of acidic deposition in the nation (Herlihy et al. 1993). The natural resources that appear most sensitive to and at greatest potential risk from acidic deposition are aquatic ecosystems, aquatic-dependent species, and high-elevation red spruce forests. Research conducted under the auspices of the NAPAP concluded that regions in the U.S. most at risk from continued acidic deposition are located along the Appalachian Mountain chain stretching from the Adirondacks in New York to the southern Blue Ridge in Georgia.

Within the SAA area, 54% of stream miles have high sensitivity to acid deposition, 18% have medium sensitivity, and 27% have low sensitivity. Published scientific evidence (Lynch and Dise 1985) indicates that some streams in the region have become increasingly acidic. Projections for the future suggest that many additional streams could become more acidic in the decades to come. The northern part of the assessment area is more vulnerable than the southern part, because of its location relative to sources of acid deposition and because of climate factors such as length of growing season. The aquatic team also noted that headwaters in rugged terrain are the most susceptible streams to acid deposition in the region. The findings of the Aquatic Team provide a better understanding of the integrity of the southern Appalachians and potential stressors from acid deposition. The information may provide a foundation for developing an ecosystem management strategy to address this environmental integrity issue.

Watersheds are also used as a basis for ecosystem management (GAO 1994). The spatial connectedness of the southern Appalachians watershed, over time, will play a vital role in supporting the existing health of the region and provide a foundation for potential recovery efforts (Doppelt et al. 1993). The SAA region contains parts of 72 8-digit U.S. Geological Survey hydrologic unit codes (huc) (Seaber et al. 1984)(Figure 12). The median U-index value on a per-watershed basis is 22%, with a range of about 1% to 52%. The frequency distribution of per-watershed U-index values (not shown) is a way to identify watersheds with relatively high or low amounts of human

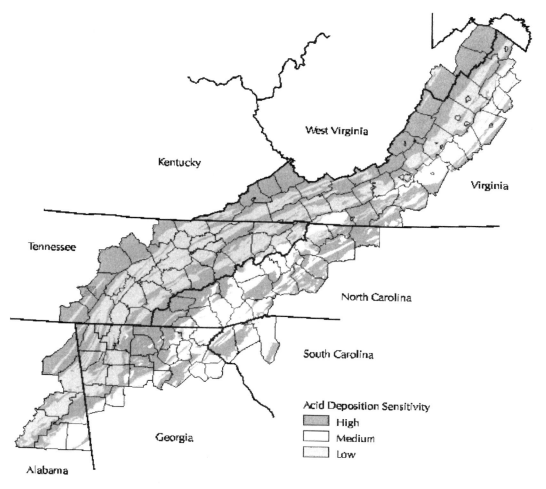

FIGURE 11   Stream sensitivity to acid deposition based on geology of region.

use. These watershed statistics are potentially related to stream water quality within watersheds. This possibility is a topic for follow-up investigations.

Aquatic ecosystems interconnect with both terrestrial landscapes of watersheds and atmospheric flows and inputs. Realistic assessment of aquatic ecosystem condition requires consideration of all of these parts and their interactions, not just a narrow view of the water in the stream. The SAA's aquatic resources assessment encourages integration of water, land and air factors that influence aquatic ecosystem integrity by gathering information on diverse effects and stressors, and by illustrative examples of useful ways to summarize this information. While the percentage of degraded streams in the study area cannot be estimated accurately with available information, evidence documented in the aquatic resources report provides some estimates.

State assessments of designated uses for aquatic life, drinking water, recreation, and other uses show that approximately three quarters of all drainages in the SAA area have at least 6% of their streams over committed. Additionally, 13 basins have greater than 20% of their stream miles not fully supporting uses (SAMAB 1996b). Because most states' monitoring programs cover only a small fraction of waters, and their monitoring network locations are not chosen to represent all streams in the SAA area, we can consider the range of 6 to 20% degraded streams to be an estimate

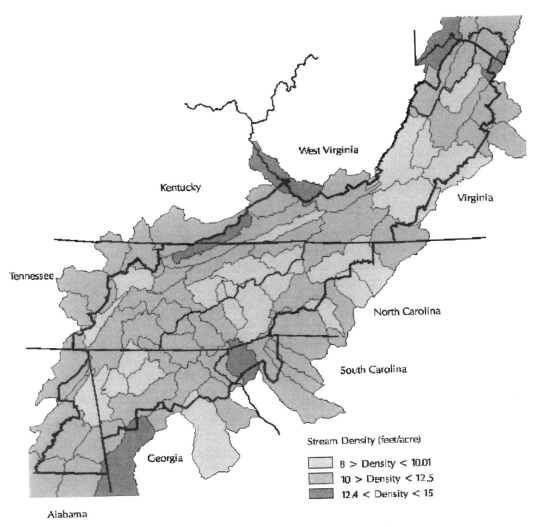

FIGURE 12   Steam density based on hydrologic unit codes.

for the larger streams. Second, studies of selected portions of the SAA area, using fish community biological samples of smaller drainages in several basins (SAMAB 1996b), suggest that over 70% of locations sampled show moderate or severe fish community degradation. Third, a statistical sample of stream habitat condition overlapping portions of the study area in Virginia and West Virginia suggests that about 50% of stream miles in the area studied show habitat impairment compared to relatively unimpacted reference conditions (SAMAB 1996b). Because these estimates are inadequate to represent the entire SAA area and indicate a wide possible range of degradation, a comprehensive statistical sample of streams in the SAA study area is suggested to determine the extent of degraded streams with known confidence.

Instream habitat for fish and other aquatic life is essential for healthy aquatic systems. Human activities on the landscape can adversely affect aquatic habitats in many ways. Sediment from erosion and loss of vegetation from denuded stream banks are prime examples. Stream habitat assessments use a variety of both qualitative and quantitative approaches. These methods focus on stream substrates, organic matter essential to stream food chains, such as leaf litter and large woody debris, stream channel form (geomorphology), and riparian and bank structure. Hankin and Reeves

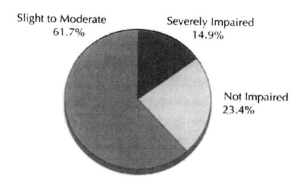

FIGURE 13   Percent distribution of stream habitat conditions.

(1988); Plafkin et al. (1989); Meador and others (1993); Dolloff et al. (1993); Harrelson et al. (1994); and Rosgen (1994) provide representative examples of stream habitat assessment methodologies. State and federal resource agencies are increasing emphasis on habitat assessment as one essential component that characterizes the condition of stream systems (Fausch et al. 1988; Rankin 1995; North Carolina Department of Environment, Health, and Natural Resources 1995; Dissmeyer 1994).

Stresses to aquatic habitats in the SAA area are considered substantial. Growth of urban areas, widespread agricultural activities, road building, and other human activities have the potential to increase the extent and severity of aquatic habitat degradation for streams. A significant portion of streams in the SAA area are likely to evidence habitat degradation, based on studies of subsets of the SAA area. Qualitative visual habitat assessments of 235 sites in the Holston and Hiwassee drainages show 15% of the sites sampled were severely impaired, 62% slightly to moderately impaired, and 23% not impaired (Figure 13). Qualitative visual habitat assessments of 178 statistically selected sites in the Mid-Atlantic Highlands Assessment (MAHA) area (including the SAA area in Virginia and West Virginia and some areas outside the study area in Pennsylvania, Maryland, and West Virginia) estimate that 50% of stream miles have impaired physical habitat (Gerritsen et al. 1995). Approximately 37% of stream miles in the Blue Ridge ecological region of the MAHA area and 60% of stream miles in the Ridge and Valley ecological region of the MAHA are impaired, due to habitat factors.

A consistent and comprehensive picture of aquatic stream habitat condition is not currently available for the SAA area. Also, many of the habitat condition data now available are based on qualitative visual estimates with different agencies using incomparable methods. Reliable aquatic habitat status and trend information will be necessary to successfully protect and restore stream systems in the southern Appalachians. Hydrologic changes that result in alterations in the amounts, duration, timing, frequency, and rate of change of stream flow should also be addressed as a critical component of stream habitat condition (Richter et al. 1995).

Natural and human activities on the landscape have the potential to significantly influence water quality and aquatic ecological integrity (Hunsacker et al. 1993). Humans currently manage or otherwise have changed most of the landscape of the SAA area. The entire landscape of a watershed can affect aquatic resources (Hunsacker and Levine 1995). Additionally, areas close to streams and other watercourses can dominate important factors that influence aquatic ecosystem integrity, such as vegetation along streams and erosion from stream banks (Steedman 1988). Landscape information for the SAA area, developed from satellite imagery, provides part of the basis to begin relating important landscape factors to instream conditions of chemistry, habitat integrity, and ecological condition (Roth and others 1995). Geographic information systems have the potential to integrate these and other data, which can provide improved management of nonpoint source pollution (Lee et al. 1991).

The 17 classes in the land cover analysis were aggregated into 7 classes that are believed to have utility for discerning influences of landscape on water resource integrity: forest, wetlands,

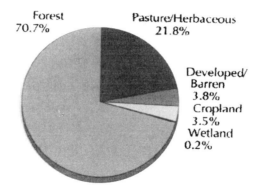

FIGURE 14 Distribution of aggregated land cover classes important for water resource integrity.

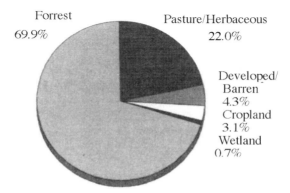

FIGURE 15 Riparian land cover by aggregated classes for SAA riparian zones identified as 100 feet from watercourses.

agriculture–pasture and herbaceous, agriculture–crops, developed and barren, water, and indeterminate (clouds, shadows, etc.). The area that was covered by each of these aggregated classes was calculated using analysis boundaries with the most relevance to aquatic systems. These include ecoregions, hydrologic units, and smaller areas defined by the overlap of ecoregions and hydrologic units; for example, the Ridge and Valley within the French Broad drainage. Aggregated land cover classes thought to strongly influence water resource integrity are distributed in the study area as follows: forest, 70.7%; pasture/herbaceous, 21.8%; cropland, 3.5%; developed/barren, 3.8%; and wetlands, 0.2% (Figure 14).

Aggregated land cover classes for the riparian zone of the entire study area are distributed as follows: forest, 69.9%; pasture/herbaceous, 22.0%; cropland, 3.1%; developed/barren, 4.3%; and wetlands, 0.7%. Figure 15 shows the distribution of land cover classes for riparian areas within 100 feet (30 meters) of watercourses for the entire study area. Forest cover in the riparian zones of the study area ranges from less than 25% to 100%. Figure 16 shows subdivisions of the SAA study area that are defined by portions of hydrologic units within ecological regions. Each area is classed according to the fraction of forest cover in the riparian zone.

Instream habitats for aquatic life are very dependent on natural bank and riparian zone vegetation. Intact riparian zones provide numerous critical ecological functions (Gregory et al. 1991). They stabilize stream banks and prevent bank erosion while providing inputs of organic matter that constitute the base of stream food chains. They provide structure for important habitat types, such as undercut banks, root cover, and large woody debris for fish and other organisms. They provide essential shade and temperature regulation for many fish, such as trout. If properly planned and

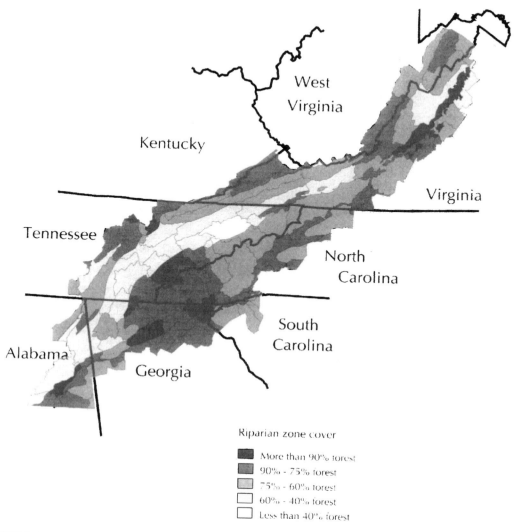

FIGURE 16   Riparian zone forest cover by aggregating ecoregion and hydrologic units to identify aquatic resources integrity problems based on percentage of forest cover in riparian zone.

managed, they can serve as filters to reduce sediment input from upland erosion (Barling and Moore 1994). Managed and regularly harvested forested near-stream zones (beyond the intact zone of natural vegetation) can also potentially reduce nutrient inputs (National Association of Conservation Districts 1994). Recommendations and regulations for stream bank and riparian area protection, such as best management practices, vary widely from state to state as do recommended riparian zone sizes. All streams need well-established riparian buffers of natural vegetation to attain and maintain their biological integrity (National Association of Conservation Districts 1994).

Assessments of riparian zones covering large geographic areas are not generally available. Remote sensing (satellite data) and GIS technologies currently make wide area inventories of riparian conditions practical (Hunsaker and Levine 1995; Roth et al. 1995; Steedman 1988). Since the land cover classification for the SAA was produced using satellite data with 30-meter resolution, only larger watercourses are detected. The location of all smaller waterways is assumed to correspond to the reach file stream tracings (EPA's Reach File 3 or RF3). These GIS coverages were combined to define riparian zones within the ecoregion and hydrologic unit analysis boundaries. The aggregated land cover classes relevant to aquatic systems are summarized within the 100-foot (30-meter) buffer zone using a combination of ecoregions and hydrologic boundaries. This information can guide priorities for future more detailed studies relating riparian factors, patterns and management practices to instream conditions (Zucker and White 1996).

Increased attention to instream habitats, riparian areas, and landscape influences on aquatic ecosystems will be essential to guide and evaluate continued efforts to restore and maintain the integrity of aquatic systems in the southern Appalachians.

ATMOSPHERIC TEAM RESULTS

Different images are formed in people's minds when the words "air pollution" are mentioned. For some people they may picture smog hanging over a city, smoke from the stack of a power plant, or a dark cloud following a vehicle; but modern society is dependent on the combustion of coal, oil, and gas in order to generate electricity, and for industrial uses to produce products for people. Furthermore, large quantities of gasoline are used to transport people and products. The combustion of fossil fuels generates energy, as well as toxic gases and small particles. These pollutants are transformed in the atmosphere and are transported downwind from where they were generated, often for long distances. Air pollutants are of a great concern because they are known to adversely effect human health, and there is also a concern because they are known to impact natural resources. The southern Appalachians are one area in the U.S. where there have been repeated concerns voiced that air pollution could be adversely impacting vegetation, soils, water quality, aquatic life, and visibility.

Recently, an analysis of air quality in the southern Appalachian Mountains was completed, and the following draws heavily from the Atmospheric Report of the Southern Appalachian Assessment (SAMAB 1996c). The topics which will be addressed include: (1) which air pollutants are of greatest concern and where do the pollutants originate; (2) what impact might ground-level ozone be having on trees in the southern Appalachians; (3) how much acidic deposition is deposited from the atmosphere in the southern Appalachians; and (4) how good is visibility in the southern Appalachians.

The Pollutants of Concern — Pollutants emitted directly into the atmosphere are referred to as primary pollutants. In high enough concentrations, primary pollutants can cause damage to vegetation and soils. A century of emissions of sulfur dioxide in the Copper Hill region in eastern Tennessee is an example of where extensive long-term damage has been done to the environment by a primary pollutant. The Tennessee Valley Authority and others have devoted large amounts of resources in reclaiming the Copper Hill region. There are still areas where the impacts can be seen today, even though there has been successful reclamation in portions of the Copper Hill region (Muncy 1986).

The pollutants which are of greatest concern in the southern Appalachians result from the transformation of primary pollutants in the atmosphere. These secondary pollutants include sulfates, nitrates, and ground-level ozone, and they are formed from the emissions of sulfur dioxides, oxides of nitrogen, and volatile organic compounds. The secondary pollutants that are impacting the southern Appalachians are formed from primary pollutant emissions both within and outside of the region. For example, sulfur dioxide emissions are emitted in the region (as in eastern Tennessee), but it is believed that emissions from the Ohio Valley and northern Georgia and Alabama may be contributing significant amounts of sulfates to the southern Appalachians (Figure 17).

The primary sources of pollution are emitted from numerous source categories, but emissions from certain categories dominate. For example, nationally, electrical utilities emit about 85% of the sulfur dioxide (Figure 18); whereas, the most important sources of oxides of nitrogen are on-road vehicles (32%) and electrical utilities (33%) (Figure 19). Interesting to many people is the fact that the main source of volatile organic compounds are gases emitted by trees during the growing season.

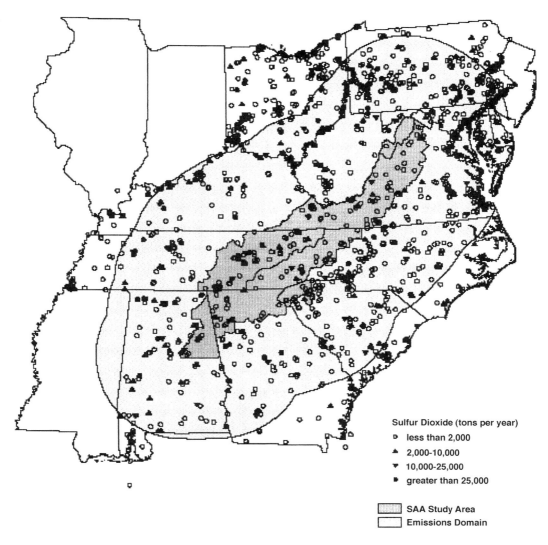

FIGURE 17 Location of point sources of sulfur dioxide — 1995. The "emissions domain" is a 250-km boundary around the SAA area. (Source: SAMAB 1996c.)

Ground-Level Ozone Exposures in the Southern Appalachians — Ozone, a chemical composed of three oxygen atoms linked together, is a highly beneficial gas in the upper atmosphere because it protects against ultraviolet radiation. At ground-level, however, ozone can kill living tissue that it contacts. Ground-level ozone is naturally occurring, but elevated concentrations occur in the southern Appalachians primarily because large human-made releases of oxides of nitrogen react with naturally occurring volatile organic compounds. High ozone exposures can occur on hot, sunny days when these two compounds are abundant in the atmosphere.

Air regulatory agencies use equipment to measure the amount of ground-level ozone in the atmosphere. The hourly average concentrations are used to calculate the amount of ozone exposure that a plant receives during a growing season. One ozone statistic used by scientists and resource managers is the cumulative exposure index called the W126 (Lefohn and Runeckles 1987, Lefohn et al. 1995). The W126 is calculated by putting a weight on all of the hourly average concentrations and then adding the values together for the growing season (April through October). The cumulative

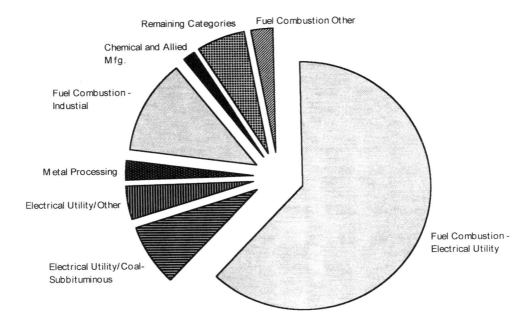

FIGURE 18 National sulfur dioxide emissions by principal source categories, 1994. (Source: EPA 1995.)

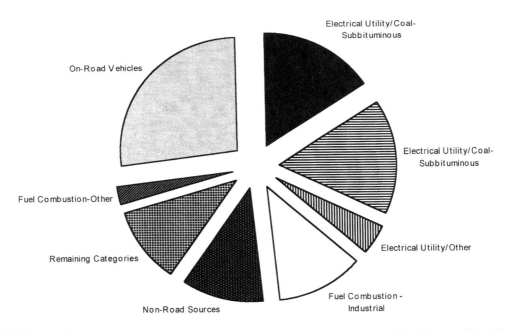

FIGURE 19 National nitrogen oxide emissions by principal source categories, 1994. (Source: EPA 1995.)

ozone exposure that plants receive does vary between years because the meteorology and the amount of primary pollutants present during a growing season will vary yearly. For example, 1989 was a cool year with frequent rainfall and consequently the ozone exposures were considered to be low (Figure 20). Most of the southern Appalachians had W126 values which ranged between 23.8 and 66.5 ppm-hours, and many monitoring locations had a lack of hourly average ozone concentrations

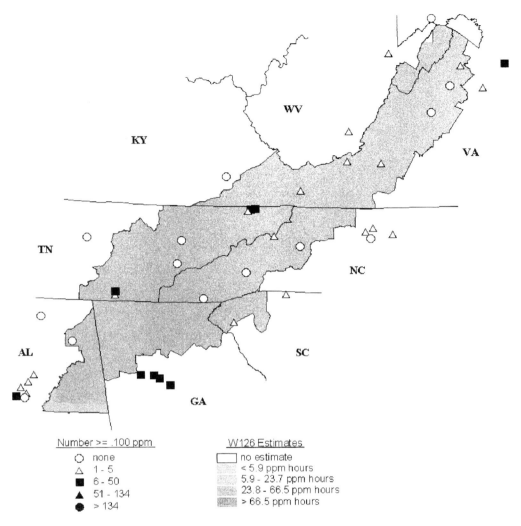

FIGURE 20 W126 ozone exposure results for the SAA area, 1989.

above 0.10 ppm. On the other hand, 1988 was one of the hottest and driest years in the 1980s and ozone exposures were considered to be high. This is especially true for the northern portion of the Southern Appalachian Assessment area, where a large amount of the area was predicted to have a W126 value greater than 66.5 ppm-hours. The remaining area had W126 values which ranged between 23.8 and 66.5 ppm-hours. Examining the number of hours greater than or equal to 0.10 ppm at monitoring sitess in 1988 reveals that most of monitoring site had a frequent occurrence of high hourly average ozone exposures (Figure 21).

Ozone can kill plant tissue when it enters openings in leaves called stomates. The stomates are used by plants to exchange gas with the air surrounding the leaves. Scientists and other specialists can identify injury to a plant from ozone by using a checklist of characteristic symptoms. Symptoms of ozone injury are likely to be commonly found in and throughout the southern Appalachians in most years (Dowsett et al. 1992, SAMAB 1996c). For example, field studies at Great Smoky Mountains National Park have reported 90 species with ozone symptoms. The presence of ozone symptoms is an indicator that the plant has had a physiological response (called an injury) to the ozone which entered the leaf (called the dose). Though ozone injury is important, often resource

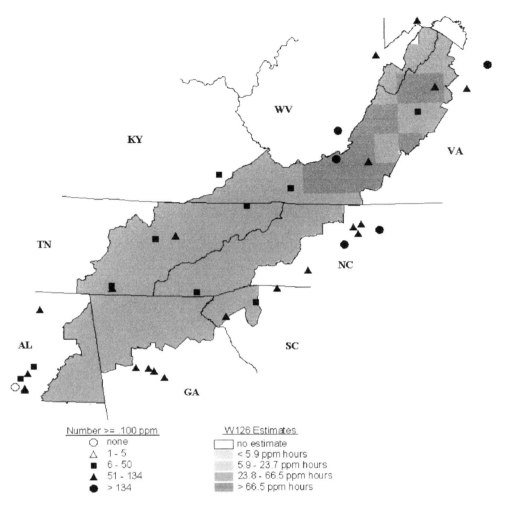

FIGURE 21   W126 ozone exposure results for the SAA area, 1988.

managers are more concerned if trees will be damaged, i.e., the ozone dose is large enough to reduce the intended use or value of a species or ecosystem (Tingey et al. others 1991).

An approach has been developed to identify which areas of the southern Appalachians, between 1983 and 1990, may have had damage from ozone exposures (SAMAB 1996c). The first step in the process involved compiling and analyzing data from previous studies where mature trees or tree seedlings were exposed to ozone under controlled conditions. The results from this effort identified four response categories (Table 5). Both the W126 index value and the frequency of hourly average ozone concentrations above 0.10 ppm were used to identify which categories of species were at risk to growth loss (damage). Grid cells which met the W126 criteria for a particular category and not the requirement for the number of hours greater than or equal to 0.10 ppm for the same category were assigned the category which matched the number of occurrences greater than or equal to 0.10 ppm. Applying the rules in Table 5 indicates where ozone exposures may have been large enough to cause damage, but environmental factors influence the openings of the stomates. Ozone cannot penetrate into a leaf without the stomates opened. Soil moisture is considered to be an important environmental variable which influences the uptake of ozone by a plant (EPA 1986). During periods of low soil moisture, such as occurred in the southern Appalachians

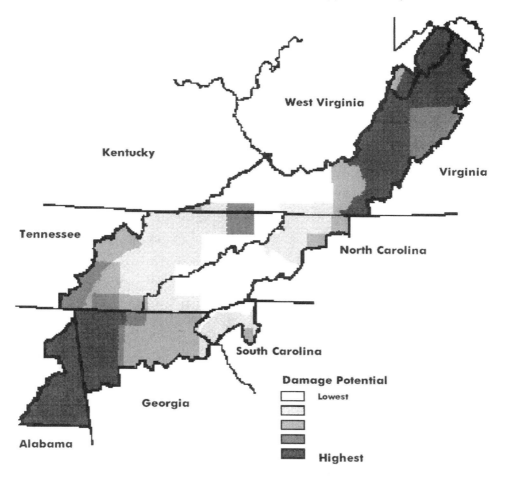

FIGURE 22   Areas with the greatest frequency of potential ozone damage, 1983 to 1990. (Source: SAMAB 1996c.)

between 1985 and 1988, it is believed ozone exposures may have a minimal contribution to growth reductions which may be observed (SAMAB 1996c).

Areas with the greatest potential for ozone damage are those where both soil moistures were adequate for the stomates to remain open, and the ozone exposures were large enough to cause damage to sensitive species. Figures 20 and 21 both show that the W126 values were large enough to cause damage to Level 1 and Level 2 species, but for most of the years (1983 to 1990) these reductions were not anticipated for Level 2 species because there were less than 51 occurrences of hourly average ozone concentrations above 0.10 ppm. Furthermore, large amounts of the southern Appalachians in 1985 through 1988 data had low soil moisture, so those areas were assigned the minimal category because ozone was unlikely to penetrate into the leaves and cause damage (SAMAB 1996c). Figure 22 depicts the results of integrating all of the years of ozone data and soil moisture. The northern and southern portion of the assessment area had the highest scores and were most likely to be the areas with the greatest amount of ozone damage to Level 1 species.

The results from SAMAB (1996c) present a picture of what impact ground-level ozone may be having across the landscape. There are exceptions to the findings. For example, there were probably areas in 1985 through 1988 that had adequate soil moisture because soil availability is also influenced by the site conditions. Furthermore, these results may not be applicable to trees found above 3000 feet in elevation because significant amounts of soil moisture are available through moisture obtained from

TABLE 5
Ozone Exposure Levels Associated with Forest Tree Response

Forest tree response category[a]	W126	Hours $\geq$ 0.10 ppm
Minimal	<5.9	<6
Level 1 (only highly sensitive species affected, e.g., black cherry)	$\geq$5.9	$\geq$6
Level 2 (moderately sensitive species affected, e.g., tulip poplar)	$\geq$23.8	$\geq$51
Level 3 (all species affected, even those normally resistant, e.g., red oak)	$\geq$66.6	$\geq$135

[a] Level 2 includes Level 1 species, and Level 3 includes species included in Levels 1 and 2.

From SAMAB 1996c.

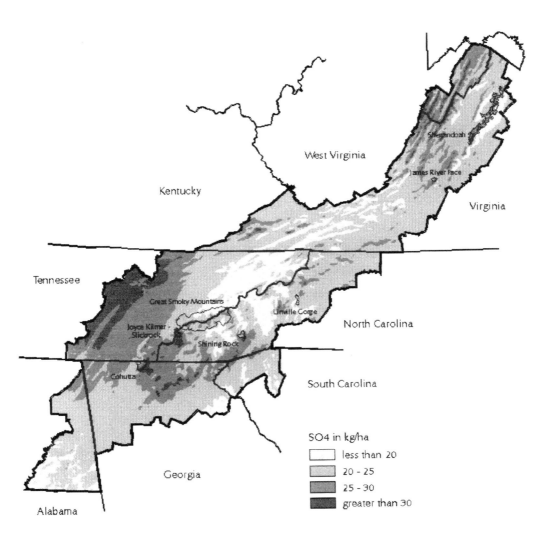

FIGURE 23 Modeled distribution of mean wet sulfate loading (in kg/ha/year) during the period 1983 to 1990. Outlined are political boundaries and Class I area parks and wilderness. (Source: SAMAB 1996c.)

clouds (SAMAB 1996c). Finally, Winner et al. (1989) have reported that visible injury to foliage increased with elevation, even though the number of elevated hourly ozone concentrations did not increase. Thus, the authors speculated that a particular species growing at a high elevation site may be more sensitive to ozone exposures than the same species growing at a low elevation.

The National Ambient Air Quality Standard (NAAQS) currently focuses on the number of times monitored ozone concentrations exceeds an hourly average of 0.12 ppm. As of 1996, only one high elevation, forested ozone monitor site had a violation of the ozone NAAQS. Results reported in SAMAB (1996c) indicated there was a potential risk for damage to sensitive tree species found in the southern Appalachians. It is believed that current form of the NAAQS should be changed and based upon a cumulative index value. It is probably also necessary to include the number of hourly average ozone concentrations above 0.10 ppm since experimental studies for tree species had growth losses only with ozone exposures with frequent concentrations above 0.10 ppm (SAMAB 1996c).

Acidic Deposition in the Southern Appalachians — During the 1980s, many people in the U.S., and other countries were disturbed by reports from the media that "acid rain" was killing forests, lakes, and streams. In response to the concerns, a large amount of research was conducted on the impact of acidic deposition on forest and aquatic ecosystems through the National Atmospheric Precipitation and Assessment Program. The chemicals of interest in studies of acidic deposition include hydrogen ion (pH), sulfates, nitrate, and ammonium. The total amount of acidic deposition includes measurements of the amounts which are dry and wet. Dry deposition measurements are difficult to obtain and therefore are only measured at a few places in the U.S. The wet deposition portion includes rain, snow, sleet, hail, fog and cloudwater. Weekly samples of wet deposition are obtained at about 200 sites in the U.S. The wet deposition collections of rain, snow, sleet, or hail are perhaps the most used data sets when studies are conducted on deposition rates. Lynch et al. (1996) have developed a method to extrapolate the wet deposition data beyond the monitoring sites. The method uses a statistical technique that includes wet acidic deposition data, rainfall data, and the influence of topography to predict annual deposition rates and hydrogen ion concentration. Average wet sulfate deposition loading throughout the southern Appalachians ranges between 20 and 40 km per hectare per year (kg/ha/year) (Figure 23). High wet sulfate loading is found in the Cumberland Plateau region in eastern Tennessee. Wet sulfate deposition loading is highest also at the highest peaks in the southern Appalachian Mountains, which includes many of the Class I wildernesses and national parks. The modeling results for wet nitrate deposition loading (Figure 24) is generally in the 5 to 10 kg/ha/year, with some localized areas of higher loading.

Figure 23 estimates only a portion of the total sulfate deposition. Annual estimates of dry sulfate deposition in many areas are equal to the wet estimates. Measurements of wet sulfate deposition do not include those coming from clouds or fog that come in contact with the mountains. At high elevations there are significant inputs of sulfates from the clouds. For example, Figure 23 shows that most of Great Smoky Mountains National Park receives an average of 20 to 30 kg/ha/year, but studies conducted (during 1986 through 1989) at the Park (at 5742 feet in elevation) estimate the total annual deposition of sulfates at 48 kg/ha. Most of the estimated annual sulfate deposition was deposited from clouds or fog (Johnson and Lindberg 1992).

Acid deposition is of significant concern because it may adversely impact the long-term health of forest and aquatic ecosystems. The southern Appalachians are composed primarily of weather-resistant rock, and many areas are sensitive to acid inputs. Sulfates, nitrates, and ammonium are deposited annually, and the bedrock is not able to fully buffer the acid inputs. Furthermore, some soils will retain sulfates that are deposited, and the sulfates may be released in the future. The continued additions of acid inputs results in (1) soil acidification, (2) leaching of base cations from soils, and (3) surface water acidification. Initial research was concerned with the long-term effects of chronic acidification, but recently there has been greater attention on episodic acidification and nitrogen saturation (SAMAB 1996c). Acidification results in negative resource changes in pH, acid

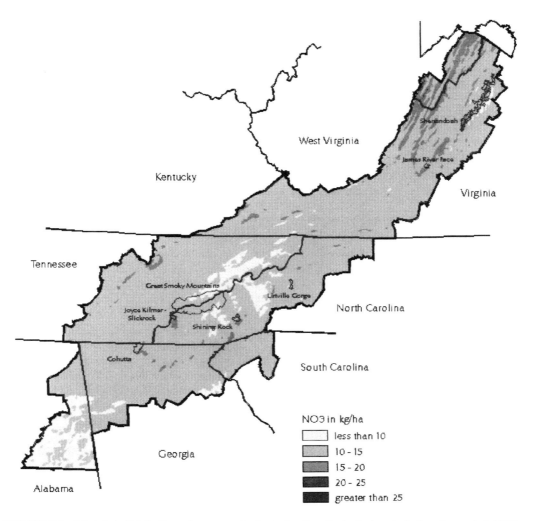

FIGURE 24   Modeled distribution of mean wet nitrate loading (in kg/ha/year) during the period 1983 to 1990. Outlined are political boundaries and Class I area parks and wilderness. (Source: SAMAB 1996c.)

neutralizing capacity (ANC), and aluminum concentrations. Increases in aluminum concentrations are of concern because aluminum is known to penetrate into the roots more readily than the calcium and magnesium which are essential for tree growth. Furthermore, increase in aluminum in streams is known to be toxic to fish and other aquatic organisms.

The concerns from acidic deposition were great enough for political leaders in the U.S. to amended the Clean Air Act in 1990. The additional laws are expected to significantly reduce emissions of sulfur dioxide by the year 2004. Electrical utilities have begun making the necessary reductions mandated by law. The 1995 results from the initial reductions are very encouraging (Figure 25) with 10 to 25% reductions in sulfates, measured at wet deposition monitoring sites, occurring throughout most of the southern Appalachians (Lynch et al. 1996).

Visibility in the Southern Appalachians — People visit the southern Appalachians for many reasons, and viewing scenery is one of the most cited reasons for visiting national forests and parks. Discussions on visibility often focus on the distance (called the standard visual range) a person can see a dark object, such as a distant mountain or ridge, against the sky. Furthermore, studies in

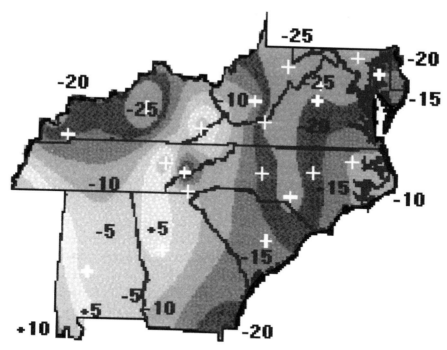

FIGURE 25   Percent departures of 1995 annual sulfate ion concentration from predictions made for 1983 to 1994. (Source: Modified from Lynch et al. 1996.)

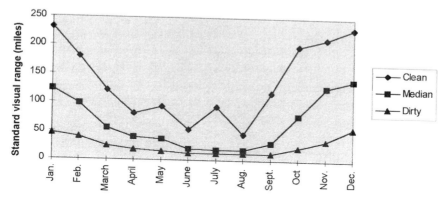

FIGURE 26   Monthly changes in standard visual range (miles) at Shining Rock Wilderness (December 1988 to September 1992). The "clean" value is where the standard visual range is greater than the value 10% of the time, the median value is where the standard visual range is greater 90% of the time.

visibility are concerned with how well the object can be seen. For example, are the fall colors obscured, or can a person see the texture formed by the forest tree canopy cover?

How far a person can see will vary depending on the time of year. Visibility monitoring has been conducted for many years at wilderness and national parks in the southern Appalachians. One technique used is to take daily pictures of the same mountain, such as was accomplished at Shining Rock Wilderness in western North Carolina. Examination of the data (Figure 26) reveals good visibility was captured in 10% of the photographs between the months of September and March. Figure 27 shows an example of good visibility. Notice the distant ridge can be seen in the center of the picture between the two mountains. Figure 28 is a picture taken at the same location as Figure 27, but the distant ridge and the mountain on the right can barely be seen. Furthermore, the appearance of the colors and texture of the forest has been significantly reduced. Scenes of poor visibility occur frequently during the months of June through September (Figure 28) when visitation to National Forests and Parks are the greatest. Studies of visitor use at Great Smoky Mountain National Park have revealed that the most popular recreational activity is sight-seeing and being able to see clearly from vistas (Peine and Renfro 1988). Visitors consider "clean air" to be among the top-ranked features of importance at the Park. The average visitor ranked "viewing landscapes" as "very important" (Morse 1988) to their overall experience.

Visibility conditions at Shining Rock Wilderness are not unique, but are similar to visibility measurements taken throughout the southern Appalachians (Figure 29). Examination of camera data for the worst 2 months in the summer shows that visibility is poor throughout the southern Appalachians (Table 6), but on an annual basis visibility is probably worst in the West Virginia, followed by northern Alabama, Georgia, and Virginia (SAMAB 1996). In the eastern U.S., annual average natural background visibility is considered to be 93 ± 30 miles (Trijonis et al. 1991). Currently the annual average visibility in the southern Appalachian is estimated to be 20 miles (Sisler et al. 1993).

Many people who live within or visit the southern Appalachians believe the reduction in visibility is probably natural. Sometimes people respond by saying, "Well, they are called the Smoky Mountains for a reason." The name "Smoky Mountains" may refer to the small vertical cloud formations that look like smoke rising from the mountains. Ground level clouds do block the view of some of the highest mountains in the southern Appalachians. For example, clouds, rain storm, or snow fall reduce visibility at Shining Rock Wilderness about 38% of the time (ARS 1995). A portion of the mountains within the southern Appalachians is also referred to as the Blue Ridge Mountains. The name is given because of the blue appearance they have during the summer months. The blue haze (which now is rarely seen) was a result of volatile organic compounds released by trees during the summer months.

The visibility reductions that are most commonly seen are best described as a whitish or grayish veil that obscures or blocks distant scenes. This type of visibility reduction indicates that fine particles (less than 2.5 μm in size) are responsible. The analysis of fine particle data from Great Smoky Mountains (Figure 30) and Shenandoah National Park shows that sulfates are the most abundant fine particle when visibility is poor. Sulfates which result from the sulfur dioxide transformation in the atmosphere are considered to be the largest single human-caused contributor to visibility reduction in the southern Appalachians (SAMAB 1996c). The sulfates are also known to cause greater amounts of visibility reductions in an atmosphere with high humidity, because sulfate readily binds to water molecule. The size particle formed by sulfates and water molecules is known to cause a greater amount of light scattering than sulfates by themselves.

Visibility is worst in the southern Appalachians during the summer months for several reasons. First, the greatest demands for electricity occurs during the summer in order to cool homes and businesses. Therefore, the greatest emissions of sulfur dioxide occur during the summer within and near the southern Appalachians. During the summer months also relative humidity is greatest and photochemicals smog (ozone) in the atmosphere may increase the formation of sulfates (Husar et al. 1994). Finally, during the summer months there is a large number of days when air stagnation occurs. The lack of air movement allows sulfates and other pollutants to increase in the southern Appalachians.

The current scientific evidence clearly indicates that sulfates from electrical utilities are primarily responsible for reductions in visibility (Trijonis et al. 1991; National Research Council 1993; Sisler et al. 1993; and IMPROVE 1994) (Table 7). Further reductions in sulfur dioxide emissions are anticipated under the Clean Air Act Amendments of 1990. The Environmental Protection Agency (EPA) has used atmospheric models to predict that summer standard visual range will increase by about 4 miles during the summer months. Despite this improvement the EPA has concluded "… there will still be perceptible man-made regional visibility impairment …" (EPA 1993b).

FIGURE 27 Photograph of Shining Rock Wilderness showing good visibility. Visibility is this good 10 to 25% of the time. Estimated standard visual range is 112 to 168 miles.

FIGURE 28 Photograph of Shining Rock Wilderness showing poor visibility. Visibility is poor 10 to 25% of the time. Estimated standard visual range is 16 to 22 miles.

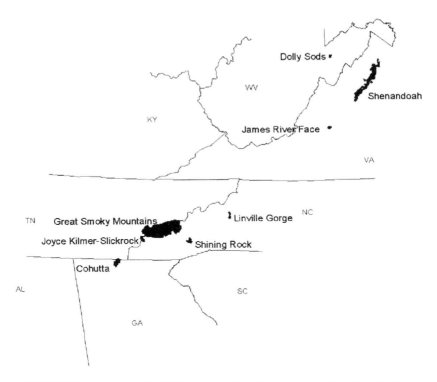

FIGURE 29   Location of visibility monitoring sites in the southern Appalachians.

TABLE 6
Median Camera-Based Standard Visual Range Estimates in Miles for the
Summer and Winter Seasons (for the Combined Years 1987 through 1993)

Class I Wilderness	Summer July/Aug.(miles)	Winter Dec./Jan.(miles)
James River Face, VA	15	66
Cohutta, GA	15	76
Dolly Sods, WV	17	NA
Joyce Kilmer-Slickrock, NC	17	151
Linville Gorge, NC	19	87
Shining Rock, NC	19	138

Note: These seasons represent the worst and best visibility conditions in the southern Appalachians.

Source: SAMAB 1996c.

SOCIAL, CULTURAL, ECONOMIC TEAM RESULTS

Krannich and Luloff (1991) define resource-dependent communities as areas "where the economic, social, and cultural conditions of community life are intertwined with, and ultimately dependent upon, the production of a natural resource commodity or commodities". This definition has been

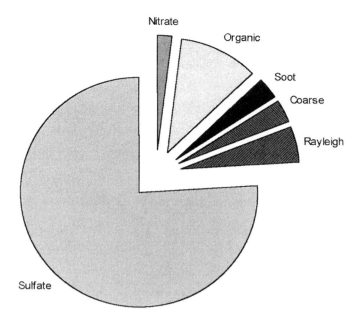

FIGURE 30 Annual visibility extinction budgets derived from aerosol measurements for Great Smoky Mountains National Park (March 1988 to February 1994). The chart clearly shows the predominant role of sulfate in visibility reductions when visibility is considered dirty. (Source: SAMAB 1996c.)

TABLE 7
Percentage Contribution by Source Category to Pollutants Which Affect Visibility in the Eastern United States

Source category	SO_x	Organic particles	VOC	Elemental carbon	Suspended dust	NH_3	NO_x
Electrical utilities	78	—	—	—	—	—	39
Diesel-fueled mobile sources	1.5	—	—	47	—	—	16
Gasoline vehicles	1	34	31	29	—	—	26
Petroleum and chemical industries	4.5	—	11	—	—	—	—
Industrial coal combustion	7	—	—	—	—	—	—
Residential wood burning	—	20	13	15	—	—	—
Fugitive dust (on/off-road traffic)	—	—	—	—	100	—	—
Feedlots and livestock waste management	—	—	—	—	—	66	—
Miscellaneous	8	46	45	9	—	34	19

Note: SO_x — sulfur oxides; VOC — volatile organic compounds; NH_3 — ammonia; No_x — nitrogen oxides.

Source: National Research Council 1993.

indicative of the southern Appalachians because of the vast natural resources available in the region. The southern Appalachian community has utilized the natural resources available to them in a number of important industries. The forested land is used for timber and wood product manufacturing while rivers are being utilized for fisheries and electric generating facilities.

TABLE 8
Percentage Shares of Employment, Employee Compensation, and Total Industrial Output Across 10 Sectors of the Southern Appalachian Region Economy, 1977 and 1991

Major sectors	Employment		Employee compensation		Total output	
	1977	1991	1977	1991	1977	1991
Agriculture, forestry, and fisheries	1.8	3.6	0.8	0.6	3.0	3.3
Mining	1.4	0.6	3.6	1.1	3.4	2.9
Construction	7.6	7.4	8.8	6.6	8.0	9.2
Manufacturing	35.8	22.6	38.4	30.4	52.2	39.8
Transportation, communication, and utilities	3.7	3.8	5.4	5.2	5.3	5.8
Wholesale and retail trade	14.8	20.4	11.6	15.2	8.9	9.6
Finance, insurance, and real estate	3.0	5.2	3.3	4.6	8.4	8.8
Services	15.3	21.6	11.4	19.1	9.2	13.3
Government enterprises	14.7	13.8	16.2	16.9	1.5	7.0
Special industries	1.9	1.0	0.6	0.3	0.0	0.3
Total	100	100	100	100	100	100

Source: ES202 Data, 1993, Bureau of Labor Statistics; U.S. Department of Labor; County Business Patterns, 1993; Bureau of the Census; U.S. Department of Commerce; Regional Economic Information System (REIS), 1993; Bureau of Economic Analysis, U.S. Department of Commerce Bureau of Economic Analysis (estimates) and other surveys, 1993, U.S. Department of Commerce).

Analysis of the social, cultural, and economic situation in the SAA area covered a broad spectrum of issues and concerns. One of the primary concerns, identified early in the community participation process of the SAA, confronted the question of social pattern changes that were affecting the region's approximately 6 million citizens. Specifically, the community groups were concerned about the stability of the local economy and identification of changing employment trends. This information may help the community devise sustainable strategies to ensure economic growth for the region. To answer this question, the Social, Cultural, Economic Team (SCE Team) focused attention on a number of employment trends in various occupations.

To grapple with the complexity of forecasting economic trends the SCE Team analyzed the percentage of employment, employee compensation, and total output across 10 critical sectors of the economy identified in Table 8. Employment figures were used as indicators in order to understand the importance of the natural resources in relation to the economy of the area. The analysis focused on county level data and provides the best indication of how specific sectors of the economy can impact a given county's economic standing.

A recent study, conducted by Science Applications International Corporation (SAIC) for the Environmental Protection Agency, used a geographical information system approach to identify linkages between natural resources available in a county and the percentage employment resulting from natural resources use (SAIC 1996). The work first characterized the economic sectors of the entire SAA region. United States census data was then used to identify Standard Industrial Classification (SIC) codes of businesses in each county. This allowed an analysis of county dependence on specific natural resources, based on SIC code databases, to be visually analyzed.

Similar efforts have been performed by the U.S. Department of Agriculture. Figure 31 reflects data used in the SAA database. Peggy Cook and Karen Mizer, in the Rural Development Research Report Number 89, identified natural resource dependence for many of the counties identified in the southern Appalachians. The SEC Team technical report process focused on a broad analysis of the region, but studies such as the SAIC's and U.S. Department of Agricultures may provide useful future analysis of community dependence on specific natural resources at the county level. This information can then be incorporated into a sustainable management plan for the region.

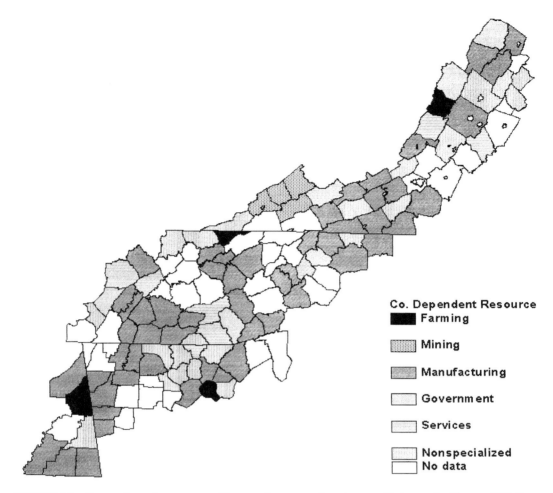

FIGURE 31 County dependence on specific natural resources based on rural/urban code. (Source: U.S. Dept. of Agriculture, Rural Development Research Report Number 89).

The economic force most prevalent in the SAA area is in manufacturing. This economic sector employs about 23% of the labor force in the region. Additional resource dependent industries include timber, mining and agriculture (Malchlis 1990). However, the analysis, covering the period from 1977 to 1991, showed a decrease in the number of manufacturing jobs in the region. The analysis also portrays a significant percentage increase in the service industry as well as the agriculture, forestry, and fisheries sector of the economy. Two additional natural resource industries can be included in the Southern Appalachian Assessment: recreation and tourism. The expansion in these two primary sectors indicates a continued relationship between the resources of the Appalachians and the communities inhabiting the region. The growth in recreation and tourism is reflected in the 30% increase in the service industry during the past 15 years.

One major issue which is closely linked with the resource dependency of the local community is that the community is often dependent on extra-local organizations that control decisions regarding natural resource use (Cramer 1993). This can result in economic upheavals to the local community due to the arrival or departure of resource-related activities (Cramer 1993). The relationship aspect is most pronounced in resource-dependent communities that maintain limited economic diversity (Weber et al. 1988).

TABLE 9
Trends in Percentage of People Participating in Recreation Activities in the Nation, the South and the Southern Appalachian Assessment region in 1972, 1982 and 1992

Activity	1972[a]		1982[b]		1992[c]	
	Nation	South	Nation	South	Nation	South
Primitive camping	5	3.6	10	7	12.9	12.9
Developed camping	11	8.0	17	14	17.4	18.6
Picnicking	47	44.1	48	40	49.3	49.5
Sightseeing	37	35.2	46	41	56.1	54.3
Off-road driving	5	3.3	11	9	17.0	18.2
Hunting	3	2.8	12	15	13.8	14.3
Fishing	24	26.7	34	39	35.3	37.0
Bicycling	10	9.3	32	27	27.8	30.6
Horseback riding	5	6.1	9	8	9.5	10.4
Day hiking	5	2.8	14	9	20.1	19.7
Pleasure walking	34	28.4	53	49	62.2	62.7
Sailing	3	2.0	6	4	4.6	4.8
Canoeing/kayaking	3	2.2	9	10	11.1	10.4
Nature study	17	13.4	12	8	37.2	38.7
Non-pool swimming	34	32.2	32	30	36.7	37.7
Water skiing	5	5.9	9	10	11.1	10.4

[a] Source: Outdoor Recreation Survey, 1973.
[b] Source: Nationwide Recreation Survey, 1983.
[c] Source: National Survey for Recreation and the Environment, 1992.

Changes in forestry policies may reflect the upheaval from manufacturing to service-based industries. The multiple-use forest policies enacted during the 1970s, due in large part to increased public concern for the environment and additional time spent at recreational activities, has changed the management practices in the national forests. Forest managers now consider sustainable timber yields in respect to other potential forest uses. Table 9 identifies trends in recreational use during the past 2 decades.

The SCE Team also measured the economic demand for outdoor recreational activities. One analysis method relied on the net economic value or consumer surplus to determine the value of recreation in the southern Appalachians (SAMAB 1996d). The net economic value was determined using the travel cost method. This approach analyzed the expenditure of recreation participants for each trip for a given activity. The results are displayed in Table 10.

The economic base identified by the SCE Team indicates a strong connection between natural resource use and future employment opportunities. The value of protecting the recreational aspects in the area may provide the necessity of long-term economic growth strategies in association with ecosystem management strategies for the SAA region. Growth in the service industry occupations and decline in manufacturing and timber harvesting provide a foundation for implementing a comprehensive approach to ecosystem management practices which incorporate social, cultural, and economic indicators into a community driven strategy for natural resource management.

CONCLUDING REMARKS

Ecosystem "health and integrity" are descriptive parameters that are subjective, subject to change over time, and often difficult to quantify (Callicott 1995). Without adequate environmental data

TABLE 10
Net Economic Values for Selected Outdoor Recreation Main Activities in the Southern Appalachians

Activity	Sample size (number)	Net Economic Value	
		Per day (dollars)	Per trip (dollars)
Camping	756	6.05	27.62
Cold water fishing	147	27.02	54.04
Day hiking/trail walking	107	5.79	11.58
Family gathering/picnics	276	73.90	73.90
Pool/outdoor swimming	323	28.54	28.54
No main activity	295	35.52	71.03
Sail boating	153	11.79	22.37
Sightseeing/driving	294	14.04	14.04
Warm water fishing	614	10.06	20.12
Motorboating/skiing	248	33.89	33.89
River rafting/canoeing	156	126.04	252.08
Bass fishing[a]	903	32.40	
Deer hunting[a]	1068	40.29	
Wildlife watching[a]	943	29.00	

[a] These values were obtained from Waddington et al. using the 1991 U.S. Fish and Wildlife survey and the contingent valuation method.

Source: Public Areas Recreation Visitor Survey, 1985 to 1987.

and assessment criteria, a description of environmental "health" is generally guesswork. The SAA was a tremendous success story of federal, state, and private organizations working together. The approach provided information to all interested stakeholders while integrating a vast amount of scientific information. Final reports for each of the four resources teams and a summary report of the SAA are accessible on the internet home page of the Southern Appalachian Man and the Biosphere Program (SAMAB).

The information and reports produced by the SAA will be of interest to a diverse group of stakeholders. For example, we believe that the SAA project will provide necessary information for ecosystem management projects, such as baseline data for future gap analyses on the 31 rare communities and associated species identified for the SAA area to basic economic information needed by regional planners. The next step of the assessment will be for local stakeholders to use the information provided by the SAA to help build a sustainable future for the region.

ACKNOWLEDGMENTS

The Southern Appalachian Assessment is the product of hundreds of dedicated individuals, and we could not hope to begin to list the names and contributions of individual participants. Recognition, instead, should be given to the local community, both individual and interested parties, who supported and provided direct interaction in developing the scope of issues and concerns for the assessment. Acknowledgment of the local, state, and regional planning entities that provided much-needed data and support is also appropriate. Finally, recognition is given to all of the federal agencies which provided personnel and financial support to complete the assessment products.

Additional credit should be given to all SAA participants for their cooperative spirit and teamwork approach required to produce the assessment of such a remarkable ecosystem. The products resulting from this endeavor are applauded by all the authors as communities seek to identify cooperative methods to protect the integrity and health of our natural resources.

APPENDIX

Part I — Questions developed by the Terrestrial Team with local community input.

1. Based on available information and reference material, what plant and animal species occur in the SAA area, and what are their habitat associations?
2. What are the status, trends, and spatial distributions of terrestrial habitats, wildlife and plant populations for:
   A. Federal Threatened and Endangered species?
   B. Viability is of concern species?
   C. Rare communities?
   D. Wildlife species that are hunted, viewed, or photographed?
   E. Species for which there is high management and public interest?
   F. Species with special or demanding habitat needs?
3. What habitat types, habitat parameters, and management activities are important for maintaining viable populations of the species on the "short list" of plants and animals?
4. Based on our current knowledge of ecological land unit capabilities in the Region Appalachians, what are the conditions needed to:
   A. Recover Threatened and Endangered species?
   B. Conserve populations of viability concern species?
   C. Maintain existing species and community diversity?
   D. Provide suitable populations on national forests?
5. What changes or trends in forest vegetation are occurring in response to human-caused disturbances or natural processes?
6. What are the potential effects of the presence or absence of fire on forest health?
7. How is the health of the forest ecosystems being affected by native and exotic pests?
8. How are current and past management practices affecting the health and integrity of forest vegetation in the southern Appalachians?

Part II — Questions developed by the Aquatic Team with local community input.

1. What is known about the current status and apparent trends in water quality, aquatic habitat, and aquatic species within the southern Appalachian study area?
2. What management factors are important in maintaining aquatic habitat and water quality?
   A. What is the extent of riparian area?
   B. What is the composition of riparian area?
3. What laws and policies for the protection of water quality, streams, wetlands, and riparian areas are in place, and how do they affect aquatic resources, other resources, and human uses within the SAA?
4. What are the current and potential effects on aquatic resources from various activities?
   A. What species are at most risk?
   B. What habitat types are at most risk?
5. What are the status and apparent trends in water usage and supplies within the SAA, including water rights and uses on national forest system land?

Part III — Questions developed by the Atmospheric Team with community input.

1. What are the major air pollutants which could impact the southern Appalachians, and what areas receive the greatest exposure?
2. What is the current concentration of particulate matter in the air of the southern Appalachians?
3. How good is visibility in the southern Appalachians, and how does air pollution affect visibility?
4. To what extent are aquatic resources in the Southern Appalachian Assessment area being affected by acid deposition?
5. What impact does ground-level ozone have on forests?

Part IV — Questions developed by the Social, Cultural, Economic Team are identified under respective subteam categories.

Human Dimensions Subteam questions developed with community.

1. How has the social pattern of southern Appalachian communities changed over the past two decades?
2. How has the changing social pattern of the southern Appalachians affected management of natural resources in the region, and what future effects of social trends can we predict?
3. How might management of natural resources impact the economic and social status of local communities in the region, particularly communities near major tracts of public land?
4. To what extent have interests or publics outside of the southern Appalachians affected the status and management of the region's ecosystems and public land?
5. What are the important attitudes and values that southern Appalachian residents hold toward natural resources and ecosystem management?
6. With particular emphasis on tourism and extractive and other resource-dependent industries, what are the important economic trends in the southern Appalachians?
7. What are the status of and the priorities for management of private land by non-industrial owners in the region?

Timber Supply and Demand Subteam questions developed with community.

1. What are the supplies of and demands for wood products in the southern Appalachians?
2. Where and how does the wood-products industry depend on National Forest System timber in the southern Appalachians?
3. What are the relationships among timber production, employment, and income in the southern Appalachians?
4. What national forest land is tentatively suitable for timber production in the region and how can assessment findings be incorporated in further analysis of timber suitability?

Outdoor Recreation Subteam questions developed with community input.

1. What opportunities are there for public land in the southern Appalachians to provide unique or unsatisfied forest-related recreation demands?
2. How has the recreating public within traveling distance of public land changed in the past 10 years and what are the predicted future changes?
3. What are the supplies of and the demands for major types of recreation settings and activities within the area?

4. How is the changing social context within the southern Appalachians likely to affect future recreation demands on public lands?
5. How do recreation opportunities affect the lifestyle and local culture of the area?

Roadless and Wilderness Subteam questions developed with community.

1. Where are roadless areas on the national forests in relation to existing wildernesses on national forest and national park land and primitive area on state and private land in the southern Appalachians?
2. What is the Forest Service doing to maintain or enhance the health and integrity, including scientific, educational, scenic, or historic values, of roadless areas and wildernesses?
3. Are major population centers and the culture, background, beliefs, and values of the people affecting wilderness areas? If so, how?

REFERENCES

Air Resource Specialists, Inc. (ARS). 1995. Historical data summaries and permanent photographic archive for Shining Rock Wilderness (1988–1992). Fort Collins, CO.

Allen, T.F.H., R.V. O'Neill, and T.W. Hoekstra. 1987. Interlevel relations in ecological research and management: some working principles from hierarchy theory. *J. Appl. Syst. Anal.* 14:63–79.

Allen, T.F.H. and T.W. Hoekstra. 1992. *Toward a Unified Ecology.* Complexity in Ecological Systems Series. Columbia University Press: New York.

Anderson, R.L., P.E. Avers, T. King, et al. 1994. Dogwood Anthracnose and its Spread in the South. Protection Report R8-PR 26. Atlanta, Ga: U.S. Department of Agriculture, Forest Service, Southern Region. p. 10.

Bailey, R.G., P.E. Avers, and T. King, Eds. 1994. Ecoregions and Subregions of the United States (map). U.S. Geological Survey, Washington, D.C.

Bailey, R.G. 1995. Descriptions of the Ecoregions of the United States. Misc. Pub. 1391. USDA Forest Service, Washington, D.C.

Barling, R.D. and I.D. Moore. 1994. Role of buffer strips in management of waterway pollution: a review. *Environ. Manage.* 18:543–558.

Barry, R.G. 1981. *Mountain Weather and Climate.* Methuen: New York.

Boone, D.L. and G.H. Aplet. 1994. Sustaining biodiversity in the Southern Appalachians, in *The Living Landscape,* The Wilderness Society: Washington, D.C. Vol. 4.

Callicott. 1995.

Costanza, R., B.G Norton, and B.D. Haskell. 1992. *Ecosystem Health: New Goals for Environmental Management.* Island Press, Washington, D.C.

Cramer, L.A., J.J. Kennedy, R.S. Krannich, and T.M. Quigley. 1993. Changing forest service values and their implications for land management decisions affecting resource-dependent communities. *Rural Sociol.* 58(3):475–491 pp.

DeGraff, R.M. and J.H. Rappole. 1995. *Neotropical Migratory Birds: Natural History, Distribution, & Population Change.* Cornell University Press: New York.

Dissmeyer, G. 1994. Evaluating the Effectiveness of Forestry Best Management Practices in Meeting Water Quality Goals or Standards. Misc. Pub. 1520. Atlanta, GA: U.S. Department of Agriculture, Forest Service, Southern Region. 179 pp. [+Appendix].

Dolloff, C.A., D.G. Hanken, and G.H. Reeves. 1993. Basinwide Estimation of Habitat and Fish Populations in Streams. Gen. Tech. Rep. SE-83. Asheville, NC: U.S. Department of Agriculture, Forest Service, Southeastern Forest Experiment Station 25 pp.

Doppelt, B., M. Scurlock, C. Frissell, J. Karr. 1993. *Entering the Watershed: A New Approach to Save America's River Ecosystems.* Island Press, Washington, D.C.

Dowsett, S.E., R.L. Anderson, and W.H. Hoffard. 1992. Review of selected articles on ozone sensitivity and associated symptoms for plants commonly found in the forest environment. R8-TP 18. U.S. Department of Agriculture, Southern Region, Forest Pest Management, Atlanta, GA. 22 pp.

EPA (U.S. Environmental Protection Agency). 1986. Air quality criteria for ozone and other photochemical oxidants. Research Triangle Park, NC: U.S. Environmental Protection Agency, Office of Air Quality Planning and Standards; EPA-600/8-84/020a-e. Available from: NITIS, Springfield, VA. 306 pp.

EPA (U.S. Environmental Protection Agency). 1992. Securing Our Legacy: An EPA Progress Report 1989–1991. U.S. Environmental Protection Agency. EPA 175 R-92-001. Washington, D.C.

EPA (U.S. Environmental Protection Agency). 1993a. Geographic Targeting: Selected State Examples. 1993. U.S. Environmental Protection Agency, EPA-841-B-93-001, Washington, D.C.

EPA (U.S. Environmental Protection Agency). 1993b. Effects of the 1990 Clean Air Act Amendments on visibility in class I areas: An EPA report to Congress. Office of Air Quality Planning and Standards Reports EPA-452/R-93-014. Research Triangle Park, NC. 92 pp.

EPA (U.S. Environmental Protection Agency). 1994a. Toward a Place-Driven Approach: The Edgewater Consensus on an EPA Strategy for Ecosystem Protection. U.S. Environmental Protection Agency, Ecosystem Protection Workgroup, March 15, 1994 Draft.

EPA (U.S. Environmental Protection Agency). 1994b. Assessment Framework. Environmental Monitoring and Assessment Program. Environmental Protection Agency, EPA/620/R-94/016, Cincinnati, OH. 45 pp.

EPA (U.S. Environmental Protection Agency). 1994c. Landscape Monitoring and Assessment Research Plan — 1994. U.S. Environmental Protection Agency, EPA/620/R-94/009, Washington, D.C. 53 pp.

EPA (U.S. Environmental Protection Agency). 1995. National air pollutant emission trends, 1900–1994. U.S. Environmental Protection Agency, Office of Air Quality Planning and Standards; Research Triangle Park, NC; EPA-454/R-95-011. 201 pp.

EPA (U.S. Environmental Protection Agency). 1996. Proposed Guidelines for Ecological Risk Assessment. Public Review Draft. U.S. Environmental Protection Agency, Risk Assessment Forum.

Etnier, D.A. and W.C. Starnes. 1991. An analysis of Tennessee's jeopardized fish taxa. *J. Tenn. Acad. Sci.* 66(4):129–133.

Fausch, K.D., C.L. Hawkes, and M.G. Parsons. 1988. Models that Predict Standing Crop of Stream Fish from Habitat Variables: 1950–85. Gen. Tech. Rep. PNW-213. Portland, OR: U.S. Department of Agriculture, Forest Service, Pacific Northwest Research Station.

GAO (General Accounting Office). 1994. Ecosystem Management — Additional Actions Needed to Adequately Test a Promising Approach. United States General Accounting Office, B-256275, Washington, D.C. 87 pp.

Gawel, J.E., et al. 1996.

Gerritsen, J., M.T. Barbour, J.S. White, and J.E. Lathrop-Davis. 1995. Physical Habitat Assessment in the Mid-Atlantic Highlands Using the Rapid Assessment Approach: 1993 Mid-Atlantic Highlands Assessment Data Preliminary Report. Wheeling, WV. U.S. EPA Region 3.

Gonzalez, R.C. and R.E. Woods. 1992. *Digital Image Processing.* Addison-Wesley: Reading, MA.

Gregory, S.V., F.J. Swanson, W.A. McKee, and K.W. Cummins. 1991. An ecosystem perspective of riparian zones: focus on links between land and water. Bioscience 41:540–551.

Grumbine, R.E. 1994. What is ecosystem management? *Conserv. Biol.* 4(4):27–38.

Hankin, D.G. and G.H. Reeves. 1988. Estimating total fish abundance and total habitat area in small streams based on visual estimation methods. *Can. J. Fish. Aquat. Sci.* 45:834–844.

Harrelson, C.C., C.L. Rawlins, and J.P. Potyondy. 1994. Stream Channel Reference Sites: An Illustrated Guide to Field Technique. Gen. Tech. Rep. RM-245. Fort Collins, CO: U.S. Department of Agriculture, Forest Service, Rocky Mountain Forest and Range Experiment Station.

Herlichy, A.T., P.R. Kaufmann, M.R. Church, et al. 1993. The effects of acidic deposition on streams in the Appalachian Mountain and Piedmont Region of the Mid-Atlantic United States. *Water Resour. Res.* 29:2687–2701.

Hunsaker, C.T., D.A. Levine, S.P. Timmins, et al. 1993. Landscape Characterization for Assessing Regional Water Quality. In *Ecological Indicators.* Vol. 2. McKenzie, Hyatt, and McDonald, Eds. New York: Elsevier Applied Science: 997–1006.

Hunsaker, C.T. and D.A. Levine. 1995. Hierarchical approaches to the study of water quality in rivers. *BioScience.* 45:193–203.

Husar, R.B., J.B. Elkins, and W.E. Wilson. 1994. U.S. visibility trends, 1960–1992. In: Air and Waste Management Association 87th Annu. Meeting an exhibition. Cincinnati, OH. 18 pp.

Interagency Monitoring of Protected Visual Environments (IMPROVE). 1994. Visibility protection. Available from Cooperative Institute for Research in the Atmosphere, Colorado State University, Fort Collins, CO. 8 pp.

Johnson, D. and S. Lindberg, Eds. 1992. Atmospheric deposition and forest nutrient cycling. New York: Springer-Verlag. 707 pp.

Krannich, R.S. and A.E. Luloff. 1991. Problems of Resource Dependency in U.S. Rural Communities. In Gilg, A., Ed., *Progress in Rural Policy and Planning.* London: Belhaven Press. 5–18 pp.

Krummel, J.R., R.H. Gardner, G. Sugihara, R.V. O'Neill, and P.R. Coleman. 1987. Human Use Index and Fractals in Land-Space Analysis.

Laitos, J.G. and J.P. Tomain. 1991. *Energy and Natural Resources Law.* West Publishing: St. Paul, MN.

Lee, M.T., J-J. Kao, and Y. Ke. 1991. Integration of GIS, remote sensing, and digital elevation data for a nonpoint source pollution model. In Remote Sensing and GIS Applications to Nonpoint Source Planning. U.S. Environmental Protection Agency-Region 5, Northeastern Illinois Planning Commission, Chicago, IL: 79–85.

Lefohn, A.S. and V.C. Runeckles. 1987. Establishing a standard to protect vegetation-ozone exposure/dose consideration. *Atmospheric Environ.* 21: 561–568.

Lefohn, A.S., H.P. Knudsen, and D.S. Shadwick. 1995. Using kringing to estimate the 7-month (April-October) SUM06 and W126 indices for the southeastern United States from 1983 until 1993. Prepared for Tennessee Valley Authority. 76 pp.

Likens, G.E., C.T. Driscoll, and D.C. Buso. 1996. Long-term effects of acid rain: response and recovery of a forest ecosystem. *Science,* 272:244–245.

Lynch, D.D. and N.B. Dise. 1985. Sensitivity of Stream Basins in Shenandoah National Park to Acid Deposition. U.S. Department of Interior, Geological Survey. Richmond, Va.

Lynch, J., V. Bowersix, and J. Grimm. 1996. Trends in precipitation chemistry in the United States, 1983–94 — An analysis of the effects in 1995 of Phase 1 of the Clean Air Act Amendments of 1990, Title IV. U.S. Geological Survey Open-File-Report 96-0346.

Malchlis, G.E., J.E. Force, and R.G. Balice. 1990. Timber, Minerals, and Social Change: An Exploratory Test of Two Resource-Dependent Communities. Rural Sociology. 55(3):411–424.

Materna, J. 1984. Impact of atmospheric pollution on natural ecosystems. In M. Treshow, Ed., *Air Pollution and Plant Life,* John Wiley: New York. 397–416 pp.

Meador, M.R., C.R. Hupp, T.F. Cuffney, and M.E. Gurtz. 1993. Methods for characterizing stream habitat as part of the National Water-Quality Assessment Program. Open-file report 93-408. Raleigh, NC: U.S. Geological Survey. Landscape Patterns in a Disturbed Environment. *Oikos* 48:321–324.

McMinn, J.W. 1991. Biological Diversity Research: An Analysis. USDA Forest Service, GTR SE-71, Southeastern Forest Experiment Station, Asheville, NC.

Morse D. 1988. Air Quality in National Parks. USDI National Park Service, Air Quality Division, Natural Resources Programs, Nat. Res. Rep. 88-I. Denver, CO. pp 3. 1–13.

Morton, P.A. 1994. Charting a new course: national forests in the southern Appalachians. In *The Living Landscape.* The Wilderness Society: Washington, D.C. Vol. 5.

Muncy, J.A. 1986. A plan for revegetation completion of Tennessee's Copper Basin. Tennessee Valley Authority, Division of Land and Economic Resources: Norris, TN. 32 pp.

National Association of Conservation Districts. 1994. Riparian Ecosystems in the Humid U.S.: Functions, Values and Management. Washington, D.C. [Conference proceedings]. Atlanta, GA. March 15 to 18, 1993.

National Research Council. 1993. *Protecting Visibility in National Parks and Wilderness Areas.* Committee on Haze in National Parks and Wilderness Areas. National Academy of Science. National Academy Press: Washington, D.C. 446 pp.

North Carolina Department of Environment, Health, and Natural Resources. Division of Environmental Management, water quality section, biological assessment group. 1995. *Standard Operating Procedures: Biological Monitoring.* Raleigh, NC.

Norton, B. 1992. A New Paradigm for Environmental Management. Pp. 23–41 in R. Constanza et al., Eds. *Ecosystem Health.* Island Press, Washington, D.C.

Noss, R.P., E.T. LaRoe, and J.M. Scott. 1995. Endangered Ecosystems of the United States: A Preliminary Assessment of Loss and Degradation. U.S. Department of the Interior, National Biological Service. February 1995.

O'Neill, R.V., D.L. DeAngelis, J.B. Waide, and T.F.H. Allen. 1986. A Hierarchical Concept of Ecosystems. Princeton University Press: Princeton NJ.

O'Neill, R.V., J.R. Krummel, R.H. Gardner, G. Sugihara, B. Jackson, D.L. DeAngelis, B.T. Milne, M.G. Turner, B. Zygmunt, S. Christensen, V.H. Dale, and R.L. Graham. 1988. Indices of landscape pattern. *Landscape Ecol.* 1:153–162.

Overbay, J.C. 1992. Ecosystem management. In Proc. Natl. workshop: Taking an Ecol. Approach to Management. April 27 to 30, Salt Lake City. WO-WSA-3. Washington, D.C. USDA Forest Service, Watershed and Air Management.

Overton, W.S. 1977. A strategy of model construction. Pp. 50–73 in C.A.S. Hall and J.W. Day, Eds. *Ecosystem Modeling in Theory and Practice: An Introduction with Case Histories.* John Wiley & Sons, New York.

Pacific Meridian Resources, Inc. 1995.

Peine, J.D. and J.R. Renfro. 1988. Visitor use patterns at Great Smoky Mountains National Park. USDI National Park Service, Southeast Region Res./Resour. Manage. Rep. 90. Atlanta, GA 93 pp.

Peper, J.D., A.E. Grosz, T.H. Kress, et al. 1995. Acid deposition sensitivity map of the Southern Appalachian Assessment Area: Virginia, North Carolina, South Carolina, Tennessee, Georgia, and Alabama. Washington, D.C.: U.S. Geological Survey on-line digital data series, open file report: 95–810.

Plafkin, J.L., M.T. Barbour, K.D. Porter, et al., Eds. 1989. Rapid Bioassessment Protocols for use in Streams and Rivers: Benthic Macroinvertebrates and Fish. EPA/444/4-89-001. U.S. Environmental Protection Agency, Office of Water, Washington, D.C.

RAF (Risk Assessment Forum). 1992. Framework for Ecological Risk Assessment. U.S. Environmental Protection Agency, EPA/A630R92-001. Washington, D.C.

Rankin, E.T. 1995. Habitat Indices in Water Resource Quality Assessments. Pp. 181–208 in Davis, W.S. and T.P. Simon, Eds. *Biological Assessment and Criteria: Tools for Water Resource Planning and Decision Making.* Lewis Publishers: Boca Raton, FL.

Rapport, D.J., H.A. Regier, and T.C. Hutchinson. 1985. Ecosystem behavior under stress. *Am. Nat.* 125:617–640.

Ratsep, R. 1990. Potential Biological Variables for Monitoring the Effects of Pollution in Forest Ecosystems. p. 54. Nordic Council of Ministers and Ecology Institute, Lund University, Lund, Sweden.

REO (Regional Ecosystem Office). 1995. Ecosystem Analysis at the Watershed Scale (Version 2.1). The Revised Federal Guide for Watershed Analysis. Portland, OR. p. 187.

Regier, H.A. 1993. The notion of natural and cultural integrity. Pp. 3–18 in Woodley, S., J. Kay, and G. Francis, Eds., *Ecological Integrity and the Management of Ecosystems.* St. Lucie Press, Boca Raton, FL.

Richter, B.D., J.V. Baumgartner, J. Powell. 1995. A method for assessing hydrologic alteration within ecosystems. *Conserv. Biol.,* In Press.

Riitters, K.H., B.E. Law, R.C. Kucera, A.L. Gallant, R.L. DeVelice, and C.J. Palmer. 1992. A selection of forest indicators for monitoring. *Environ. Monitoring and Assessment* 20:21–33.

Riitters, K.H., R.V. O'Neill, and K.B. Jones. 1997. Assessing habitat suitability at multiple scales: a landscape-level approach. *Biol. Conserv.* 81:191–202.

Rosgen, D.L. 1994. A classification of natural rivers. Catena 22:169–199.

Roth, N.E., J.D. Allan, and D.L. Erickson, 1995. Landscape influences on stream biotic integrity assessed at multiple spatial scales. *Landscape Ecol.*

SAIC (Science Applications International Corporation). In Press. A Portrait of the Natural Resource Dependency of Communities in the Southern Appalachian Area. EPA Contract No.: 68-W5-0054, WA No. 1–4.

SAMAB (Southern Appalachian Man and the Biosphere). 1988. Interagency cooperative agreement for the establishment and operation of the Southern Appalachian Man and the Biosphere Cooperative. Gatlinburg, TN.

SAMAB (Southern Appalachian Man and the Biosphere). 1996a. The Southern Appalachian Assessment Summary Report. Report 1 of 5. U.S. Department of Agriculture, Forest Service, Southern Region. Atlanta.

SAMAB (Southern Appalachian Man and the Biosphere). 1996b. The Southern Appalachian Assessment Aquatics Technical Report. Report 2 of 5. U.S. Department of Agriculture, Forest Service, Southern Region. Atlanta.

SAMAB (Southern Appalachian Man and the Biosphere). 1996c. The Southern Appalachian Assessment Atmospheric Technical Report. Report 3 of 5. U.S. Department of Agriculture, Forest Service, Southern Region. Atlanta.

SAMAB (Southern Appalachian Man and the Biosphere). 1996d. The Southern Appalachian Assessment Social-Cultural-Economic Technical Report. Report 4 of 5. U.S. Department of Agriculture, Forest Service, Southern Region. Atlanta.

SAMAB (Southern Appalachian Man and the Biosphere). 1996e. The Southern Appalachian Assessment Terrestrial Technical Report. Report 5 of 5. U.S. Department of Agriculture, Forest Service, Southern Region. Atlanta.

Seaber, P.R., F.P. Kapinos, and G.L. Knapp, 1984. Hydrologic Units Maps. p. 63. Water-Supply Paper #2294, U.S. Geological Survey, Denver, CO.

Smith, W.H. 1984. Ecosystem pathology: a new perspective for phytopathology. *For. Ecol. Manage.* 9:193–219.

Sisler, J.F., D. Huffman, and D.A. Latimer. 1993. Spatial and temporal patterns and the chemical composition of haze in the United States — and analysis of data from the IMPROVE network, 1988–1991. Report submitted to the National Park Service (William C. Malm, Principal Investigator) and the U.S. Environmental Protection Agency (Marc L. Pitchford, Principal Investigator). 141 pp.

Steedman, R.J. 1988. Modification and assessment of an index of biotic integrity to quantify stream quality in Southern Ontario. *Can. J. Fish Aquat. Sci.* 45:492–501.

Suter, G.W. 1990. Endpoints for Regional Ecological Risk Assessments. *Environ. Manage.* 14(1):9–23.

Suter, G.W. and L.W. Branthouse. 1993. *Ecological Risk Assessment.* Lewis Publishers: Boca Raton, FL.

Tingey, D.T., W.E. Hogsett, and E.H. Lee, et al. 1991. An evaluation of various alternative ambient ozone standards based on crop yield loss data. Pp. 272–288 in Berglund, R., D. Lawson, and D. McKee, Eds., Transaction of the Tropospheric Ozone and Environment Specialty Conference. Air and Waste Management Association. Pittsburgh, PA.

Treshow, M. 1984. Diagnosis of air pollution effects and mimicking symptoms. Pp. 97–112 in M. Treshow, Ed., *Air Pollution and Plant Life,* John Wiley, New York.

Trijonis, J.C., W.C., Malm, and M. Pitchford, et al. 1991. Visibility: existing and historical conditions — causes and effects. In: Acidic Precipitation Assessment Program, Acidic Deposition: State of Science and Technology. Vol. III, Report 24. 129 pp.

Troyer, M.E. and M.S. Brody. 1994. Managing Ecological Risks at EPA: Issues and Recommendations for Progress. U.S. Environmental Protection Agency. EPA/600/R-94/183.

Turner, M.G. 1989. Landscape ecology: the effect of pattern on process. *Annu. Rev. Ecol. Syst.* 20:171–197.

USMAB Directorate on Biosphere Reserves. 1989. Guidelines for the Selection and Management of Biosphere Reserves (draft). Available from USMAB Secretariat, Department of State, Washington, D.C.

Waring, R.H. 1984. Imbalanced Ecosystems — Assessments and Consequences. Pp. 112–117 in G.I. Agren, Ed., State and Change of Forest Ecosystems — Indicators in Current Research, Rep. Num. 13, Swedish University of Agriculture and Science, Department of Ecology and Environmental Resources.

Weber, B.A., F.N. Castle, and A.L. Shiver. 1988. Performance of Natural Resource Industries. In Brown, D., Reid, N, Bluestone, H., McGranahan, D., and Mazie, S., Eds., Rural Economic Development in the 1980s: Prospects for the Future. Washington, D.C.: U.S. Department of Agriculture, Economic Research Service.

Weins, J.A., N.C. Stenseth, B. Van Horne, and R.A. Ims. 1993. Ecological mechanisms and landscape ecology. *Oikos* 66:369–380.

Whittaker, R.H. 1972. Evolution and measurement of species diversity. *Taxon* 21:213–251.

Wickham, J.D., R.V. O'Neill, K.H. Riitters, T.G. Wade, and K.B. Jones. 1997. Sensitivity of selected landscape pattern metrics to land-cover misclassification and differences in land-cover composition. *Photogrammetric Eng. Remote Sensing* 63:397–402.

Winner, W.E., A.S. Lefohn, I.S. Cotter, et al. 1989. Plant response to elevation gradients of O_3 exposures in Virginia. Proceedings.

Zucker, L.A. and D.A. White. 1996. Spatial modeling of aquatic biocriteria relative to riparian and upland characteristics. In proceedings, Watershed 96. June 8 to 12, 1996. Baltimore, MD.

8 Environmental Monitoring

Elizabeth R. Smith, Charles R. Parker, and John D. Peine

CONTENTS

Introduction 167
   The Role of the Ecosystem Manager in Environmental Monitoring......................................168
   A Regional Framework 170
   Scale in Environmental Monitoring..171
   The History of Environmental Monitoring in Southern Appalachia 173
Overview of Current Monitoring in the Region..174
   Intensive-Site Monitoring..174
   Monitoring within Administrative Areas 174
   Regional Scale Monitoring 175
Specific Environmental Monitoring Programs in Southern Appalachia......................................176
   The Great Smoky Mountains National Park 176
   The SAMAB Forest Health Monitoring Demonstration..180
   Risk Assessment Report Card for Ecosystem Managers 182
SAMAB's Role in Future Environmental Monitoring...183
References 184

INTRODUCTION

Despite years of observation, our understanding of both ecosystem process and the effects of environmental perturbation on ecosystems is far from complete. Recognition of the many interactions that influence system dynamics has resulted in the development of increasingly sophisticated monitoring systems that often lack the temporal or spatial resolution necessary to capture the range of variability required to predict response to various management options. Likewise, the introduction of new environmental stresses confounds comprehension of cause–effect. Because of this, ecosystem management is a difficult concept to define in terms of particular steps towards a specific result. Adoption of an ecosystem approach to resource management will require continued environmental monitoring to increase and modify current understanding of ecosystem function and thus the impacts of management decisions. Future management can then be adapted to enhance the benefits derived from the system. Because of this, environmental monitoring is recognized as a critical tool for ecosystem management (Slocombe 1993; Grumbine 1994). Through a careful review of what has and has not worked in the past, combined with careful design towards desired end-products that includes establishing linkages across scales, future monitoring should further increase our understanding of human influences on ecosystem dynamics as well as provide guidance for improved resource management.

Hellawell (1991) defines environmental monitoring as "intermittent surveillance carried out in order to ascertain the extent of compliance with a predetermined standard or the degree of deviation from an expected norm." This differs from a survey in that a survey is done without any preconception of what the findings will be. Monitoring is typically established with one of three objectives: (1) determining the degree of compliance to environmental regulations, e.g., pollution monitoring; (2) testing an hypothesis to establish cause–effect in order to define regulatory standards; and (3) trend monitoring where large-scale changes are anticipated as a result of multiple activities (e.g., regional air pollution impacts and climate change) (Spellerburg 1991). Monitoring to enhance ecosystem management adds another objective, that of evaluating and predicting the effectiveness of prescribed management options through improved understanding of ecosystem function and response.

THE ROLE OF THE ECOSYSTEM MANAGER IN ENVIRONMENTAL MONITORING

Ecosystem monitoring is one of the most difficult responsibilities of the ecosystem manager. The purpose of such an initiative is rarely specific and is therefore obtusely defined. Alternative design and implementation strategies are endless. The potential for bias toward vested scientific interests at the expense of a balanced perspective is always present. The sustainability of the program over the long term is at best problematic. The ecosystem manager should provide informed leadership in order to insure that the monitoring program being conducted is clearly relevant, well designed, engages the best scientific perspectives, leverages other research and monitoring initiatives, periodically synthesizes data and reports on findings, and insures sustainability over the long term. In order to accomplish these goals, the following key elements should be taken into consideration.

Defining Program Objectives.   Monitoring programs may be based on providing answers to specific management questions, describing basic ecosystem processes and functions, and/or detecting impacts from prevalent perturbations such as those discussed elsewhere in this volume (pests and pathogens, exotic species, species reintroduction, boundary effects, air pollution and/or climate change). Some monitoring programs are designed to utilize the accumulating database to build models to predict ecosystem response to perturbations such as those listed. The more specific the objectives, the higher the probability for efficiency in operation of the program. On the other hand, the more generic the purpose, the more likely that basic trends in ecosystem dynamics will be portrayed in such a way that unforeseen influences may be detected.

Whatever the rationale for the monitoring strategy, the informed ecosystem manager should be very aware of the implications of the chosen path, what opportunities are afforded by the strategy chosen, and what is foregone concerning characterization of ecosystem and related processes and the ability to characterize and predict the impacts of perturbations.

Placing the Program within a Regional Context.   The ecosystem manager is inevitably in contact with a wide network of institutions, some of which are likely to be engaged in ecosystem monitoring activity. One of the manager's most important roles in ecosystem monitoring is to ensure that those involved in the monitoring activities under her/his jurisdiction are fully aware and engaged with those pursuing relevant monitoring activity in the region.

The other aspect of this concern is to always place specific monitoring activities in a broader regional context. External influences are invariably of concern as are the regional context of the natural ecosystem of primary interest. If all available resources for monitoring are restricted to measurement within the boundaries of property directly under the manager's control, the dynamics of external influences will not be understood.

Conduct of Resource/Risk Assessments.   An understanding of ecosystem dynamics and risks provides key insight as to how to best utilize the invariably limited funds available for resource monitoring. As described by Peine and Berish in Chapter 19 of this volume, an instructive risk assessment can be readily conducted by assembling experts in various aspects of the ecosystem of concern and its utilization. In the case of the southern Appalachians, the ecosystem at greatest risk is the high elevation spruce–fir forest. Gypsy moth is the pest of greatest concern. Air pollution (acid rain, ozone, and visibility loss) is a pervasive problem. Encroachment along the boundary of public lands is a major issue as well. Climate change in the montane landscape is of long-term concern. These dynamics alone provide important insight concerning design of a monitoring system in the region.

Designing the Program. As the objectives of the monitoring program become resolved, innumerable choices must be made concerning the specifics of program design. Program elements might include paired research watersheds; sensitive ecosystems, ecotones, and species potentially at risk; and indicators of ecosystem processes and functions such as nutrient cycling; and indicator species for population dynamics and productivity. Strategies of what to monitor and at what scale should be primarily the purview of the scientists engaged to design the program, but the manager should be involved as well. Some elements of the program may have educational and/or political relevance beyond the scientific interests of the technocrats. It is important for the manager to judge whether the scope of the work proposed can be sustained in the long run, given the inevitable fluctuations likely to occur in resource allocation for the program. The more the manager becomes a stakeholder in the design process, the more likely the individual will defend the program as perspectives and priorities change in overall property management.

Forming Scientific Partnerships. The manager should realize that there is a fine line between research and resource monitoring and that there will always be more questions about ecosystem dynamics than resources to study them. Therefore, the ecosystem manager needs to provide leadership in cultivating partnerships with research institutions like national laboratories and universities to participate in the monitoring activities bringing additional expertise and resources to the program. Some key initiatives that managers can take to attract researchers is to designate paired research watersheds where collaborative research and monitoring activity are encouraged to take place and permanent plots and transects are carefully monumented; provide transportation, housing and laboratory space on site; and maintain well documented and easily accessible databases on the program.

Peer-Reviewed Protocols. If the data from the program are to be accepted by entities outside the agency sponsoring the monitoring program, rigorous peer review of the specific procedural protocols must be conducted. The manager should be sure that this base is covered.

Protect the Resources from Adverse Impacts from the Monitoring Activity. This is a less than obvious responsibility of the ecosystem manager. Unregulated research and resource monitoring activity can be quite destructive, particularly on a microscale within research watersheds, where many people are likely to be engaged in a large variety of activities with little familiarity with the big picture. One person's sampling procedure may be another's disturbance adversely impacting the opportunity to accurately characterize trends in resource relationships being monitored. Vehicular and foot traffic on steep mountain slopes, snow fields, alpine meadows, or on desert varnish can be very destructive. Permits for monitoring should be required to insure regulation of activity to minimize disturbance.

Monumenting Permanent Plots and Transects. Another seemingly mundane activity that befalls the ecosystem manager is to insure that all permanent plots and transects are monumented utilizing standardized protocols. Time and again, those "permanent plots" that a scientist installs fully intending to revisit get lost as the individual's priorities and interests change, leaving the manager literally with a lost investment in the ability to measure trends.

Data Management. This element is just as critical as the actual data collection itself. Data documentation and appropriate storage are very important. Protocols for quality assurance and quality control must be run on the data as well. Utilizing a geographic information system greatly facilitates access and interpretation. Spatial portrayal enhances information utility in circumstances when a decision must be quickly made such as whether or not to let a naturally set fire burn.

Integrative analysis of monitoring trends is greatly enhanced by a comprehensive data management system. The manager needs to insist that adequate funds are available for such activity.

Periodic Synthesis and Reporting of Findings.   Sometimes the obvious is missed. All the energy going into data collection and the processing of field samples and entering data becomes so all consuming that little time and energy is left to perform the task of conducting an integrated analysis of all elements of the program to define baseline conditions and trends in population dynamics and ecosystem processes. This task needs to be conducted in a timely fashion on a scheduled periodic basis to insure that the manager gets the information flow needed.

A REGIONAL FRAMEWORK

The first step in designing a regional framework for environmental monitoring is to carefully evaluate what indicators to use to answer the questions in which resource managers are interested. One way to do this is to use a conceptual model (see Figure 1). The conceptual model helps to identify what information is needed from the monitoring system by specifying what components of the ecosystem are of value to various stakeholders, then identifying the system inputs that affect these components. As system inputs are identified, specific parameters to indicate trends in these components are developed. Then, depending on regional variability in both time and space for these measures, needed sampling intensity and remeasurement frequency can be determined. Clear identification of information necessary to predict future ecosystem response through a conceptual model can justify resources and commitment needed to ensure success of the effort. Commitment to needed remeasurements is critical; many monitoring systems have failed to produce critical information on ecosystem dynamics simply because measurements were not continued for a long enough period of time to capture trends.

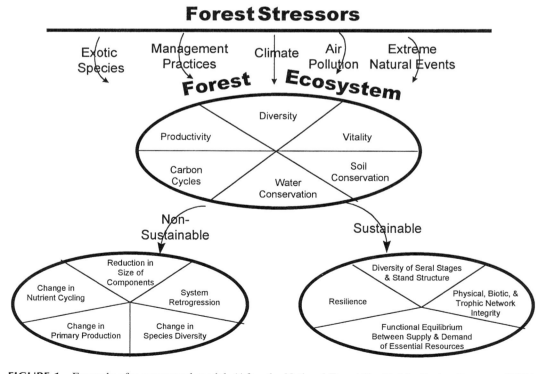

FIGURE 1   Example of a conceptual model. (After the National Forest Health Monitoring Program, 1996.)

Next, one of the most important steps in formulating a regional monitoring framework is to look at existing monitoring programs, regardless of which agency or institution maintains them, to capitalize on information that is already available. Further identification and prioritization of gaps in understanding and data through a regional resource assessment is critical if new monitoring systems are to maximize usefulness and efficiency. Input from other interested parties, including other agencies, institutions, and the public, will ensure that valued ecosystem components are included in the monitoring system. The framework should also clearly identify intended uses of the data, as well as links and interactions across scales. Methodologies, from sampling, to analysis and reporting, should be documented. Data management should be anticipated and maintained throughout the process. Geographic information systems should be the organizing tool, with a network of management and cross reference systems.

The framework and design of the monitoring system should maintain a degree of flexibility so that new methods and technologies can be incorporated when they become available. Once designed, the framework should also be reviewed for scientific validity and to ensure that stated goals can be met.

Once underway, data collected within the system should be periodically synthesized and integrated towards an assessment of current condition, with results communicated with other managers, scientists, and the public.

SCALE IN ENVIRONMENTAL MONITORING

Within any ecosystem, there exist communities or species that function on different spatial or temporal scales and thus have different requirements for monitoring. A holistic evaluation of an ecosystem requires assessment across these scales, recognizing interactions that occur from one scale to the next. A monitoring system designed for such a purpose should include both extensive monitoring, to establish range and extent of condition, and intensive monitoring, to monitor rare or sensitive indicator species, or to identify process. Both types of monitoring are important, and in ecosystem assessment, neither will suffice alone. However, establishment of a nested network of both intensive and extensive monitoring systems may not be feasible because of limited resources. Through interagency cooperation and careful planning of a regional monitoring framework, both types of monitoring information can be made available to all those involved in ecosystem management.

Both the temporal and spatial scales used in environmental monitoring depend on the question being posed (Figure 2). The size of the population or area being monitored and the frequency of disturbance that effects change are typically determinants. Practically, established monitoring programs reflect a trade-off determined by balancing the amount of information needed with the available financial support. Thus, studies that seek to obtain a great deal of information on physiological process generally are limited to a small spatial scale. These studies have typically been established in what is considered a "regionally representative" area or within a sensitive ecosystem. Unfortunately, because of their limited spatial scale, these studies cannot represent the range of natural variability, limiting their usefulness when small variations in ecosystem characteristics effect large changes in response.

At the other extreme of the spatial scale is extensive monitoring across a region. This type of monitoring attempts to establish extent of observed symptoms or characteristics, and may serve as a way of identifying new environmental problems before large-scale damage occurs. Additionally, extensive monitoring may provide insight into existing problems such that new cause–effect hypotheses are generated that can be tested on a smaller scale. However, because the objectives of this type of monitoring include representing a large spatial area or population, the number of variables sampled is necessarily limited due to cost. Thus, choice of parameters examined is critical and must effectively represent the ecosystem changes anticipated.

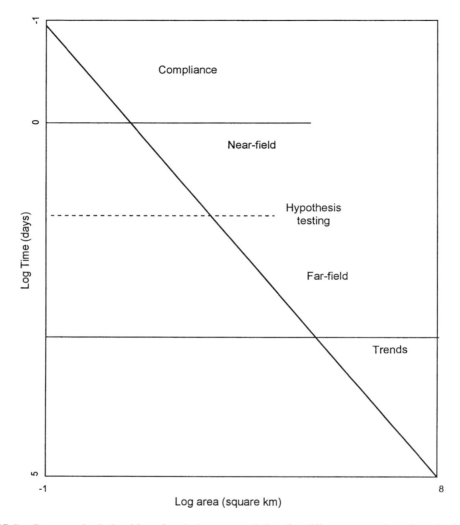

FIGURE 2 Conceptual relationships of scale in space and time for different categories of monitoring. The temporal scale represents the duration of sampling (with respect to each perturbation), and the spatial scale represents the potential sampling area. (After Spellerberg, I.F., *Monitoring Ecological Change,* Cambridge University Press, Cambridge, 1991.)

Environmental monitoring in the southern Appalachian region might be divided into three overlapping spatial classes. The first would include intensive, small spatial-scale monitoring aimed at identifying ecological process or a cause–effect relationship, or alternatively, monitoring a very small, rare population. The second, overlapping with both the first and third, would be environmental monitoring within some administrative jurisdiction. The third is regional scale monitoring that crosses administrative boundaries, the least-used type of monitoring in the region.

Temporal scale is another important consideration when establishing any type of environmental monitoring. While it is rare that short-term monitoring at any spatial scale will effectively address the objectives of the study or project, long-term support for monitoring is difficult to obtain. Yearly variations in weather patterns, for example, cause large changes in patterns of pollution deposition, insect and disease spread, and overall vigor of tree species, yet similarly frequent fluctuations in political support often result in termination of monitoring projects before sufficient information has been collected. Spellerburg (1991) identifies four classes where monitoring should be long term: (1) where slow processes, such as succession are monitored; (2) where rare events need to be identified; (3) where processes studied are subtle; and (4) where complex phenomena are being monitored. Within the southern Appalachian region, several long-term monitoring projects have provided crucial understanding of ecological processes. Unfortunately, there are also examples of monitoring that has been discontinued that might have offered critical insight into mitigation of current environmental problems.

Under an ecosystem management approach, consideration of both spatial and temporal scales used in environmental monitoring gains in importance. With our acknowledgment that ecosystem processes occur at many scales comes a realization that no one scale is correct for environmental monitoring. True ecosystem management incorporates a hierarchy of both spatial and temporal scales as we try to better understand changes that occur in response to management actions. Similarly, boundaries for areas to be monitored should follow ecological lines, a community, watershed, or ecoregion, for example.

THE HISTORY OF ENVIRONMENTAL MONITORING IN SOUTHERN APPALACHIA

Environmental monitoring in southern Appalachia has changed significantly towards development of today's systems designed to address concerns of ecosystem viability and resiliency. Before the 20th century, virtually no monitoring of any type was done within the region. A view of limitless resources resulted in exploitation of forest and soil resources and pollution and modification of water resources. In the 1930s the region was in a state of extensive degradation and federal legislation finally recognized the need for both protection (creation of the Great Smoky Mountains National Park and National Forest lands) and reforestation (efforts of the Tennessee Valley Authority [TVA] and the Civilian Conservation Corps). The first long-term research by the U.S. Forest Service (USFS) on the effects of forest management on ecosystem properties was initiated in 1937 at Coweeta, NC, identifying and demonstrating improvements in management techniques.

As the region began to recover, the first regional efforts at monitoring changes in forest status were initiated by TVA and the USFS using temporary plots. Later in the late 1950s, both agencies began using Continuous Forest Inventory (CFI) programs (USFS Forest Inventory and Analysis [FIA] and TVA County-Wide Surveys) to monitor forest growth by measuring permanent inventory plots periodically. Similarly, TVA, charged with improved navigation and flood control of the Tennessee River, began monitoring physical characteristics throughout the watershed.

In the 1980s, concern over the potential damage being wrought by air pollution on forests, lakes, and river systems of the nation resulted in a 10-year research program, the National Acid Precipitation Assessment Program* (NAPAP). Many small-scale studies designed to quantify dose–response were established. NAPAP also reviewed data from large-scale surveys such as FIA, but found little evidence of changes in forest productivity, largely because these systems were designed to monitor changes in timber volume rather than forest health. Despite little evidence of damage, concern remained that subtle changes were occurring on a regional scale. However, past monitoring did not provide enough information to identify when changes were outside the range of natural variability. NAPAP served to point out how little we know about the health of our ecosystems as a whole, especially in quantifying range and extent of ecosystems characteristics and concerns. Today, environmental monitoring systems typically include monitoring at many scales as essential components.

Two current case studies illustrating different types of monitoring useful for ecosystem management are presented later in this chapter. Environmental monitoring in the Great Smoky Mountains National

---

* The National Acid Precipitation Assessment Program, created as a result of the Acid Precipitation Act of 1980, was a 10-year research and assessment program to examine the causes, effects, and control options associated with acidic deposition. The program coordinates the work of federal agencies that compose the Acid Deposition Task Force and integrates that work with state, private sector, and international organizations. The Program has been reauthorized by the 1990 Clean Air Act Amendments.

Park is conducted to determine degree of compliance with mandated guidelines for National Park Service protection of significant resources as well as to determine the effectiveness of management actions. Also, watershed studies in the Park are used to determine biological response to pollutant input as information for future environmental protection legislation. The Southern Appalachian Man and the Biosphere (SAMAB) Forest Health Monitoring Demonstration is one of the first examples of regional scale monitoring across political boundaries to establish range and extent of current and future forest health problems.

OVERVIEW OF CURRENT MONITORING IN THE REGION

In the southern Appalachians, many kinds of monitoring are being undertaken. These efforts generally have been implemented independently without regional coordination, and therefore lack cohesiveness. Despite this, important ecological insights have been obtained from this work. Below, we review the nature and scope of monitoring in the region, and the agencies that have undertaken the work.

INTENSIVE-SITE MONITORING

There are many examples of process-level monitoring within southern Appalachia. Much of our current understanding of ecological processes is based on fundamental long-term studies undertaken at Coweeta Hydrologic Laboratory, the Oak Ridge National Environmental Research Park, and Great Smoky Mountains National Park, each designated as one of the Southern Appalachian Man And the Biosphere's (SAMAB) Man and the Biosphere core units (see Chapter 5). Monitoring in the Great Smoky Mountains National Park includes both intensive and extensive componentsprojects, with the objectives of both (1) identifying processes and establishing their spatial extent, and (2) serving as an early-warning system for this administrative area. Monitoring in the Park is described in detail in the following section.

At Coweeta Hydrologic Laboratory, there has been continuous monitoring of stream flow, climate, and forest growth since 1934, thus providing one of the longest ecological data records within the region. Coweeta's mission is "to evaluate, explain, and predict how water, soil, and forest resources respond to management practices, natural disturbances, and the atmospheric environment; and to identify practices that mitigate impacts on these watershed resources" (Swank and Crossley 1986). Research at Coweeta combines short-term (5 years or less) studies with long-term studies of response to natural and anthropogenic disturbance on a watershed level. Coweeta was designated as a part of the Long-Term Ecological Research (LTER) program in 1980.

Walker Branch Watershed, part of the Oak Ridge National Environmental Research Park, was established in 1967 to provide an intensive research site aimed at better understanding changes in ecological process in response to environmental stress. Sponsored by the U.S. Department of Energy, Walker Branch has three primary objectives: (1) providing base-line values for unpolluted natural waters; (2) contributing to our knowledge of cycling and loss of chemical elements in natural ecosystems; and (3) enabling the construction of models for predicting the effects of man's activities on the landscape (Van Hook 1989). Studies of varying length have focused on ecological process, identifying trace contaminants, forest micrometeorology, atmospheric deposition, and detailed acidic deposition effects on canopy processes and soil chemistry. Current work at Walker Branch is examining the possible effects of changes in weather patterns brought about by climate change.

MONITORING WITHIN ADMINISTRATIVE AREAS

Environmental monitoring within administrative areas typically is done to ensure that the responsible agency is complying with federally mandated environmental protection, although data collected within these programs have often been used for larger-scale analyses because of their regional representativeness or historical record. Examples of monitoring within administrative areas include systems established in the Great Smoky Mountains National Park (see following section) as well as other National Park Service lands, the Oak Ridge Reservation, the Tennessee Valley Watershed, and the U.S. Forest Service National Forest System.

Monitoring on the Oak Ridge Reservation (which includes the National Environmental Research Park) is done primarily because work carried out to meet the changing defense, energy, and environmental needs of the U.S. has frequently produced hazardous wastes, requiring documented environmental monitoring. In 1980, 13,590 acres of the reservation were designated as the Oak Ridge National Environmental Research Park. The purpose of this park is to provide opportunities for both education and research while demonstrating compatibility between energy technology development and a quality environment (Oak Ridge National Laboratory 1994). A database of wetland and plant information and a plant and animal reference collection are maintained.

The Tennessee Valley Authority, originally mandated to provide sustainable development opportunities for the Tennessee River and surrounding region, has carried out environmental monitoring of air, water, and land resources since its creation in 1933. Periodic reinterpretation of the agency's mandate has resulted in changes in level and type of monitoring, but much valuable information on the region's resources has been collected. Monitoring of air pollutants and their effects on the region's vegetation have been done since the 1950s, when TVA first began power generation using coal-fired steam plants. These studies have focused on gradient effects of pollutants on sensitive species, changes in ecosystem dynamics as a result of specific pollutant inputs, and the behavior of pollutants when they are released into the atmosphere. Because part of TVA's original mission was reforestation and demonstration of effective land management, monitoring on TVA lands has included the objective of evaluating changes in the ecosystem as a result of specific land management activities. Monitoring of the Tennessee River system is some of the most intensive that TVA does currently, including physical and chemical characteristics of the river and its biological populations. Watershed monitoring varies in intensity throughout the system, because of both limitations in funding and the specific needs of various locations. Much of the environmental monitoring that occurs within the river system uses a "representative watershed" approach to obtain estimates of parameters for the entire region.

The U.S. Forest Service similarly conducts environmental monitoring within administrative units such as national forests. Again, the objective here is to obtain information on forest development so that effective management decisions can be made. This monitoring includes gathering information on forest tree species and stand structure, wildlife habitat and populations, and water quality. Unfortunately sampling design is usually determined by the administrative unit, so there is little consistency from unit to unit. Generally, forest monitoring data applicable to ecosystem management are available in the USFS Continuous Inventory of Stand Conditions (CISC).

REGIONAL SCALE MONITORING

Environmental monitoring on a regional scale within southern Appalachia is probably the least utilized, due to a lack of support for cross-jurisdictional work requiring a great deal of effort and cooperation. Examples include regional air pollution monitoring such as the Southern Oxidants Study (SOS Strategic Planning Committee 1992) and the National Atmospheric Deposition Program (Eaton et al. 1987). Although the objective here is to obtain estimates of atmospheric deposition across the region, monitoring equipment is so costly that only a few sites are actually located within the southern Appalachian region. While data from these sites indicate the magnitude of variation within the region, the number of sites is insufficient to determine the effect of specific site characteristics on values obtained, thus limiting estimates of regional deposition.

The other major monitoring example is the Environmental Monitoring and Assessment Program (EMAP) — Forest Health Monitoring (FHM) (FHM National Office 1996) that initially aspired to provide statistics on environmental health on a national level once the program was fully established.

Otherwise, there are surveys done on a regional scale, including the U.S. Forest Service's Forest Inventory and Analysis, and TVA's Forest Industry Survey. The objectives for these programs have not been aimed at collecting information on environmental condition; however data from these programs have been used successfully to augment environmental studies to a limited degree.

SPECIFIC ENVIRONMENTAL MONITORING PROGRAMS IN SOUTHERN APPALACHIA

THE GREAT SMOKY MOUNTAINS NATIONAL PARK

The National Park Service began a nation-wide natural resource inventory and monitoring program in the late 1980s (National Park Service 1992). The national program recognized at the outset that no one correct way to conduct environmental monitoring exists, and that the needs of parks vary widely across the nation. Therefore, a system of prototype monitoring parks was envisioned in which a wide range of monitoring approaches would be developed and tested. The most successful concepts then would be adopted for use in other parks. Ten prototype parks or clusters of parks are planned. The Great Smoky Mountains National Park (hereafter referred to as the Park) was selected as one of the first four prototypes in 1991. The manner in which the Park's monitoring program was established may be instructive for others wishing to set up long-term environmental monitoring programs. Although monitoring had occurred in the Park for a number of years (Peine et al. 1985), it had always been on an *ad hoc* basis. There was never a commitment from Park management, financial or otherwise, to sustain any of these monitoring efforts. However, when the national Inventory and Monitoring program began to solicit proposals for the prototype program, the Park quickly responded. An initial proposal was developed that was judged to be comprehensive, defensible, and responsive to the needs of Park management. It included many aspects of the prior monitoring efforts, as well as many new elements. It was estimated to cost nearly $1,000,000 annually. Budget restrictions forced several redrafts, with each draft featuring the elimination of different elements of the initial program. Each component that was dropped, i.e., salamanders, small mammals, birds, lichens, and paired research watersheds, resulted in a weakening of the overall cohesiveness of the original concept. Finally, a proposal was submitted that was less than half the cost of the first draft. The conceptual plan had been considerably altered in the process, numerous ideas had been debated and dropped, and many important elements had been eliminated. Therefore, the final proposal represented a compromise between what was perceived by the scientists and resource managers as an excellent, comprehensive program, and what was perceived by program managers as a proposal that stood a reasonable chance of being funded.

The Park is the largest (211,029 ha) protected area in the eastern deciduous forest biome. It is renowned as a center of temperate biodiversity of many groups of organisms, including flowering plants, millipedes, spiders, insects, and amphibians. It is for these reasons, along with the historic cultural resources present, that it has been designated an International Biosphere Reserve and a World Heritage Site, and has been selected as one of the first four prototype parks for the National Park Service's Inventory and Monitoring program. The Park faces many threats to its integrity. Among the most immediate are the effects of air pollution, the introduction of exotic species, and habitat fragmentation on the periphery. This has led to the development of a strategy to monitor biodiversity at a hierarchy of scales. The scales are species, community, watershed, ecosystem, and landscape. Program goals are to track, and eventually to predict trends in biodiversity at these scales. This is designed to alert Park management to significant trends in time to undertake research into the causes and then to take corrective actions as warranted. For example, if trends outside the accepted norms for natural variation are noted in the biodiversity indicators of an important group, of several groups simultaneously, or across two or more levels of the hierarchy, Park management may respond by soliciting studies into the causes and possible consequences of the trends. If the results warrant, management actions may be instituted to counter the trends. In addition, the Park serves as an important baseline reference site of relatively pristine conditions for the region and the nation.

At the species level, monitoring is focused on species of special management concern, such as rare species of plants, heroic species such as black bears and white-tailed deer, and the unique southern Appalachian brook trout. There are approximately 520 species of plants, including vascular plants, bryophytes, and lichens, which are considered rare in the Park. Of these, about 170 species appear on state and/or federal lists. Fifteen of these species are now believed to be extinct from the Park. In order to prioritize the species to be monitored, a list was prepared using the Natural Heritage rarity rankings in the Park's Biological and Conservation Data System. Balanced with the perceived threats to each species, an initial selection of 30 species was made (Rock and Langdon 1993). Baseline monitoring for each species includes site characterization, associated species, threats, and population density. Primary focus is on detecting and tracking population declines and increases. Periodic revisits are scheduled that vary in frequency from several per year to once every several years. High priority is assigned to any species that appears to be in decline. Subsequent visits accumulate data toward development of predictive, demographic population models; the identification and efficacy rating of pollinators; and testing the impact of management activities. As an example, rare plants are linked to wildlife, especially white-tailed deer, by monitoring the impact of herbivore grazing on plants. Monitoring of rare plants in Cades Cove, which has a high density of deer, has led to a recent management decision to evaluate the effectiveness of deer exclosures on plant survivorship.

Monitoring of white-tailed deer is restricted to Cades Cove (Delozier and Stiver 1994a). The deer population there has been the subject of several studies over the years, and has been the target of management actions to reduce the size of the herd (Wathen and New 1989, and references therein). Over-browsing by deer is believed responsible for the elimination of at least two species of plants from the Park, Virginia chain fern (*Woodwardia virginica*) and purple fringeless orchid (*Plantanthera peramoena*). In addition, an attempt is underway to reintroduce the endangered red wolf to the Park. The monitoring of white-tailed deer is intended to address the issues of population density, over-browsing, and significance as a prey base for wolves within this one small area of the Park. Two methods are used to estimate population density and provide an index of herd health (an indicator of over-population stress). Density is estimated using roadside night counts, and herd health is estimated using abomasal parasite counts.

Black bear monitoring (Delozier and Stiver 1994b) is divided into two distinct components of different geographic extent. The most intensive monitoring is concentrated in the western end of the Park. Here, individual bears are captured, tagged, tattooed, weighed, sexed, and examined for parasites. A variety of body measurements are taken, along with blood, hair, tooth, and fecal samples. Selected animals are fitted with radio collars. In the winter, denning females are located and their reproductive status is determined. Home ranges are determined by tracking radio collared animals at intervals throughout the year.

Parkwide monitoring of bears includes a bait station survey and a hard mast survey. Bait stations are established each July along routes selected to represent the complete range of habitat and elevation within the Park at a density of approximately one bait site per three square kilometers. Bait stations are spaced at least 0.9 km apart along each route. A bait station consists of three partially opened cans of sardines hung with nylon string at a height of approximately 3 m from a tree. After 5 nights, the stations are revisited and checked for visitation by bears (missing or chewed cans). The hard mast survey is conducted in conjunction with the states of Georgia, North Carolina, and Tennessee, using the Whitehead tree count survey (Whitehead 1969) during August and September each year. Hard mast is the most important food of black bears, and the availability of mast affects food habits, movements, habitat preference, and reproductive success.

Brook trout monitoring includes annual estimates of population density and biomass, using three-pass depletion electrofishing. Fifteen permanent monitoring locations are sampled annually. Another 10 sites are sampled each year on a rotating basis from the complete list of known brook trout populations (Moore and Kulp 1995). In addition, distribution surveys are conducted each year on a rotating basis to determine the upper and lower extent of the species in all streams. Eventually, all brook trout streams will be included in the rotation. The combination of quantitative population monitoring with distribution monitoring, and annual permanent site monitoring with less frequent rotating site monitoring permits the Park to maintain an extensive program using limited funds.

Aquatic macroinvertebrates and fish other than brook trout are monitored at the community scale. Fish community monitoring locations are established at intervals along the lengths of major drainages (Abrams Creek, Little River, Cataloochee Creek, and Hazel Creek), from near the Park boundary to the upper limit of nonsalmonid species occurrence. (Only rarely do species other than brook and rainbow trout occur above about 800 meters elevation.) At each location fish are quantitatively sampled using three-pass depletion electrofishing (Moore and Kulp 1995). Some macroinvertebrate sites are in stream reaches that do not have fish populations, but most are collocated with brook trout and fish community monitoring sites to provide linkages among invertebrate and fish community data. Aquatic macroinvertebrates are sampled using rapid bio-assessment protocols (Parker and Salansky 1995), which provide reproducible, semiquantitative data. Macroinvertebrates are sampled annually at 25 permanent locations, including most permanent fish monitoring locations, and another ±15 sites are sampled each year on a rotating basis in the same manner that brook trout are sampled.

Watershed monitoring takes place at the Noland Divide Research Watershed, a south-facing 17.4-ha watershed on the headwaters of Noland Creek at 1575 m. The watershed is dominated by red spruce (*Picea rubra*) and Fraser fir (*Abies fraseri*), American beech (*Fagus grandifolia*), yellow birch (*Betula alleghaniensis*), and fire cherry (*Prunus pensylvanica*). Most of the firs have been killed by the exotic balsam woolly adelgid, *Adelges piceae*. The Integrated Forest Study documented levels of nitrate input at an adjacent site higher than at any of 11 other sites in North America (Johnson and Lindberg 1992, Lovett and Lindberg 1993). When the inventory and monitoring program was begun in 1992, two flumes were installed on the paired streamlets that drain the watershed. The core elements of the watershed program will be maintained in the long term, including basic physical, chemical, and biotic monitoring of the streams, and chemical monitoring of precipitation. Selected parameters (discharge, temperature, pH, conductivity) are continuously monitored with probes and data loggers. Weekly grab samples are collected for detailed chemical analyses (anions and cations). Monthly 24-hour drift samples are collected to monitor aquatic macroinvertebrate species composition, life histories, and phenology, as well as stream system dynamics. Objectives of the watershed scale program are to provide monitoring that is directly related to air pollution effects and to contribute to the extensive research program into nutrient cycling at the Noland Divide site (Cook et al. 1994). Ultimately, the research programs led to the establishment of the monitoring program at this site, and now the monitoring program contributes data to the ongoing research.

The combined aquatic monitoring program consists of the brook trout and fish community monitoring, aquatic macroinvertebrate monitoring, the Noland Divide Research Watershed program, and a parkwide water quality monitoring network. This represents the highest degree of integration among scales in the Park resource monitoring program, incorporating population, community, and watershed components with a landscape scale element. Water quality is used to link the levels of the hierarchy, from atmospheric inputs, through the watershed scale, to the fish and macroinvertebrate communities, to the brook trout populations. Water quality monitoring occurs quarterly at each of 70 sites throughout the Park. Many of these sites are collocated with the brook trout, fish community, and macroinvertebrate monitoring sites, although others are located along the elevation gradient. The sites are distributed among the five water quality regions of the Park as defined by Nodvin et al. (1995). This also shows the flexibility of the prototype inventory and monitoring program. The extensive water quality monitoring program was not a part of the original Park program, but grew out of a short-term assessment of acidification of Park streams. When the results of that work were made available, Park managers decided to redirect resources. Some funds were diverted from the intensive watershed research to a subset of the extensive parkwide monitoring network developed by the acidification study.

The vegetation monitoring component has the primary goal of detecting changes in the vegetation through time. This will be done by establishing a parkwide network of permanent plots that encompass the geographic extent and include the major plant communities of the Park. The basic monitoring design will consist of approximately 300 plots, the major portion of which will be randomly located within ten watersheds, with a subset of plots subjectively located to ensure full representation of Park vegetation types. Individual plots will be scheduled for remeasurement at 5-year intervals. The nested plot design is based partially on the methodologies of the North Carolina Vegetation Survey* and on previous work in the Park. The nested subplots, beginning with 10 cm × 10 cm through 10 m × 10 m, allow this component to operate at a variety of scales. The design also permits the Park program to integrate with other programs such as the North Carolina Vegetation Survey, using similar plots in adjacent areas of the southern Appalachians. A secondary goal of this element is to provide anchor locations for gathering data on other taxonomic groups (i.e., , amphibians, birds, invertebrates), and thereby attribute detailed habitat data to those groups.

Landscape-scale monitoring will be accomplished primarily by remote sensing to analyze changes in vegetation within the Park and land use patterns around the Park. The national inventory and monitoring program is undertaking vegetation mapping in all natural areas parks in the nation. The Nature Conservancy's community classification is being used to identify plant associations with a combination of aerial photography and field verification. The Smokies represent perhaps the greatest challenge to the program because of its topographic relief and extraordinary diversity of plant communities. Because of this, the Park is one of the prototypes chosen for the mapping effort. When the vegetation mapping is completed, it will serve as a baseline for identifying vegetation changes in future years. The Southern Appalachian Assessment (Chapter 7) has provided a regional assessment of land use that includes the vicinity of the Park. This will be used as a baseline for future assessments of changes in land use, the degree of habitat fragmentation, and the recognition of existing and potential corridors for wildlife movement between the Park and surrounding natural areas. Vegetation and land use mapping are examples of activities undertaken and funded by sources outside the Park that contribute significantly to the Park's national resource inventory and monitoring program.

Air Quality

Air quality is an important aspect of long-term monitoring in the Park, although it is funded and operated separately from the monitoring program described above. The air quality monitoring program has been jointly funded by the Park and the National Park Service Air Quality Program, with cooperation from the state of Tennessee, the Environmental Protection Agency, the Tennessee Valley Authority, and the Oak Ridge National Laboratory. Unlike the inventory and monitoring program, the air quality monitoring program is not base funded and has to depend on project money each year. The program monitors a wide range of parameters in three general categories: visibility and particulates; ozone; and acid precipitation.

The Park Service has been monitoring visibility and particulates in parks across the nation since 1982 (Malm and Sisler 1987). The Look Rock visibility monitoring station was established in the Park in 1980 (Peine et al. 1985) and operated continuously since then. The program was enhanced in 1988 when the Park Service entered into a cooperative effort with the Environmental Protection Agency, known as the Interagency Monitoring of Protected Visual Environments or IMPROVE (Joseph et al. 1986). Visibility is monitored by photography, transmissiometry, and nephelometry, and aerosols are monitored by size-fractionated filtration. The data from the visibility monitoring program are used by the Park in evaluating the potential for significant deterioration of air quality standards when new or modified emitters apply for permits.

---

* Peet, R.K., T.R. Wentworth, and P.S. White. 1990. A flexible, multipurpose method for measuring vegetation. Unpublished draft.

Ground level ozone has been monitored in the Park at Look Rock since 1980, at Cove Mountain since 1986, and at Clingmans Dome since 1992 (J. Renfro, pers. commun.). These monitoring stations provide an elevation transect from 823 m at Look Rock, to 1243 m at Cove Mountain, and to 2015 m at Clingmans Dome. At each site, monitors track ambient ozone concentrations continuously. The diurnal pattern of ozone concentration, in which peaks occur in late afternoon and minima occur just before dawn, are muted to virtually eliminated at higher elevations in the Park (Neufeld et al. 1992). This means that high elevations receive higher cumulative ozone exposure than low elevations. Studies of the biological effects of ozone on native species of plants (Neufeld et al. 1992; Renfro 1992) have documented the nature and extent of ozone induced damage to more than 95 species. Tall milkweed is monitored annually for ozone damage in the Park.

Acid precipitation monitoring is an outgrowth of the National Acid Precipitation Assessment Program of the 1980s (National Acid Precipitation Assessment Program 1991), and the Integrated Forest Study (Johnson and Lindberg 1992). Three sites are monitored by the Park: a low elevation (640 m) site at Elkmont; and high elevation protected (1737 m) and exposed sites (1920 m) at Noland Divide. In addition to precipitation and meteorological monitoring, sulfur and nitrogen are monitored in their various forms (e.g., sulfate, nitrate, and ammonium) and in the manner in which they are deposited (wet or dry). The monitoring of sulfur and nitrogen provide the greatest linkage between the air quality monitoring and the inventory and monitoring programs in the Park. The air quality data document some of the highest known deposition rates for nitrate, ammonium, and sulfate in the East, and the inventory and monitoring program has determined that nitrogen is being exported from the Noland Divide watershed (Nodvin et al. 1995). These results have contributed to the recent increase in concern among scientists and policy makers about the role of atmospheric nitrogen in forest ecosystems (Thomson 1996).

THE SAMAB FOREST HEALTH MONITORING DEMONSTRATION

The national Forest Health Monitoring (FHM) Program is mandated by Congress* to provide a means of annually assessing the nation's forest resources with regard to possible stresses such as air pollution, climate change, nutrient depletion, and large-scale pest and pathogen threats. No other regional scale monitoring program looking at ecosystem function and health has ever been established. However, the need for such a program is clearly recognized if only to establish what portion of our forests are truly affected by air pollution and what extent of damage, if any, there is. Initially, the program was jointly administered by the U.S. Forest Service (USFS)(the Forest Health Monitoring Program) and the Environmental Protection Agency (EPA)(Environmental Monitoring and Assessment Program — EMAP — Forests), with the hope that it would be a multiagency effort once plots were established and annual measurements implemented. Currently, administrative responsibility lies primarily with the USFS, with cooperators including the National Association of State Foresters, 7 federal agencies, 24 state forestry or agricultural agencies, and 19 universities.

Specific objectives of the FHM Program include:

1. Estimate with known confidence the current status, changes, and trends in selected indicators of forest ecosystem condition on a regional basis
2. Identify associations between changes of trends in indicators of forest ecosystem condition and indicators of natural and human-caused stressors, including changes in forest extent and distribution
3. Provide information on the health of the nation's forest ecosystems in annual statistical summaries and periodic interpretive reports for use in policy and management decisions

---

* Forest Ecosystems and Atmospheric Research Act, 1988; Food, Agriculture, Conservation, and Trade Act, 1990; Clean Air Act, 1990.

4. Identify mechanisms of ecosystem structure and function through long-term monitoring of ecosystem processes at intensively monitored sites representing major forest ecosystems
5. Improve the effectiveness and efficiency of forest health monitoring through directed research
6. Integrate forest health monitoring with other EMAP resource groups in order to complete multi-ecosystem assessments

The SAMAB FHM Demonstration, begun in 1991, was the first project implemented using ecological boundaries rather than on a state-by-state basis. The collaboration of SAMAB-member agencies made the demonstration an example of multiagency cooperation, where, through the collaboration of the SAMAB member agencies, valuable forest resource information was collected more quickly and cheaply than if this had been a single agency undertaking. Funding for the SAMAB FHM Demonstration was provided equally by TVA and the EPA. Other SAMAB partners provided in-kind contributions such as the National Park Service making seasonal housing available for field crews and the National Forest System providing contracting services.

Information collected in the FHM Program is more far-reaching than traditional forest invento- ries, and assessments of ecosystem health are done using a "suite of indicators" rather than depending on a single variable like growth. The program also incorporates several levels of monitoring. The first level is detection monitoring that incorporates the basic sampling design and measures variables such as growth, crown condition, and presence of insects and disease. The second level is evaluation monitoring, employed in areas where problems are observed through detection monitoring. In evaluation monitoring, the number of plots or the number of variables measured is increased to ascertain a probable cause. The third level is intensive site ecosystem monitoring, where hypotheses regarding cause/effect are examined through intensive, continuous measurement at biologically representative sites. FHM also has a research on monitoring techniques (ROMT) component that investigates the functionality and feasibility of new monitoring techniques.

The sampling design used by FHM at the detection level is a systematic grid system that uses a fixed-radius cluster plot design where each plot represents approximately 158,000 acres. This sampling intensity will provide forest health information on a *regional* scale, for *regional*-scale stresses such as air pollution damage, climate change, and large-scale pest problems. This grid is insufficient for smaller, localized areas such as counties, or for detection of problems that are small- scale or occur only on a subset of the population. The sample design can be easily supplemented with additional grid points to provide data necessary to evaluate trends on a smaller-than-regional scale.

For the SAMAB FHM Demonstration, considered a research project designed to test indicators, data collected include site characteristics; tree species, size, condition, and canopy position; soil chemical and physical characteristics; foliage condition and chemistry; annual growth and wood chemistry from tree increment cores; vegetation structure to evaluate biodiversity and habitat condition; measures of photosynthetically active radiation (PAR) as a surrogate for canopy density; lichen community and chemical analyses; and evaluation of adjacent air pollution bioindicator plants.

The SAMAB FHM Demonstration utilized the detection monitoring grid through a "¼ inter- penetrating" scheme, where ¼ of the total number of plots was to be established in each of the first 4 years. Justification for this strategy was that as many of the variables collected on the plots are not likely to change quickly (i.e., site and soil characteristics), plot establishment and initial measurements could be spread out, and a "running average" could be used to estimate regional conditions. Annual remeasurements would then be done on those variables that are especially sensitive to change (i.e., assessments of insect defoliation), while those less sensitive are measured less frequently. It was anticipated that a total of approximately 130 plots in forested areas in the SAMAB region would be established over a 4-year period. Measurement of one plot averaged one day with a six-person crew, causing initial costs to be extremely high. Following plot establishment, however, it was anticipated that fewer measures would be taken (for example, PAR, bioindicator plants, and vegetation structure), thus reducing time and costs associated with visiting the plots.

In a functional mode, the FHM Program relies heavily on help from collaborators. State forestry agencies, the USFS, and the TVA have, in the past, provided field personnel to establish plots and collect routine measurements. EPA and the USFS have provided personnel for training and quality assurance/quality control. The Natural Resources Conservation Service has provided soil scientists to help with soil sampling and characterization. Other examples of collaboration include research into forest health indicator development — TVA and the U.S. Fish and Wildlife Service have worked closely with the USFS and EPA toward resolving issues associated with proposed indicators through both on-site and off-site research. The SAMAB FHM Demonstration collaborators included TVA's Resource Group, the National Park Service–Great Smoky Mountains National Park, The U.S. Forest Service–National Forests in North Carolina, EPA's EMAP–Forests Program, the Natural Resources Conservation Service, and universities in the region through the hiring of undergraduate and graduate students as field crew personnel.

The benefits of a such regional monitoring system are numerous. Data collected within the FHM Program will be available to anyone who has a need for it. Every agency with responsibility for conservation of natural resources recognizes the need for information reflecting the current status of these resources, especially given that most agencies now espouse ecosystem management. Large-scale assessment data provide the context for smaller-scale studies, improving understanding of landscape processes, health of migratory wildlife, and spread of insects, disease, and even fire. It will also allow periodic assessments of the status of, and trends in, forest health that are necessary for responsible resource management within the region and allow response to potential problems in a timely manner.

Unfortunately, like many large-scale monitoring projects, support for the SAMAB FHM Demonstration has declined. The last plots were established in 1992. No additional funding has been provided, leaving one half the plots yet to be established. The FHM Program has seen many changes, especially in funding, which today primarily comes from the USFS rather than EPA. Implementation has resumed on a state-by-state basis, although no detection activities are planned for the SAMAB region for the immediate future.

RISK ASSESSMENT REPORT CARD FOR ECOSYSTEM MANAGERS

The case of the evolving environmental monitoring program at Great Smoky Mountains National Park provides an instructive perspective as to the risks taken by a manager directing the development of such a new and innovative program. The current effort is very expensive, costing $500,000 per year. Park personnel believe the monitoring program funds should be doubled in order to have a more balanced program. This would represent approximately 10% of the total operations budget for the nation's most visited national park. Application of the Risk Assessment Report Card for the Park ecosystem managers is judged to be as follows:

Vision: B

The program is one of four prototype programs selected by the National Park Service. The program, however, has been more narrowly defined than potentially necessary by not encompassing other ongoing and previously initiated monitoring programs, such as the extensive air quality program, bird monitoring transects, hog exclosures, and the spruce–fir–lichen and small mammal monitoring plots. Also, the regional context of the Park has been largely ignored.

Resource risk: C+

As documented in other chapters in this volume, several species in the Park are at risk, as is the entire spruce–fir ecosystem. However, the overall biodiversity of the Park is at relatively low risk compared to that of other national parks, such as the Florida Everglades.

Resource conflicts: C+

Over the long term, the fundamental question will be raised concerning the amount of funds being used for monitoring vs. proactive resource management activities. Also, the choice of monitoring activity is likely to change over time. This will not be a problem as long as the original designers of the monitoring program remain in the employ of the Park, but as they are replaced, new ideas and perspectives will invariably be introduced. This basic human nature is inevitable. It is demonstrated in the design of the Park monitoring program presented here, since many then-ongoing monitoring programs started by others were not utilized when the new program was designed in 1992.

Procedural protocols: A

Documentation and peer review of protocols of the subject program have been completed.

Scientific validity: B

The scientific validity of the subject program has been peer reviewed. However, a strategy for periodic analysis of data, integrated synthesis, creation of predictive models, and reporting of results has not been determined.

Legal jeopardy: A

A scientifically valid information base will add credibility to the more informed ecosystem manager's decisions.

Public support: B

Public support has not been an issue for the subject program. Most Park visitors are not that aware of it.

Adequacy of funding: C

Park officials engaged in the program claim that the program is being funded at only 50% of its optimum level.

Policy precedent: B

The subject program is implicit in the enabling federal legislation for the national park system but would more likely expand to other parks if it became a congressional mandate for the agency.

Administrative support: A

Support by the Park administrative services personnel has been extraordinary.

Transferability: B

Part of the rationale of designating prototype parks is that the protocols developed can be applied in other parks. However, each landscape/ecosystem is somewhat unique, therefore inevitably requiring development of independent monitoring protocols.

The end result of this ambitious monitoring initiative will be a much better understanding of the resource conditions allowing the ecosystem manager to make much more informed decisions.

SAMAB'S ROLE IN FUTURE ENVIRONMENTAL MONITORING

Environmental monitoring in southern Appalachia has made great progress since it first became a tool for resource management in the region. Improvements include (1) a shift from single resource monitoring toward more holistic ecosystem monitoring; (2) recognition of the need for monitoring at several scales; and (3) more utilization of ecological boundaries over political ones. The SAMAB

Cooperative has played a significant role in making these improvements through education and in bringing agencies together to work collaboratively toward goals of mutual interest. There is still much that could be done to further improve monitoring as a tool for ecosystem management, and SAMAB can continue to play an important role as we learn the most effective ways to protect the region while continuing to provide the goods and services valuable to society.

One of the most important accomplishments achieved by SAMAB over its years of existence was the Southern Appalachian Assessment (SAA) (Berish et al. Chapter 7 of this volume). SAMAB member agencies worked together to assemble, evaluate, and analyze existing data available from the many environmental monitoring efforts that have been established in the region over the years. This work both gave a clearer picture of what we know about current environmental health in the region and identified areas where there are significant gaps in data and understanding. The results of the SAA will be used by SAMAB to guide future research and monitoring efforts.

SAMAB can provide mechanisms to convey the results of the SAA, as well as any future research and monitoring, to the public. This is an important step toward implementing ecosystem management, as privately owned lands are also part of the regional ecosystem. Many environmental problems can be more effectively addressed through a community-based approach, allowing local communities to prioritize issues and thus secure buy-in from residents who play a role in the decision-making process. SAMAB also should work to help the public communicate with its member agencies about what society values in terms of resource management, recognizing that these values may change over time and that people are a part of the region's ecology.

SAMAB can play a role in furthering environmental monitoring in other areas, including improved collaboration among the agencies, establishment of linkages across scales, and development and implementation of monitoring standards. One particular area in which SAMAB can potentially improve environmental monitoring through the region is by improving coordination among the biosphere reserve units. Originally envisioned as a cluster of research units that together provide insight into regional ecosystem dynamics, the biosphere reserve units have never developed a joint plan where research and monitoring activities are complementary. This could provide a tremendous opportunity for the region, as collaboration among these geographically distinct units offers the potential for improving spatial representation and thus ability to extrapolate results.

By improving collaboration among biosphere reserve units and identifying other regional monitoring needs, SAMAB can initiate the process of developing a regional framework for monitoring. In efforts that cross jurisdictional boundaries, SAMAB member agencies can work together as a team, similar to the way the SAMAB FHM Demonstration developed and began to be implemented. Communication about the design of monitoring programs can help ensure that there is consistency among programs so that data may be combined to represent larger spatial or temporal ranges. Minimum standards for data quality, storage, and retrieval should be recommended and adopted for all SAMAB agencies, and data sharing should be encouraged wherever possible.

REFERENCES

Cook, R. B., J. W. Elwood, R. R. Turner, M. A. Bogel, P. J. Mulholland, and A. V. Palumbo. 1994. Acid-base chemistry of high-elevation streams in the Great Smoky Mountains. *Water, Air, Soil Pollut.* 72:331–356.

Delozier, K. and W. Stiver. 1994a. White-tailed deer population monitoring. Great Smoky Mountains National Park. 3 pp.

Delozier, K. and W. Stiver. 1994b. Black bear population monitoring protocols. Great Smoky Mountains National Park. 9 pp.

Eaton, W. C., E. L. Tew, C. E. Moore, and D. A. Ward. 1987. Summary Report for the National Atmospheric Deposition Program/National Trends Network (NADP/NTN) Site Visitation Program. EPA 600-87-936. Research Triangle Institute, Research Triangle Park, NC.

Forest Health Monitoring National Office. 1996. *Forest Health Monitoring.* Unpublished report. USDA Forest Service, Research Triangle Park, NC. 26 pp.

Grumbine, R. E. 1994. What is ecosystem management? *Conserv. Biol.* 8(1):27–38.

Hellawell, J. M. 1991. Development of a rationale for monitoring. in Goldsmith, F. B., Ed., *Monitoring for Conservation and Ecology.* Chapman and Hall: London. 271 pp.

Johnson, D. W. and S. E. Lindberg, Eds., Eds. 1992. *Atmospheric Deposition and Forest Nutrient Cycling: a Synthesis of the Integrated Forest Study.* Springer-Verlag: New York. 707 pp.

Joseph, D. B., J. Metsa, W. C. Malm, and M. Pitchford. 1986. Plans for IMPROVE: A Federal program to monitor visibility in class I areas. In Proceedings of the 1986 APCA International Specialty Conference on Visibility Protection, Research and Policy Aspects, Pittsburgh, pp. 113–125.

Lovett, G. M. and S. E. Lindberg. 1993. Atmospheric deposition and canopy interactions of nitrogen in forests. *Can. J. For. Res.* 23:1603–1616.

Malm, W. C. and J. F. Sisler. 1987. Sources of visibility reducing haze at Shenandoah National Park. In Proc. 80th Annu. Meeting of the Air Pollution Control Association, New York, June 1987. Paper 87-40A.4.

Moore, S. E. and M. A. Kulp. 1995. Fisheries monitoring protocols. Great Smoky Mountains National Park. 71 pp.

National Acid Precipitation Assessment Program. 1991. 1990 Integrated Assessment Report. National Acid Precipitation Assessment Program Office of the Director: Washington 520 pp.

National Park Service. 1992. Natural Resources Inventory and Monitoring Guideline. NPS-75. Washington, D.C. 37 pp.

Neufeld, H. S., J. R. Renfro, W. D. Hacker, and D. Silsbee. 1992. Ozone in Great Smoky Mountains National Park: dynamics and effects on plants. In R. L. Berglund, Ed. *Tropospheric Ozone and the Environment II.* Pittsburgh: Air and Waste Management Association. Atlanta, November 1992, CONF-9211233, pp. 594–617.

Nodvin, S. C., T. Flum, J. Shubzda, H. Rhodes, M. Malick, E. Williams, H. VanMiegrot, S. E. Lindberg, E. F. Pauley, A. K. Rose, and N. Nicholas. 1995. A comprehensive water quality monitoring program for the Great Smoky Mountains National Park. Tech. Rep. Ser., National Biological Service, Washington, D.C.

Oak Ridge National Laboratory. 1994. Oak Ridge Reservation Annual Site Environmental Report for 1993. ES/ESH-47.

Parker, C. R. and G. K. Salansky. 1995. Benthic macroinvertebrate monitoring protocol manual. Great Smoky Mountains National Park. 67 pp.

Peine, J. D., C. Pyle, and P. S. White. 1985. Environmental monitoring and baseline data management strategies and the focus of future research in Great Smoky Mountains National Park. Res./Resour. Manage. Rep. SER-76. U.S. Department of the Interior, National Park Service. 114 pp.

Renfro, J. R. 1992. Ozone fumigation plant testing and sensitivity results at Great Smoky Mountains National Park: 1987–1991. In J. D. Peine, Ed. Forum on Air Quality Management for Class I Areas in the Southern Appalachians. Gatlinburg, TN, March 1992. Report Section Mc.

Rock, J. H. and K. R. Langdon. 1995. Rare plant monitoring protocols manual. Great Smoky Mountains National Park. 67 pp.

Shubzda, J., S. E. Lindberg, C. T. Garten, and S. C. Nodvin. 1995. Elevational trends in the fluxes of sulfur and nitrogen in throughfall in the Southern Appalachian Mountains: some surprising results. *Water, Air Soil Pollut.* 85:2265–2270.

Slocombe, D. S. 1993. Implementing ecosystem-based management. *Bioscience* 43(9):612–622.

Southern Oxidants Study Strategic Planning Committee. 1992. Southern Oxidants Study Strategic Plan. Informal document. SOS Office of the Director, North Carolina State University, Raleigh, NC.

Spellerberg, I. F. 1991. *Monitoring Ecological Change.* Cambridge University Press: Cambridge. 304 pp.

Swank, W. T. and D. A. Crossley, Jr., Eds. 1986. *Forest Hydrology and Ecology at Coweeta.* Ecological Studies Vol. 66. Springer Verlag: New York.

Thomson, V. E. 1996. Southern Appalachian Clean Air Partnership. USDA Forest Service and USDI National Park Service, R8-TP 30. 105 pp.

Van Hook, Robert, Ed. 1989. *Walker Branch Watershed.* Springer-Verlag: New York.

Wathen, W. G. and J. C. New, Jr. 1989. The white-tailed deer of Cades Cove: population status, movements and survey of infectious diseases. National Park Service, Res./Resour. Manage. Rep. SER-89/01. Southeast Regional Office, Atlanta, Georgia. 139 pp.

Whitehead, C. J. 1969. Oak mast yields on wildlife management areas in Tennessee. (Unpublished report.) Tennessee Game and Fish Commission, Nashville. 11 pp.

9 The Role of Indicator Species: Neotropical Migratory Song Birds

Theodore R. Simons, Kerry N. Rabenold, David A. Buehler, Jaime A. Collazo, and Kathleen E. Franzreb

CONTENTS

Introduction 187
Importance of the Southern Appalachians 190
Evidence and Causes of Declining Songbird Populations...190
An Ecosystem Approach to Forest Bird Conservation 193
Insights from On-Going Research..201
Conclusions 203
Risk Assessment Report Card 206
References 207

INTRODUCTION

Southern Appalachian forests support some of the richest avian diversity in North America, including some 75 species of neotropical migrants, birds that perform the remarkable feat of making much of the Western Hemisphere their home. This diverse group includes the swallows, kingbirds, and other flycatchers that feed in the air on flying insects. The Eastern Kingbird is a typical species. It breeds in forested areas, primarily in the Eastern U.S. and winters in Central America and Northern South America (Figure 1). Species such as tanagers glean insects from forest foliage and also feed extensively on fruit. Other groups include the vireos, orioles, thrushes, and even the tiny hummingbirds.

But the largest and most striking members of this group of birds are the wood warblers, some 50 closely related species of what can best be referred to as "quintessential" songbirds. These brightly colored songsters occupy an astonishing diversity of habitats. The Blackburnian Warbler inhabits the spruce–fir forests as far north as boreal Canada. Black and White Warblers glean insects from branches of the tallest trees in mature deciduous forests but nest on the ground. Worm-eating Warblers are specialists at prying insects out of the protective covering of curled up leaves, while Chestnut-sided Warblers are shrub nesting specialists of disturbed sites and forest edges.

Neotropical migrants predominate in the breeding bird community of eastern deciduous forests. In some parts of the southern Appalachians, up to 80% of the breeding bird community is comprised of these species (Figure 2). These approximately 75 species use ground, shrub, and especially canopy nests, and about 80% of them are insectivores (Figure 3). Recent concern over the status of these birds has been prompted by surveys showing widespread population declines.

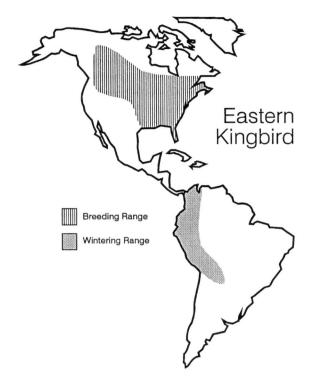

FIGURE 1   Breeding and wintering ranges of the Eastern Kingbird, a typical neotropical migratory passerine. (From Rappole, J.H. et al. USDI/FWS, U.S. Government Printing Office, Washington, D.C., 1983.)

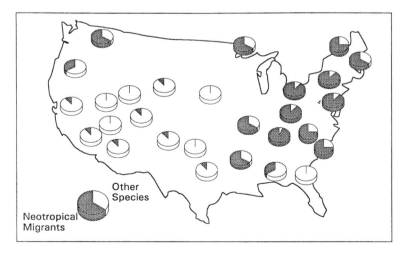

FIGURE 2   Proportion of neotropical migrants in the breeding bird community. (From MacArthur, R.H., *Proc. Natl. Acad. Sci. U.S.A.,* 43, 293–295, 1957.)

FIGURE 3 Population trends, nest locations, and foraging ecology of southern Appalachian neotropical migrants.

IMPORTANCE OF THE SOUTHERN APPALACHIANS

The southern Appalachian region is significant to forest birds for a variety of reasons:

- It is a an internationally recognized refugium of temperate forest biodiversity. Great Smoky Mountains National Park (GSMNP) is a designated World Heritage Site and an International Biosphere Reserve, recognized primarily for its biological diversity. The steep, complex topography harbors species richness along extreme temperature and moisture gradients.
- It contains the largest remaining stands of virgin forests in the eastern U.S. Over 30% of the forests of GSMNP are considered to be high in primary forest attributes, representing perhaps as much as 80% of the primary forest remaining in the eastern U.S. (Davis 1993). These forests provide a rare opportunity to study the unique characteristics of undisturbed forest ecosystems.
- A substantial number of ecosystems and species are at risk The most threatened ecosystem is the high elevation spruce–fir forests that have been decimated by exotic insects and air pollution. There are 120 species of vascular plants recognized as rare enough to be of managerial concern. A similar number of bryophytes, lichens, and fungi are also considered rare at the regional, national, or global level. At least 22 species of breeding birds are considered of serious management concern due to significant reductions in populations or habitats (Hunter 1993).
- The region contains the largest block of protected forested landscape in the eastern U.S. Over five million acres of protected lands in the region include a matrix of National Forests, federally designated wilderness areas, state lands, Tennessee Valley Authority reservoirs, and National Park Service lands (Figure 4).

EVIDENCE AND CAUSES OF DECLINING SONGBIRD POPULATIONS

We know that several species have been in decline for some time. In the western U.S., Golden-cheeked Warblers and Black-capped Vireos, and in the east Bachman's and Kirtland's Warblers are listed as endangered species. In these birds, population problems could usually be traced to extremely limited and specialized breeding and/or wintering habitat, and it is generally believed that their populations have historically always been low.

More recently, a larger problem has been detected. The U.S. Fish and Wildlife Service has conducted the Breeding Bird Survey in the U.S. since 1966. A survey consists of 50 3-minute point counts (censuses from a fixed point in which an observer records all the birds seen or heard during a set period of time) along a 25-mile roadside route. Routes are randomly assigned and run once each year during the peak of the breeding season. Using an all-volunteer force, the program conducts 2000 to 3000 surveys a year along some 50,000 miles of secondary roads. Population trends for neotropical migrants from 1978 to 1987 are summarized in Figure 5 (Robbins et al. 1989b). Of the species classified as neotropical migrants, 71% declined during the period. Of the 44 species showing negative trends, 20 exhibited statistically significant declines. Declines for some species, such as the Bay-breasted Warbler, were precipitous, averaging 16% per year. More typical are species like the Kentucky Warbler whose populations decreased at rates of 2 to 3% per year.

A more recent analysis of data from the Breeding Birds Survey from 1966 to 1992 (Peterjon et al. 1995) indicates that, continent-wide, as many species of neotropical migrants have shown increasing as decreasing populations. Nevertheless, widespread population declines were evident in many species over the past 15 years, particularly in the eastern U.S. (Askins et al. 1990). The southern Appalachians have shown consistent negative trends for both neotropical migrants and woodland birds over the 26 year period (Figure 6).

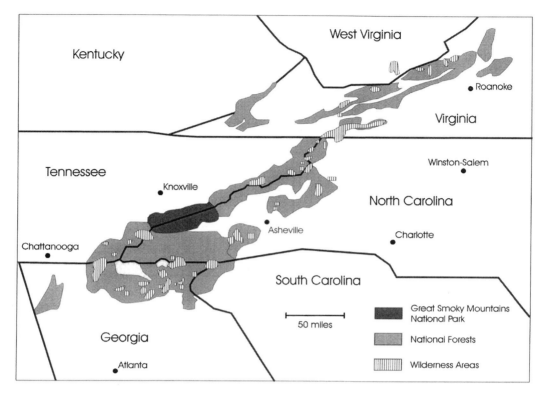

FIGURE 4 Protected areas in the southern Appalachians.

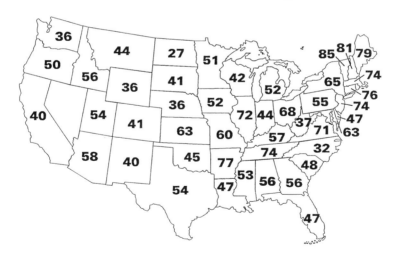

FIGURE 5 Percent of migratory bird species showing population declines from 1978 to 1987. (From Robbins, C.S. et al., *Wildl. Monogr.,* 103, 1989.)

Habitat changes on both the breeding and wintering grounds are thought to be responsible (Sherry and Holmes 1995). In the tropics, logging and land clearing for agriculture and ranching are reducing habitats at rates of 1 to 4% per year and, in some countries, these rates are rising.

The effects of winter habitat loss on population are amplified due to the simple fact that tropical wintering habitats comprise but a fraction of the land area available in North America for breeding.

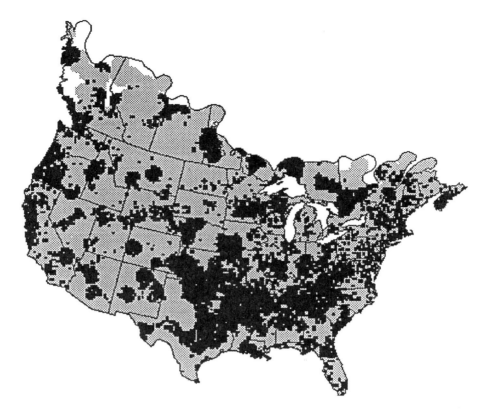

FIGURE 6 Mean trends for populations of neotropical migrant bird species during the period 1966 to 1992, based on the Breeding Bird Survey. Dark areas indicate negative population trends, light areas indicate positive population trends. (From Peterjon, B.J. et al., in Our Living Resources: a Report to the Nation of the Distribution, Abundance, and Health of U.S. Plants, Animals, and Ecosystems. U.S. Department of the Interior, National Biological Service, Washington, D.C., 1995.)

Nevertheless, it is important to keep in mind that forested habitats on the breeding grounds have also changed enormously over the past 3 centuries. Most of the bottomland hardwood forest of the Mississippi Valley and old growth forest in the east and west are gone. As managed forests become less diverse, they provide reduced habitat for forest birds (Thompson et al. 1995). Many of the remaining forested habitats have been severely fragmented by human activities. This has created additional pressures on many woodland birds that historically nested in the interior of large forest tracts.

The Wood Thrush, a so-called "area-sensitive" species (Robbins et al. 1989a), is a good example. The southern Appalachians are the center of abundance for the species which breeds in deciduous forests throughout the eastern U.S. (Figure 7). Populations have shown a steady decline over much of the species range during the past 20 years, based on a variety of indices (Figure 7). Research on the breeding and wintering grounds points to several consequences of habitat fragmentation that may explain these trends. Temple and Cary (1988) and Robinson et al. (1995) have shown that nest parasitism by Brown-headed Cowbirds is higher in fragmented forests. Small patches also harbor more potential predators such as squirrels, raccoons, and crows that generally avoid forest interiors. Wilcove et al. (1985) have shown that birds nesting in large contiguous forests suffered lower rates of nest predation than those nesting in forest fragments.

Much of this research, has been coordinated under the Partners in Flight Program (Finch and Stangel 1993), a cooperative interagency effort involving state, private, and federal land management and conservation organizations in the U.S. and Latin America. The program has focused attention on these birds in a hemisphere-wide effort to stop population declines.

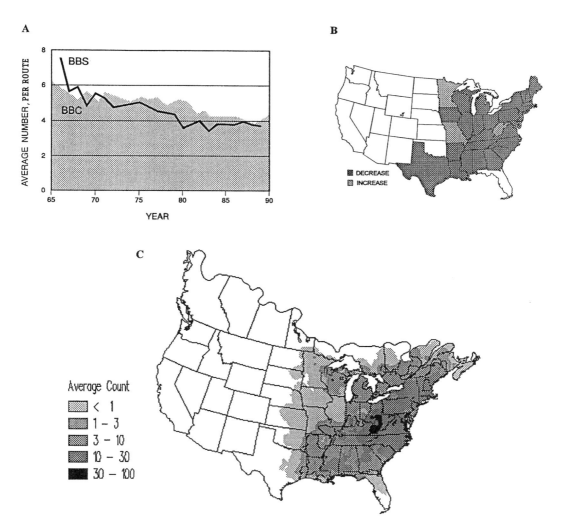

FIGURE 7 Status of breeding wood thrush populations in the U.S. Data from the Breeding Bird Census (BBC) and Breeding Bird Survey (BBS) indicate declining population trends (A) across most of the species breeding range in the eastern U.S. (B). Population densities reach their maximum in the southern Appalachians, based on BBS census data (C).

It is becoming clear that the population declines shown by many neotropical migrants and other forest birds are probably the result of a combination of factors. A wide variety of research efforts are underway in the southern Appalachians to help conserve forest bird populations (Table 1, Figure 8). These research projects were initiated independently, but efforts have been made to share data through Partners in Flight and use standardized methodologies where possible. These studies could serve as a framework for an ecosystem-scale program to address the conservation needs of forest birds across the southern Appalachians.

AN ECOSYSTEM APPROACH TO FOREST BIRD CONSERVATION

These collaborations have resulted in some standardization of methods and objectives, but they have also highlighted the need for a regional-scale approach to understanding the overall importance of the southern Appalachians to forest birds and the effect that land use practices are having on

TABLE 1
On-Going Avian Research in the Southern Appalachians

Principal investigator	Project title	Location	Partners
Buehler (UT)	Forest Avian Diversity	Cherokee N.F.	USFS
Collazo (NBS)	Forest Bird Productivity	Nantalahala N.F.	USFWS, USFS
Franzreb (USFS)	Effects of Timber Harvest on Cove Hardwood Birds	North Carolina National Forests	USFS, NC
Simons (NBS) Rabenlod (Purdue)	Old Growth Bird Community Studies	Great Smoky Mountains NP	NPS Purdue
Simons (NBS)	Wood Thrush Productivity	Great Smoky Mountains NP	NPS

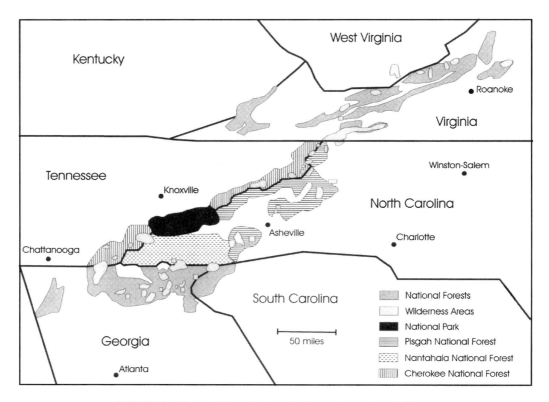

FIGURE 8   Forest bird study sites in the southern Appalachians.

their populations. An approach (Simons et al. 1995) for an ecosystem-scale research program has been developed that could serve as a model for understanding a number of important land management issues in the region. The proposed program would consist of two components:

- The establishment of standardized population and productivity monitoring protocols on "control" sites within GSMNP to allow comparisons to regional and national monitoring programs such as the Breeding Bird Survey
- A landscape scale "case study" to develop landscape models of how habitat conditions on protected and managed, public and private lands in the southern Appalachian region are affecting forest bird populations

A review of the objectives of the proposed ecosystem program and some of the findings to date will illustrate how an ecosystem-scale study of forest bird populations could enhance conservation efforts in the future.

Simons et al. (1995) have examined several questions related to the establishment of a population monitoring program within GSMNP, which would serve as a control site for comparison to more disturbed sites in the region. Because forest habitats in the Park are older and relatively more stable than surrounding areas, understanding bird population trends in the Park would provide important insights into larger-scale population trends and the relative significance of factors affecting populations on the wintering grounds, such as habitat loss, vs. those affecting populations on the breeding grounds, such as forest fragmentation. Some of the questions such an approach could address include:

- Do changes in bird populations within GSMNP mirror those observed on a regional scale (suggesting that populations are responding to conditions on the wintering grounds)?
- Do populations within the National Parks remain stable while regional populations decline (suggesting that populations are responding to conditions on the breeding grounds)?
- Does the productivity of neotropical migrants within these protected areas exceed that required for population stability (suggesting that the National Parks may serve as population "source" areas at a local or regional scale) (Simons and Farnsworth 1995)?

Work to establish GSMNP as a control site for regional scale studies began in 1991. Initial efforts were focused on calibrating and testing methodologies for long-term population monitoring to determine the costs and benefits of various population monitoring techniques, the appropriate scale for a Park-based monitoring program, and the relationship between bird population and habitat variability. We initially looked at three common methods of quantifying breeding bird populations; point counts, spot mapping, and mist netting. Based on the results of that calibration (Figure 9), we selected 10-minute fixed radius point counts as the primary sampling method. We then used point counts to estimate the breeding bird populations at five pairs of old growth–second-growth cove hardwood sites in the vicinity of Gatlinburg, TN. Permanent census points were established at each of the sites, and replicated censuses were conducted in May and June from 1992 to 1994. Results provided estimates of the relative abundance of 56 species of breeding birds at these sites and the sampling variability inherent in those estimates. Old growth sites showed higher breeding bird species diversity than second growth sites, presumably a reflection of the more even distribution of a larger number of species at old growth sites and the structural complexity caused by the large trees and tree-fall gaps characteristic of old growth forests (Figure 10). An analysis of the natural annual variability of census data (Figure 11) was used to evaluate the trade-off between statistical error rates and sample size requirements for a range of species (Figure 12). This approach can be used to ensure that future monitoring programs will be capable of meeting their objectives.

Forest breeding bird communities could serve as one of several models to examine how land use patterns and land management practices are affecting biological diversity in the southern Appalachians. Other potential indicators of regional forest health include the black bear, which is recognized as a management indicator species by the U.S. Forest Service (Clark and Pelton, Chapter 10 this volume) and salamanders, whose diversity in the southern Appalachians exceeds that anywhere else in North America (Duellman and Treub 1986). Clearly no one species or community can serve as a reliable indicator of ecosystem health, and legitimate questions have been raised concerning the management indicator concept (Doak and Mills 1994, Harrison 1994). Nevertheless, by carefully selecting a balanced sample of indicator species and communities, land managers should be able to track changes in biotic diversity and abundance that are being driven by changes in land use.

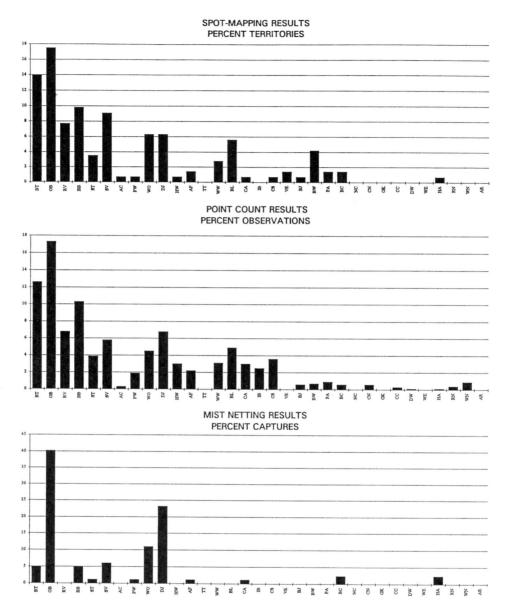

FIGURE 9 A comparison of census results from the Roaring Fork study site in Great Smoky Mountains National Park; 10-min, 50-m radius point counts proved to be the most efficient method for sampling forest bird communities in these habitats. (From Simons T.R. et al., Characterization of deciduous forest breeding bird communities of Great Smoky Mountains National Park. Final Report to the National Park Service, 1995.)

Accomplishing that goal for forest birds will require a coordinated landscape scale program focused on two major questions.

- How is the diversity, abundance, and productivity of avian populations related to habitat features such as habitat abundance, structural complexity, and the spatial arrangement of habitat types?
- How are these habitat features affected by land use and management?

Bird species codes used in Figures 9 through 11

Code	Species	Scientific name	Code	Species	Scientific name
AC	American Crow	*Corvus brachyrhynchos*	MD	Mourning Dove	*Zenaida macroura*
AF	Acadian Flycatcher	*Empidonax virescens*	MW	Magnolia Warbler	*Dendroica magnolia*
AR	American Redstart	*Setophaga ruticilla*	NC	Northern Cardinal	*Cardinalis cardinalis*
BB	Black-throated Blue Warbler	*Dendroica caerulescens*	NF	Northern Flicker	*Colaptes auratus*
BC	Black-capped chickadee	*Parus atricapillus*	NO	Northern Oriole	*Icterus galbula*
BG	Blue-gray Gnatcatcher	*Polioptila caerulea*	NR	Northern Raven	*Corvus corax*
BH	Broad-winged Hawk	*Buteo platypterus*	NW	Nashville Warbler	*Vermivora ruficapilla*
BJ	Blue Jay	*Cyanocitta cristata*	OB	Ovenbird	*Sieurus aurocapillus*
BL	Black-and-white Warbler	*Mniotilta varia*	OO	Orchard Oriole	*Icterus spurius*
BN	Brown Thrasher	*Toxostoma rufum*	PA	Parula Warbler	*Parula Americana*
BO	Barred Owl	*Strix varia*	PI	Pine Siskin	*Carduelis pinus*
BR	Brown Creeper	*Certhia familiaris*	PW	Pileated Woodpecker	*Dryocopus pileatus*
BT	Black-throated Green Warbler	*Dendroica virens*	RB	Red-bellied Woodpecker	*Melanerpes carolinus*
BW	Blackburnian Warbler	*Dendroica fusca*	RC	Red Crossbill	*Laxia curvirostra*
CA	Carolina Wren	*Thryothorus ludovicianus*	RG	Rose-Breasted Grosbeak	*Pheucticus ludovicianus*
CB	Brown-headed Cowbird	*Molothrus ater*	RK	Red-Tailed Hawk	*Buteo jamaicensis*
CC	Carolina Chickadee	*Parus carolinensis*	RN	Red-breasted Nuthatch	*Sitta canadensis*
CE	Cedar Waxwing	*Bombycilla cedrorum*	RO	American Robin	*Turdus migratorius*
CG	Common Grackle	*Quiscalus quiscalus*	RT	Rufous-sided Towhee	*Pipilo erythrophthalmus*
CH	Chestnut-sided Warbler	*Dendroica pennsylvanica*	RU	Ruffed Grouse	*Bonasa umbellus*
CN	Canada Warbler	*Wilsonia canadensis*	RV	Red-eyed Vireo	*Vireo olivaceus*
CO	Cooper's Hawk	*Accipiter cooperii*	SA	Swainson's Warbler	*Limnothlypis swainsonii*
CS	Chimney Swift	*Chaetura pelagica*	SI	Swainson's Thrush	*Catharus ustalatus*
CU	Black-billed Cuckoo	*Coccyzus erythropthalmus*	SS	Song Sparrow	*Melospiza melodia*
CW	Cerulean Warbler	*Dendroica cerulea*	ST	Scarlet Tanager	*Piranga olivacea*
DJ	Dark-eyed Junco	*Junco hyemalis*	SU	Summer Tanager	*Piranga rubra*
DW	Downy Woodpecker	*Picoides pubescens*	SV	Solitary Vireo	*Vireo solitarius*
EK	Eastern Kingbird	*Tyrannus tyrannus*	TT	Tufted Titmouse	*Parus bicolor*
EP	Eastern Phoebe	*Sayornis phoebe*	TW	Tennessee Warbler	*Vermivora peregrina*
ES	Starling	*Sturnus vulgaris*	VE	Veery	*Catharus fuscescens*
EW	Eastern Wood-pewee	*Contopus virens*	WE	Worm-eating Warbler	*Helmitheros vermivorus*
GC	Gray Catbird	*Dumetella carolinensis*	WI	White-eyed Vireo	*Vireo griseus*
GF	Great Crested Flycatcher	*Myiarchus crinitus*	WN	White-breasted Nuthatch	*Sitta carolinensis*
GK	Golden-crowned Kinglet	*Regulus satrapa*	WO	Wood Thrush	*Hylocichla mustelina*
GO	American Goldfinch	*Carduelis tristis*	WT	White-throated sparrow	*Zonotrichia albicollis*
HA	Hairy Woodpecker	*Picoides villosus*	WV	Warbling Vireo	*Vireo gilvis*
HB	Ruby-throated Hummingbird	*Archilochus colubris*	WW	Winter Wren	*Troglodytes troglodytes*
HO	House Wren	*Troglodytes aedon*	YB	Yellow-billed Cuckoo	*Coccyzus americanus*
HT	Hermit Thrush	*Catharus guttatus*	YC	Yellow-breasted Chat	*Icteria virens*
HW	Hooded Warbler	*Wilsonia citrina*	YR	Yellow-rumped Warbler	*Dendroica coronata*
IB	Indigo Bunting	*Passerina cyanea*	YS	Yellow-bellied Sapsucker	*Sphyrapicus varius*
KW	Kentucky Warbler	*Oporonis formosus*	YT	Yellow-throated Warbler	*Dendroica dominica*
LF	Least Flycatcher	*Empidonax minimus*	YV	Yellow-throated Vireo	*Virea flavifrons*
LW	Louisiana Waterthrush	*Seiurus motacilla*	YW	Yellow Warbler	*Dendroica petechia*

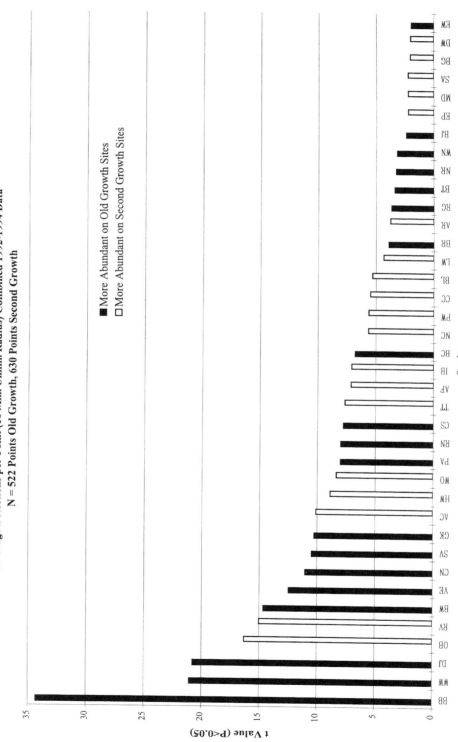

FIGURE 10 Significant differences in relative abundance at old growth and second growth sites. Average detections per point (10 min. unlim. radius) combined 1992–1994 data. N = 522 points old growth, 630 points second growth.

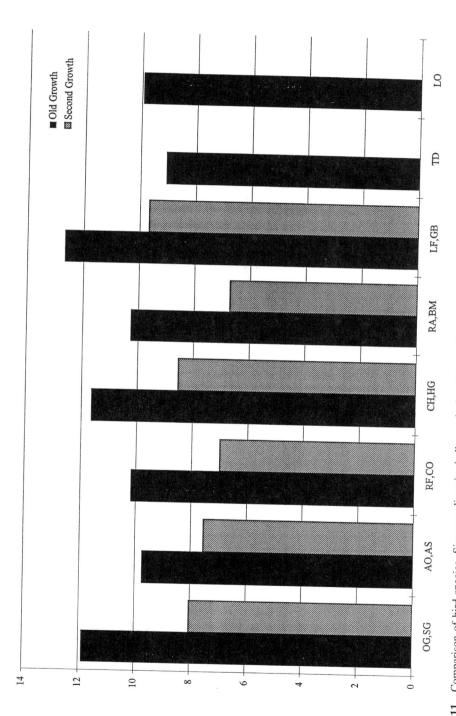

FIGURE 11 Comparison of bird species. Simpson diversity indices calculated from old growth and second growth study sites in Great Smoky Mountains National Park. Sites: (OG) combined old growth sites, (SG) combined second growth sites, (AO) Albright Grove old growth, (AS) Albright Grove second growth, (RF) Roaring Fork, (CO) Cherokee Orchard, (CH) Chimneys, (HG) Husky Gap, (RA) Ramsay Cascade, (BM) Brushy Mountain, (LF) Laurel Falls, (GB) Grassy Branch, (TD) Thomas Divide, (LO) Laurel Falls old growth.

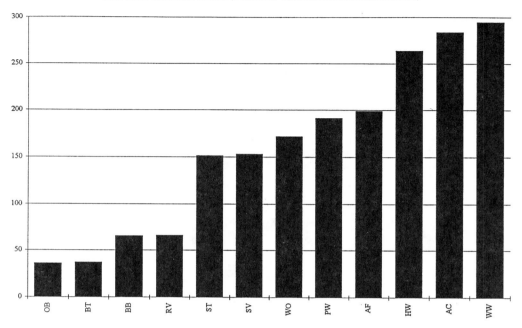

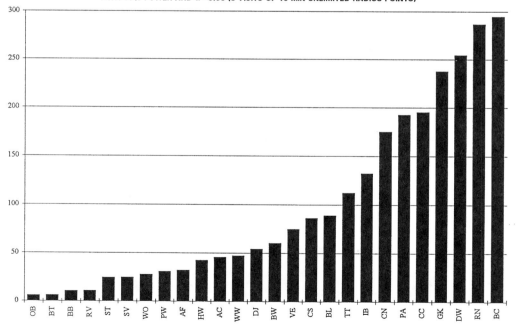

FIGURE 12   Estimated sample sizes (number of point counts) required to detect population changes from one year to another. Estimates based on point counts from 1992 to 1994 in Great Smoky Mountains National Park.

Some of the hypotheses to be tested through such a program might include:

- Old-growth forests support higher levels of breeding bird diversity, abundance, and productivity than younger but floristically similar stands
- Differences in habitat suitability between old growth and younger stands can be attributed to the greater structural complexity of old growth forests
- Breeding bird diversity and productivity are correlated with the abundance and connectivity of suitable habitat in southern Appalachian landscapes
- Rates of productivity will be lower and Brown-headed Cowbird nest parasitism will be higher in landscapes characterized by small, isolated habitat patches and patches with high edge-to-area ratios
- Local trends in bird abundance, diversity, productivity, and parasitism within patches of contiguous forest will be associated with distance to the patch boundary

INSIGHTS FROM ON-GOING RESEARCH

A comparison of preliminary results from our adjacent study sites provides a glimpse of the insights that such an ecosystem-scale approach might provide. The Southeastern Working Group of Partners in Flight has identified 22 high priority species in the region (Hunter 1993) (Table 2). These species were identified based on evidence of declining populations, specialized habitat requirements, or regional trends in habitat loss.

Preliminary results from our study sites indicate some differences between the bird communities on managed and unmanaged forests in the region. For example, comparison of spot-mapping results from cove hardwood/oak–hickory forests in GSMNP and adjacent North Carolina National Forests indicate that breeding bird densities for most high priority species are higher in the Park. The patterns were apparent for both old growth (Figure 13) and second growth (Figure 14) sites. A comparison of point count censuses from each of our study sites presents a more complex picture. (Figure 15). About half of the high priority species found on each of the sites showed higher indices of abundance within the Park, which may reflect higher habitat quality on less-disturbed sites. Populations of other species, such as Worm-eating and Hooded Warblers, appear to fare better on more disturbed sites where forest management practices presumably create preferred habitats.

Nesting productivity is another measure of habitat quality. The Wood Thrush has become a model species whose nesting success appears to be closely linked to levels of forest fragmentation (Robinson et al. 1995). On-going studies of Wood Thrush nesting success in GSMNP indicate relatively high levels of productivity, suggesting that the contiguous forests in the Park may be serving as a regional population source (Pulliam 1988) for the species (Simons and Farnsworth 1995). Data from 1993 indicate slightly higher Wood Thrush productivity in the Park than on the Cherokee National Forest (Table 3). Similar monitoring at a regional scale will be necessary to determine relationships between forest management practices and the diversity and abundance of forest birds.

The message from our preliminary research is clear: an effective program for conserving avian diversity in the southern Appalachians must be based on ecosystem-scale data that integrate a variety of influences across the regional landscape. A successful landscape scale program must include: (1) a commitment to long-term studies, (2) a standardized regional-scale habitat map, (3) fully standardized sampling protocols for birds and their habitats, and (4) controlled, hypothesis-driven studies of bird/habitat relationships that will provide an understanding of how habitat quality, the spatial characteristics of habitats, and land management practices influence habitat suitability for forest birds. The results of this program could serve as one of several complementary indicators of the health of southern Appalachian ecosystems.

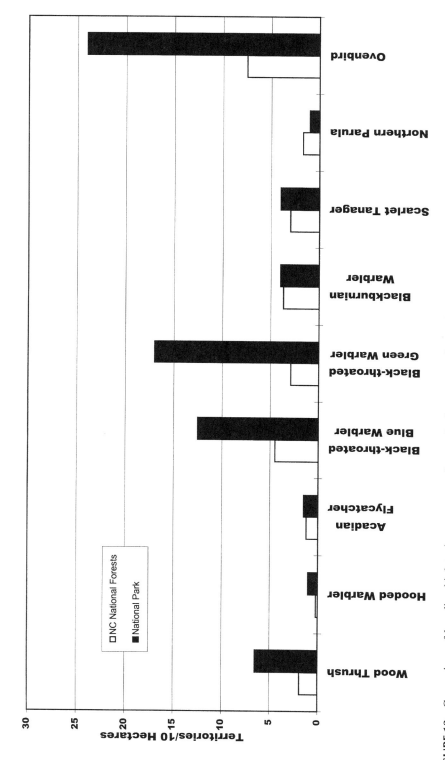

FIGURE 13 Comparison of breeding bird territory spot-mapping results from mature forest sites on North Carolina National Forests and old growth sites in Great Smoky Mountains National Park.

TABLE 2
High Priority Southern Appalachian Bird Species,
Based on Analysis by Hunter (1993)

Rank	Species	Rank	Species
1	Chestnut-sided Warbler	12	Eastern Wood-Pewee
2	Swainson's Warble	13	Yellow-throated Vireo
3	Louisiana Waterthrush	14	Black-throated Green Warbler
4	Wood Thrush	15	Blackburnian Warbler
5	Golden-winged Warbler	16	Kentucky Warbler
6	Cerulean Warbler	17	Scarlet Tanager
7	Worm-eating Warbler	18	Gray Catbird
8	Hooded Warbler	19	Blue-winged Warbler
9	Acadian Flycatcher	20	Northern Parula
10	Black-throated Blue Warbler	21	Prairie Warbler
11	Canada Warbler	22	Ovenbird

TABLE 3
Wood Thrush Productivity on Study Sites in Great Smoky Mountains National
Park and the Cherokee National Forest in 1993

Site	# Active nests 1993	# Successful nests	# Chicks fledged	Fledglings/successful nest
National Park	54	26	95	3.65
Cherokee National Forest	13	7	23	3.30

The components of such a program are beginning to take shape. Prioritization of species and habitats through Partners in Flight has greatly refined our ability to set appropriate research and conservation priorities. Numerous smaller studies such as ours have helped determine the relevant questions and methods to apply to a regional-scale study. Finally, digital map and information databases generated through the Southern Appalachian Assessment (see Berish et al., Chapter 7 of this volume) are providing an information base at an appropriate scale for ecosystem research and monitoring.

CONCLUSIONS

The diversity of southern Appalachian forest birds creates a dilemma for land managers seeking to conserve declining or sensitive species. The difficulty stems from the need to set objectives in an appropriate management context. Species of concern occur across a wide range of forest successional stages and management regimes, so that simple prescriptions are usually not possible. Protecting old growth, managing for snags, reducing clearcuts, or preserving large forest tracts will benefit some species, while management that creates edge and early successional habitats will benefit others.

Thompson et al. (1992, 1995) have discussed these trade-offs and the need to develop a hierarchical approach to management that scales down from the continental to the habitat-stand level. Large scale assessments of population trends and habitat requirements provided through programs such as Partners in Flight (Hunter 1993) and the Breeding Bird Survey (Peterjon et al. 1995) provide the best guidelines for evaluating regional and continental priorities. Because land use practices on private land tend to favor early successional and edge species, the best opportunities to manage for

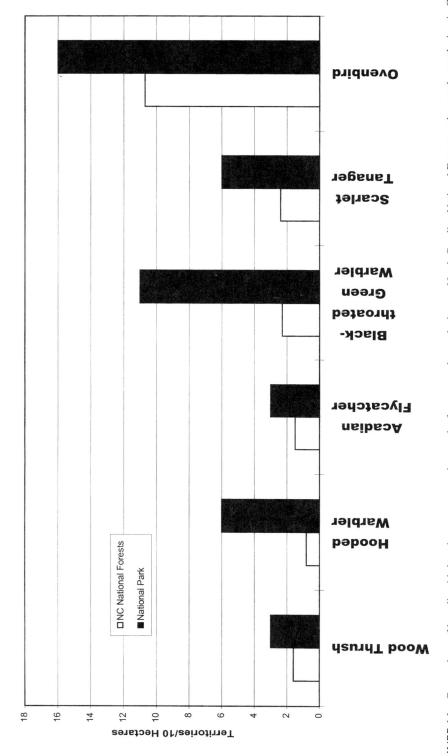

FIGURE 14 Comparison of breeding bird territory spot-mapping results from second growth sites on North Carolina National Forests and second growth sites in Great Smoky Mountains National Park.

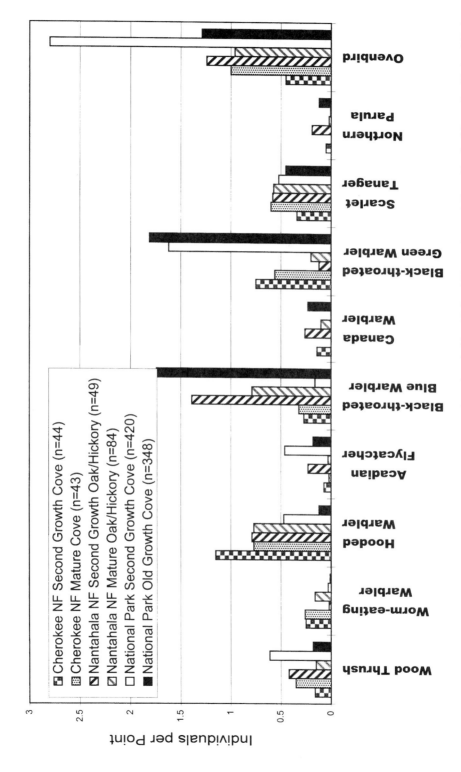

FIGURE 15 Comparison of breeding bird point count censuses on southern Appalachian National Forest sites and Great Smoky Mountains National Park.

late successional and forest interior species may often occur on public lands. The extent to which public lands should be managed to buffer or compensate for land use practices on private lands remains a major unanswered question of land management policy.

Finally, a successful ecosystem-scale program to conserve avian diversity in the southern Appalachians must have a strong public education component. The challenge of conserving these birds is shaped by their complex life histories and a web of interrelated social, economic, and ecological factors. Given the numerous environmental threats facing the region, one might legitimately ask, why worry about forest birds? Certainly we could invoke the "canary in the coal mine" argument and point to migratory bird population declines as a symptom of the impending free-fall in global biodiversity resulting from the unprecedented destruction of natural habitats that is currently taking place world-wide.

We could also argue that these birds perform an important ecological function. Warblers and related species have, in some circumstances, been shown to be important regulators of forest insects (Holmes 1990), but we would be hard pressed to convince anyone that loss of these birds will trigger an ecological collapse, even in an area as threatened by exotic insects as the southern Appalachians.

We believe that one of the most important reasons for directing conservation efforts toward these birds is their tremendous capacity to educate and inspire. A Black Poll Warbler, weighing less than a 25 cent piece, can fly from New England to Venezuela in 60 hours (McNair and Post 1993). The trip can include a 2000-km over-water flight at an altitude of 5000 meters, the metabolic equivalent of a person running 4 minute miles for 80 hours straight (Nisbet et al. 1963, Greenberg and Lumpkin 1991). The compelling stories these birds tell about the interconnection and interdependence of ecosystems ultimately provides the best incentive for preserving the southern Appalachian habitats on which they depend. Public education may well be the most important component of any strategy to protect the ecosystem because, ultimately, the political will to protect habitat for these birds will not derive from a wealth of ecological data but from their simple beauty, their remarkable life histories, and because knowing about them, enriches our lives.

RISK ASSESSMENT REPORT CARD

[*Editor's note:* The risk to managers taking action to enhance habitat for neotropical migratory birds centers around the lack of understanding of why some species are in such dramatic decline. Since their life cycle is so complex, it is difficult to ascertain the crux of the problem and how relevant breeding habitat is to the big picture. The Partners in Flight program provides a key focal point for understanding the context for taking action.]

Vision: B

Development of a hierarchical approach to management that evaluates management needs from the continental to the habitat-stand level is in its very early stages. Land management policies that relate management priorities on public lands to land use trends on private lands are lacking.

Resource risk: D

Several neotropical migrant species, particularly those associated with mature forests, are experiencing serious decline in the eastern U.S.

Socioeconomic conflicts: C–

Natural resource utilization versus conservation conflicts on public lands and private property remain unresolved although the value of old growth forest is becoming more accepted.

Procedural protocols: B+

Protocols for large-scale monitoring are available.

Scientific validity: A–

Understanding of species biology and habitat requirements exceeds that for most other groups of vertebrates.

Legal jeopardy: C

Legal challenges via the Endangered Species Act and Migratory Bird Treaty Act likely.

Public support: A

Public interest and support in bird conservation has historically always been high.

Adequacy of funding: B

Commitments of research funding through land management agencies, states, and Partners in Flight have been adequate. Funding for land management and conservation have been limited.

Policy precedent: C

Land management policies that relate management priorities on public lands to land use trends on private lands are lacking.

Administrative support: B+

Land management agencies participating in Partners in Flight have been strongly supportive of program objectives.

Transferability: B+

Research and monitoring techniques are broadly transferable. Necessary land management practices are often site and species specific.

[*Editor's note:* Here's hoping that stewards of forests are as inspired as the author by the beauty and courage of these magnificent creatures and gain resolve from the haunting calls of the wood thrush and oven bird while visiting an eastern forest.]

REFERENCES

Askins, R. A., J. F. Lynch, and R. Greenberg. 1990. Population declines in migratory birds in eastern North America. In D.M. Power, Ed. *Curr. Ornith.,* 7: 1–57.

Davis, M.B. 1993. Old growth in the east. The Cenozoic Society: Richmond, VT. 149 pp.

Doak, D.F. and L.S. Mills. 1994. A useful role for theory in conservation. *Ecology* 75: 615–626.

Duellman, W.E. and L. Treub. 1986. *Biology of Amphibians.* McGraw Hill: New York.

Finch, D.M. and P.W. Stangel, Eds. 1993. Status and Management of Neotropical Migratory Birds. USDA Forest Service, Gen. Tech. Rep. RM-229. Fort Collins, CO. 422 pp.

Greenberg, R. and S. Lumpkin. 1991. Birds Over Troubled Forests. Smithsonian Environmental Research Center: Edgewater, MD.

Harrison, S. 1994. Metapopulations and conservation. In Edwards, P.J., R.M. May, and N.R. Webb, Eds. *Large-Scale Ecology and Conservation Biology.* Blackwell Scientific, London.

Holmes, R.T. 1990. Ecological and evolutionary impacts of bird predation on forest insects: an overview. *Stud. Avian Biol.* 13:6–13.

Hunter, W.C. 1993. Species and habitats of special concern within the southeast region. In Status and Management of Neotropical Migratory Birds. Proc. national training workshop, Estes Park, CO. September 1992.

Keast, A. 1980. Spatial Relationships Between Migratory Parulid Warblers and Their Ecological Counterparts in the Neotropics. In A. Keast and E.S. Morton, Eds. *Migrant birds in the Neotropics: Ecology, Behavior, Distribution, and Conservation.* Smithsonian Institution Press: Washington, D.C.

MacArthur, R.H. 1957. On the relative abundance of bird species. *Proc. Nat. Acad. Sci. U.S.A.* 43:293–295.

McNair, D.B. and W. Post. 1993. Autumn migration route of Blackpoll Warblers: evidence from southeastern North America. *J. Field Ornithol.* 64(4):417–425.

Nisbet, I.C.T., W.H. Drury, and J. Baird. 1963. Weight loss during migration. Part I: deposition and consumption of fat by the Blackpoll Warbler, *Dendroica striata. Bird Banding* 34:107–138.

Peterjon, B.J., J.R. Sauer, and S. Orsillo. 1995. Breeding bird survey: population trends 1966–92. In LaRoe, E.T., G.S. Farris, C.E. Puckett, P.D. Doran, and M.J. Mac, Eds. Our Living Resources: a Report to the Nation of the Distribution, Abundance, and Health of U.S. Plants, Animals, and Ecosystems. U.S. Department of the Interior, National Biological Service, Washington, D.C. 530 pp.

Pulliam, H.R. 1988. Sources, sinks, and population regulation. *Am. Nat.* 132: 652–661.

Rappole, J. H., E. S. Morton, T. E. Lovejoy, and J. L. Ruos. 1983. Nearctic avian migrants in the Neotropics. USDI/FWS. U.S. Government Printing Office: Washington, D.C.

Robbins, C.S., D.K. Dawson, and B.A. Dowell. 1989a. Habitat area requirements of breeding forest birds of the middle Atlantic states. *Wildl. Monogr.* 103. 34 pp.

Robbins, C.S., J.R. Sauer, R.S. Greenberg, and S. Droege. 1989b. Population declines in North American birds that migrate to the neotropics. *Proc. Natl. Acad. Sci. U.S.A.* 86:7658–7662.

Robinson, S.K., F.R. Thompson III, T.M. Donovan, D.R. Whitehead, and J. Faborg. 1995. Regional forest fragmentation and the nesting success of migratory birds. *Science* 267: 1987–1990.

Sherry, T.W. and R.T. Holmes. 1995. Summer vs. winter limitation of populations: what are the issues and what is the evidence? In Martin, T.E. and D.M. Finch, Eds. *Ecology and Management of Neotropical Migratory Birds.* Oxford University Press: New York.

Simons, T.R., K. Rabenold, and G.L. Farnsworth. 1995. Characterization of deciduous forest breeding bird communities of Great Smoky Mountains National Park. Final report to the National Park Service. Gatlinburg, TN. 93 pp.

Simons, T.R., and G.L. Farnsworth. 1995. Evaluating Great Smoky Mountains National Park as a population source for Wood Thrush. 1994 Annual report to the National Park Service. 18 pp.

Temple, S.A. and J.R. Cary. 1988. Modeling dynamics of habitat-interior bird populations fragmented landscapes. *Conserv. Biol.* 2(4):340–347.

Thompson, F.R. III, W. D. Dijak, T. G. Kulowiec, and D. A. Hamilton. 1992. Breeding bird populations in Missouri Ozark forests with and without clearcutting. *J. Wildl. Manag.* 56:23–30.

Thompson, F.R., III, J.R Probst, and M.G. Raphael. 1995. Impacts of silviculture: overview and management recommendations. In Martin, T.E. and D.M. Finch, Eds. *Ecology and Management of Neotropical Migratory Birds.* Oxford University Press; New York.

Wilcove, D. S. 1985. Nest predation in forest tracts and the decline of migratory songbirds. *Ecology* 66:1211–14.

10 Management of a Large Carnivore: Black Bear

Joseph D. Clark and Michael R. Pelton

CONTENTS

Historical Perspective..209
Black Bear Ecology 211
    Habitat 211
    Population Dynamics 213
Research and Management 214
Ecosystem Management and Black Bear Conservation..215
    Case Study: the Southern Appalachian Black Bear Study Group 215
    Bear Conservation Issues..216
    Toward an Ecosystem Management Approach..219
Risk Assessment for Managers...221
References 222

HISTORICAL PERSPECTIVE

The black bear *(Ursus americanus)* is considered by many to be the embodiment of southern Appalachia. It is one of the most popular wildlife species in the region; a survey of visitors in Great Smoky Mountains National Park suggests that the chance of viewing black bears is a primary reason for visiting the park (Burghardt et al. 1972). Historically, black bears occurred in large numbers in southeastern highlands and played an important role, both physically and spiritually, to native Americans. The southeastern Indians treated the black bear with great respect and admiration (Hudson 1982). When a bear was killed, the flesh was often eaten as a symbol of courage, strength, and wisdom. According to Cherokee folklore, bears would gather on Clingman's Dome, the highest point in the Great Smoky Mountains, to talk and dance just before going into hibernation for the winter (Connelly 1968). As a material resource, bear fur was used as robes or bedding, the claws were worn as amulets, and the fat was prized for cooking, waterproofing, and as oil for the skin and hair. Much of today's fried cuisine that we commonly associate with southern culture actually originated with the Indians who used bear oil as the medium for this type of cooking.

During the period of European settlement (16th through 19th centuries), bears were an important resource and were commonly mentioned in the writings of early explorers to the region. On an expedition to Virginia in 1671, Captain Abraham Wood wrote that they killed a bear almost every day, often up to three per day (Raybourne 1987). Bear meat, skins, and oil were valued commodities during this period and there was a thriving market for the trade of this species. There were no regulations prohibiting the killing of bears at that time and, as a result, exploitation occurred year-round. Meshach Browning, hunting in the Alleghenies, reportedly killed 300 bears, 2000 deer *(Odocoileus virginianus)*, and countless wolves *(Canis rufus)*, panthers *(Felis concolor)*, and wildcats

0-57444-053-5/99/$0.00+$.50
© 1999 by CRC Press LLC

(Lynx rufus) during the early 1800s (Connelly 1968). Such feats were not uncommon during that period.

Along with this unprecedented exploitation, habitat destruction was occurring at a quickening pace. The old growth forests of chestnut *(Castanea denta)*, white *(Quercus alba)* and black oak *(Quercus velutina)*, beech *(Fagus grandifolia)*, and hickory *(Carya* spp.*)* that the early colonists encountered (Van Doren 1955) were logged for homesteads, cooperage, tanbark, heating, and to clear land for fields and livestock. The most extensive logging took place from 1900 through the 1920s for railroad construction, mine ties, and charcoal (SAMAB 1996a). Initially, this timber cutting probably was beneficial to bears by increasing the production of shade-intolerant soft mast species such as blackberry *(Rubus* spp.*)*, pokeberry *(Phytolacca americana)*, and blueberry *(Vaccinium* spp.*)*. However, these benefits were short lived. As the forests gave way to ax and plow, bears were forced into the most rugged and inaccessible reaches of the Appalachian Highlands. Shrinking habitat made bears even more susceptible to killing by hunters, trappers, and homesteaders.

The fact that these animals are today associated with rugged mountain habitats is largely by default rather than preference. These areas are relatively less productive than the settled bottomlands, and bears often visited farmsteads to feed on crops and livestock. As a result, homesteaders used all available means to eliminate the bear from the southern Appalachians. Sharpshooters killed bears with flintlocks and long rifles, bears were trapped with large steel traps and deadfalls (giving rise to places with names like Bear Trap Hollow and Bear Pen Gap), and a special breed of dog, the Plott hound, was even developed in the southern Appalachians for the sole purpose of hunting bears. Bounties on bears were enacted by local authorities; the bounty system existed until as late as 1972 in Virginia (Raybourne 1987).

In only about 200 years, most of the virgin forests in the southern Appalachians had been cleared. Families were forced from their farms because of eroded, overworked soils, whereby they then moved to other land where this cycle was repeated. Other large mammals such as the elk *(Cervus canadensis)*, bison *(Bison bison)*, red wolf, and cougar also disappeared. Bear numbers in the southern Appalachians reached their lowest ebb in the mid-1800s to early 1900s, when they could be found only in the most remote areas.

During the 1930s and 1940s, the precarious situation for black bears in the southern highlands slowly began to right itself. Those abandoned farms gradually began to revert back to forest and, as a result, early successional stage species such as white-tailed deer, cottontail rabbits *(Sylvilagus floridanus)*, and bobwhites *(Colinus virginianus)* flourished. As those young forests matured, bear numbers slowly began to respond. At the same time, the system of national parks and national forest lands in the region was rapidly being expanded by the federal government. The most dramatic of these land acquisitions was the establishment of Great Smoky Mountains National Park in the 1930s. The protection of those approximately 520,000 acres may have had more to do with the survival of the black bear in the southern Appalachians than did any other single event.

Just as forests in the southern highlands were beginning to recover, another event resulted in yet another setback for the black bear. A blight of the American chestnut all but eliminated this tree from upland forests in the East. First noticed on trees in New York in 1904, the disease spread rapidly, with outbreaks occurring in North Carolina and South Carolina as early as 1923 (Kuhlman 1978). This tree was once the dominant species in southern Appalachian uplands and, unlike oaks, is a consistent nut producer from year to year. Although chestnut, too, was heavily logged prior to the blight, its rapid growth and early nut-producing capabilities resulted in it being an important second-growth source of food for bears. After the blight eliminated the chestnut, bears had to settle for inconsistent acorn and hickory nut crops as their primary food. Undoubtedly, the blight greatly decreased the carrying capacity of southern highland habitats for these animals. As a result, bear demographics in the region today are significantly influenced by wide annual fluctuations in acorn production (Pelton 1996).

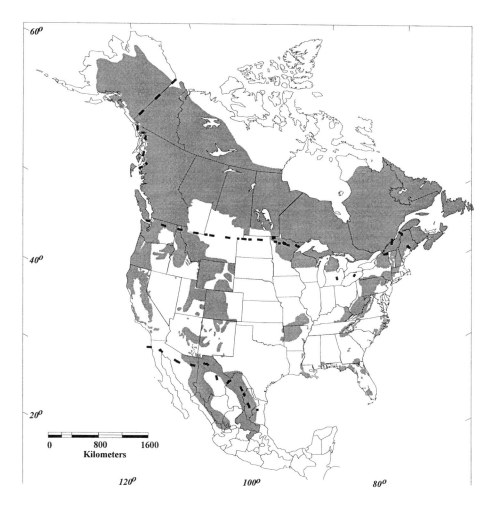

FIGURE 1   Black bear distribution in North America. (From Van Manen and Pelton, unpublished data).

Following that setback, bear numbers in the southern Appalachians gradually increased. Following World War II, most states had elevated the status of the bear to that of a game animal, which afforded it some legal protection from exploitation. Due to this protection, a recovering land base, and better knowledge of bear ecology, the species has recovered on most public lands.

BLACK BEAR ECOLOGY

Habitat

The range of the black bear is increasingly fragmented from New England southward through the mid-Atlantic and southeastern states (Figure 1). Bear populations in the southern Appalachians are largely centered on publicly owned land, principally national forests and national parks. However, public lands make up only about 5 million of the 37 million acres in the region, or 16% of the land base (K. Hermann, U.S. Environmental Protection Agency, unpubl. data).

Although the quantity of bear habitat in the region is limited, the quality of that habitat is good, due to the long growing season and rich diversity of foods. Black bears have high energy requirements but have an inefficient digestive system; therefore, food probably is the most critical habitat component. Bears need low fiber foods that have high digestible energy and high soluble nutrients (Eagle and Pelton 1983). They require certain amino acids in their diet which they fulfill by eating protein, mostly in the form of insects. In addition, their denning adaptation requires that they meet an entire year's nutritional needs within a period of 6 to 8 months.

Bear foraging ecology in the Appalachians can be classed into three major seasons: early spring, late spring–early fall, and late fall. After fasting in winter dens for several months, bears emerge in the early spring to find only scarce amounts of poorly digestible foods and, consequently, many continue to loose weight. Squawroot *(Conopholis americana)*, a parasitic plant of oaks, is an especially important spring food item, becoming available as early as late March. During late May, serviceberries *(Amelanchier arborea)*, young grasses, and insects are the primary food. Not until late spring and summer, when a variety of readily digestible foods become available, do bears begin to gain back some of the body weight they have lost. Blueberries, huckleberries *(Gaylussacia* spp.), blackberries, and black cherries *(Prunus serotina)* are progressively important foods as summer wanes (Eagle and Pelton 1983). Those foods commonly occur in forest openings created by timber harvesting, in clearings caused by natural disturbances, along roads, and at abandoned homesites.

Late fall has been identified as the most critical period for foraging bears. A tremendous amount of energy is required to prepare for the denning period and cub production. Most of that energy is derived from acorns, which are high in fats and carbohydrates, and body weight increases dramatically during late fall. Consequently, the mature oak–hickory forest type probably is the single most important habitat component in the region.

Bears drink regularly but water also can be important for thermoregulation by enabling bears to cool themselves during summer. Likewise, thickly vegetated areas of rhododendron *(Rhodendron* spp.) or mountain laurel *(Kalmia latifolia)* provide shade and, thus, a refuge from heat.

During denning, black bears must live off energy stored as fat. The less energy required by bears to maintain their body temperature during cold winter weather, the greater the probability of survival and successful cub rearing upon den emergence (Johnson and Pelton 1981). Well insulated den sites, therefore, should have advantages over poorly insulated ones. Bears in southeastern uplands may have more stringent denning habitat requirements than bears in more northern latitudes even though the winter climate is less severe. Snow serves to conceal and insulate winter dens in northern ranges but, in the Southeast, heavy winter rains can cause flooding of dens and bears lose heat much more rapidly when they are wet. Bears, therefore, tend to choose den sites that are well insulated, dry, and secure.

Bears have been known to den in a wide array of sites, including tree or rock cavities, in brushpiles, underneath logs, on open ground nests, and in road culverts. Female bears tend to make particular use of standing or fallen trees. Trees probably provide better insulation and offer more security than other available den types. Tree dens have been shown to reduce energy loss by up to 40% as compared to ground nests used by bears (Pelton 1996). Trees used as dens average about 100 cm in diameter and 250 to 300 years of age. Tree dens are more common at higher elevations because the trees are less accessible to loggers and are more prone to ice and lightning damage, eventually resulting in the formation of a cavity.

Escape cover and vehicle access are two habitat components that can be especially critical for bears in the Appalachian highlands. The region is rich in cultural history and, consequently, traditions are strong, the life-style is independent, and the use of hounds for hunting bears and other wildlife is commonplace. That combination of culture and a relatively dense bear population has produced efficient hunters and, in some cases, poachers. Consequently, road densities are a special concern because hunting and poaching efficiency increases with improved vehicle access. Habitat suitability is increased if road densities are kept low or if logging roads are closed after the timber has been harvested. Escape cover is another special requirement for regulating hunting efficiency and for reducing poaching. Rhododendron and laurel thickets, sapling-stage clearcuts, and wetlands can serve as areas where bears can take refuge. Dispersal corridors are especially important in the region where habitats and bear populations are fragmented.

POPULATION DYNAMICS

Black bears are a relatively long-lived species with relatively low mortality rates, reproductive rates, and population densities. Bear densities have been estimated at 0.53 to 1.58 bears per square kilometer in Great Smoky Mountains National Park in Tennessee (Coley 1995), and 0.71 to 1.49 bears per square kilometer at Shenandoah National Park in Virginia (Carney 1986). These densities are generally higher than bear densities in other regions of North America; however, densities in the southern Appalachians outside these two national parks (particularly private land) are lower, largely because of intolerance by humans. The black bear population in Great Smoky Mountains National Park, where the most extensive demographic analyses have been performed, have trended upward since 1973 when mark–recapture data were first collected. One subpopulation within the Smokies has increased on average by about 9% annually, based on mark–recapture population estimates (Coley 1995). McLean and Pelton (1994) reported stable population growth in two national forests adjacent to the Smokies and reported a 5% annual increase within the boundaries of the Park. Carney (1986) reported Shenandoah National Park bear population trends as stable. Bait-station surveys, a technique designed to detect gross changes in bear population densities, have trended upward in areas outside Great Smoky Mountains National Park as have hunter kill data in the region (Southern Appalachian Black Bear Study Group, unpublished data).

Mortality of black bears typically is low, particularly as the animals increase in age. Annual mortality rates for black bears in Shenandoah National Park have been estimated at 30% for cubs, 54% for yearlings, 39% for 2-year olds, and 21% for older bears (Carney 1986). In Great Smoky Mountains National Park, annual mortality and emigration was 39% for cubs, 7 to 13% for subadults, and 16 to 39% for adults (Coley 1995). Most mortality is human-induced and includes hunting, poaching, and road kills.

Black bears breed in summer and give birth while denning in January and February. The cubs accompany the adult females the entire year and again den with her the following winter. The following spring, the cubs (now yearlings) leave the adult female, who then again comes into estrus and breeds. Thus, unless a litter is lost, bears typically produce cubs every other year. Bears in the southern Appalachians usually breed at age 3 and produce their first litter around their 4th birthday. In the Great Smoky Mountains, the average number of cubs counted from 74 litters was 2.2, ranging from 1 to 4 (Coley 1995). In Shenandoah National Park, average litter size from cub counts was 2.0 (Carney 1986). An average of 2.7 placental scars were present from an analysis of 108 reproductive tracts from Georgia, Tennessee, and North Carolina (Carlock et al. 1983); the number of cubs recruited into the population would be lower, however, due to intrauterine and post-partum mortality.

Dispersal is an important demographic parameter that reduces localized bear numbers and enables genetic interchange to take place. Very little is known about dispersal in black bears except that it primarily occurs in the subadult male segment of the population. Apparently, subadult females are able to establish home ranges within the ranges occupied by their mothers, whereas subadult males are forced to disperse in order to establish home ranges that do not conflict with those of adult males (Rogers 1987; Schwartz and Franzmann 1992). Although subadult and adult mortality rates presented above likely include dispersal, actual rates of dispersal have not been estimated in the southern Appalachians.

Bear populations in the region are influenced by a variety of biological factors, the most important being annual hard mast production. Production of acorns is the major force behind the population dynamics of the species. Acorn production in southeastern uplands is high (up to 200 kg/ha) but extremely variable (Powell and Seaman 1990). A poor acorn crop, particularly white oak, has a marked effect on adult and cub survival, movements and dispersal, breeding, and cub production. Total acorn failures have been known to result in increased hunter harvests, heavy highway mortality, and almost complete reproductive failure in bear populations (Pelton 1989).

RESEARCH AND MANAGEMENT

Pelton (1980) states that four developments during the 1960s to 1970s greatly improved the ability of biologists to collect data on bear populations: (1) the Aldrich spring-activated snare which enabled more practical backcountry bear trapping, (2) safer and more effective immobilization drugs, (3) better tooth sectioning and aging techniques, and (4) more sophisticated and reliable radio telemetry systems. Since that time, the development of habitat modeling techniques using Geographic Information Systems (GIS) and improved taxonomic analyses with DNA fingerprinting can be added to the above list. All these methods have been used to study black bears in the southern Appalachians.

Research in the region has largely centered on public lands. Bear research in Great Smoky Mountains National Park began in 1968 by University of Tennessee (UT) researchers and today represents the longest continuous bear research effort in the world. Many of the techniques commonly used to study black bears today were pioneered by students from UT. Over the years, bear research in the Smokies has concentrated on such varied topics as food habits, movements, habitat use, denning ecology, reproduction, behavior, bear–human interactions, population monitoring, population demographics and modeling, and habitat modeling using GIS. Great Smoky Mountains National Park has served not only as the important core of the bear population in the southern Appalachians, but also the testing grounds for many of the research and management techniques presently being used. This relatively roadless wilderness, free of timber harvest activities and hunting, has served as an important control and test population where the regionally applied bait-station survey was developed, where trapping techniques were modified or new ones created, and where other aspects of handling and monitoring bear populations were evaluated. Similar research at the UT has been conducted on the adjacent Pisgah, Nantahala, and Cherokee National Forests.

In Virginia, extensive research has been conducted by Virginia Polytechnic Institute and State University (VPI) students (Kasbohm and Vaughan 1992). At Shenandoah National Park, they have investigated bear habitat use and population dynamics and have evaluated the effects of the gypsy moth *(Lymantria dispar)* on bear movements. VPI researchers continue to study the relationship between bear nutrition and reproduction using captive bears. Research is currently underway to investigate the population dynamics of black bears in the George Washington National Forest in Virginia. Also at VPI, an extensive taxonomic study of the bears of the southeastern U.S. is being conducted. This effort involves the use of DNA fingerprinting and other techniques to determine the genetic similarities between bear populations in the region. North Carolina State University, the University of Georgia, Clemson University, various state wildlife agencies, and a number of other institutions have also been involved in black bear research in the southern Appalachians over the years. Together, these researchers have generated a wealth of data on the ecology of bears in the region.

Most management for black bears in the region is effected by state wildlife agencies, the National Park Service, and the U.S. Forest Service. Depending on available habitat, population level, and management philosophy, however, management goals vary considerably. Bear management programs administered by these agencies generally consist of harvest regulation, law enforcement, habitat management, population monitoring, and nuisance bear assistance.

In states where bear populations are sparse (e.g., Kentucky, Maryland, and South Carolina), harvests are light or the species is totally protected. In other states with large bear populations and strong hunting traditions, the regulation of hunting is a major management activity (e.g., Virginia, North Carolina). Virginia bear harvests usually range from 600 to 700 annually. In North Carolina, bear harvests since 1982 usually are from 200 to 300 annually, but have ranged as high as 600 (1992).

Hunting regulations, consisting of setting season lengths, timing, method of take, and bag limits, vary widely by state. The use of hounds is generally considered to be one of the least conservative methods of bear hunting and, consequently, only those states with the largest bear populations (North Carolina, Tennessee, Virginia, West Virginia) generally permit their use. There is a recent trend toward regulations that protect the adult female segment of the population. Females den earlier than males and later hunting seasons (late November and December) have been shown to reduce the proportion of adult females harvested. Consequently, many states in the region have moved their hunting seasons to this period after the females have denned. In addition, a system of "sanctuaries" or large areas closed to hunting have been established in some states. These provide a nucleus of breeding females that are protected from hunting. No hunting over bait or trade in bear parts (i.e., gall bladder, teeth, claws, skins) is permitted in southern Appalachian states.

Laws pertaining to bears are primarily enforced by the state wildlife management agencies, but federal wildlife law enforcement officials (e.g., National Park Service, U.S. Fish and Wildlife Service) also play a valuable role. Bear poaching is common in the southern Appalachians and a sting operation was completed in 1988 in an attempt to reduce the illegal harvest. In this operation, nicknamed "Operation Smokey," some 43 individuals in Tennessee, Georgia, and North Carolina were arrested and charged with 130 state and federal violations involving poaching and selling bear parts (Warburton 1992). The demographic consequences of illegal take on the bear populations of the southern Appalachians is largely unknown.

Habitat management for bears is most effectively accomplished working in concert with the U.S. Forest Service because that agency is the largest public landowner and it conducts most of the habitat manipulation work. State wildlife officials work with their local U.S. Forest Service ranger districts to review timber sales to make sure that core areas and corridors are protected, denning habitats are sufficient, there is an adequate source of hard mast, and that vehicular access is regulated. Other habitat management activities include burning and creating openings for fruit production, and the planting of fruit trees and shrubs.

Bear numbers are monitored each year by checking the hunter kill; many states require checking and will extract a tooth for aging and collect a reproductive tract to estimate litter sizes. A systematic survey of bear visits to bait sites (i.e., the bait-station survey) is conducted annually in almost all the states in the region. This survey provides a relative index of bear population levels so that trends can be assessed. All states in the region report stable to slightly increasing bear populations. In addition, surveys of hard mast (i.e., acorns, hickory nuts) have been conducted annually in the region for a number of years and many states now are conducting annual surveys of soft mast (i.e., fruits and berries).

Tennessee and Virginia have been actively involved in bear reintroduction efforts. In Tennessee, bears have been moved from the Great Smoky Mountains National Park to the Big South Fork National River and Recreation Area, along the Kentucky–Tennessee border on the Cumberland Plateau, where bears had been previously extirpated. Similar efforts have been underway in Virginia to stock areas of suitable bear habitat where bears are nonexistent or where densities are low.

Finally, most of the southern Appalachian states, along with both Shenandoah and Great Smoky Mountains national parks, have nuisance bear programs. For many agencies, this is where the bulk of the human and monetary resources for bear management are invested. Such programs generally consist of education, preventative measures such as electric fencing for bee yards, and trapping and relocation of problem bears. Nuisance bear programs and policies vary by agency, ranging from aggressive to nonexistent, depending on the level of damage and agency philosophy. Several agencies (e.g., Great Smoky Mountains National Park, the city of Gatlinburg, and the Tennessee Wildlife Resources Agency) are working together to develop better solid waste management programs in municipal areas where problems occur.

ECOSYSTEM MANAGEMENT AND BLACK BEAR CONSERVATION

Case Study: the Southern Appalachian Black Bear Study Group

During the 1960s a number of state wildlife agencies in the southern Appalachians reported decreasing legal harvests of bears. Because monitoring of the harvest was the only mechanism for tracking bear population trends at that time, it was assumed that those decreased harvests likely reflected a declining population. This population decline was probably due to a combination of both increased access to bear habitat on national forests and increased hunting efficiency, facilitated by the use of more sophisticated, high-tech equipment such as CB radios, four-wheel drive vehicles, and radio-collared hunting dogs.

Due to their large effective population size and extensive range, it has long been recognized that the southern Appalachian bear population was a contiguous resource shared among several states. Concerns over the apparent dwindling population led states of North Carolina and Tennessee to take significant management actions. In 1971, the North Carolina Wildlife Resources Commission established a network of bear sanctuaries scattered throughout the Pisgah and Nantahala national forests to protect breeding age females. Because of the relatively narrow band of habitat available to bears in east Tennessee in and around the 600,000-acre Cherokee National Forest, the Tennessee Wildlife Resources Agency completely closed the bear hunting season from 1970 to 1973.

Up until the late 1960s and early 1970s, little detailed research had been conducted on black bears anywhere in North America, much less the southern Appalachians. Techniques for capture, handling and monitoring bears were in their infancy. Data were scattered and scanty among unpublished reports from state wildlife agencies. Although it was recognized that the bear population was a shared resource, the southern Appalachian population was only being superficially studied and monitored through four separately evolving and independent efforts. Consequently, there was no history of standardization or consolidation of regional data on this species, making interpretations of data confusing and coordinated management extremely difficult.

In 1974 at the 2nd Eastern Workshop on Black Bear Management and Research, biologists from the southern Appalachian states and the University of Tennessee met informally to discuss the mutual problem of dwindling bear numbers. The Tri-State Black Bear Study was conceived at that meeting and, in 1976, three cooperating states (North Carolina, Georgia, and Tennessee) and the UT's Department of Forestry, Wildlife and Fisheries began a 5-year cooperative project. Major objectives were to pool and standardize data and to initiate new research and management programs that would put more adequate population monitoring into place. In 1983, results of the Tri-State Black Bear Study were published; this 286-page compendium consolidated, analyzed, and summarized all available data on the regional bear population. From that report a series of multiorganizational management recommendations were formulated giving future direction for a long-term bear management program for the southern Appalachians.

With the official inclusion of South Carolina, the National Park Service, and the U.S. Forest Service within the group in 1990, the Tri-State Black Bear Study became known as the Southern Appalachian Black Bear Study Group (Bear Study Group). This multiagency group is currently responsible for conducting a comprehensive monitoring and management program for bears in the region. Field data are collected, assimilated, and analyzed biannually by this group. Besides annual harvests, data from bait-station surveys, hard and soft mast crops, road kills, incidental kills, management actions (i.e., nuisance bear complaints), and trapping efficiency from ongoing research studies are collectively analyzed. Problems associated with this species are now consolidated at the regional level rather than separately in each individual state. More recently, the states of Virginia, West Virginia, and the Virginia Polytechnic Institute and State University have expressed interest in becoming part of Bear Study Group.

BEAR CONSERVATION ISSUES

Perhaps the most important influence on bear populations in southern Appalachia, both now, and in the future, is human population growth. The human population of the southern Appalachians grew by almost 28% from 1970 to 1990, compared to only 16% for the entirety of those seven states that make up the region. The region is projected to grow an additional 12.3% by 2010, with most of that growth in northern Georgia, eastern Tennessee, western North Carolina, and northern Virginia (SAMAB, 1996b). The influx of people to a region where bear populations are already severely fragmented will significantly reduce habitat for these animals. Bear habitat reduction and fragmentation continues on a large scale due to urban sprawl, increased development for summer or retirement homes, and road building, often directly adjacent to prime bear range (Figure 2). Bears that have historically used habitats where those developments have occurred likely will come into close contact with humans. Open garbage dumps and unprotected apiaries, for example, inevitably attract bears. This often results in conflicts and the subsequent reduction in bear numbers.

In the southern Appalachians, only about 16% of the total land base remains in public ownership as national forests, national parks, national wildlife refuges, or state wildlife management areas. That represents a relatively secure land base, the majority of which consists of upland habitats. The bulk of the land base in the region, however, is in private ownership which most often results in the elimination of bears. The sad truth is that the continued urban and suburban growth of the region probably precludes the existence of bears in those areas. For now, the current level of fragmentation has not prevented occasional movements or dispersal, thus maintaining the genetic viability of the regional population. However, without some level of management in the future the risk of crossing unsuitable habitats will likely become formidable.

As an example, black bears now occupy over 5.9 million acres in the mountains of Georgia, North Carolina, South Carolina, and Tennessee; this area includes four national forests and one national park. Although this landscape is forested, mountainous habitat, consisting primarily of publicly owned lands, it is regularly disrupted by unsuitable habitats such as major highways and human development. Bears, particularly males and translocated nuisance animals, continue to traverse these human barriers, often at great risk. One of the most noteworthy barriers, I-40 between Newport, TN and Asheville, NC, serves as an obvious deterrent to bear movements between the adjacent Pisgah National Forest and Great Smoky Mountains National Park. Many bears are killed annually trying to cross; a few successfully cross this "ribbon of risk". For others, the risks of crossing are simply too great for any rewards that could be gained (i.e., breeding or feeding) and thus, they remain isolated from other bears and habitats. Corridor management must be a high priority to maintain the integrity and stability of bears in the region. The long-term demographic, social, and genetic consequences of this fragmentation are yet unknown.

The rural character of the southern Appalachians is steadily giving way to more trade- and service-oriented activities leading to a proliferation of housing projects (primary or secondary homes) and strip development along state or county highways adjacent to occupied bear habitat. As more humans settle in the region, the problems of fragmentation, isolation, and bear–human interactions will intensify, not only along the ribbons of risk but also along any interface between bear habitat and human habitat in the region. These interactions will become particularly difficult during years of hard mast shortage when bears are forced to forage over larger areas. Dealing with all the issues surrounding bear–human interactions in the future can only be handled through a multiagency regional approach such as that established by the current Bear Study Group.

Currently, the southern Appalachian population is generally regarded as healthy and stable to increasing. Although a number of factors have been attributed to this increase over the past 20 to 30 years, the two most relevant and influential factors likely have been the maturation of the southern Appalachian oak forest (natural) and the creation of sanctuaries, either de facto or planned. Approximately one million of the 5.9 million acres of occupied habitat in Georgia, North Carolina, South Carolina, and Tennessee are currently regarded as bear sanctuaries. The sanctuary concept has likely built resiliency into the regional population, allowing it to sustain occasional poor hard mast years without extensive losses. The extensive network of sanctuaries in the southern Appalachians is a direct product of the coordination made possible through the Bear Study Group.

Within the occupied range of black bears in the southern Appalachians, a range of scenarios occur that dictate a variety of management strategies. Certainly the large protected core of the region, Great Smoky Mountains National Park, is the focus of many of these special management situations. The juxtaposition of this protected population with areas that are becoming rapidly

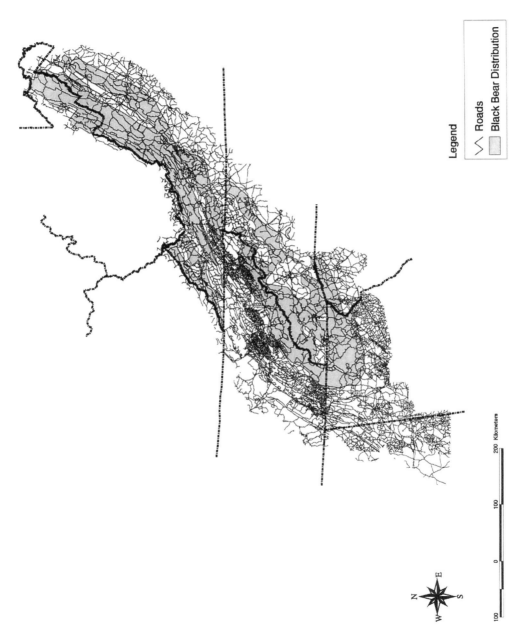

FIGURE 2   Black bear distribution in relation to class 1 and 2 roads in the southern Appalachians.

developed, areas that were once historic, low elevation fall feeding sites for bears, results in severe seasonal nuisance problems. The interface between these two starkly contrasting habitats is becoming increasingly well-defined. Conversely, where a protected population is juxtaposed with areas where bears are hunted (e.g., national forest lands), such nuisance situations usually do not occur. Most of the other sanctuaries in the region are buffered by adjacent national forests.

The relatively narrow band of mountainous, national forest habitat in Tennessee also sharply contrasts with the more expansive occupied habitat in North Carolina. Each situation requires different management strategies, for example, shorter and later hunting seasons in Tennessee versus more liberal seasons in North Carolina. Whereas Tennessee and North Carolina allow hunting with dogs, Georgia allows only still hunting. Also, bears residing in different habitats with varying levels of human intrusion react differently in terms of their movement ecology and population dynamics. Consequently, harvest sex/age ratios and success rates often reflect these environmental variables. Much can ultimately be learned from these different scenarios, and they warrant monitoring. However, the bottom line is the regional health of the species as reflected by landscape-level parameters. A regional approach facilitated through the Bear Study Group decreases the possibility that a subset of the whole is accepted as the rule when, in fact, it is the exception.

Furthermore, there are continuing improvements in methods of bear harvest (the use of CB radios, all-terrain vehicles, dog radio collars) and growing economic incentives to kill bears (i.e., illegal trade in skins, gall bladders, claws, and teeth). Although the extent of bear poaching in the southern Appalachians is unknown, serious local population losses can and do occur. The above issues are all exacerbated due to the inherently low biotic potential of black bears and because the historic stable food source of the species (chestnuts) has been replaced by an unstable substitute (acorns). This situation is further complicated by mast failures and oak mortality caused by the southerly range expansion of the gypsy moth. Recent reseach from Virginia, however, suggests that soft mast foods may help ameliorate acorn crop failures (Kasbohm et al. 1996). Nevertheless, long-term changes to bear habitats that, at the time, seem small, can pose a significant cumulative threat. Such habitat changes may soon test the resilience of the southern Appalachian bear population.

TOWARD AN ECOSYSTEM MANAGEMENT APPROACH

The Bear Study Group is an excellent example of a successful long-term monitoring program coordinated by a multiagency cooperative effort. The group has functioned for over 20 years, compiling an increasingly valuable data set. In retrospect, the Bear Study Group has practiced many of the principles of ecosystem management and may be one of the earliest and best examples of interjurisdictional wildlife management in the region.

For example, the Bear Study Group embraces the concept of sustainability, one of the key principles of the ecosystem management philosophy. When harvests began to decline, it was recognized that the goal of sustainability could not be achieved without a coordinated, multiagency effort. Being a landscape-level species, bears have to be managed across jurisdictional lines, even though management strategies, tools, and objectives may vary. Nevertheless, managers have to find common ground and work together; this has been effectively accomplished through the Bear Study Group.

Besides the political difficulties due to the cross-jurisdictional nature of the southern Appalachian bear population, the species has to be managed on the ecosystem level, at an appropriate biological scale. Unlike many species of animals and plants, this could not be done within a watershed, river basin, or even a physiographic region, but had to be accomplished across entire physiographic regions. For example, bear population demographics (e.g., natality, population growth, dispersal) in the Great Smoky Mountains National Park have a major effect on those in the adjacent Pisgah National Forest and both have to be considered in tandem. This landscape-scale management strategy was an immense challenge to the Bear Study Group but is a necessary component of successful ecosystem management.

Finally, sound science has always been the pillar of the Bear Study Group. The group has always based decisions on biological data, along with social, cultural, economic, and political factors. This, of course, is a key principle of ecosystem management.

The Bear Study Group is a model of interagency cooperation and landscape-level planning. Although the Bear Study Group has been extremely successful, a necessary future direction should be to become better integrated with other resource managers and users in the region. Bear biologists have often been accused of "single species myopia" (i.e., nearsightedness) and, in the future, we will have to do a better job of balancing the needs of bears with those of other species, ecological communities, and the public. This is the ultimate step towards true ecosystem management. Given the changing needs of society, the emphasis on biological diversity and ecological communities, and the policy changes within state and federal agencies towards more holistic management, the Bear Study Group cannot afford to isolate itself from these other natural resource managers and issues.

One positive step towards a more coordinated approach has been the Southern Appalachian Assessment (SAMAB, 1996b; also see Berish et al., Chapter 7 of this volume). This assessment is an evaluation of current environmental conditions, trends, and impacts of the region to assist those involved with resource planning for ecosystem management. Independent technical reports address terrestrial resources, aquatic resources, air quality, and social/cultural/economic issues. The assessment is a cooperative effort between the U.S. Forest Service, U.S. Environmental Protection Agency, Tennessee Valley Authority, U.S. Fish and Wildlife Service, U.S. Geological Survey, and the National Park Service.

The initial goal of the assessment was to compile existing data on southern Appalachian ecosystems, evaluate those data, and determine the status and future trends for natural resources in the region. One of the major products of this assessment is a working database on the region, much of it in GIS format, that is accessible and available to anyone. Although the initial work centered principally on locating and evaluating existing data, there is much potential for using this database to address other issues such as projecting human population growth, predicting spread of exotic pests, or for modeling wildlife–habitat relationships. If, for example, working models of bear habitat needs were in place, environmental perturbations or habitat management options could be evaluated for black bears on a regional basis. If other models were available for other species, the trade-offs resulting from management actions (e.g., the development of a road) could be evaluated for each species and a decision made based on the desired objective. Some of this work is already underway. A bear habitat suitability map based on road density information was initially developed as part of the assessment. Following that, a more detailed model developed by van Manen (1994) for the Great Smoky Mountains is being applied to the entire region, based on landscape data obtained during the assessment (R. Tankersley, University of Tennessee, pers. commun.). This would be a valuable tool to identify critical habitats to address the extensive losses and fragmentation currently occurring in the region; the Bear Study Group could play a significant role in this process. However, the issue of habitat destruction is much larger than just the Bear Study Group. These habitat losses affect far more than just bears and they must be addressed in a forum that involves all the affected land management agencies and the public. However, it is oftentimes difficult to find a mechanism where this interagency coordination can take place. SAMAB might be an effective conduit to address such a broad regional problem. The Bear Study Group, working through SAMAB, could play a valuable role to insure that the needs of bears are balanced against those of cerulean warblers *(Dendroica cerulea)*, brook trout *(Salvelinus fontinalis)*, spruce–fir, mountain bogs, and Smoky madtoms *(Noturus baileyi)*.

Finally, public education about bears can reap enormous rewards. In the southern Appalachians, bears do not commonly occur in areas of even light human habitation. Bears have been known to flourish in areas of heavy human use, however, if their presence is accepted by the public (G. Alt, Pennsylvania Game Commission, pers. commun.). There may be opportunities to expand bear range on private lands, but only if human attitudes toward bears are changed. Again, the Bear Study Group, possibly coordinated through SAMAB, could develop a regional program to address this need.

To many, ecosystem management implies that single species management is no longer appropriate. We could not disagree more. Valuable contributions to ecology and wildlife science have come from both systematists (i.e., top down) and species biologists (i.e., bottom up). We believe that ecological knowledge is best advanced if both scientific disciplines are used simultaneously. It is important to have a good understanding of how all the pieces fit together and function as a whole, but it is equally important to have thorough knowledge of the workings of each of the individual components. In many cases, it is much more efficient to individually study the component pieces. Because bears pose some unique management problems and as long as bears are harvested and we measure that harvest in terms of numbers of *bears* and not kilograms of *biomass*, it will be vital to understand species biology and to manage bears on a species level. That can only be done if we have a measure of annual mortality, recruitment, population growth, etc.

Furthermore, principles of ecosystem management imply passive management to some. Again, we urge caution. Black bears primarily occur in the southern Appalachians on public land. On most of those lands, timber harvest, road building, wildland recreation, hunting, fishing, and other activities routinely take place. Many keystone species are no longer present and a number of exotic pests (plant, animal, and disease pathogens) have greatly altered and are altering ecosystems. These are not naturally regulating ecosystems (if such a thing exists) and, consequently, we believe that sustainability and conservation of biological diversity can only be achieved with planned, scientifically based, active management.

The black bear is a historically and biologically vital component of southern Appalachian ecosystems. Management of this landscape-level species presents special problems that, under today's environmental pressures and political climate, can only be accomplished with a management approach that is well integrated with the public, effected across jurisdictional lines, and studied across scientific disciplines. However, despite the "single species bashing" often associated with the ecosystem management philosophy, the black bear deserves equal consideration along with all the other possible management objectives for the region. While we applaud the holistic efforts to take a broader perspective, let's not substitute farsightedness for nearsightedness, but rather strive to keep all ecosystem aspects in clear focus.

RISK ASSESSMENT FOR MANAGERS

To summarize, we recommend that the Bear Study Group attempt to integrate with other resource managers in the region to address the broader habitat issues. Habitat loss and fragmentation is the primary threat to bears in the region, but it also is a major impact on a wide range of other natural resources in the southern Appalachians. That might best be addressed through the framework of SAMAB. Second, we suggest that there is some potential to increase bear numbers on private lands, but only if public attitudes are dramatically changed. Again, this might be addressed by the Bear Study Group. Our *Risk Assessment Report Card for ecosystem managers* reads as follows:

Vision: A

The interagency management approach for black bears represents one of the oldest and most well-established programs in the region. We have a good idea of the threats facing bears and have a ready-made infrastructure that could be used to deal with it.

Resource risk: B

Although the bear population has recovered in recent years and appears stable or increasing in most areas, the population is limited largely to public lands. Shrinking habitats outside those publicly owned areas likely will have serious consequences to the population in the not-so-distant future.

Resource conflicts: B

These bear management actions would enhance a number of other resources (wildlife, water, air quality).

Socioeconomic conflicts: C

Although bears can be a valuable resource for tourism, bears can conflict with typical southern Appalachian land uses. Bears are known to cause nuisance problems, and land uses such as urban development and road building can be incompatible with bear management. Additionally, although timber harvest can be beneficial to bears by increasing soft mast production, certain types and levels of harvest can be detrimental and can cause conflicts.

Procedural protocols: A

Protocols for management and monitoring are in place.

Scientific validity: A

We have a good general knowledge of what needs to be done and we are actively developing specific tools for identifying critical habitat components.

Legal jeopardy: A

Not likely to be challenged.

Public support: B

Although bears are generally classed as a charismatic species and receive high public support, nuisance problems can and do occur. This often results in negative feelings toward bears by landowners.

Adequacy of funding: C

Funding for bear management is sorely inadequate but is better than for many other wildlife species. Funds for large-scale habitat protection, however, are lacking.

Policy precedent: A

Habitat protection and acquisition common policy.

Administrative support: A

Strong administrative support from state and federal agencies.

Transferability: B

Although bear management shares many commonalities across regions, there are some unique problems associated with bear management in the southern Appalachians (i.e., fragmentation, unique habitat resources).

REFERENCES

Burghardt, G. M., R. O. Hietala, and M. R. Pelton. 1972. Knowledge and attitudes concerning black bears by users of the Great Smoky Mountains National Park. *Int. Conf. Bear Res. Manage.* 2:255–273.

Carlock, D. M., R. H. Conley, J. M. Collins, P. E. Hale, K. G. Johnson, A. S. Johnson, and M. R. Pelton. 1983. The tri-state black bear study. Tenn. Wildl. Resour. Agency Tech. Rep. No. 83-9. Knoxville. 286 pp.

Carney, D. W. 1986. Population dynamics and denning ecology of black bears in Shenandoah National Park, Virginia. Proc. East. Workshop Black Bear Res. Manage. 8:243–245 (abstract only).

Coley, A. B. 1995. Population dynamics of Black Bears in Great Smoky Mountains National Park. M. S. thesis. University of Tennessee, Knoxville. 180 pp.

Connelly, T. L. 1968. *Discovering the Appalachians.* Stackpole Books: Harrisburg, Pa. 223 pp.

Eagle, T. C. and M. R. Pelton. 1983. Seasonal nutrition of black bears in the Great Smoky Mountains National Park. *Int. Conf. Bear Res. Manage.* 5:94–101.

Hudson, C. 1982. *The Southeastern Indians.* University of Tennessee Press: Knoxville. 574 pp.

Johnson, K. G., and M. R. Pelton. 1981. Selection and availability of dens for black bears in Tennessee. *J. Wildl. Manage.* 45:111–119.

Kasbohm, J. W. and M. R. Vaughan. 1992. Virginia Polytechnic Institute and State University status report. *East. Black Bear Workshop* 11:87–88.

Kasbohm, J. W., M. R. Vaughan, and J. G. Kraus. 1996. Effects of gypsy moth infestation on black bear reproduction and survival. *J. Wildl. Manage.* 60:408–416.

Kuhlman, E. G. 1978. The devastation of American chestnut by blight. Pages 1–3 *in* MacDonald, W. L. et al., Eds., Proc. American Chestnut Symp. West Virginia University, Morgantown.

McLean, P. K. and M. R. Pelton. 1994. Estimates of population density and growth of black bears in the Smoky Mountains. *Int. Conf. Bear Res. Manage.* 9:253–261.

Pelton, M. R. 1980. Black bear. Pages 504–514 *in* J.A. Chapman and G.A. Feldhamer, Eds. *Wild Mammals of North America; Biology, Management, Economics.* Johns Hopkins University Press: Baltimore.

Pelton, M. R. 1989. The impacts of oak mast on black bears in the southern Appalachians. Pages 7–11 *in* C. E. McGee, Ed., Proc. Workshop on Southern Appalachian Mast Manage. University of Tennessee, Knoxville.

Pelton, M. R. 1996. The importance of old growth to carnivores in eastern deciduous forests. In M. B. Davis, Ed., *Eastern Old Growth Forests:* Island Press. 420 pp.

Powell, R. A. and D. E. Seaman. 1990. Production of important black bear foods in the southern Appalachians. *Int. Conf. Bear Res. Manage.* 8:183–187.

Raybourne, J. W. 1987. Pp. 104–117 *in* Restoring America's Wildlife: 1937–1987. U.S. Department of Interior, Fish and Wildlife Service. U.S. Government Printing Office, Washington, D.C.

Rogers, L. L. 1987. Factors influencing dispersal in the black bear. Pages 75–84 in B. D. Chepko-Sade and Z. T. Halpin, Eds. *Mammalian Dispersal Patterns: the effects of Social Structure on Population Genetics.* University Chicago Press: Chicago.

SAMAB (Southern Appalachian Man and the Biosphere). 1996a. The Southern Appalachian Assessment Summary Report. Report 1 of 5. U.S. Department of Agriculture, Forest Service, Southern Region. Atlanta. In Press.

SAMAB (Southern Appalachian Man and the Biosphere). 1996b. The Southern Appalachian Assessment Aquatics Technical Report. Report 4 of 5. U.S. Department of Agriculture, Forest Service, Southern Region. Atlanta. In Press.

Schwartz, C. C. and A. W. Franzmann. 1992. Dispersal and survival of subadult black bears from the Kenai Peninsula, Alaska. *J. Wildl, Manage.* 56:426–431.

Van Doren, M., Ed. 1955. *Travels of William Bartram.* Dover Publications: New York. 414 pp.

van Manen, F. T. 1994. Black Bear Habitat use in Great Smoky Mountains National Park. Ph.D. dissertation. University of Tennessee, Knoxville. 212 pp.

Warburton, G. S. 1992. North Carolina status report. *East. Black Bear Workshop* 11:64–68.

Wathen, W. G. 1983. Reproduction and Denning of Black Bears in the Great Smoky Mountains. M.S. thesis, Univ. Tennessee, Knoxville. 135 pp.

11 Species Repatriation: Red Wolf

Chris F. Lucash, Barron A. Crawford, and Joseph D. Clark

CONTENTS

Historical Background 225
Translocation Ecology 226
The Southern Appalachians 227
Releases of Red Wolves in Great Smoky Mountains National Park..................................228
    Background...228
    Experimental Release 229
    Subsequent Releases 229
Wolves and Livestock Depredation 236
Obstacles to Reestablishment 239
    Limited Land Base 239
    Small Staff...239
    Hybridization with Coyotes 240
    Taxonomic Status 240
Keys to the Future Success of the Reintroduction Effort 241
Risk Assessment for Managers...243
References 245

HISTORICAL BACKGROUND

The red wolf (*Canis rufus*) once ranged throughout the southeastern U.S. from central Texas to the Atlantic coast and from the Gulf of Mexico northward into the Ohio River Valley, southern Pennsylvania, and southeastern Kansas (Nowak 1979). Indiscriminate killing, bounties, and habitat destruction expedited the initial decline of the red wolf. Further disruption caused by lumbering practices, mineral exploration, and agriculture forced red wolves into open habitats, thereby increasing contact and conflict with humans and livestock while simultaneously creating favorable conditions for the more adaptive coyote (*C. latrans*). As red wolf numbers decreased, coyotes moved into the vacated areas. Government predator control efforts inadvertently selected for the more easily caught red wolf over the more elusive coyote (Pimlott 1965). As a result, hybridization with coyotes increased because red wolves found it increasingly difficult to locate conspecifics (McCarley 1962; Paradiso and Nowak 1971; Nowak 1979), largely due to the breakdown of habitat barriers (Mayr 1963). The few remaining red wolves were left only in areas of marginal habitat in forested bottoms and coastal marshes of southwest Louisiana and southeast Texas (Riley and McBride 1972).

The uniqueness of the red wolf and its drastic decline has been documented by a number of researchers (McCarley 1962; Nowak 1967; Paradiso 1968; Paradiso and Nowak 1971; Pimlott 1965; Pimlott and Joslin 1968). The U.S. Fish and Wildlife Service (FWS) officially recognized the red wolf as an endangered taxon in 1967 when this canid became listed as an endangered species. Subsequently, it received federal protection with the passage of the Endangered Species Act in 1973.

An Interim Red Wolf Recovery Team was appointed in 1974 to direct a field crew already in place. It became apparent that a "wolf–coyote hybrid swarm" had formed in central Texas and was spreading eastward (Carley 1975). It was also apparent that red wolves were becoming the minority canid in the few places where it still occurred. As a result, efforts were directed to capture all remaining red wolves in the wild for inclusion in a captive breeding program.

From fall 1973 to summer 1980, over 400 wild canids were captured in southeastern Texas and southwestern Louisiana and examined. Of these, only 43 were admitted to the breeding certification program as probable red wolves (Carley 1975; McCarley and Carley 1978). Examination of offspring produced from the controlled breeding provided additional evidence of taxonomic status for these select canids. That screening program, plus some inadvertent mortality, left only 14 animals to become the founding stock for all red wolves in existence today (Nowak et al. 1992).

A Recovery Plan for the red wolf was originally approved in 1982 and a revised Recovery Plan (U.S. Fish and Wildlife Service 1990) was integrated into a Species Survival Plan and approved in October 1990. This plan suggests that at least three mainland restoration sites will be needed to maintain a viable population of 220 red wolves in the wild. Furthermore, the plan recommends that another 330 wolves remain in captive breeding programs to safeguard genetic diversity and for use to establish additional wild populations (U.S. Fish and Wildlife Service 1990).

To determine the feasibility of restoring red wolves into parts of their former range and to develop translocation techniques, experimental releases were conducted during 1976 and 1978 on Bulls Island, Cape Romain National Wildlife Refuge in South Carolina (U.S. Fish and Wildlife Service 1990). The releases were considered successful and were valuable in providing information for future reintroduction efforts (Carley 1979). In fall 1987, wolves were released into the 48,000-ha Alligator River National Wildlife Refuge (ARNWR) in eastern North Carolina. Continued success and population expansion has established ARNWR as the first permanent red wolf recovery site.

TRANSLOCATION ECOLOGY

The translocation of wildlife species (transport by people from one place to another) as a conservation strategy has increased in recent years (Griffith et al. 1989). Griffith et al. (1989) conducted a survey of translocations of birds and mammals to document current activities, identify factors associated with success, and suggest guidelines for future study. From releases made from 1973 to 1986, the researchers estimated nearly 700 translocations were conducted by a variety of government and private institutions each year and that approximately 44% had been successful. They found translocations of native game animals were more successful than those of threatened or endangered species, herbivore translocations were more successful than those of carnivores, and wild-captured animals were more successfully translocated than captively raised animals. They also found early breeders with large reproductive potential were more successful when translocated and animals released into the core of their former range in excellent habitats were most successful, as were species without competitors. Other factors that contributed to successful translocations were the use of multiple release sites and if that species was increasing and numbers were moderate to high, rather than translocations of species that had already reached critically low levels.

For birds and mammals, most translocations do not succeed, and the proportion may be even lower for reptiles and amphibians (Dodd and Seigel 1991). In addition to the conclusions reached by Griffith et al. (1989), Dodd and Seigel (1991) make a number of general recommendations to enhance success of translocation efforts. First, it is important to know the cause of the initial decline. Translocation programs should only be attempted if the causes of the decline are reasonably well understood and those problems have been eliminated. Second, it is important to know the habitat, demographic, and biophysical constraints of the animal being translocated. Although seemingly obvious, several projects have failed in part because of lack of attention to those details. A thorough habitat analysis should be undertaken when possible. This includes the availability of food, cover, and water but also movement corridors and potential for conflicts with humans or other species. Third, population genetics and social structure should be considered. There needs to be sufficient numbers and variability of founding individuals to avoid the deleterious effects of inbreeding and genetic drift. Furthermore, outbreeding should be considered; the translocation of individuals from naturally isolated deems can result in reduced fitness if gene complexes resulting from local adaptation are broken up through hybridization (Reinert 1991). Additionally, information on the social structure of the translocated species is essential, even though chances of success may not always be maximized by releasing sex ratios and age structures typical of natural populations. Dodd and Seigel (1991) also warn that translocation can be a mechanism for disease transmission and that precautions taken can increase success and reduce risks to native populations. Finally, long-term monitoring is emphasized. Dodd and Seigel (1991) state that most translocations are poorly monitored following release and that evaluations of success are lacking.

THE SOUTHERN APPALACHIANS

In 1989, the FWS turned to the southern Appalachians as a second site for reestablishment of red wolves. The southern Appalachians are considered to be those federal lands consisting of the Chattahoochee National Forest in north Georgia, the Nantahala and Pisgah National Forests in North Carolina, the Cherokee National Forest in Tennessee, and the Great Smoky Mountains National Park (hereafter referred to as the Park) in Tennessee and North Carolina. As such, the southern Appalachians represent the largest federally owned land base east of the Mississippi River.

This land base offers great potential for restoration and management of various threatened and endangered wildlife species. Possible indigenous species for management or restoration include the peregrine falcon (*Falco peregrinus*), red-cockaded woodpecker (*Picoides borealis*), northern flying squirrel (*Glaucomys sabrinus coloratus*), mountain lion (*Felis concolor*), elk (*Cervus canadensis*), fisher (*Martes pennanti*), and red wolf. The last group of species requires particularly large tracts of land for home ranges and for dispersal of offspring.

The beauty of the mountains and the scenic rivers in the southern Appalachians attract millions of visitors who are generally supportive of wildlife enhancement programs. They come to the area to view, photograph, and hunt wildlife in conjunction with other activities which include hiking, camping, canoeing and rafting, fishing, off-road vehicle use, and mountain biking. That diversity of recreational interests can be difficult to accommodate and presents significant challenges to managers.

The public land within the southern Appalachian area is managed as four national forests containing a number of wilderness and state wildlife areas, and one national park. Three U.S. Forest Service offices, three state wildlife agencies, and one National Park Service (NPS) office oversee the work force charged with the management of natural resources of the area. The FWS regularly consults with many of these agencies in addressing specific concerns pertaining to the management of threatened and endangered species, especially the reestablishment of such species. This results in no less than six agencies from eight independent offices charged with wildlife resource management.

To complicate matters, each agency must comply with specific internal mandates, respond to the pressures of user and political groups, and be sensitive to the interests of private landowners within the general boundaries of public lands. This can cause some division among the agencies and obscure otherwise clear biological decisions for wildlife management. For example, a state wildlife management area inside a national forest may be managed by the state agency for hunting. An adjacent national park, by contrast, may be set aside to solely protect, preserve, and restore species. These areas, mandated with dramatically different wildlife management objectives, are separated by boundary lines invisible to wildlife.

In light of the great potential for success, and despite the potential for conflict and the enormous challenges presented, the FWS, in cooperation with the NPS, proposed to release red wolves into the Park and, while doing so, assess the feasibility of restoring wolves to other areas of the southern Appalachians.

RELEASES OF RED WOLVES IN GREAT SMOKY MOUNTAINS NATIONAL PARK

BACKGROUND

The Park was selected as a possible release site for a number of reasons. The Park consists of 2072 square kilometers of forested ridges and drainages that are largely accessible only by foot. There is strong regional and national interest in the Park as evidenced by its high visitation, over 9 million visits annually, the most of any national park in the country. The vast majority of visitors are interested in viewing and experiencing the area's unique features and wildlife. Hunting is prohibited, thus decreasing the chances of intentional or accidental taking of a wolf within the Park. The Park is in proximity to or bordered by four national forests, an Indian reservation, and an extensive system of lakes and adjacent lands managed by the Tennessee Valley Authority. The Park contains some of the richest floral diversity found in North America with over 1500 vascular and 2400 nonflowering plants (King and Stupka 1950). In addition to plants, animal diversity is high. A number of large- and medium-sized mammals are found in the Park, including white-tailed deer (*Odocoileus virginianus*) and the exotic wild boar (*Sus scrofa*), providing a potential prey base for the wolves. The NPS has already reintroduced other threatened and endangered species such as river otter (*Lutra canadensis*), peregrine falcon, and the Smoky madtom (*Noturus baileyi*). Additionally, natural and human-assisted immigrations by beavers (*Castor canadensis*) and coyotes have occurred in the past decade.

To gather information and better prepare for the potential conflicts resulting from wolves, the FWS put into effect a three-phase plan prior to any permanent releases. These phases included a coyote study, public information and education, and a 1-year experimental release (Parker 1990). Prior to wolf stocking, the FWS was concerned with the presence of coyotes in the release area. Hybridization had played an integral part in the decline of the red wolf, leading to its gradual extirpation from the wild, and there was the possibility that it could occur again if red wolves were reestablished. However, the specific circumstances of hybridization between coyotes and red wolves during the mid-1970s were not well understood and no information was available from the ARNWR project because the range of coyotes had not yet expanded to that area. To collect information and begin addressing the potential problem, the FWS contracted with the University of Tennessee to assess the density and ecology of coyotes in the Park. This assessment was conducted from January 1990 to May 1991 and provided base-line information, including coyote home range size, habitat use, and the development of a practical method to monitor coyote abundance (Crawford 1992).

The second phase of the restoration plan consisted of the initiation of a public education and information exchange program by the FWS and NPS. Visitors to the Park and area residents were soon informed about the project through NPS interpretive programs, pamphlets, and local news coverage. An essential part of the education process involved a documentary about the red wolf produced by WBIR-TV of Knoxville, TN for the "Heartland Series." Also, an educational package developed by the FWS, NPS, WBIR-TV, and the Southern Appalachians Man and the Biosphere Cooperative (SAMAB) was distributed free of charge to over 800 schools, media, and resource organizations. This package contained the regional Emmy-winning documentary video "Front Runner," a teacher's work guide and a wolf activities poster. "Front Runner" aired on regional stations, was updated, and additional copies were prepared for distribution upon request.

A series of meetings with federal and state agencies, local citizens, and various civic and special interest groups were held. These meetings served as a sounding board where significant details of the general proposal and management policy could be discussed and criticized prior to implementation. This led to the formation of a communication committee representing a variety of interests to continually serve as a liaison with wolf project personnel. Their concerns were addressed in the final proposal and the 1-year experimental release.

This release, the third phase of the plan, would provide an opportunity to demonstrate a tangible commitment to solve any problems, allow field personnel to demonstrate their ability to manage the wolves, work cooperatively with livestock owners, and develop a sense of trust with the local communities surrounding the Park. The objectives of the experimental release were to collect information on major human-related issues common to the southern Appalachian area, the primary issue being the potential threat to livestock, and continue gathering information on red wolf/coyote interactions. Depredation threats to livestock would be assessed in Cades Cove where a private tenant leases approximately 330 hectares (825 acres) of pasture land for a 500+ head, cattle-breeding operation. The FWS agreed that if this experiment failed, the project would not continue. To assure the public that this was just an experiment, all wolves were to be returned to captivity after approximately 1 year, during which wild reproduction would be prevented. To allow normal breeding behavior and prevent reproduction, the adult male was vasectomized before release.

EXPERIMENTAL RELEASE

On 12 November, 1991, after 9 months of acclimation, one family of wolves consisting of an adult pair and two female pups, were released in the Cades Cove area of the Park (Figure 1). Initially, these animals were intensively monitored but the number of radio-locations collected was gradually reduced to four daily as movements became more predictable. As expected, the wolves generally restricted their movements to Cades Cove. However, on four occasions wolves strayed off Park property. On two occasions the wolves were captured within several hours and the other two times they returned on their own within 24 hours.

Although these wolves had spent most of their lives in captivity, field observations of the animals' condition and activities indicated they were able to secure food. Scat analyses revealed that items most frequently consumed were white-tailed deer and raccoons (*Procyon lotor*).

The wolves were sighted on numerous occasions throughout the experiment by visitors and project personnel. Scavenging in the picnic area by the wolves after dark presented a minor problem similar to experiences with skunks (*Mephitis mephitis*), raccoons, and black bears (*Ursus americanus*). Although the wolves generally presented no serious conflicts with human use, the adult male was captured 3 months after release and returned to captivity because he repeatedly demonstrated a high tolerance of humans at close distances. It is believed that older wolves, born and raised in captivity, develop behavior reflecting their routine interactions with human keepers; this 8-year old male was a good example. The adult female was also tolerant, but to a lesser extent. She roamed free for the duration of the experiment. In contrast, the two pups developed an increasing wariness of humans as they spent more time in the wild.

The adult pair of wolves was observed on several occasions aggressively pursuing coyotes, indicating an attempt to exclude coyotes from the core area of their range. However, after the capture of the adult male wolf, two or more coyotes dominated the adult female wolf in aggressive interactions.

At the end of the experiment, the family of wolves was recaptured and project personnel presented these results to the NPS and the communication committee. The decision was made to proceed with a full attempt at repatriation, continuing with the same study objectives as the experimental release, but without inhibiting reproduction and with no scheduled plans to return the wolves to captivity.

SUBSEQUENT RELEASES

In October 1992, six wolves: two adults and four juveniles (three males, one female), were released in Cades Cove. The following December, a second family of six wolves of like composition were released 11 km from the Cove, in the Tremont area of the Park (Figure 2).

The wolf group at Cades Cove quickly displaced resident coyotes and confined their movements to the Cove area. Preferred prey species of the Cove wolves included white-tailed deer, raccoons, cottontail rabbits (*Sylvilagus floridanus*), and ground hogs (*Marmota monax*). The adult wolves also killed a 32-kg wild boar within the first month of release (Lucash and Crawford 1994).

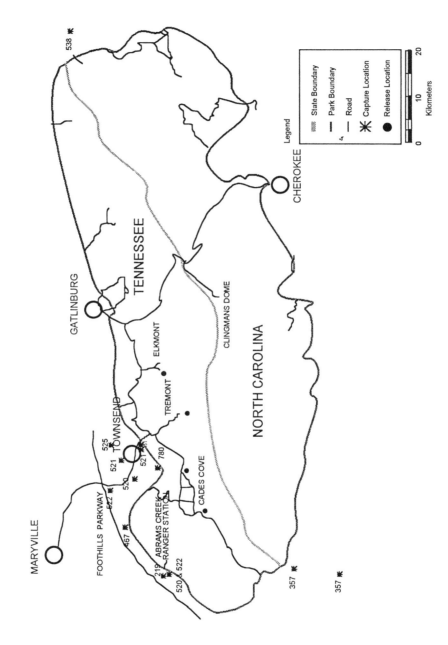

FIGURE 1   Release sites and red wolf capture locations outside the Park from November 1991 to May 1996.

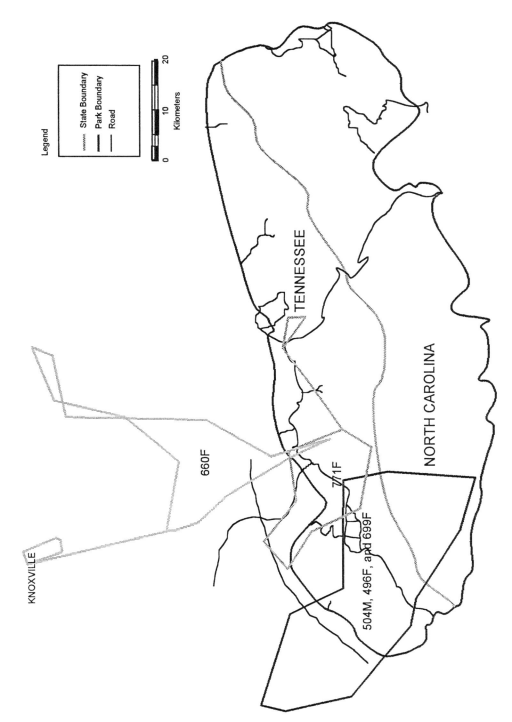

FIGURE 2   Movements of red wolves 504M, 496F, 699F, 711F, and 660F on Park and nonfederal property during 1995.

TABLE 1
Red Wolves That Traveled off Park Property from November 1991 to May 1996

Wolf	Birth date	Release date	Capture date	Distance from park (km)	Comments
219M[1]	4/19/83	11/12/91	01/21/92	1	Permanent captivity
467F [1]	4/26/91	11/12/91	01/15/92	3	Transfer to Island project
538M [1]	4/25/92	10/15/92	02/09/94	7	Given second chance
		04/22/96	Not captured	0	Free in Park
357M[2]	4/29/89	12/09/92	01/28/93	3	Given second chance
		02/08/93	03/07/93	11	Given third chance
		04/11/93	Not captured	0	Died (Table 2)
520M[2]	4/14/92	12/09/92	12/21/92	1	Given second chance
		01/04/93	02/08/93	3	Transfer to ARNWR
521M[2]	4/14/92	12/09/92	12/27/92	4	Given second chance
		01/04/93	01/30/93	1	Transfer to ARNWR
522F [2]	4/14/92	12/09/92	01/16/93	1	Given second chance
		01/25/93	01/28/93	1	Given third chance
		03/30/96	05/15/96	4	Permanent captivity
525M[2]	5/15/92	12/09/92	01/27/93	1	Given second chance
		09/09/93	12/14/93	20	Permanent captivity
592M[2]	4/18/93	Wild born	Not captured	3	Wolf returned to Park
593F [2]	4/18/93	Wild born	03/13/94	56	Transfer to ARNWR
451M[2]	4/25/91	06/28/94	08/29/94	<1	Given second chance
		09/21/94	10/18/94	<1	Blind (PRA); euthanized
711F [2]	4/25/94	06/28/94	Not captured	<1[a]	Died (Table 2)
660F [3]	5/05/93	04/10/95	Not captured	30[a]	Fate unknown (Table 3)
496F [1]	4/15/91	07/24/95	Not captured	20[a]	Died (Table 2)
504M[1]	5/01/91	07/24/95	Not captured	20[a]	Died (Table 2)
699F [1]	4/10/94	07/24/95	Not captured	10[a]	Died (Table 2)
780M[1]	4/07/95	07/24/95	04/10/96	1	Temporary captivity
782F [1]	4/07/95	07/24/95	Not captured	<1[a]	Not collared, died (Table 2)

Note: Release areas: 1 = Cades Cove, 2 = Tremont, 3 = Elkmont.

[a] These wolves were not captured; distances indicate wolf's furthest location from the Park.

In contrast to the Cades Cove wolves, the Tremont wolves showed very little affinity to the forested mountains of the release area or for one another. Within the first 8 weeks after the release, all six wolves traveled significant distances from the release site, five eventually leaving the Park (Table 1, Figure 1). During February 1993, three of the four juveniles continued to leave the Park even after multiple recaptures and releases. By March 1993, all four juveniles had to be returned to the holding pen and three were eventually transported to ARNWR. The fourth juvenile was held in captivity for later release. The juvenile wolves were not the only Tremont wolves leaving park property. The adult male traveled south of the Park along Cheoah Lake on two occasions and was captured in late January and again in early March. He was released with the adult female in early March, and the pair settled into an area just east of Cades Cove (Lucash and Crawford 1994).

During mid-April 1993, both the Cove and Tremont adults produced litters of pups. The Cove female had four pups (one male, three females) inside an abandoned hay barn in the center of the Cove. The labyrinth of tunnels beneath several stacked layers of large hay bales served as the den. The Tremont female had three pups (one male, two females) beneath a large rock outcropping in a hollow near the headwaters of Laurel Creek. Both litters of pups were captured at 10 weeks of

TABLE 2
Mortality of Red Wolves in Southern Appalachians Recovery Project
from November 1991 to May 1996

Wolf	Birth date	Release date	Death date	Death location	Distance from park	Cause of death
337F	5/05/88	12/09/92	06/26/93	Cades Cove	GSMNP	Intraspecific aggression
357M	4/29/89	12/09/92	10/11/93	Cades Cove	GSMNP	Poison — ethylene glycol
496F	4/15/91	07/24/95	10/11/95	Hwy. 129	Park Line	Intestinal blockage
501M	5/01/91	04/10/95	12/25/95	Cades Cove	GSMNP	Interspecific aggression
504M	5/01/91	07/24/95	12/25/95	Cades Cove	GSMNP	Undetermined
594F	4/18/93	Wildborn	08/08/93	Crib Gap	GSMNP	Interspecific aggression
595M	4/18/93	Wildborn	07/31/93	Cades Cove	GSMNP	Parvovirus
596F	4/18/93	Wildborn	07/31/93	Cades Cove	GSMNP	Parvovirus
597F	4/18/93	Wildborn	07/31/93	Cades Cove	GSMNP	Parvovirus
598F	4/18/93	Wildborn	07/28/93	Cades Cove	GSMNP	Parvovirus
699F	4/10/94	07/24/95	11/22/95	Dalton Gap	GSMNP	Illegally taken (gun shot)
711F	4/25/94	06/28/94	02/17/96	Dry Valley	0.4 km	Undetermined
782F	4/07/95	07/24/95	03/01/96	Townsend	0.5 km	Undetermined

age, surgically implanted with radio-telemetry transmitters, and released at the capture site within several hours. This was our first wild-born generation of red wolves, for a total of 16 free-ranging animals. This was a major step in the progression of the program, but it was to be short-lived.

Before the end of the year, seven wolves had died of various causes (Table 2). Four pups died from an infectious disease, one adult female and one female pup died from injuries received from aggressive activity with other wolves or coyotes, and one adult male had ingested poison.

In late June 1993, the Tremont adult female was found dead. The necropsy indicated that wounds inflicted by other large canids was the cause of death. This female had been making nightly hunting forays into the Cove and had likely encountered one or more of the Cove pack. Aggressive interactions involving breeding adults are not uncommon in the wild and are occasionally fatal (U.S. Fish and Wildlife Service, unpublished report). For several weeks, the adult male and the three pups were given supplemental food. Supplemental feeding was discontinued when biologists felt confident that the adult male was acquiring food for the pups.

Within a period of only a few days in July 1993, all four pups from the Cove litter died. Necropsy of one pup carcass implicated parvovirus. Given the age of the pups and short span of time in which all pups died, it is likely that the parvovirus infection was responsible for the death of the other three pups.

In response to a potential parvovirus outbreak affecting the Tremont pups, project personnel decided to recapture and inoculate the litter against parvovirus, distemper, rabies, and other canid diseases. During early August, the Tremont adult male was captured to facilitate capture of the pups. Within two days, two of the three pups had been captured. On the following day, the third pup, a female, was found dead. In the absence of the adult male, one or more coyotes had moved into the temporarily vacant territory and killed the third pup. On 9 September, the Tremont adult male, one juvenile and the remaining pups were released after receiving the vaccination series.

On October 12, 1993 the Tremont adult male was found dead near the center of his home range. The cause of death was organ failure resulting from ingestion of ethylene glycol, the main ingredient in automotive antifreeze. Federal investigators were unable to determine if the poisoning was intentional or accidental.

With the death of the Tremont adult male, the juvenile male and pups apparently lost family cohesiveness. The juvenile male left the Park and was captured near Newport, TN in mid-December

1993. The female pup left the Park immediately after the capture of her sibling and was not captured until the following March, near Madisonville, TN. During the 5 months that the female pup was outside the Park she covered approximately 120 km², including a number of farms. Project personnel contacted all landowners involved and no wolf sightings or complaints were reported.

With the approach of winter 1993 to 1994, the Cades Cove juveniles were entering their first breeding season. Generally, red wolves do not breed until 22 months of age or later. As expected, the juveniles began to travel independent of their parents and dispersed out of their natal home range, indicative of having reached sexual maturity. On February 4, one juvenile male was located 5 kilometers from the Park on U.S. Forest Service property, along the Appalachian Trail. He was captured on February 9, and returned to an acclimation pen. Another juvenile male occupied an area adjacent to his natal home range and remained well within the Park. However, he was observed consorting with a coyote and was captured on 15 February. The juvenile female was observed on several occasions traveling with a coyote during March. She was captured on 1 April, and placed in captivity with the juvenile male from the Tremont release.

On April 16, 1994, the Cove livestock operator witnessed wolves killing two young calves and reported seven other calves missing since his last count, 2 weeks prior. As a result of these circumstances, the remaining free-ranging wolves in the Cove were captured and placed in captivity. Project and Park personnel discussed options that would provide long-term protection for young calves and debilitated cattle at risk to predation. No wolves were to be released into the Cove until these options/strategies were decided upon and implemented. Selected options were the construction of a high-tensile, high-voltage electrified fence to enclose 32 ha as a calving corral, and changes to the Cove operator's lease to provide tighter management of the cattle herd. Both the corral and the lease changes were to be in place prior to the initial peak of the Cove calving season in March 1995, and the increased demand for prey by wolves raising litters of pups later in the spring.

On June 28, 1994, an adult pair of wolves and their five pups were released from the Tremont acclimation pen. In an attempt to localize the wolves' movements, the release took place while the pups were still at a very young age and relatively immobile. These wolves remained in the general release area for several months. During early fall, the adult male of this pack was frequently observed traveling along the paved roads near the Park entrance at Townsend and scavenging refuse from a campground just outside the Park. The wolf seemed unusually confused, tolerant of approaching vehicles, and reluctant to leave the roadway. Unsuccessful attempts were made by project personnel to haze (harass) this animal from the roads. The wolf was captured, held for 2 weeks in the Tremont pen and released again — only to repeat the same behavior. He was captured a second time and permanently confined to captivity. He was later diagnosed with progressive retinal atrophy, a condition believed to be genetically transmitted, in which inflammations in the wall of the retina build up scar tissue, eventually leading to total blindness. The Tremont adult female and three of the five pups were captured during late winter 1995, placed in the Tremont pen with a new adult male, and released again in April 1995. The female used the hollow inside of a large boulder for a den and produced a litter of at least two pups in late April. The pack remained in the release area for 8 weeks but by late June began a gradual move to the Cove, which was still void of a resident breeding pair.

An adult pair of wolves that had been acclimated to the Elkmont area (Figure 1) for 2.5 months was also released in April 1995. Wolves had not been released nor had they occupied the Elkmont area prior to this release. The initial movements of the wolves released into this new area were expected to be wide-ranging and unpredictable, especially given the uncertainty of prey densities in the Elkmont area and the relatively short acclimation period. Biologists intended to rely on the timing of release to coincide with whelping as a means of restricting these initial movements. Unfortunately, the female did not produce pups and she spent the majority of time well north of the Park until early August, when her transmitter ceased to function. Although the wild-born female apparently avoided contact and conflict with residents, attempts were made to capture her due to

TABLE 3

Livestock Depredation Claims and Reimbursements Made from the Red Wolf Recovery Project in the Southern Appalachians from November 1991 to May 1996

Date	Claim	Location	Distance from park	Settlement
12/21/91	1 chicken	Cades Cove GSMNP	0	$ 3.00
01/21/92	3 turkeys	Happy Valley, TN	1 km	Not requested
04/04/92	1 calf	Cades Cove, GSMNP	0	$ 250.00
12/24/92	1 calf	Cades Cove, GSMNP	0	$ 360.00
01/27/93	3 chickens	Happy Valley, TN	0.5 km	$ 10.00
03/05/93	1 calf	Cades Cove, GSMNP	0	$ 360.00
03/18/93	3 calves	Cades Cove, GSMNP	0	$ 1,080.00
04/08/93	1 calf	Cades Cove, GSMNP	0	$ 360.00
05/21/93	1 calf	Cades Cove, GSMNP	0	$ 360.00
06/01/93	1 calf	Cades Cove, GSMNP	0	Denied
12/05/93	1 cow	Cades Cove, GSMNP	0	Denied
01/30/94	1 calf	Cades Cove, GSMNP	0	$ 360.00
02/02/94	1 calf	Cades Cove, GSMNP	0	$ 247.00
03/07/94	1 calf	Cades Cove, GSMNP	0	$ 360.00
04/17/94	9 calves	Cades Cove, GSMNP	0	$ 3,240.00
12/04/95	3 calves	Cades Cove, GSMNP	0	Denied
01/03/96	1 cow	Cades Cove, GSMNP	0	Denied
01/12/96	1 calf	Cades Cove, GSMNP	0	$ Pending
04/02/96	1 calf	Cades Cove, GSMNP	0	Pending
05/14/96	1 calf	West Millers Cove,TN	4 km	Pending

the distance she remained from the Park and other red wolves. The Elkmont male spent several weeks traveling throughout the southwestern third of the Park and eventually settled in the Cove during late May.

In late July 1995, an entire pack of wolves, consisting of two adults, one juvenile female and five pups, were released into the Cove. This pack (named "Whitetail") had been transferred to the Park in early April from private land in eastern North Carolina, adjacent to the ARNWR project. The initial plans for these wolves included release in early May, following the completion of the protective calving corral, and preceding the inevitable arrival of the Tremont wolves into the Cove. Although this new pack of wolves would also have a very short acclimation, biologists felt the wolves' initial exploratory movements would be minimized by the immobility of the young pups and the attractiveness of the Cove's abundance of prey. Unfortunately, the completion of the corral and the release was delayed until late July. By then, the Elkmont adult male had settled in the central Cove and the Tremont adult male had solidified his occupation of the western Cove. Several days following release, the Whitetail adults moved into the Cove and the Tremont male was located very near or with them. After that location, the Whitetail adults separated, moved out of the Cove area, and were not located together again. During the next few weeks it became evident the pack had lost cohesion. The adults and the yearling female ranged widely throughout the southwestern third of the Park and onto state and private property to the west and north of the Park. Although the five pups were too young to be fitted with radio collars, several observations of individual pups at opposite ends of the Cove indicated they were not in contact with the adults. Four of the five pups were captured during late summer and held until large enough to wear radio collars. The adults continued the same pattern of movement, independent of one another for the short remainder of their lives.

The Whitetail adults and juvenile female all died during the fall of 1995 (Table 2). During October, the adult female starved to death due to an upper intestinal blockage. The juvenile female was illegally shot and killed along a little used section of the Appalachian Trail inside the Park during November, and the adult male, otherwise in excellent physical condition, died on Christmas Day of an unknown cause. The fifth pup was found dead and reported by a landowner just outside the Park in Townsend, during early March 1996.

Three other wolves died during late winter 1996 (Table 2): the adult male released at Elkmont was killed either by dogs or coyotes near the Cove in January, and a juvenile female from the Tremont pack and the fifth Whitetail pack pup, an untransmittered female, were found dead of unknown causes on private property near the Park's north boundary in Townsend, TN.

To replace the losses of wolves from mortality, five adult wolves were transferred to the Park from ARNWR and one of the island propagation sites at Cape Romain National Wildlife Refuge (CRNWR) near Charleston, SC. These five wolves were introduced to existing free-ranging and captive wolves at the Park to form five free-ranging breeding pairs by late April, 1996. Three of the five free-ranging females produced litters by early May, 1996, adding 14 pups to the 12 monitored adults to the Park's red wolf population.

WOLVES AND LIVESTOCK DEPREDATION

The perceived economic threat of a large predator to livestock operators is perhaps the greatest political barrier to reestablishing a self-sustaining red wolf population to the southern Appalachians. Such concerns from local residents were expressed through representatives of state agricultural agencies and farm bureaus. The major worry was depredation of livestock and the project's financial responsibility to the livestock owner. Other issues were repercussions for accidentally killing a wolf, private landowners' legal ability to protect domestic animals against wolf attacks, and the commitment and ability of personnel to monitor and manage errant wolves. Although much initial resolution of these concerns was attempted, the remaining apprehensions could only be addressed by releasing the wolves and repeatedly demonstrating that conflict situations could be effectively dealt with. This would require a continual process of reevaluating and fine-tuning policies as circumstances warranted.

At the Cades Cove cattle operation, some 200 to 240 calves are born during a 10-month period each year and are allowed to roam freely. All the pastures contain woodlots or are bordered by them, where the cattle go for cover to give birth. Most depredations occur in or along these woodlots. Before release of the wolves, the livestock owner and Park personnel speculated that five to ten calves were taken each year by black bears, coyotes, and other scavengers, but no actual counts were ever made. The FWS recognized the opportunity to use the Park's leased livestock operation as a proving ground for depredation prevention and reimbursement policies. The Cove operation was viewed as a worst-case scenario where the threat of depredation would be greatest due to the constant presence of wolves, extensive woodlots, and extended calving period. The livestock operators outside the Park would have the opportunity to observe depredation activity and the project's response policies.

Prior to the experimental release, an indemnity fund was established to compensate for domestic animal depredation by red wolves. The fund was comprised of donations to the National Fish and Wildlife Foundation from the National Parks and Conservation Association and other private sources. Any person suffering legitimate losses from red wolves would be fully compensated. Administration of the $25,000 fund was kept at the field level, requiring only co-signatures from the Park superintendent and the FWS red wolf coordinator. This insured expedient reimbursement with minimal administrative complications.

Since the first release of red wolves in the Park in 1991, approximately 1000 calves have been born in Cades Cove. Of 28 claims for red wolf depredation of cattle reported by the Cove livestock operator, 22 have been paid for an approximate total of $7,200.00. Of these, 14 were for missing calves less than 1 week of age. No healthy, weaned cattle were taken. Ten dollars has been paid for three chickens taken outside the Park (Table 3).

During the experimental release, numerous sightings were made of coyotes and wolves in the pastures pursuing young calves. Four depredations were reported. In one case, coyotes were observed consuming a recently killed calf; in the other three, claims for missing calves were filed. In two of these cases, the wolves had been continuously monitored and were not in the area where the depredations occurred. Observations by field personnel and the livestock owner suggested that coyotes were responsible for these incidents. Based on circumstantial evidence, the wolves may have been responsible for one calf depredation. During this period one chicken and three domestic turkeys outside the Park also were taken by the adult male wolf. Reimbursements for the calf and chicken were paid to the owners, but offers for reimbursement were declined by the owners of the turkeys.

Close monitoring of the wolves and the cattle operation continued with the second and subsequent releases of wolves into Cades Cove. The number of observations of coyotes fluctuated relative to season, and the number, age, and breeding status of wolves in the Cove. Since wolves were continuously present in the Cove, they were suspect in all depredations. The lack of evidence often resulted in reimbursement, except when observations and other evidence implicated coyotes or other animals as the offenders.

Ten paid claims were based on the operator's supposition that a calf had been born sometime during the previous night and had disappeared, though he had never actually seen it. These disappearances were based on evidence of a lone cow with a turgid udder, bawling for several hours. Although it is common for even the most efficient cattle breeding operations to suffer an average of 3 to 5% mortality during calving (F. Hopkins, University of Tennessee Veterinary Teaching Hospital, pers. commun.), the Cove operator was not required to show proof that the reported missing calf had ever been alive, nor was he required to report calf mortality from natural causes. Payments for these ten claims were made when wolves were judged the most likely suspects based on movement patterns and the lack of any other plausible explanation. Reimbursement amounts were based on current average price per pound ($0.55 to $0.75) for weaned calves at average expected weights of 200 kg (450 lbs). This lenient policy was implemented to illustrate the project's commitment to the operator and to insure that reimbursement adequately covers lost economic potential. Payment was denied for six claims based on one or more of the following: the preponderance of evidence (although scant in some cases) indicating coyotes; necropsy concluded the cow or calf was dead prior to the wolves' attendance, or when the livestock operator did not adhere to cattle management guidelines within his lease.

Livestock and hunting groups expressed concerns that released wolves and their offspring would be indistinguishable from coyotes. These groups feared the consequences of accidentally taking a red wolf while hunting or protecting their livestock. To alleviate this, project officials classified all wolves in the Smokies program as experimental/nonessential. This classification allows more freedom when customizing the regulations to fit the specific demands of the release areas (Parker and Phillips 1991).

Regulations under the experimental/nonessential designation ensure that accidental taking of red wolves will not be prosecuted, provided that the activity resulting in the taking was legal, and the taking is reported. The policy of intensive radio-monitoring wolves allows biologists to contact landowners when a wolf moves onto private property. This precludes the likelihood of the landowner misidentifying the wolf and, if necessary, gains biologists access for removal of the wolf. The public education efforts continue to concentrate on using relative size and the obvious radio-collars to distinguish red wolves from other wild canids. In all cases the circumstances will be investigated for evidence of misrepresentation or intent.

As a result of input from the Tennessee Farm Bureau, a harassment clause was written into the regulations enabling landowners to protect their property from the threat of damage by wolves as long as it is noninjurious to the wolf (Henry 1991). Landowners were instructed to immediately contact project personnel before taking any further action. If capture attempts by project personnel fail and damage threats continue, personnel and/or the stock owner(s) would then be permitted to destroy the wolf. However, the North Carolina and Tennessee Farm Bureaus continued to request that private landowners be given the legal right to kill red wolves at any time on their property. To prevent erosion of public support and still provide protection to nonoffending wolves, regulations were revised and further relaxed in 1995. These changes include allowing any private landowner or person having the landowner's permission to take red wolves found on his/her property when the wolves are in the act of killing livestock or pets (Henry 1995). To date, the Smokies project has not encountered a situation where a harassing red wolf has been injured or killed on private property.

Understandably, the concerns regarding the commitment and ability of the FWS to continually manage the wolves under the outlined agreements could not be completely alleviated by written or verbal assurances. Our intent, therefore, was to establish a sense of trust through example in the experimental period and nurture this trust throughout the duration of the project.

Since the experimental release in 1991, 18 wolves have left Park property, with 12 requiring recapture. Eleven of these 12 wolves were successfully recaptured, using leg-hold traps or immobilizing darts (Table 1). On one of these 12, the transmitter ceased functioning before the wolf could be captured. The majority of captures outside the Park occurred within four kilometers of the Park boundary (Figure 1) and many occurred within 24 hours of initiation of capture efforts. Recaptures seem to have caused minimal inconvenience for involved residents. Approximately 80 private residents have been contacted in association with our attempts to capture wolves. All residents and landowners have been cooperative and granted access to their property. The majority of individuals were forthright in their support for the program and in many instances requested that we allow the wolf to remain on their property. Several landowners, though expressing ambivalence or mild support for the project in the Park, requested that red wolves be kept off of their property, regardless of the threat of or actual depredation.

The remaining 5 of 18 wolves that have left the Park occupied ranges which encompassed both Park and nonfederal properties (Figure 2). These wolves were constantly on the move and regularly returned to the Park. There was no indication that they were preying on domestic animals, and we received no reports of their presence on private property. As a result of these behavior and movement patterns, biologists felt these wolves did not warrant recapture. One young male returned to the Park on his own. Two females were found dead just outside the Park of unknown causes; the other three wolves eventually returned to the Park. Unfortunately, two of these three died of natural causes and the third was illegally shot inside the Park.

It was expressed by opponents of the project and acknowledged by project personnel that there would inevitably come a time when biologists would lose contact with or not be able to capture some wolves. Currently, 7 of the possible 19 free-ranging wolves have had unknown fates (Table 4). Two had failed transmitters; five were pups documented at birth or at several weeks of age but never captured. Efforts were made to capture these five pups at 8 to 12 months of age, when they would be large enough to wear transmitter collars. These five pups were from two litters born to the same female in consecutive years. During trapping efforts, each member of the respective pack was captured two or more times, with no sign of the targeted pups. Thus, it is doubtful that 5 of the 7 wolves with unknown fates are alive.

It is reasonable to expect the number of wolves with unknown fates to rise in proportion with a growing wolf population. The FWS and project personnel remain committed to removal of wolves that settle on private property, and we have demonstrated a strong ability to do so. Since the experimental period, the 37 wolves that have been involved in releases have been captured 67 times for movement, behavioral problems, or technical problems with transmitters. The den sites of every female that whelped in the wild have been located to document size and sex ratios of the litters. As the project expands, however, we recognize our limited ability to respond to and manage all potential situations. In situations involving wolves on private property, we feel the most recent

TABLE 4
Red Wolves in Southern Appalachians with Unknown Fates
from November 1991 to May 1996

Wolf	Birth date	Release date	Last contact	Cause of lost contact
660F	5/05/93	04/10/95	07/31/95	Radiotransmitter malfunction
707M	4/25/94	06/28/94	06/28/94	Never captured, presumed dead
712F	4/25/94	06/28/94	06/28/94	Never captured, presumed dead
771M	5/10/94	03/30/96	04/10/96	Radiotransmitter malfunction
787F	4/25/95	Wildborn	07/04/95	Observed in wild midsummer; never captured
788F	4/25/95	Wildborn	07/04/95	Observed in wild midsummer; never captured
802?	4/25/95	Wildborn	07/04/95	Observed in wild midsummer; never captured

amendment to protection regulations in April 1995 further demonstrates our willingness to give landowners' concerns priority over wayward wolves.

OBSTACLES TO REESTABLISHMENT

In accessing the southern Appalachians for red wolf restoration, impediments regarding availability of personnel, funding, and land are identified. Additionally, there are complicated biological and legal issues facing the red wolf and other imperiled species that must be overcome for successful reestablishment to occur.

LIMITED LAND BASE

Even before the experimental release period, we were concerned that the Park may not be large enough to support a sustainable population of red wolves; the movements of wolves since that time continue to strengthen our concerns. Thirteen wolves have left Park property on a minimum of 25 occasions since November, 1991. An additional five wolves occupied ranges which regularly took them on and off Park property. Furthermore, many wolves released elsewhere eventually moved into the Cades Cove area. This area supports the highest prey populations and is the largest contiguous area in the Park in early succession. The prey base in the majority of the Park, which are generally mature forests, may not be sufficient.

On the other hand, releases have taken place only in Cades Cove, the Tremont area, and Elkmont areas. All three of these locations are in close proximity (3 to 6 km) to the Park boundary. Releases in more central areas of the Park have not been conducted because of the difficulty of radio telemetry there and because wolves could leave for more remote and inaccessible National Forest land south of the Park.

Seven wolves have made at least ten sallies to or through U.S. Forest Service property and there are no formal agreements allowing dispersing wolves to remain there. Currently, wolves not returning to the Park require retrieval. Until an agreement is in place, additional releases that place wolves in closer proximity to National Forest lands would be futile.

SMALL STAFF

Wolves will need to be regularly radio monitored to track movements, interactions with coyotes, validate depredation claims, detect mortality, and document reproduction. Although we were familiar with the time and effort required to maintain contact with wolves through experience at ARNWR, telemetric monitoring of these highly mobile animals in mountainous habitats presented increased logistical difficulties due to weather patterns and topography. In the mountains, poor visibility and high winds often make radio-telemetry flights dangerous, limiting aerial tracking opportunities. Tracking can be accomplished from vehicles or by foot, but the rugged and remote terrain decreases the effective range of the equipment and greatly increases the time required to maintain regular contact with wolves.

To date, the project has effectively consisted of three personnel, all serving in the field as well as conducting administrative, public relations, and clerical activities. The crew is on call 24-hours/day, 7 days/week. A six- to eight-person crew, stationed throughout the study area, would be needed to manage a future population of 60 to 100 wolves, given current monitoring protocols.

HYBRIDIZATION WITH COYOTES

Clarifying the circumstances that promoted hybridization and then minimizing or preventing it from occurring in reestablished red wolf populations is a high priority. Detailed information on red wolf/coyote interactions is difficult to obtain, but we were able to make some observations of interactions. During the experimental release period, when the adult pair of wolves encountered coyotes, the wolves generally were the aggressors. However, when the adult male wolf was returned to captivity, two or more coyotes tended to dominate the adult female wolf. In contrast to the first release, sightings of coyotes significantly decreased after the second release. The adult wolves used in the second release spent a greater amount of time together than the experimental pair, possibly indicating a stronger bond and perhaps explaining the decrease in observed coyote activity. Very few observations of aggressive interactions between coyotes and wolves have been made since this second release. Several observations were made during early spring 1994 of juvenile wolves traveling with or being followed by single coyotes. Later that same spring a 2-year-old female wolf gave birth to a litter of pups shortly after she was captured in the Cove. Observations of her traveling with a coyote strongly suggested that the coyote sired the litter. The only unpaired male wolves in the general area of this female were her two litter mates, but they were not located near her during the peak of the breeding season. When the remaining wolves were removed from the Cove in late Spring 1994 due to calf depredation, the number of observations of coyotes in the Cove increased.

From this and additional information from ARNWR, we believe that paired wolves will attempt to exclude coyotes from a portion of their home range, two or more coyotes will react aggressively to individual wolves, red wolves may prefer conspecific mates but unpaired wolves will consort and breed with coyotes if no conspecific is available, and coyotes will kill lone red wolf pups. As a result, we recommend a reintroduction strategy to maximize red wolf pairing by saturating a given area with released wolves. Coyote control also may be an option. This decreases opportunities for wolf–coyote pairing, thereby promoting red wolf genetic integrity. This approach is already in practice at ARNWR.

TAXONOMIC STATUS

The taxonomic status of red wolves has been the subject of much recent controversy. The presence of this canid in the southern U.S. was first documented in 1791 (Harper 1942) and later given a scientific name by Audubon and Bachman (1851). Although the uniqueness of the red wolf was recognized in works by Goldman (1937) and Atkins and Dillon (1971), the first extensive scientific work to determine its taxonomic status was performed by Nowak (1979). In that study, he measured a number of cranial characteristics of red wolves, coyotes, and gray wolves and concluded, based on multivariate analyses, that the red wolf was a distinct species. The finding of a unique electrophoretically determined allele with a distribution congruent with geographical distribution of the remaining red wolf population by Ferrel et al. (1980) further supported species status. Nowak (1992) states that cranial characteristics for the species began to change dramatically after 1930, however, due to hybridization with coyotes. Animals were later collected in southeast Texas that were thought to represent the purest red wolf strain in existence; of these, 14 individuals became the founders of today's captive breeding program.

Based on mitochondrial DNA analyses, however, Wayne and Jenks (1991) challenged Nowak's (1979) conclusions. In that paper, Wayne and Jenks (1991) were unable to find segments of red wolf mitochondrial DNA that would distinguish it from gray wolves or coyotes, yet they found gene sequences that were characteristic of both coyotes and gray wolves. From that analysis, Wayne and Jenks (1991) and Wayne (1992) offered various possible interpretations of the data, but clearly favor the hypothesis that the red wolf was a hybrid between coyotes and gray wolves and was not a distinct species. A companion paper by Gittleman and Pimm (1991) criticized the red wolf recovery program based on that hybrid status.

Several rebuttals were subsequently published. Nowak (1992), Phillips and Henry (1992), and Dowling et al. (1992) criticized both Wayne and Jenks' (1991) and Gittleman and Pimm's (1991) conclusions, based on some erroneous interpretations of historical data, the low number of genetic sequences compared, the exclusive use of genetic characters to identify species and their hybrids, and behavioral, ecological, and morphological differences noted in red wolves. In a subsequent paper, Wayne and Gittleman (1995) published interpretations based on nuclear DNA analysis data (Roy et al. 1994a, b) further supporting their hypothesis that the red wolf was originally a hybrid between the gray wolf and the coyote with subsequent extinction of the gray wolf in the southern U.S. However, other interpretations of the nuclear DNA evidence point out sampling problems and suggest that the present red volves are descendants of a relic population of wolves. Furthermore, the hybrid origin interpretation fails to account for the rest of the other available evidence on the taxonomic question, while species or subspecies origins are also supported by the DNA evidence and are consistent with the other taxonomic evidence. To date, the preponderance of evidence supports species status, but academic deliberations will undoubtedly continue as new information becomes available.

This debate over red wolf taxonomy and systematics has profound implications for the recovery program. Wayne (1992) and Gittleman and Pimm (1991) originally argue that the red wolf should not be awarded the same status as truly distinct species and Gittleman and Pimm (1991) further state that recovery efforts for red wolves are misguided. They argue that no distinct species is being saved, that the red wolf does not occupy a unique ecological role in areas where coyotes are present, and that any lessons about reintroducing species are likely to be obscured by hybridization with the coyote.

Within the larger arena of conservation biology, however, we feel that these taxonomic questions are largely moot. It is undisputed that a large canid was originally present in the southern Appalachians that has since gone extinct, that the present red wolf population contains the genetic remains of that canid, and, despite claims to the contrary, these animals do occupy a niche different from that of the coyote. Differences in food habits are a clear indication of that (Phillips and Henry 1992), and Wayne and Gittleman (1995) later admit that red wolves may play an ecological role in some habitats that coyotes cannot fill. From an ecological standpoint, the loss of natural predators of white-tailed deer and other large and medium-sized mammals in the southern Appalachians has had significant ecological consequences. We believe that the status of endangered species recovery programs should not be judged based on taxonomic status alone, but that ecological role should play a large part. The Endangered Species Act clearly provides for this, as evidenced by recent genetic augmentation efforts on behalf of the Florida panther. We agree with Wayne and Gittleman's (1995) recently revised conclusions that "biologists must look beyond the taxonomic classification of an endangered hybrid or subspecies; they should also take into account its unique function in an ecosystem or possession of special traits that cannot be reproduced…."

KEYS TO THE FUTURE SUCCESS OF THE REINTRODUCTION EFFORT

One of the most critical and complex issues affecting red wolf recovery is the public perception of the wolf itself and the consequences of reestablishing the species. There is a need for large-scale efforts to inform and involve the public, especially local residents. The biological concept of a predator and its role in a healthy ecosystem is misunderstood by a large portion of the general public; this is especially true in the case of the wolf. Myth, misunderstanding, exaggeration of historical events, and livestock depredation have accumulated through the centuries to give the wolf a larger-than-life reputation as a vicious and evil killer, devoid of redeeming qualities. Education of the public with accurate biological information focusing on predator/prey relationships and general wolf ecology is necessary to overcome the ignorance-based fear and hatred still present in the general populace. Effective education could be achieved through school systems, hunting/outdoor clubs, business and agriculture organizations, and public forum.

An accurate understanding of the animal, though crucial, is only the beginning. Landowners have come to view the presence of an endangered species as a curse. They fear regulations protecting that species will not only affect trapping, hunting, and access, but may prohibit other current uses of the area. Recently, great strides have been made in allowing managers flexibility under the Endangered Species Act through the experimental population provision, to reduce and avoid the inconvenience or losses the public has come to expect regarding endangered species. However, to effectively tailor a restoration and management strategy to an area is a complicated process. Timely and thorough scoping is necessary to identify and respond to all interests and concerns of the local, regional, and national public. Constant information exchange is necessary to establish and maintain program stability within the general limits defined by the affected public. This aspect of the program is necessary prior to and throughout the release stages.

The majority of concerns voiced by various private individuals, organizations, and agencies can be generally summarized as questions of responsibility. Who is going to be responsible for the program, throughout and beyond the recovery stages? What guarantees are there that private individuals or groups will not have to bear a disproportionate share of the burden imposed by the recovery of the red wolf? As a federally protected endangered species, the red wolf comes under the aegis of the FWS, the primary federal agency charged with endangered, nonmarine, species recovery. The federal regulations governing the reestablishment of experimental populations will promote interstate consistency. Amendment of federal and state protection regulations will be critical to allow control of animals that come into conflict with human activities. The willingness and ability to accomplish this has already been demonstrated within the program.

Although likely to be in general agreement with endangered species recovery, other agencies potentially involved with the program need clear answers to questions of long-term responsibility. The cost for implementing such programs may exceed the means of a single agency. This may require multiple-agency funding at the federal and state levels, as well as assistance from private sources. In the case of the red wolf, the federal agencies involved must commit to long-term plans that shield state agencies from much of the financial burden of coordination and management and yet remain open and responsive to state input. State wildlife agencies must play a critical role in the red wolf project. State concerns are attuned to local interests and resource management needs and include potential conflicts with public use of national forest, state, and private land. The need for active state agency involvement is unavoidable when fostering management of the species outside federal properties. All affected agencies must be involved in the decision making process.

Before initiating any large-scale restoration project, all agencies and interests involved must develop comprehensive strategies to avoid and resolve potential conflicts. This task will likely present a considerable challenge. Wolf repatriation is an issue involving biology, socioeconomic, factors politics, and emotion, and thus cannot be effectively addressed without the participation of all groups interested in the responsible management of the area.

To promote genetic diversity and promote natural conspecific mate selection, we ideally recommend that a large area be saturated with red wolf packs. Releases representing the 14 founder lines would be desirable. Large contiguous tracts of federal property are essential for wolves to travel and disperse freely. In a situation with no human-related management limitations, repatriation could best be accomplished through simultaneous mass releases of paired wolves. Saturating an area with wolves would displace the majority of resident coyotes, as breeding pairs of red wolves are more inclined to be territorially exclusive. The resultant reproduction would provide additional wolves for mate selection and increase the effective core area through dispersal.

However, this approach is not completely feasible, due to the unpredictable behavior of individual wolves and the logistical difficulties of managing such a large number of animals within the given human-related limitations. The most feasible alternative is slow, controlled releases of breeding pairs or family groups with intensive monitoring and management, combined with limited coyote control, until a sufficient core can be established. This is a painstaking and labor-intensive process which may take a decade or more. A large field crew will be needed to minimize conflict with humans and their related interests. Part of the crew's responsibilities would be to continue to inform and work with the affected public toward more public-involved and cooperative solutions to perceived wolf/human conflicts — such as predator-resistant domestic animal management. The members of such a crew, representing several agencies, could potentially have other duties as well. This arrangement would enhance multiple-agency and public involvement, disperse program cost, and benefit other resource programs. Coordination of field crews by one agency would be necessary to maintain management consistency and focus.

The federal landholdings of the southern Appalachians are not contiguous, and wildlife species do not recognize political boundaries. Federal and state boundaries will need to be philosophically dismantled. Travel corridors between federal, state, and leased or permitted private property will be necessary to permit wolves to disperse and encounter conspecifics from other areas, fostering necessary genetic exchange. This would also benefit existing species such as black bears and potential populations of elk and mountain lions.

There is enough suitable habitat in the southern Appalachians to potentially achieve a self-sustaining red wolf population. Although habitat is simply a biological need for the wolf, the presence of a red wolf population in a given area is far more than a simple issue for the communities and agencies involved. There are numerous requirements of the public that must be met before they are willing to support or even tolerate the presence of restored large carnivores. There are also risks, some real, some merely perceived, that managing agencies and the public must be willing to take in order to achieve such a goal.

RISK ASSESSMENT FOR MANAGERS

Based on the factors linked with success of translocations identified by Griffith et al. (1989) and discussed earlier in this chapter, the repatriation of red wolves to the southern Appalachians will be an immense challenge. The characteristics of the species and the scenario of the translocation are inconsistent with virtually all of the attributes identified by Griffith et al. (1989). However, that does not automatically rule out success. The project has been exceptionally well researched and planned and, consequently, satisfies all of the recommendations of Dodd and Siegel (1991). Our *Risk Assessment Report Card* reads as follows:

Vision: A+

There is a lot of good background information from other areas and the pilot study in the Smokies has yielded much valuable information for making management decisions.

Resource risk: D

During the repatriation process the red wolf population is highly vulnerable to disease, coyote cross-breeding, illegal shooting, and dispersal outside public lands where potential for conflict is great.

Resource conflicts: A

The reestablishment of red wolves will have some positive benefits to ecosystem function. Predators to white-tailed deer have long been absent from the southern Appalachians and possible reduction in numbers of the exotic European wild boar would be desirable.

Socioeconomic conflicts: C

Red wolf depredations do occur, although the perceived risk is probably much greater than the actual risk. Nevertheless, this has resulted in some significant socioeconomic and political conflicts.

Procedural protocols: C

Although procedures for successful repatriation of captive species have not been well documented, the pilot study and the effort at ARNWR have yielded some information and that has been used to develop restoration guidelines.

Scientific validity: C

The conditions for the initial decline of red wolves were not well documented and, therefore, procedures to eliminate those conditions are lacking. As a result, repatriation strategies have been developed based on limited information and, at this point, may be based more on art than science.

Legal jeopardy: C

The hybrid issue as well as other issues relative to the Endangered Species Act remains controversial.

Public support: B+

There appears to be strong public support for the project; most landowners in the area also seem supportive given the current depredation policies.

Adequacy of funding: C

Manpower and funds for successful repatriation are yet unknown.

Policy precedent: A

The proposed action is well within the mandates of most of the affected agencies.

Administrative support: B–

Affected agencies are all supportive of the project although there needs to be a more proactive strategy for sharing responsibilities.

Transferability: B

The mechanics of this action are probably unique, although the political and administrative structure necessary to accomplish the task would be transferable to other locales.

Achieving the goal of a recovered population of red wolves in the southern Appalachians will require a comprehensive approach that will address the variety of concerns and possible conflicts mentioned above and those yet to surface. This task will be extremely difficult to accomplish. Indeed, achieving concurrence among six or more agencies on one management plan, for a controversial species, may seem nearly impossible. It will demand an unprecedented level of interagency cooperation, but it is the only reasonable means to establishing and maintaining a viable population of this large carnivore. We have the opportunity to restore an integral component to the southern Appalachians ecosystem, and with it, create a blueprint for implementing landscape-scale management of natural resources. If the red wolf project is successful in the southern Appalachians, the mechanism and hope for restoring and managing other landscape species also will be established.

REFERENCES

Atkins, D. L. and L. S. Dillon. 1971. Evolution of the cerebellum in the genus *Canis. J. Mammal.* 52:96–107.

Audubon, J. J. and J. Bachman. 1851. The viviparous quadrupeds of North America, vol. 2, New York, NY. 334 pp.

Carley, C. J. 1975. Activities and Findings of the Red Wolf Recovery Program from late 1973 to July 1, 1975. U.S. Fish and Wildlife Service, Albuquerque, NM. 215 pp.

Carley, C. J. 1979. Status summary: The red wolf (*Canis rufus*). Endangered Species Report No. 7, U.S. Fish and Wildlife Service, Albuquerque, NM. 36 pp.

Crawford, B. A. 1992. Coyotes in Great Smoky Mountains National Park: Evaluation of Methods to Monitor Relative Abundance, Movement, Ecology and Habitat Use. M.S. thesis, University of Tennessee, Knoxville, TN. 84 pp.

Dodd, C. K. and R. A. Seigel. 1991. Relocation, repatriation, and translocation of amphibians and reptiles: are they conservation strategies that work? *Herpetologica* 47:336–350.

Dowling, T. E., W. L. Minckley, M. E. Douglas, P. C. Marsh, and B. D. DeMarais. 1992. Response to Wayne, Nowak, and Phillips and Henry: use of molecular characters in conservation biology. *Conserv. Biol.* 6:600–603.

Ferrell, R. E., D. C. Morizot, J. Horn, and C. J. Carley. 1980. Biochemical markers in a species endangered by introgression: the red wolf. *Biochem. Gen.* 18:39–47.

Gittleman, J. L. and S. L. Pimm. 1991. Crying wolf in North America. *Nature* 351:524–525.

Griffith, B., J. M. Scott, J. W. Carpenter, and C. Reed. 1989. Translocation as a species conservation tool: status and strategy. *Science* 245:477–480.

Goldman, E. A. 1937. The wolves of North America. *J. Mammal.* 18(1):37–45.

Harper, F. 1942. The name of the Florida wolf. *J. Mammal.* 23:339.

Henry, V. G. 1991. Endangered and threatened wildlife and plants; determination of experimental population status for and introduced population of red wolves in North Carolina and Tennessee. *Fed. Reg.* 56(213):56325–56334.

Henry, V. G. 1995. Endangered and threatened wildlife and plants; final rule for nonessential experimental populations of red wolves in North Carolina and Tennessee. *Fed. Reg.* 60(71)18940–18948.

King, P. B. and A. Stupka. 1950. The Great Smoky Mountains — their geology and natural history. *Sci. Month.* 71:31–43.

Kleiman, D. G. 1989. Reintroduction of captive mammals for conservation. *Bioscience* 39:152–161.

Mayr, E. 1963. *Animal Species and Evolution.* Harvard University Press: Cambridge, MA. 797 pp.

McCarley, H. 1962. The taxonomic status of wild *Canis* (Canidae) in the South-central United States. *Southwest. Nat.* 7(3–4):227–235.

McCarley, H. and C. J. Carley. 1979. Recent changes in the distribution and status of wild red wolves *Canis rufus.* Endangered Species Report No. 4, U.S. Fish and Wildlife Service, Albuquerque, NM. 38 pp.

Nowak, R. M. 1967. The red wolf in Louisiana. *Defenders Wildl. News* 45(1):82–94.

Nowak, R. M. 1979. North American Quatery *Canis.* Museum of Natural History, University of Kansas Monograph 6. 154 pp.

Nowak, R. M. 1992. The red wolf is not a hybrid. *Conserv. Biol.* 6:593–595.

Nowak, R. M., M. K. Phillips, B. G. Henry, C. W. Hunter, and R. C. Smith. 1995. The origin and fate of the red wolf. Pages 409–416 in L. N. Carbyn, S. H. Fritts, and D. R. Sipe, Eds. Ecology and Conservation of the Wolf in a Changing World. Canadian Circumpolar Institute, University of Alberta, Edmonton.

Paradiso, J. L. 1968. Canids recently collected in east Texas, with comments on the taxonomy of the red wolf. *Am. Midl. Nat.* 80(2):529.

Paradiso, J. L. and R. M. Nowak. 1971. A report on the taxonomic status and distribution of the red wolf. U.S. Department of the Interior. Spec. Sci. Rep. — Wildl. No. 145. Washington, D.C.

Parker, W. T. 1990. A proposal to reintroduce wolves into the Great Smoky Mountains National Park. Red Wolf Management Series Tech. Rep. No. 7, U.S. Fish and Wildlife Service, Asheville, NC. 33 pp.

Parker, W. T. and M. Phillips. 1991. Application of the experimental population designation to recovery of endangered red wolves. *Wild. Soc. Bull.* 19:73–79.

Phillips, M. K. 1994. Reestablishment of red wolves in the Alligator River National Wildlife Refuge, North Carolina. September 14, 1987, to September 30, 1992. Red Wolf Management Series Tech. Rep. No. 10. U.S. Fish and Wildlife Service, Atlanta, GA. 28 pp.

Phillips, M. K. and V. G. Henry. 1992. Comments on red wolf taxonomy. *Conserv. Biol.* 6:596–599.

Pimlott, D. H. 1965. A study of the status and ecology of the red wolf in the South-central United States; progress report. Department of Zoology, University of Toronto, Ontario. 6 pp.

Pimlott, D. H. and P. W. Joslin. 1968. The status and distribution of the red wolf. *Trans. North American Wildlife and Natural Resources Conference* 33:373–384.

Reinert, H. K. 1991. Translocation as a conservation strategy for amphibians and reptiles: some comments, concerns, and observations. *Herpetologica* 47:357–363.

Riley, G. A. and R. T. McBride. 1972. A survey of the red wolf *Canis rufus*. U.S. Department of the Interior. Spec. Sci. Rep., Wildl. No. 162. Washington, D.C. 15 pp.

Roy, M. S., D. J. Girman, and R. K. Wayne. 1994a. The use of museum specimens to reconstruct the genetic variability and relationships of extinct populations. *Experientia* 50:1–7.

Roy, M. S., E. Geffen, D. Smith, E. Ostrander, and R. K. Wayne. 1994b. Patterns of differentiation and hybridization in North American wolf-like canids revealed by analysis of microsatellite loci. *Mol. Biol. Evol.* 11(4):553–570.

U.S. Fish and Wildlife Service. 1990. Red wolf recovery plan. Atlanta, GA. 110 pp.

Wayne, R. K. 1992. On the use of morphologic and molecular genetic characters to investigate species status. *Conserv. Biol.* 6:590–592.

Wayne, R. K. and J. L. Gittleman. 1995. The probematic red wolf. *Sci. Am.* 273 (1):36–39.

Wayne, R. K. and S. M. Jenks. 1991. Mitochondrial DNA analysis implying extensive hybridization of the endangered red wolf, *Canis rufus*. *Nature* 351:565–568.

12 Management of Isolated Populations: Southern Strain Brook Trout

Stanley Z. Guffey, Gary F. McCracken, Stephen E. Moore, and Charles R. Parker

CONTENTS

Introduction 247
History and Status of Brook Trout Populations in the Southern Appalachians 248
    The Distribution of Brook Trout...248
    Brook Trout Decline in the Southern Appalachians...249
    Great Smoky Mountains National Park...250
    Effects of Rainbow Trout...250
Distinctiveness of Southern Appalachian Brook Trout and the Impact
of Hatchery Introductions 252
    Stocking History..252
    The Native Southern Appalachian Brook Trout 252
    Molecular Population Genetics..253
    Genetic Inventory of Brook Trout Populations 254
    Effects of Hybridization..255
    Heterogeneity among Native Southern Appalachian Populations 256
Management and Conservation of Southern Appalachian Brook Trout.......................257
The Southern Appalachians as a Center of Biological Diversity 259
Risk Assessment Report Card for Ecosystem Managers 259
References 261
Glossary..265

INTRODUCTION

For many people, trout and clear, free-flowing trout streams are synonymous with mountain wildness. From this perspective preservation of wild trout requires preservation of wildness, and loss of wild trout is indicative of environmental deterioration (Leopold 1949). Jordan and Evermann (1896) expressed this perception:

> The members of this genus [*Salvelinus*] are by far the most active and handsome of the trout, and live in the coldest, cleanest, most secluded waters. No higher praise can be given to a salmonid than to say it is a charr.

The history of brook trout distribution in the southern Appalachians over the past century supports the reality of this relationship between trout well-being and environmental quality. Since 1900, stream mileage occupied by brook trout in the southern Appalachians has decreased by about 79% as a consequence of human activities (Bivens 1985). While most of this decline occurred prior to 1970 and brook trout populations have fared much better in recent years (Habera and Strange 1993), land use and fisheries management practices still impact native brook trout. Their current restricted habitat makes the remaining populations highly vulnerable to extinction (Nagel 1991; Platts and Nelson 1988). Concerns for the preservation of southern Appalachian brook trout are heightened by findings from molecular population genetics. Southern Appalachian brook trout are genetically distinct from populations throughout the taxon's main range and warrant recognition as a distinct evolutionary unit (Stoneking et al. 1981; McCracken et al. 1993; Hayes et al. 1996). The molecular studies also show high levels of differentiation among native southern Appalachian populations, indicating that population extinction will also result in the irretrievable loss of genetic diversity.

In this chapter we examine the biology and ecology of brook trout in the southern Appalachians from the perspectives of biodiversity management, genetics, and conservation. In the following sections we discuss (1) the history and status of brook trout populations in the southern Appalachians; (2) the distinctiveness of the southern Appalachian brook trout and the impact of hatchery introductions; and (3) the sustainable management of their remaining populations. Our approach highlights the necessity of examining the complex historical, ecological, and genetic dimensions of biodiversity management concerns in order to devise sound ecosystem management policies. In a final section, southern Appalachian brook trout are presented as a model for organisms that exist in numerous, small, isolated populations. Demographic, genetic, and conservation implications of this metapopulation structure are considered in reference to other species in the region that have distributions and population structures similar to brook trout. We conclude by presenting the southern Appalachians as a center of biodiversity for brook trout and other organisms and discuss implications for biodiversity management in the region. We emphasize the situation in the Appalachian mountain region of Tennessee and North Carolina, particularly in Great Smoky Mountains National Park (hereafter refered to as "the Park"), where southern brook trout have been most extensively studied.

HISTORY AND STATUS OF BROOK TROUT POPULATIONS IN THE SOUTHERN APPALACHIANS

THE DISTRIBUTION OF BROOK TROUT

The brook trout or brook charr, *Salvelinus fontinalis*, is the only salmonid native to the southern Appalachians. The native range of brook trout extends from northern Quebec, westward to Manitoba, and along eastern North America south to northern Georgia (MacCrimmon and Campbell 1969). In its northern range, brook trout inhabit a wide variety of stream, riverine, and lacustrine habitats, including some anadromous populations (Hendricks 1980). In the southern part of its range, brook trout are confined to high elevation first and second order streams of the Appalachian mountains (Jones 1978; Strange 1979; Bivens 1985; Habera and Strange 1993; Flebbe 1994). Brook trout, like most salmonids, have been successfully reared in hatcheries for over a century and have been widely stocked throughout the world (MacCrimmon and Campbell 1969).

Prior to extensive human disturbance the limits of the southern range were abiotically determined, principally limited by maximum summer air temperature and maximum ground water temperature. McCrimmon and Campbell (1969) demonstrated that the southern range of native brook trout is approximately delimited by the 21°C mean July isotherm. However, air temperature is not the major determinant of the thermal characteristics of stream habitats. The temperature of ground water sources is the major determinant of the temperature of high elevation streams in the southern Appalachians (Meisner 1990). For brook trout, the 15°C ground water isotherm occurring at 35–39°N latitude is thought to be the major limit to brook trout distribution in the southern Appalachians (Meisner 1990; Power 1980).

In this southern range, the 21°C mean July isotherm and the 15°C groundwater isotherm are functions of both latitude and elevation. Calculating 15°C groundwater temperature as approximately 1.5°C plus mean annual air temperature, Meisner (1990) demonstrates the lower stream boundary for brook trout rises from sea level at 39° north latitude to about 600 meters at 35°N in northeastern Georgia. This corresponds to an estimate of 1°C change in temperature in 188 meters of elevation or 110 kilometers of latitude (Meisner 1990). In the absence of human impacts suitable habitat will extend below this groundwater isotherm as a result of aspect shading and shading by riparian vegetation. However, as a consequence of human activities few southern Appalachian brook trout populations are presently found near this theoretical boundary (Kelly et al. 1980; Bivens 1985; Habera and Strange 1993). In most locations brook trout in the southern Appalachians are confined to first- and second-order stream segments above at least 900 meters in elevation (Habera and Strange 1993).

BROOK TROUT DECLINE IN THE SOUTHERN APPALACHIANS

Logging, road and railroad construction, land clearing for agriculture, over-fishing, and the introduction of nonnative salmonids have all contributed to the habitat compression of southern brook trout (Kelly et al. 1980; Bivens 1985). Timber harvest and clearing for agriculture directly influence the thermal characteristics of streams by removing shading vegetation (Swift and Messer 1971). These activities, as well as associated road and railroad construction, also can result in stream siltation leading to changes in stream faunas (Hansen 1971). Although the effects of these changes on brook trout feeding ecology have not been investigated in the southern Appalachians, brook trout are visual predators and increased turbidity would likely reduce feeding success. Siltation also negatively impacts brook trout reproduction by smothering eggs in spawning sites and by covering or filling in spawning sites, rendering them unusable. Road construction in the high elevations of Great Smoky Mountains National Park has revealed another source of negative impacts, the exposure of acidic rock strata (Huckabee et al. 1975). Leachate from argillaceous slates, phyllites, and schists of the Anakeesta Formation (King et al. 1968) can reduce Park stream pH from 6.7 to 4.5 or lower, and mobilize heavy metal sulfide constituents of the Anakeesta rocks (Morgan et al. 1976). These inputs can render stream segments sterile (Green 1975; Huckabee et al. 1975). Even after pH recovery, the resultant metal hydroxide precipitates inhibit benthic macroinvertebrate recolonization (Morgan et al. 1976). Ironically, rainbow trout (*Oncorhynchus mykiss*) are more sensitive to low pH than are brook trout (Curtis et al. 1989), and this can inhibit rainbow trout encroachment in acidified streams where brook trout are not altogether eliminated (Bivens 1985).

During the early part of this century, logging and associated activities appear to have been the primary factors responsible for reduction of brook trout range in the southern Appalachians. Powers (1929) and King (1937) indicated that brook trout distribution at the time the Park was established (1936) was limited to areas that had little or no logging. By 1930, extensive logging in Tennessee south of the Park had reduced brook trout populations to a few stream segments in Monroe County and eliminated brook trout from Polk County (Bivens 1985). Studies of logging history and native brook trout decline are lacking for most of the southern Appalachian region; however, it is likely that such studies would confirm the relationship between logging intensity and brook trout decline.

These landscape level impacts have diminished over the past century, at least on public lands. Timber harvest is prohibited on National Park and National Forest wilderness lands. On public and private lands where these activities take place, harvest methods exist and are often employed, that can minimize impacts on aquatic systems including brook trout populations (Dixon 1975). However, although many habitats have largely recovered, brook trout have generally not regained range lost due to prior impacts. Indeed, brook trout continued to decline after the end of extensive logging and forest recovery (Seehorn 1979; Kelly et al. 1980). Widespread introductions of foreign salmonids, brown trout (*Salmo trutta*) from northern Europe and rainbow trout from the Pacific northwest, are strongly implicated in this continued decline (King 1937; Seehorn 1979; Kelly et al. 1980; Bivens 1985; Bivens et al. 1985; Larson and Moore 1985; Habera and Strange 1993).

GREAT SMOKY MOUNTAINS NATIONAL PARK

The dimensions of these various impacts on southern Appalachian brook trout have been most thoroughly examined in Great Smoky Mountains National Park. King (1937) reported that prior to 1900, brook trout were abundant in streams of the Park above 2000 feet (610 meters), with some populations extending to 1600 feet (490 m). By the time of his surveys in 1935, brook trout were generally limited to stream segments above 3000 ft (910 m) where extensive logging had not taken place, a loss of over 160 stream miles (260 km) or 55% of the estimated range in 1900 (Kelly et al. 1980). Surveys in the 1950s indicated brook trout had not regained lost habitat after the end of logging in 1935, but had instead declined an additional 15% (Lennon 1967). These surveys observed that stream segments previously inhabited by brook trout were inhabited by rainbow trout (Lennon 1967) that had been stocked into the lower sections of most Park streams between 1910 and 1930 (Kelly et al. 1980). By the 1970s brook trout occupied about 123 miles (198 km) of Park streams (Kelly et al. 1980), only a slight decline from the 1950s. However, the number of stream segments with sympatric brook trout and rainbow trout populations had increased (Kelly et al. 1980). Brown trout invading from stocked stream segments outside the Park had become important by the 1970s as well, with reproducing populations sympatric with rainbow trout in 50 miles of low elevation Park streams (Kelly et al. 1980).

The studies cited above identify two broad categories of impacts, operating more or less sequentially, which negatively impacted brook trout in the Park: logging and associated activities, and stocking with nonnative salmonids. As in areas outside the Park, logging directly impacted brook trout streams through removal of shading vegetation and increased surface runoff and siltation (King 1937). Road, railroad, bridge, and dam construction in support of timber harvest directly impacted brook trout streams as well (King 1937). Even after the timber had been removed, landscape and stream recovery were delayed by the frequent fires that swept through slash and remaining vegetation on cut-over lands (King 1937). Even more direct during this period of destructive exploitation, although perhaps of lesser consequence, were the impacts of destructive fishing practices. Fishing was not regulated and the use of baits, nets, and dynamite insured successful fishing (King 1937). With the establishment of Great Smoky Mountains National Park in 1936, logging was terminated and some regulation of fishing was instituted, but brook trout populations did not recover with the landscape (Lennon 1967).

EFFECTS OF RAINBOW TROUT

King (1937) suggested that the upstream invasion by rainbow trout of stream segments from which brook trout had been extirpated was a consequence of rainbow trout's broader habitat tolerances, higher fecundity, and higher growth rate. The thermal maximum and optimum of rainbow trout are higher than those of brook trout (Coutant 1977; Peterson et al. 1979) and they are generally more tolerant of silt and pollution than are brook trout (King 1937). Thus, they could successfully occupy recovering habitat prior to brook trout. Presumably this priority of arrival, coupled with higher fecundity and growth rate, inhibited the downstream invasion by brook trout after habitat recovery. Lennon (1967) ascribed the failure of brook trout to recover lost habitat, and the continued loss of habitat to rainbow trout, to consequences of life in small headwater streams. Fish in these populations are smaller than those found in larger downstream segments, and fecundity is lower because of the scarcity and small size of spawning sites. Lennon (1967) also believed these small headwater populations had higher incidence of disease than larger brook trout and rainbow trout populations living in larger lower elevation segments, a factor working to the further demographic detriment of brook trout. Empirical support for most of these hypothesized effects is strong, but experimental evidence for the mechanisms involved in the replacement of brook trout by invading rainbow trout has been more difficult to obtain and results difficult to interpret (Fausch 1988).

It is well documented that southern Appalachian brook trout populations south of central Virginia have lost and have continued to loose habitat to rainbow trout (Kelly et al. 1980; Moore et al. 1984; Bivens et al. 1985; Larson and Moore 1985). Allopatric brook trout populations above natural barriers which prevent upstream movement of rainbow trout have not declined significantly during the period of monitoring by Park biologists; in streams without barriers, rainbow trout populations have expanded upstream to the detriment of brook trout (Larson and Moore 1985). In sympatry, there is a negative relationship between the densities of adult brook trout and the densities of rainbow trout, suggesting competitive superiority of rainbow trout. A negative relationship also exists between the adult population size of either species and numbers of young of the year of the other species (Larson and Moore 1985), suggesting either interspecific predation or greater access to spawning sites by the higher density species. Stomach content analysis of both species in allopatry and sympatry has excluded interspecific predation as significant factors in the displacement of brook trout by rainbow trout (Habera 1987; Ensign 1988). However, removal experiments in the Park indicate that this biotic exchange is a consequence of interspecific interactions rather than the occupation by rainbow trout of empty habitat created by declining brook trout populations. Electrofishing to remove rainbow trout in streams where they are sympatric with brook trout results in increases in the biomass of brook trout and the establishment of age structures similar to that of allopatric brook trout populations (Moore et al. 1986). This is achieved even if rainbow trout are not fully eradicated, although without complete removal of rainbows, brook trout recovery is temporary (Moore et al. 1984).

Interactions between brook trout and rainbow trout involve differences in stream microhabitat utilization. In both sympatry and allopatry, brook trout prefer pools in low velocity stream segments near cover, while rainbow trout prefer higher velocity riffles and runs (Cunjak and Green 1983; Lohr and West 1992; Welsh 1994). Brook trout allopatric from rainbow trout are largely confined to high gradient headwater stream segments where water temperatures are lower. However, high gradient segments are not the preferred habitat of brook trout, but of rainbow trout. The observation that rainbow trout do not displace brook trout in low gradient high elevation streams even in the absence of barriers to upstream movement (Moore et al. 1986) supports the contention that lower water temperature of headwater streams is important for brook trout. At intermediate elevations, rainbow trout encroachment in steep gradient streams exhibits considerable annual variation as a consequence of different habitat tolerances of the two species and of annual variation in stream flow (Larson et al. 1995). Segregation of microhabitat utilization suggests the possibility of stable sympatry in at least some stream segments, although food utilization by brook trout and rainbow trout in the Park overlap significantly (Habera 1987; Ensign 1988) and populations of both species may be food limited (Chapman 1965). Encroachment of rainbow trout or their removal from sympatry with brook trout does not result in a shift in prey taken by adults of either species (Habera 1967; Ensign 1988).

Numerical displacement (Chapman 1965) involving some type of antagonistic interaction at some stage in the life cycle, though incomplete in its formulation, appears to be a component of the displacement of brook trout by rainbow trout in the southern Appalachians (Fausch 1988). Competition for spawning sites, which may be limiting (Morgan and Robinette 1978), is a factor only if sites are defended throughout most of the year or if use by rainbow trout renders them unsuitable for use by brook trout. In the Park and throughout the southern Appalachians generally, brook trout spawn in the fall and emerge in early spring. Rainbow trout spawn in the spring and emerge in early summer, and by late summer are larger than the earlier emerging brook trout. Effects of rainbow trout spawning at sites containing emerging brook trout fry have not been investigated. The effects of behavioral interactions between the two species as adults and as juveniles

have primarily been investigated in the laboratory (Cunjak and Green 1983, 1984, 1986), and the
results are equivocal (Fausch 1988; Lohr and West 1992).

DISTINCTIVENESS OF SOUTHERN APPALACHIAN BROOK TROUT AND THE IMPACT OF HATCHERY INTRODUCTIONS

STOCKING HISTORY

Stocking with hatchery-reared brook trout began in the Park in 1937 in an attempt to augment wild
populations and to recolonize habitats from which wild populations had been extirpated. From
1937 through 1939, about 200,000 hatchery reared native brook trout were stocked annually into
Park streams (Kelly et al. 1980). Heavy stocking continued from 1940 through 1947, using hatchery
strains derived from northeastern populations. Stocking was curtailed after 1947 but was not entirely
eliminated in the Park until 1975. Over that period more than 800,000 eggs, fry, fingerlings, and
adults from northern derived hatchery strains were stocked into at least 76 Park streams. Only 12
streams in the Park have no record of stocking (McCracken et al. 1993). Outside the Park, federal
fish hatcheries in Tennessee, North and South Carolina, and Virginia maintained northern derived
strains for stocking (Bivens 1985; Kriegler 1993; McCracken and Guffey 1994; Guffey 1995).
Kriegler (1993) documents stockings into at least 85 streams in Tennessee between 1951 and 1988.
Stocking with northern-derived hatchery brook trout strains throughout Tennessee was terminated
in 1988, but continues in North Carolina and Virginia.

Stocking with hatchery-reared fish was largely unsuccessful in halting the loss of brook trout
range, with stream mileage occupied by brook trout continuing to decline until at least the mid
1970s. Management efforts over the past 25 years have emphasized stream restoration and construction of barriers against rainbow trout encroachment where feasible, and removal of rainbow
trout from sympatric populations above barriers. Work in the Park has demonstrated the effectiveness, as well as the large manpower requirements, of rainbow trout removal (Moore et al. 1986).
Removal of rainbow trout from all former or potential brook trout habitat is clearly not feasible,
and for many managers, not desirable. However, rainbow trout removal from selected streams
appears to be the best strategy for establishing brook trout sport fisheries and as a hedge against
demographic extinction of small headwater populations.

THE NATIVE SOUTHERN APPALACHIAN BROOK TROUT

The distinctiveness of southern Appalachian brook trout is a part of the lore of the region (Yuskavitch
1991; Venters 1993). Known in the region as "speckled trout" or "speckles," southern Appalachian
brook trout are smaller and reputed to have more spots and be more brightly colored than northern
derived hatchery fish. Local opinion also holds they offer a more exciting angling experience and
are tastier. Fisheries biologists in the region have occasionally come to a similar conclusion. In a
National Park Service internal report, Holloway (1945) appears to have been the first fisheries
biologist to suggest differences between native brook trout and northern-derived hatchery strains.
He noted that hatchery strains were less hardy than native populations and less likely to become
established. For these reasons, Holloway recommended that stocking with hatchery strains be
discontinued in the Park. As previously discussed, stocking was sharply reduced after 1947 but
was not terminated until 1975. In 1967, another Park fisheries biologist argued that native and
hatchery trout differed morphologically and possibly in life history characteristics as well (Lennon
1967). Only with the application of molecular population genetics techniques coupled with examination of stocking history has the distinctiveness of southern Appalachian brook trout been established. Because of the mandate of the National Park Service to manage for native varieties, most
of the initial research was conducted in, and on behalf of, the Park. Subsequent work has also been

supported and conducted by state and other federal resource agencies in the southern Appalachian region.

MOLECULAR POPULATION GENETICS

In a unpublished examination of allozyme protein variation among brook trout populations, Brussard and Nielsen (1976) concluded that southern Appalachian populations might differ from northern populations at the level of subspecies or species. This work was extended by Stoneking et al. (1981), who reached a similar conclusion. Stoneking et al. (1981) observed significant allele frequency differences between three southern populations (two from Tennessee and one from the Park in North Carolina) and five northern populations (three from Pennsylvania and two from New York) at 4 of 39 allozyme loci. Stoneking et al. (1981) suggested the possibility that stocking might be a confounding factor in their study. Kriegler (1993) has confirmed this possibility, documenting that the two Tennessee populations examined by Stoneking et al. (1981) had received stocking onto native populations. Stoneking et al. (1981) considered the one Park population examined (Bunches Creek) to be the most distinctively southern of the three. Bunches Creek is one of only twelve Park streams with no record of stocking (McCracken et al. 1993).

McCracken et al. (1993) considered stocking history to explicitly test the hypothesis of genetic differences between native brook trout populations in the Park and the northern derived hatchery strains used for stocking in the Park. Five of the presumed native southern Appalachian populations with no record of stocking with hatchery fish were sampled along with three other Park streams with documented stocking onto existing populations. Northern derived hatchery strains were represented by two strains used in stocking in the region and by a wild reproducing population from a Park stream, Meigs Creek, which was devoid of brook trout prior to stocking. McCracken et al. (1993) showed that unstocked populations and hatchery strains were fixed for different alleles at one locus (CK-A2*) and had significant allele frequency differences at an additional 9 of 16 polymorphic loci that contained alternative alleles of presumed northern ancestry. The effects of hybridization were evident in the three populations where hatchery strains had been stocked onto extant populations. Homozygotes and heterozygotes were observed for both CK-A2* alleles, and the alternative "northern" alleles were present at several other loci in all of these stocked populations.

McCracken et al. (1993) estimated a mean genetic similarity of $I = 0.906$ between unstocked and northern hatchery strains, indicating a level of genetic divergence similar to that observed by Stoneking et al. (1981). For the larger sample of populations, we estimate a mean genetic similarity of $I = 0.840$ between native southern Appalachian populations, and northern and northern derived hatchery populations. These genetic similarity estimates are similar to those observed in allozyme studies of genetic differentiation between strains and subspecies of other salmonid fishes (Allendorf and Utter 1979; Loudenslager and Gall 1980; Leary et al. 1987). For example, genetic identity estimates were obtained for morphologically differentiated subspecies in the cutthroat trout complex range from 0.743 to 0.928 for three subspecies with two or more fixed diagnostic loci (Leary et al. 1987). Thorpe (1982) suggests that allopatric populations with genetic similarity below 0.85 should not be considered conspecifics, but that populations with higher values require further data for adequate determination. The range of these values for morphologically differentiated cutthroat trout subspecies, as well as data from other taxa (Nei 1987), point to the difficulties of basing taxonomic decisions solely on indices of genetic similarity (Buth 1984; Frost and Hillis 1990).

The investigation of mitochondrial DNA (mtDNA) variation also confirms the genetic differentiation between northern and native southern brook trout populations. Hayes et al. (1996) observed polymorphisms in 7 of 13 unique restriction endonucleases used to examine mtDNA variation in 11 southern Appalachian populations and five northern derived hatchery strains. The status of southern Appalachian populations as pure native or hybrid was informed by records of stocking histories and the earlier studies using the nuclear gene, allozyme markers. The seven informative

TABLE 1
Genetic Status of Brook Trout Populations

	Native	Hybrid[a]	Northern	Total
Great Smoky Mountains National Park	36	15	1	52
Tennessee[b]	48	33	14	95
North Carolina[b]	2	10	0	12
South Carolina	1	9	2	12
Georgia[c]	3	8	0	11
Total	90	75	17	182

[a] Fifteen populations (5 from the Park, 5 from Tennessee, 3 from North Carolina, 1 from South Carolina, and 1 from Georgia) did not show the diagnostic northern CK-A2 allele but were classified as hybrid on the basis of presumptive 'northern' alleles, at other loci.
[b] Outside of Great Smoky Mountains National Park.
[c] Data from Dunham et al. 1994.

restriction enzymes generated 16 distinct haplotypes, 12 of native southern Appalachian origin, and 4 from the northern derived hatchery strains. All native southern haplotypes differed from the hatchery haplotypes at a minimum of 3 restriction sites, with a mean difference between southern and the nearest northern haplotype of 4.7 restriction sites (range 3 to 6). Native southern mtDNA haplotypes differed from one another by an average of 2.6 restriction sites (range 1 to 5), and hatchery haplotypes differed from one another by an average of 1.5 restriction sites (range 1 to 3). By the method of Nei and Li (1979) these haplotype differences correspond to an estimate of 0.84% ± 0.02 mean sequence divergence between native southern populations and northern-derived hatchery strains (range 0.48 to 1.13%). Estimated mean sequence divergences were 0.41% ± 0.02 among southern haplotypes (range 0.15 to 0.80%), and 0.24% ± 0.07 among hatchery strain haplotypes (range 0.16 to 0.32%). As in the case with the allozyme estimates of genetic similarity, this mtDNA sequence divergence estimate is comparable to differences seen among subspecies in other salmonid taxa (Wilson et al. 1985; Thomas et al. 1986). Putting the issue of taxonomic assignment aside, we believe that nuclear and mtDNA genetic data as well as the biogeographical patterns are sufficiently compelling that southern Appalachian brook trout should be recognized as a distinct evolutionary unit for concerns of management and conservation.

GENETIC INVENTORY OF BROOK TROUT POPULATIONS

Employing the diagnostic allozyme markers of McCracken et al. (1993) we have now examined 171 brook trout populations: 52 from the Park, 95 from Tennessee and 12 from North Carolina outside of the Park, and 12 from South Carolina. (Table 1). These data have allowed us to evaluate the status of native southern Appalachian populations and the impacts of stocking in the region. We also have examined three hatchery strains from the Pisgah Hatchery, NC (Pisgah, Ed Ray, and Armstrong), three wild populations from Maryland, and three wild populations from Shenandoah National Park, VA. These outgroup populations are northern by the criteria of diagnostic allozyme loci. Our samples include all the known populations in Tennessee outside the Park (Strange and Habera 1995) and the majority of brook trout populations in the Park and the majority of populations in South Carolina. The North Carolina populations are all from the northwestern part of the state and are only a small fraction of the populations in North Carolina. A more extensive survey of North Carolina brook trout populations is currently underway (Jim Borawa, pers. commun.).

Overall, 87 (51%) of the 171 populations examined are apparent native southern Appalachian with no evidence for introgression of northern alleles, 17 (10%) are pure hatchery derived, and 67 (39%) are native hatchery hybrids. The Park has the highest percentage of native populations, 69%, and South Carolina the lowest, 8%. Of the brook trout populations in Tennessee, 51% are native. In Tennessee and South Carolina, hatchery-derived and hybrid populations cluster around hatcheries which previously maintained northern derived hatchery strains for stocking (Kriegler 1993; Kriegler et al. 1995; Guffey 1995; Saidak 1995). Dunham et al. (1994) examined 11 populations from Georgia and identified 4 (36%) as native southern Appalachian and the remainder as hybrid by the criterion of diagnostic alleles at the CK-A2* locus. These observations indicate that in addition to the loss of brook trout habitat since 1900, substantial erosion of indigenous genetic diversity has also occurred. The results also underscore the importance of eliminating stocking with hatchery brook trout in areas where hatchery fish can impact native populations.

Differences in the percentages of native populations in the five areas may reflect differences in stocking histories. Stocking was terminated in the Park in 1975 and later in areas managed by state agencies. Differences in habitat quality and absolute stocking effort may also have influenced the differing frequencies of native populations, but these factors are difficult to evaluate. Because of the small number of samples and the limited geographic scope of the surveys in North Carolina, the information from that state should be interpreted with caution. Additionally, many of the Atlantic drainage streams in North and South Carolina and Georgia may not have had brook trout populations prior to stocking. As most of these Atlantic drainage streams are located adjacent to the Tennessee River–Atlantic divide (Shull 1995; Dan Rankin, pers. commun.), stream capture is a possible source of native populations that must be considered.

The data from the 171 populations were consistent with McCracken et al.'s (1993) earlier survey in the Park. All populations with no known stocking history were fixed for the CK-A2* southern allele. The three northern derived hatchery populations, all populations of documented hatchery origin, and the populations from northern Virginia and Maryland were fixed for the CK-A2* northern allele and contained presumptive northern alleles at other variable loci. Populations with a known history of hatchery stocking on to extant native populations are typically segregating for both CK-A2* alleles and for presumptive northern alleles at other loci. The diagnostic utility of these other loci is problematic. Five populations from Tennessee, five from the Park, three from North Carolina, and one from South Carolina are fixed for the CK-A2* southern allele but have alternate, presumptive northern alleles at other loci. These could be interpreted as hybrids or as native populations that are polymorphic at one or more of the loci in question. Clarification of this issue may not be possible, and we prefer to err conservatively regarding the status of native brook trout, and designate those populations as southern Appalachian–northern-derived hatchery hybrids.

EFFECTS OF HYBRIDIZATION

The fixed difference at the CK-A2* locus provides the simplest means of evaluating the extent of northern derived hatchery introgression in hybrid populations. The frequency of the CK-A2* northern allele in hybrid populations varies considerably. Consistent with our evidence for a lesser impact of stocking in the Park, the frequency of the CK-A2* northern allele is lower in hybrid populations in the Park (mean = 0.16) than in hybrid populations outside the Park (0.34, 0.30, and 0.53 in Tennessee, North Carolina, and South Carolina, respectively). Correlations between levels of introgression and the number of fish stocked or the time since stocking, are not significant for either the nuclear or the mitochondrial gene markers (Hayes et al. 1996). This lack of correlation may be indicative of a nonlinear response or a consequence of our small sample size from some streams (Hayes et al. 1996). Deficiencies in the stocking record also preclude meaningful evaluation of the effects of stocking intensity (Kriegler et al. 1995).

It is clear that there has been considerable loss of pure native southern Appalachian brook trout in the region, and that attempts to ameliorate this decline by stocking with northern derived hatchery strains has further contributed to the erosion of native genetic diversity. The loss of genetic diversity is demonstrated by the fact that 39% of the 171 populations sampled in the region are now hybrid. The significance of this loss of genetic diversity is a function of the degree of differentiation between northern and southern populations and of the partitioning of genetic variance among the native populations.

HETEROGENEITY AMONG NATIVE SOUTHERN APPALACHIAN POPULATIONS

Native southern Appalachian brook trout populations exhibit low levels of heterozygosity. McCracken et al. (1993) observed an average individual heterozygosity (H_O) of 0.025 in 5 native populations examined for 34 allozyme loci. H_O was 0.112 in the three hatchery samples and H_O was 0.053 in the five hybrid populations. The larger number of populations in the Park, Tennessee, North Carolina, and South Carolina, examined at a smaller number of loci, showed a similar pattern: lowest heterozygosities in native populations, highest in hatchery strains, and intermediate in hybrid populations. These observations indicate that native southern Appalachian populations carry little genetic variation at the level of the individual population.

Despite their low levels of hetrozygosity, the native populations differed substantially in their genetic compositions. Considerably greater genetic difference was observed among native southern Appalachian populations than among northern populations with regards to differences in allele frequencies and the fixation of populations at different alleles at different loci (Morgan and Baker 1991; Perkins et al. 1993). These differences in genetic architecture between northern and southern populations reflect different histories of brook trout in the two regions. More importantly, from the perspective of biodiversity management, this partitioning of genetic variance among native southern Appalachian populations indicates that populations are not genetically interchangeable. If a goal of biodiversity management is the preservation of genetic diversity, management of brook trout in the southern Appalachians should be guided by this observed genetic heterogeneity.

An analysis of nuclear gene diversity (Nei 1973; Chakraborty 1980; Perkins et al. 1993) in native populations from the Park and Tennessee (n = 69; populations with sample size < 8 were excluded) partitions the among population variation as follows: 15.3% of the observed genetic variation is contained within the average population, 27.3% is among populations within river drainages, and 57.4% is among drainages. This indicates that about 57% of the total genetic variance in native southern Appalachian brook trout is due to differences among populations in different major drainages (South Fork Holston, Watauga, French Broad and Little Tennessee rivers), and only about 15% of the variance is shared among all populations. By contrast, in New York (n = 21) 62.5% of the total gene diversity is contained within the average wild brook trout population, 5% is among populations within river drainages, and 32.5% is among drainages (Perkins et al. 1993). Sequence variation in the more rapidly evolving mitochondrial genome conforms to this pattern and shows higher haplotype diversity among southern populations than among northern populations (Hayes et al. 1996).

The genetic distinctiveness of northern and southern brook trout populations, as well as the heterogeneity of mitochondrial and nuclear markers among southern Appalachian populations, reflects the phylogeographic history of the species. Northern populations and northern-derived hatchery strains have higher levels of heterozygosity than southern Appalachian populations, and genetic variation is partitioned differently in the two groups. These patterns of nuclear and mito-chondrial marker variation are consistent with hypotheses of different biogeographical histories of the two groups. To account for the observed genetic differences, we propose a model with the following components: (1) Fixed genetic differences between northern and southern populations evolved in populations ancestral to the two groups prior to post-glacial recolonization of the northeast, (2) Current northern populations were founded by a small number of closely related lineages, (3) Divergence among native southern Appalachian populations is a consequence of the isolation of populations in major watrersheds. Under this model, the higher heterozygosity of northern populations and the lower levels of mitochondrial and nuclear gene heterogeneity among northern populations reflects their recency of origin, their maintenance of large populations, and the persistence of gene flow, at least among some populations. Conversely, the lower heterozygosities observed in southern populations and the higher genetic heterogeneity among these populations suggests a longer duration of isolation between populations and lower effective population sizes of populations that are more isolated genetically.

Other studies of mtDNA variation among northern brook trout populations also observed limited haplotype diversity (Grewe et al. 1990; Quatro et al. 1990; Danzmann et al. 1991). These studies supports our hypothesis of relatively few lineages founding northern populations, and suggests that the low haplotype diversity we observe in northern hatchery strains is not a consequence of loss of diversity under domestication (Hayes et al. 1996). Range expansion following glacial retreat has also been hypothesized to explain low genetic heterogeneity in other North American taxa (Bernatchez and Dodson 1991; Highton and Webster 1976; Ferris et al. 1983; Hayes and Harrison 1992; Walker 1987). The presence of fixed genetic differences between native southern Appalachian populations and northern populations suggests that divergence of the two groups is deeper than the last glacial retreat.

MANAGEMENT AND CONSERVATION OF SOUTHERN APPALACHIAN BROOK TROUT

Loss of brook trout populations over the past century has motivated considerable effort to understand the causes of this decline and to prevent further losses. Establishment of National Forests and Great Smoky Mountains National Park has curtailed the loss of brook trout habitat from destructive land use practices. Adoption and enforcement of fishing regulations has largely eliminated overfishing as a factor in brook trout decline. Recognition of the consequences for brook trout populations of introduced rainbow trout has precipitated an evaluation of the appropriateness of rainbow trout stocking as a fisheries management practice. Continued encroachment of wild populations of rainbow trout on allopatric brook trout populations has motivated efforts to remove rainbow trout from some streams and to construct barriers to upstream movement in others. These mangement activities have been effective in arresting the decline of southern Appalachian brook trout and should be continued. However, the findings of molecular genetics studies suggest additional considerations which should be incorporated into management and conservation policies. Our studies indicate that (1) populations of native southern Appalachian brook trout are genetically distinct from populations in the taxon's main range, (2) native southern Appalachian brook trout exhibit significant heterogeneity among populations and among watersheds, and (3) the southern Appalachians are a center of gene pool diversity for brook trout. These characteristics increase the importance of conserving native southern Appalachian brook trout populations and indicate that traditional management approaches involving hatcheries and stock transfer should be substantially revised.

Concern for the effects of stocking on indigenous aquatic faunas has become a major focus in fisheries and conservation biology (Fergusson 1990; Allendorf 1991; Goodman 1991; Krueger and May 1991). Displacement of native taxa by introduced nonnative taxa has long been recognized as a conservation concern, and is a component of current biodiversity management policies in the southern Appalachian region. Regional differentiation of widespread taxa and the effects of hybridization on native gene pools have only recently been recognized. Genetic and ecological consequences of stocking with nonnative taxa and strains have become a major issue in the management of the diverse salmonid faunas of western North America (Frissell 1991). As is the case with trout in eastern North America, recognition of the importance of these concerns arose after their effects were manifest. Therefore, it is necessary to develop management policies which incorporate the insights of genetics, and to develop approaches to ameliorate the consequences of previous actions.

The significance of the negative effects of stocking with hybridizing hatchery strains is dependent on the extent of stocking and the rarity of the native populations. Native brook trout in the southern Appalachians may not be characterized as rare, but the observed loss of 70 to 80% of brook trout range during this century has clearly become a cause for concern. Moreover, stocking records document a half-century of extensive stocking with northern hatchery strains, and our genetic inventories observe that only about 50% of remaining brook trout populations are pure native. This documented loss of native populations coupled with our conclusion that native southern Appalachian brook trout are a distinct evolutionary lineage with substantial heterogeneity among populations indicates to us that management policies should be revised.

Preservation of habitat is the fundamental requirement of ecologically sound biodiversity management (Noss and Cooper 1994). In addition to preserving habitat we recommend six specific approaches for the conservation of native southern Appalachian brook trout:

1. Eliminate stocking with nonnative salmonids where they may compete with native brook trout.
2. Continue and expand efforts to limit the encroachment of wild rainbow trout populations on native brook trout populations.
3. Where necessary, remove rainbow trout from streams where they occur in sympatry with native brook trout. This is especially important for areas where few native populations remain.
4. Eliminate all brook trout stocking with hatchery strains.
5. Remove hatchery-derived wild brook trout populations and replenish with native populations.
6. Reestablish native brook trout populations in streams, especially higher order low elevation streams, which historically supported them but from which they have been extirpated.

Populations in larger low elevation streams could meet public demand for a brook trout fishery and serve as a preventative to demographic extinction and the long-term effects of global warming. Stocking in support of recommendations 5 and 6 should be by stock transfer from native populations in the same watershed. As approximately 57% of the genetic diversity observed among all native brook trout populations results from differences among watersheds, such a stock transfer policy would contribute substantially to the preservation of the diversity in local populations. Management of native brook trout populations emphatically should not involve establishment of native southern Appalachian hatchery strains. Strains established from single populations would preserve only a fraction of the total diversity in a region, while strains established from multiple populations would homogenize locally distinct populations.

We recognize that implementation of these recommendations is expensive, time consuming, and may often be in conflict with established management practices and with competing resource utilization values. However, we make these recommendations as appropriate guides toward the goal of preserving the diversity in native southern Appalachian brook trout. It is important to recognize too that resource managers in the region have adopted biodiversity management principles and have already incorporated most of these recommendations into their management programs. For example, trout stocking to meet public demand for a recreational trout fishery is largely confined to waters that do not support and will not impact, wild reproducing populations of any species or variety of trout, but most especially native brook trout (North Carolina Wildlife Resources Commission 1989; TWRA 1994). Similarly, stream restoration in support of brook trout conservation is informed by genetic inventories, with source populations for replenishment selected on the basis of genetic identity and geographical proximity. Fisheries resource managers have made conservation of the native southern Appalachian brook trout a policy priority that should be emulated by other land and resource managers in the region.

THE SOUTHERN APPALACHIANS AS A CENTER OF BIOLOGICAL DIVERSITY

Many plant and animal species reach their southern limit in the southern Appalachians (White et al. 1993). Affinities are primarily with boreal and northern hardwood communities (Whittaker 1956; Brooks 1971; Pelton 1984), although relationships with western North American (Anderson 1970; Wood 1970; Handley 1971) and eastern Asian (Li 1971; White 1983) biotas are recognized. Relationships between the biotas of the southern Appalachians and eastern Asia are geologically deep, with most of the related taxa originating between the Cretaceous and the Eocene periods (Hsu 1983). These taxa are remnants of an ancient northern hemisphere biota which was disrupted by late Cenozoic glaciations (White 1983). Relationships with western, and most especially northern North American biotas, are consequences of Pleistocene glacial history (Brooks 1971) and the climatic analogue between elevation and latitude (Whittaker 1956). The southern Appalachians were unglaciated throughout the Cenozoic, and the region has not been submerged since at least the early Paleozoic (King et al. 1968).

These historical contingencies, and the topographic diversity caused by complex mountain geology, have resulted in a large number of endemic and relictual taxa in the contemporary southern Appalachian biota (White 1984; White et al. 1993). Numerous factors have influenced this level of endemism and the striking biotic diversity of the southern Appalachians. Assemblage of communities in near current locations began with post glacial warming beginning about 18,000 years ago (Davis 1983). Proximity of the southern Appalachians to refugia increased the probability that entire communities, as opposed to individual species, would shift into suitable mountain habitats. At the same time, the narrow climatic zonation of the mountains provided the stage for fine-scale assembly of communities based on individual species requirements and interactions among species (Delcourt and Delcourt 1991). Finally, observed genetic differences between southern Appalachian populations of widespread taxa and populations in other regions suggest that contemporary biological communities in the different areas were assembled from populations from different Pleistocene refugia. All of these factors have contributed to the rich and diverse southern Appalachian biota of today and make its conservation and sustainable management a pressing concern.

RISK ASSESSMENT REPORT CARD FOR ECOSYSTEM MANAGERS

Formulating principles for sound ecological management of biodiversity is a goal of the southern Appalachian Assessment (Berish et al., Chapter 7 of this volume). For many biological communities and taxa we possess sufficient knowledge to at least outline the necessary principles. For other communities and taxa, such as brook trout, we now possess sufficent knowledge to identify specific actions. Development and resource extraction which affect southern Appalachian biodiversity should only be undertaken if negative impacts can be prevented. In many circumstances, development and resource utilization can take place if practices to limit negative impacts are adopted. For example, timber harvest methods that minimize surface runoff and retain undisturbed buffers around streams are compatible with preservation of native brook trout. At the same time, we must accept that some activities and some levels of activities are incompatible with biodiversity conservation in the region. Conservation of the rich and unique biota of the southern Appalachians requires that biodiversity be given at least equal status with competing resource utilization interests.

The native brook trout serves as a well studied model for the numerous taxa and populations endemic to the southern Appalachians. Conservation of these taxa presents special problems for ecosystem managers because they frequently exist as small isolated populations. Practical difficulties also arise because of the large number of taxa exhibiting these characteristics, a reality which argues for a reserve model of conservation as opposed to a multiple-use approach. Political and economic opposition is the most likely obstacle to this approach. Over the longer term, conservation of isolated endemic taxa may depend entirely on social, economic, and political changes to cope with air quality concerns and global climate change. Over the short term, the *Risk Assessment*

Report Card for Ecosystem Managers identifies the difficulties confronting brook trout management and the management of other taxa with analogous population characteristics:

Vision, brook trout; A

Fisheries managers in Great Smoky Mountains National Park and in the southern Appalachian states have been exemplary in their commitment to research, monitoring, and management of declining native brook trout populations, and have a commitment to the long-term conservation of this southern Appalachian native.

Resource risk, brook trout: B

Socioeconomic conflicts, brook trout: B

Management of native southern Appalachian brook is often perceived as being in conflict with sport fisheries. Managers are committed to meeting both the public demand for sport fishing and the management of native biodiversity. Towards that end, state and federal fisheries officials devote considerable effort to public education. Conflicts will still arise when biodiversity conservation requires the elimination of some or all alternative land uses, or the costly modification of such alternative uses.

Procedural protocols, brook trout: A

Because fisheries managers have the necessary institutional support, management policies and protocols have received considerable explicit attention and are modified and updated as data become available.

Scientific validity, brook trout: A

Southern Appalachian brook trout have received a great deal of research effort, and current and proposed policies are informed by that research.

Legal jeopardy, brook trout: B

Brook trout management policies are not likely to be challenged, at least on state and federal lands.

Public support, brook trout: A

Brook trout management policies are largely supported by the major public fishing and conservation groups, such as Trout Unlimited.

Adequacy of funding, brook trout: A–

Brook trout research and management have benefitted from considerable funding support.

Policy precedent: A

Strong policy justification on public lands. The National Park Service is mandated to manage for native biodiversity. Other federal and state management bodies are rapidly adopting biodiversity management perspectives.

Administrative support, brook trout: A

Administrative levels are evolving towards biodiversity management perspectives.

Transferability: A

A major conclusion of our analysis is that southern Appalachian brook trout can serve as a model for other taxa that have similar distributions and population structures. For the reasons outlined, this applies especially to taxa in the southern Appalachians but applies as well to other regions. Our work with brook trout has demonstrated the importance of genetic considerations when dealing with taxa found in small, isolated populations and taxa with broad geographical distributions. Our genetic analysis has confirmed the suspicions expressed by fisheries biologists in the 1930s that native southern Appalachian brook trout differ from domesticated strains used in stocking in the region. We have not explicitly addressed these differences from a taxonomic perspective, but we have demonstrated that the differences are sufficently great as to be of management concern.

REFERENCES

Allendorf, F. W. 1991. Ecological and genetic effects of fish introductions: synthesis and recommendations. *Can. J. Fish. Aquat. Sci.* 48:178–181.

Allendorf, F. W. and F. M. Utter. 1979. Population genetics. Pp. 141–160 In W. S. Hoar, D. S. Randall, and J. R. Brett, Eds. *Fish Physiology.* Vol. 8. Academic Press: New York.

Anderson, L. E. 1970. Geographical relationships of the mosses of the Southern Appalachian Mountains. Pp. 101–116 in P. C. Holt, Ed. *The Distributional History of the Southern Appalachians.* Part II: Flora. Virginia Polytechnic Institute and State University: Blacksburg.

Bernatchez, L. and J. J. Dodson. 1991. Phylogeographic structure in mitochondrial DNA of the lake whitefish (*Coregonus clupeaformis*) and its relation to Pleistocene glaciations. *Evolution.* 45:1016–1035.

Bivens, R. D. 1985. History and distribution of brook trout in the Appalachian region of Tennessee. Tennessee Wildlife Resources Agency Tech. Rep. No. 85-1. University of Tennessee, Knoxville.

Bivens, R. D., R. J. Strange, and D. C. Peterson. 1985. Current distribution of the native brook trout in the Appalachian region of Tennessee. *J. Tenn. Acad. Sci.* 60:101–105.

Brooks, M. 1971. The southern Appalachians. Pp. 1–10 In P. C. Holt, Ed. *The Distributional History of the Biota of the Southern Appalachians.* Part III: Vertebrates. Virginia Polytechnic Institute and State University; Blacksburg.

Brussard, P. F. and D. F. Nielsen. 1976. Allozyme variation in the brook trout, *Salvelinus fontinalis*: a preliminary report. Unpublished report to Great Smoky Mountains National Park, Gatlinburg, TN. 11 pp.

Buth, D. G. 1984. The application of electrophoretic data in systematic studies. *Annu. Rev. Ecol. Syst.* 15:501–522.

Chakraborty, R. 1980. Gene diversity analysis in nested subdivided populations. *Genetics.* 96:721–726.

Chapman, D. W. 1965. Food and space as regulators of salmonid populations in streams. *Am. Nat.* 100:131–138.

Coutant, C. C. 1977. Compilation of temperature preference data. *J. Fish. Res. Board Can.* 34:739–745.

Cunjak, R. A. and J. M. Green. 1983. Habitat utilization by brook char (*Salvelinus fontinalis*) and rainbow trout (*Salmo gairdneri*) in Newfoundland streams. *Can. J. Zool.* 61:1214–1219.

Cunjak, R. A. and J. M. Green. 1984. Species dominance by brook trout and rainbow trout in a simulated stream environment. *Trans. Am. Fish. Soc.* 113:737–743.

Cunjak, R. A. and J. M. Green. 1986. Influence of water temperature on behavioral interactions between juvenile brook char, *Salvelinus fontinalis*, and rainbow trout, *Salmo gairdneri. Can. J. Zool.* 64:1288–1291.

Curtis, L. R., W. K. Selm, L. K. Siddens, D. A. Meager, R. A. Carchman, W. H. Carter, and G. A. Chapman. 1989. Dole of exposure duration in hydrogen ion toxicity to brook trout (*Salvelinus fontinalis*) and rainbow trout (*Salmo gairdneri). Can. J. Fish. Aquat. Sci.* 56:33–40.

Danzmann, R. G., M. J. Joyce, and J. P. Volpe. 1991. Mitochondrial DNA variability in brook charr (*Salvelinus fontinalis*) populations sampled from the Lake Huron drainage. Unpublished report to the Ontario Ministry of Natural Resources, Ottawa.

Davis, M. B. 1983. Quaternary history of deciduous forests of eastern North America and Europe. Ann. MO Bot. Gard. 70:550–563.

Delcourt, H. R. and P. A. Delcourt, 1991. *Quaternary Ecology: A Paleoecological Perspective.* Chapman & Hall: London.

Dunham, R. A., J. DiBona, L. Robison, K. Norgen, J. Durniak, and M. Spencer. 1994. Biochemical genetics of brook trout, *Salvelinus fontinalis*, in Georgia. Unpublished report to the Georgia Department of Natural Resources, Atlanta.

Ensign, W. E. 1988. The Importance of Competition for Food Resources in the Interaction between Brook Trout (*Salvelinus fontinalis*) and Rainbow Trout (*Salmo gairdneri*). Unpublished masters thesis, University of Tennessee, Knoxville.

Fausch, K. D. 1988. Tests of competition between native and introduced salmonids in streams: What have we learned? *Can. J. Fish. Aquat. Sci.* 45:2238–2246.

Fergusson, M. M. 1990. The genetic impact of introduced fishes on native species. *Can. J. Zool.* 68:1053–1057.

Ferris, S. D., R. D. Sage, E. M. Prager, U. Ritte, and A. C. Wilson. 1983. Flow of mitochondrial DNA across a species boundary. *Proc. Nat. Acad. Sci. U.S.A.* 80:2290–2294.

Flebbe, P. A. 1994. A regional view of the margin: Salmonid abundance and distribution in the southern Appalachian Mountains of North Carolina and Virginia. *Trans. Am. Fish. Soc.* 123:657–667.

Frissell, C. A. 1991. Topology of extinction and endangerment of native fishes in the Pacific Northwest and California. Conserv. Biol. 7:342–354.

Frost, D. R. and D. M. Hillis. 1990. Species in concept and practice: Herpetological applications. *Herpetologica* 46:97–104.

Goodman, B. 1991. Keeping anglers happy has a price: Ecological and genetic effects of stocking fish. *BioScience* 41:294–299.

Green, R. L. 1975. Benthic Macroinvertebrate Communities in Great Smoky Mountains National Park Streams Influenced by Acid Drainage. Unpublished masters thesis. Tennessee Technological University, Cookeville.

Grewe, P. M., N. Billington, and P. D. N. Herbert. 1990. Phylogenetic relationships among members of *Salvelinus* inferred from mitochondrial DNA divergence. *Can. J. Fish. Aquat. Sci.* 47:984–991.

Guffey, S. Z. 1995. Allozyme Genetics of South Carolina Brook Trout. Unpublished report submitted to the South Carolina Department of Natural Resources, Clemson.

Habera, J. W. 1987. Effects of Rainbow Trout Removal on Trout Populations and Food Habits in Great Smoky Mountains National Park. Unpublished masters thesis, University of Tennessee, Knoxville.

Habera, J. W. and R. J. Strange. 1993. Wild trout resources and management in the southern Appalachian Mountains. *Fisheries* 18:6–13.

Handley, C. O., Jr. 1971. Appalachian mammal geography — recent epoch. Pp. 263–304 in P. C. Holt, Ed. *The Distributional History of the Biota of the Southern Appalachians.* Part III: Vertebrates. Virginia Polytechnic Institute and State University: Blacksburg.

Hansen, E. A. 1971. Sediment in a Michigan trout stream: Its source, movement and some effects on fish habitat. U.S. Department of Agriculture, Forest Service Research Paper NC-59.

Hayes, J. P. and R. G. Harrison. 1992. Variation in mitochondrial DNA and the biogeographic history of woodrats (*Neotoma*) of the eastern United States. *Syst. Biol.* 41:331–334.

Hayes, J. P., S. Z. Guffey, F. J. Kriegler, G. F. McCracken, and C. R. Parker. 1996. The genetic diversity of native, stocked, and hybrid populations of brook trout in the southern Appalachians. *Conserv. Biol.* 10:1403–1412.

Hendricks, M. L. 1980. *Salvelinus fontinalis* (Mitchell), brook trout. P. 114 in Lee, D. S., C. R. Gilbert, C. H. Hocutt, R. E. Jenkins, D. E. McAllister, and J. R. Stauffer, Jr., Eds., *Atlas of North American Freshwater Fishes.* North Carolina State Museum of Natural History: Raleigh.

Highton, R. and T. P. Webster. 1976. Geographic protein variation and divergence in populations of the salamander *Plethodon cinereus*. *Evolution* 30:33–45.

Holloway, A. D. 1945. Report on the fisheries of the Great Smoky Mountains National Park. Unpublished internal report to Great Smoky Mountains National Park, Gatlinburg, TN.

Hsu, J. 1983. Late Cretaceous and Cenozoic vegetation in China, emphasizing their connections with North America. *Ann. MO Bot. Gard.* 70:490–508.

Huckabee, J. W., C. P. Goodyear, and R. D. Jones. 1975. Acid rock in the Great Smokys: Unanticipated impact on aquatic biota of road construction in regions of sulfide mineralization. *Trans. Am. Fish. Soc.* 104:677–684.

Jones, R. D. 1978. Regional distribution trends of the trout resource. Pp. 1–10 in Harshbarger, T. J., Ed., Southeastern Trout Resource: Ecology and Management Symp. Proc. U.S. Forest Service, Southeastern Forest Experiment Station, Asheville, NC.

Jordan, D. S., and B. W. Evermann. 1896–1900. The fishes of North and Middle America: A descriptive catalogue of the species of fish-like vertebrates found in the waters of North America, north of the Isthmus of Panama. *Bull. U.S. Nat. Mus.* 47:1–3313.

Kelly, G. A., J. S. Griffith, and R. D. Jones. 1980. Changes in distributions of trout in Great Smoky Mountains National Park, 1900–1977. U.S. Fish and Wildlife Service Tech. Pap. No. 102.

King, P. B., R. B. Neuman, and J. B. Hadley. 1968. Geology of the Great Smoky Mountains National Park, Tennessee and North Carolina. U.S. Geological Survey Prof. Pap. No. 587.

King, W. 1937. Notes ont he distribution of native speckled trout in the streams of Great Smoky Mountains National Park. *J. Tenn. Acad. Sci.* 12:351–361.

Kriegler, F. J. 1993. The Genetics of Native and Stocked Brook Trout Populations in East Tennessee. Unpublished master's thesis. University of Tennessee, Knoxville.

Kriegler, F. J., G. F. McCracken, J. W. Habera, and R. J. Strange. 1995. Genetic characterization of Tennessee brook trout populations and associated management implications. *N. Am. J. Fish.* Manage. 15:804–813.

Krueger, C. C. and B. May. 1991. Ecological and genetic effects of salmonid introductions in North America. *Can J. Fish Aquat. Sci.* 48:66–77.

Larson, G. L. and S. E. Moore. 1985. Encroachment of exotic rainbow trout into stream populations of native brook trout in the southern Appalachian Mountains. *Trans. Am. Fish. Soc.* 114:195–203.

Larson, G. L., S. E. Moore, and B. Carter. 1995. Ebb and flow of encroachment by nonnative rainbow trout in a small stream in the southern Appalachians. *Trans. Am. Fish. Soc.* 124:613–622.

Leary, R. F., F. W. Allendorf, S. R. Phelps, and K. L. Knudsen. 1987. Genetic divergence and identification of seven cutthroat subspecies and rainbow trout. *Trans. Am. Fish. Soc.* 116:580–587.

Lennon, R. E. 1967. Brook trout of Great Smoky Mountains National Park. U.S. Fish and Wildlife Service Technical Paper 15. Department of the Interior, Washington, D.C.

Leopold, A. 1949. *A Sand County Almanac, and Sketches Here and There.* Oxford University Press: New York.

Li, H.-L. 1971. Floristic relationships between eastern Asia and eastern North America. Unnumbered publication of the Morris Arboretum, University of Pennsylvania, Philadelphia.

Lohr, S. C. and J. L. West. 1992. Microhabitat selection by brook and rainbow trout in a southern Appalachian stream. *Trans. Am. Fish. Soc.* 121:729–736.

Loudenslager, E. J. and G. A. E. Gall. 1980. Geographic patterns of protein variation and subspeciation in the cutthroat trout. *Syst. Zool.* 29:27–42.

MacCrimmon, H. R. and J. S. Campbell. 1969. World distribution of brook trout, *Salvelinus fontinalis. J. Fish. Res. Board Can.* 26:1699–1725.

McCracken, G. F. and S. Z. Guffey. 1994. Population genetics and distribution of brook trout in Great Smoky Mountains National Park. Unpub. rep. submitted to Great Smoky Mountains National Park, Gatlinburg, Tennessee.

McCracken, G. F., C. R. Parker, and S. Z. Guffey. 1993. Genetic differentiation and hybridization between stocked hatchery and native brook trout in Great Smoky Mountains National Park. *Trans. Am. Fish. Soc.* 122:533–542.

Meisner, J. D. 1990. Effect of climatic warming on the southern margins of the native range of brook trout, *Salvelinus fontinalis. Can. J. Fish. Aquat. Sci.* 47:1065–1070.

Moore, S. E., B. L. Ridley, and G. L. Larson. 1981. Changes in standing crop of brook trout concurrent with removal of exotic trout species. Great Smoky Mountains National Park. Res. Res. Manage. Rep. No. 37. Uplands Field Research Laboratory, Great Smoky Mountains National Park, Gatlinburg, TN.

Moore, S. E., G. L. Larson, and B. Ridley. 1984. A summary of changing standing crops of native brook trout in response to removal of sympatric rainbow trout in Great Smoky Mountains National Park. *J. Tenn. Acad. Sci.* 59:76–77.

Moore, S. E., G. L. Larson, and B. Ridley. 1985. Dispersal of brook trout in rehabilitated streams in Great Smoky Mountains National Park. *J. Tenn. Acad. Sci.* 60:1–4.

Moore, S. E., G. L. Larson, and B. Ridley. 1986. Population control of exotic rainbow trout in streams of a natural area park. *Environ. Manage.* 10:215–219.

Morgan, E. L., M. H. Hoff, and W. F. Trumpf. 1976. Anakeesta leachate studies and brook trout investigations. Tennessee Technological University Environmental Biology Research Center Rep. Ser. No. 8. Cookeville.

Morgan, E. L. and J. R. Robinette. 1978. Life history study of brook trout *Salvelinus fontinalis* (Mitchill). Great Smoky Mountains National Park. Gatlinburg, TN.

Morgan, R. P. and B. M. Baker. 1991. Development of genetic inventories for Maryland game fish: Brook trout. Final draft report. Appalachian Environmental Laboratory, Center for Environmental and Esturine Studies, University of Maryland, Frostburg.

Nagel, J. W. 1991. Is the decline of brook trout in the southern Appalachians resulting from competitive exclusion and/or extinction due to habitat fragmentation? *J. Tenn. Acad. Sci.* 66(4):141–143.

Nei, M. 1987. *Molecular Evolutionary Genetics.* Columbia university Press: New York.

Nei, M. and W.-H. Li 1979. Mathematical model for studying genetic variation in terms of restriction endonucleases. *Proc. Nat. Acad. Sci. U.S.A.* 76:5269–5273.

North Carolina Wildlife Resources Commission. 1989. Casting the future of trout in North Carolina: a plan for management of North Carolina's trout resources. North Carolina Wildlife Resources Commission, Division of Boating and Inland Fisheries. Raleigh, North Carolina.

Noss, R. F. and A. Y. Cooper. 1994. Saving nature's legacy. Island Press: Washington, D.C.

Pelton, M. R. 1984. Mammals of the spruce–fir forests in Great Smoky Mountains National Park. Pp. 187–192 in P. S. White, Ed. The southern Appalachian spruce–fir ecosystem: Its biology and threats. Res./Resource Manage. Rep. SER-71. U.S. Department of Interior, National Park Service. Atlanta, Georgia.

Platts, W. S. and R. L. Nelson. 1988, Fluctuations in trout populations and their implications for land-use evaluation. *N. Am. J. Fish. Manag.* 8:333–345.

Perkins, D. L., C. C. Krueger, and B. May. 1993. Heritage brook trout in northeastern USA: Genetic variability within and among populations. *Trans Am. Fish. Soc.* 122:515–532.

Peterson, R. H., A. M. Sutterlin, and J. D. Metcalfe. 1979. Temperature preference of several species of *Salmo* and *Salvelinus* and some of their hybrids. *J. Fish. Res. Board Can.* 36:1137–1140.

Power, G. 1980. The brook charr (*Salvelinus fontinalis*). Pp. 797–840 in E. K. Balon, Ed., *Charrs: Salmonid Fishes of the Genus Salvelinus.* Dr. W. Junk Publishers: The Hague, The Netherlands.

Quatro, J. M., R. P. Morgan II, and R. W. Chapman. 1990. Mitochondrial DNA variability in brook trout populations from western Maryland. *Trans. Am. Fish. Soc. Symp.* 7:470–474.

Saidak, L. R. 1995. Brook Trout (*Salvelinus fontinalis* in Tennessee: An Ecological Genetic Study Using Allozyme Electrophoresis and Random Amplified Polymorphic DNA. Unpublished masters thesis. University of Tennessee, Knoxville.

Seehorn, M. E. 1979. Status of brooktrout in the southeast. Pp. 16–17 in R. D. Estes, T. Harshbarger, and G. D. Pardue, Eds., Proc. Brook Trout Workshop. U.S. Forest Service, Southeastern Forest and Range Experiment Station, Asheville, North Carolina.

Shull, L. N. 1995. An Allozyme Analysis of Brook Trout (*Salvelinus fontinalis*) from Headwater Streams of the Blue Ridge Parkway. Unpublished masters thesis. Appalachian State University, Boone, NC.

Stoneking, M., D. J. Wagner, and A. C. Hildebrand. 1981. Genetic evidence suggesting subspecific differences between northern and southern populations of brook trout (*Salvelinus fontinalis*). Copeia 1981:810–819.

Strange, R. J. 1979. Our native trout of the south: will it survive? *Trout* 20:10–13.

Strange, R. J. and J. W. Habera. 1995. Wild trout project: 1994 Annual report. Tennessee Wildlife Resources Agency and the University of Tennessee. Pub. No. R11-2217-39-001-95.

Swift, L. W., Jr. and J. B. Messer. 1971. Forest cutting raises temperatures of small streams in the southern Appalachians. *J. Soil Water Conserv.* 26:111–116.

Thomas, W. K., R. E. Withler, and A. T. Beckenbach. 1986. Mitochondrial DNA analysis of Pacific salmonid evolution. *Can. J. Zool.* 64:1058–1064.

Thorpe, J. P. 1982. The molecular clock hypothesis: biochemical evolution, genetic differentiation and systematics. *Annu. Rev. Ecol. Syst.* 13:139–168.

TWRA. 1994. A draft strategic wildlife resources management plan for entering the 21st century. Tennessee Wildlife Resources Agency, Nashville.

Venters, V. 1993. In search of speckles. *Wildlife in North Carolina.* July:11–15.

Walker, G. L. 1987. Ecology and Population Biology of *Thuja occidentalis* in its Southern Disjunct Range. Unpublished Ph.D. thesis, University of Tennessee, Knoxville.

Welsh, T. L. 1994. Interactive Dominance: Chinook Salmon and Eastern Brook Trout (*Oncorhynchus tshawytscha, Salvelinus fontinalis*). Unpublished PhD. thesis. University of Idaho, Moscow.

White, P. S. 1983. Eastern Asian–eastern North American floristic relations: the plant community level. *Ann. MO Bot. Gard.* 70:734–747.

White, P. S. 1984., Ed. The southern Appalachian spruce–fir ecosystem: its biology and threats. U.S. Department of Interior, National Park Service. Res./Resource Manage. Rep. SER-71. Atlanta, Georgia.

White, P. S., E. R. Buckner, J. D. Pittillo, and C. V. Cogbill. 1993. High-elevation forests: Spruce-fir forests, northern hardwood forests, and associated communities. Pp. 305–337 in W. H. Martin, S. G. Boyce, and A. C. Echternacht, Eds. *Biodiversity of the Southeastern United States: Upland Terrestrial Communities.* John Wiley & Sons: New York.

Whittaker, R. H. 1956. Vegetation of the Great Smoky Mountains. *Ecol. Monogr.* 26:1–80.

Wilson, G. M., W. K. Thomas, and A. T. Beckenbach. 1985. Intra- and inter-specific mitochondrial DNA sequence divergence in *Salmo*: rainbow, steelhead, and cutthroat trouts. *Can. J. Zool.* 63:2088–2094.

Wood, C. E., Jr. 1970. Some floristic relationships between the Southern Appalachians and western North America. Pp. 331–404 in P. C. Holt, Ed. *The Distributional History of the Biota of the Southern Appalachians.* Part II: *Flora.* Virginia Polytechnic Institute and State University, Blacksburg.

Yuskavitch, J. A. 1991. Saved by the barriers. *Trout.* Summer:18–27.

GLOSSARY

Allele — A particular form of a gene at a particular locus.

Allopatric — Of a population or species occupying a different region or section of habitat from that of another population or species. See **Sympatric.**

Allozyme — An allele of an enzyme identified by electrophoretic separation and histochemical staining.

Electrophoresis — The separation of molecules in an electric field.

Genetic identity — An index of genetic similarity among populations or species, ranging from 0 when no alleles are shared, to 1 when all alleles and allele frequencies are identical.

Haplotype — A unique mitochondrial DNA lineage.

Heterozygous — Of an individual organism that possesses different alleles at a locus.

Homozygous — Of an individual organism that has the same allele at each copy of a gene locus.

Lacustrine — Found in lakes.

Locus (pl. loci) — A site on a chromosome occupied by a particular gene.

Mean sequence divergence — An estimate of the percentage difference in DNA sequences between two populations or species.

Mitochondrial DNA (mtDNA) — The genome of the mitochondria, located outside the nucleus and transmitted only by the female parent.

Polymorphic — Of a species or population that has two or more genotypes for a particular trait.

Restriction endonuclease (restriction enzyme) — An enzyme that cuts double-stranded DNA at specific short nucleotide sequences.

Restriction site — The specific DNA site cut by a restriction enzyme.

Sympatric Of two species or populations occupying the same region or section of habitat. See **Allopatric.**

Taxon (pl. taxa) — The named taxonomic units to which individuals or sets of species are assigned.

13 Control of Exotic Species: European Wild Boar*

John D. Peine and Richard Lancia

CONTENTS

Principles of Ecosystem Management: Exotic Species Control 268
  Policy 268
  Biological and Ecological Considerations 269
  Control Strategies 269
    Monitoring and Modeling 270
  Adaptive Resource Management 270
European Wild Boar at Great Smoky Mountains National Park: a Case Study 271
  Park Setting 271
  Policy Mandates 271
  Origin and Invasion 271
  Home Range and Activity Patterns 272
    Distribution, Density, and Migration 272
    Other Factors Affecting Movements 273
  Food Habits 273
  Reproductive Characteristics 274
    Rates of Increase 274
  Resource Impacts 275
    Flora 275
    Fauna 275
    Soil 276
    Disease 276
  Control Program 276
    Early History 276
    Setbacks 276
    Trapping Techniques 276
    Shooting Techniques 278
    Hog Bait Study 278
    Review of Control Activities 278
  Monitoring 279
  Hog Population Model 280
    Hog Population Model 282
    Catch per Unit Effort Model 282
  Recommendations for the Future 282
    Adequate and Stable Funding 283
    Staff Requirements 283
     284

---

* Portions of this chapter were adapted from a paper by J. D. Peine and J. A. Farmer, 1990.

0-57444-053-5/99/$0.00+$.50
© 1999 by CRC Press LLC

Fencing 284
Monitoring 284
Data Management...284
Model Verification 285
Research Agenda 285
Strategic Planning for Management...286
Review Hog Management Policy..286
Conclusions 287
Risk Assessment Report Card for Ecosystem Managers 288
Acknowledgments...288
References 288

PRINCIPLES OF ECOSYSTEM MANAGEMENT: EXOTIC SPECIES CONTROL

Control of exotic species is an extremely complex matter that involves societal values, policy decisions, economic analyses, ecological interrelationships, population dynamics, and technological capabilities. In this chapter we give an overview of the issues and describe a case history of the effort to control exotic wild pigs or European wild boar, (*Sus scrofa*) in Great Smoky Mountains National Park (hereafter referred to as the Park). The case is an excellent example of the enormous complexities associated with a attempt to eradicate or control a large, fecund mammal that has the potential to substantially alter the structure and function of natural ecosystems at a landscape spatial scale.

POLICY

Justification for initiating the invariably formidable task of exotic species control is best grounded by a clearly articulated and well-thought out mission statement at the agency or landowner level. The statement must take into consideration societal concerns as well as compatibility with the overall mission of the agency. Perpetuation of native species, historic levels of biodiversity, and maintenance or restoration of ecosystem integrity are typically guiding forces in the justification for the control of exotics. The mission statement must also be considered in conjunction with other responsibilities of the agency. A clear priority ranking of all agency missions should be established, so when financial resources are scarce the commitment to control efforts is unambiguous.

An essential step in the process of considering a control program for an exotic species is a cost–benefit analysis. Costs of control programs include activities that remove the animals, administration, monitoring and surveillance, nontarget ecosystem effects, risks associated with removal activities, and foregone economic or ecological benefits associated with the exotic species at current or higher population levels (Izac and O'Brian 1991).

Benefits reflect the potential damages avoided through eradication (Izac and O'Brian 1991), which could include agricultural, environmental, and disease costs. In the Smokies, environmental costs associated with wild pigs could include destruction of native flora and fauna (Singer et al. 1984); competition with native fauna for food; increased predation and soil erosion; altered nutrient cycling, plant growth (Lacki and Lancia 1983 1986), and plant succession; and enhanced dispersal of undesirable organisms.

Invariably, currently available data and the understanding of ecological relationships associated with exotic species control are almost certainly incomplete. Even though information available is likely to be incomplete, this shortcoming should not be a reason for ignoring cost–benefit analysis. At the least, the process of conducting the analysis is likely to better define the problem and aid in modeling which will likely lead to better management decisions. At best, scarce management dollars can be allocated in the most effective manner.

Finally, policy must also include realistic expectations associated with control efforts. In many cases it is unlikely that an exotic can be exterminated and reinvasion prevented, except perhaps in

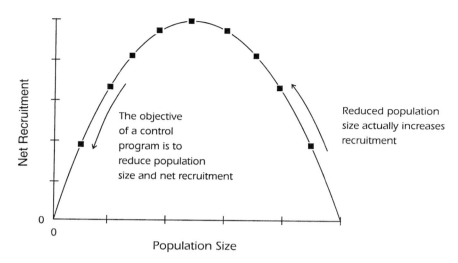

FIGURE 1 Idealized recruitment curve.

very localized areas, because costs can be prohibitively high. The alternative is a control effort that requires long-term commitment to constrain populations to acceptable levels and spatial distribution.

BIOLOGICAL AND ECOLOGICAL CONSIDERATIONS

The complexity and difficulty of control programs for exotic species require formulation of a realistic and effective control strategy. Ideally, such a program should be based on extensive research and understanding, for example, of population demography, foraging ecology, behavior, habitat requirements, and ecological relationships such as competition and predator–prey. A thorough understanding of these dynamics may suggest "vulnerabilities" in the life history strategy of the exotic that could yield effective control procedures.

Developing an effective control strategy requires understanding population demographics. Basically, population growth, and thus population size, is a function of four principle demographic rates: mortality, natality, immigration, and emigration. All four must be considered in developing an effective control program. Immigration and emigration involve the spatial distribution of animals which, for some terrestrial species can be confined by fencing. Although costly, fencing in some cases is essential to achieve eradication, unless there is the unlikely presence of an absolute natural barrier.

Mortality and natality can be viewed in the same theoretical context as sustained yield harvesting. If growth of the exotic population over time is basically sigmoidal, then recruitment of individuals added to the population, which is the combined net effect of mortality and natality, is a dome-shaped curve in relation to the stock population (Figure 1). This is the familiar sustained yield curve in stock-recruitment yield models (Ricker 1958). The fundamental demographic principle that underlies these relationships is that the recruitment per individual in the population is density dependent (Fowler and Smith 1981).

The difference between control and harvesting objectives is simply the desired location of the population on the stock-recruitment curve. In sustained yield harvesting, the objective is to maintain the population on the right side of the maximum sustained curve; whereas for control, the objective is to be as far toward the left side of the curve as possible. In other words, managers attempt to drive the population as far to the left as they can, ideally to the point of eradication, by keeping removals greater than sustainable recruitment. Effective control must also include eliminating the possibility of immigration or reinvasion.

On the left side of the curve, recruitment decreases as the size of the stock population decreases. Therefore, control efforts are designed to keep the stock population and recruitment as small as economically and ecologically possible. A corollary relationship is that as the population gets smaller, disproportionally more effort is required to reduce the population. Thus, it becomes increasingly difficult to reduce the population to smaller and smaller levels.

Therefore, a control strategy should identify a goal as either eradication or a low population that is a reasonable balance between stock population size, recruitment, and the effort required to keep the population at that level. To achieve eradication, (1) immigration must be preempted and (2) either all animals must be removed or the population reduced to the point where natural processes lead to localized extinction. Excluding immigration is also crucial to the last point because even a small number of immigrants can lead to reestablishing the population (Hone 1995). Low populations should lead to reduced damage, assuming that damage is proportional to population size (Katahira et al. 1993:273), although this might not always be the case (Hone and Stone 1989:423). Finally, eradication is possible if intensive removals and surveillance are combined when necessary with fencing (Katahira et al. 1993).

CONTROL STRATEGIES

Rarely are there easy answers to questions concerning devising the best control strategies. There is a tendency to rely on traditional methods such as trapping, poisoning, or shooting without creatively considering all options, even those that appear to be impossible, given current technology. Managers can easily get trapped into committing funds and manpower to control the population at an acceptable level in perpetuity, leaving no resources to explore potentially more effective means of control. Additionally, public tolerance of damage or acceptability of previously unacceptable, or conversely acceptable, methods could change dramatically. Thus, some resources must always be allocated to exploring new approaches and developing new technologies.

Monitoring and Modeling

Absolutely essential parts of a program of exotic species control, or any wildlife management program for that matter, are monitoring, modeling, and adaptive resource management (ARM) (Walters 1986; Walters and Holling 1990). Monitoring is simply keeping records on the control program's inputs and outputs and costs and benefits. Inputs include such things as human and financial resources used in planning, conducting, and evaluating control efforts. Outputs involve expected consequences of the control effort, such as numbers of animals removed and assessment of damage. Costs and benefits were described above.

Modeling is essentially a process of constructing hypotheses in mathematical or graphical form that explicitly state the processes that link various components in the ecosystem at a level of detail sufficient for the model's intended use. In essence, the modeling process helps organize thinking about the project. It often leads to better definition of the control effort problem and a better understanding of difficulties associated with implementing it. The model can be used to identify the most sensitive parameters that must be defined accurately, if the ecological processes underlying the control effort are to be understood. A good example of the value of a simple model is Hone and Hobards' (1980) prediction that removing 70% of a population of 1000 wild pigs twice a year would lead to near extinction in 3 years, whereas, the same proportional removal once a year would take about 9 years to eradicate the population. This model helped guide the control effort reported by Katahira et al. (1993).

ADAPTIVE RESOURCE MANAGEMENT

Adaptive resource management is a management paradigm in which the dual goals of system performance and acquiring reliable knowledge (Romesburg 1981) are accomplished simultaneously.

System performance refers to the degree of control of the exotic population. Acquiring reliable knowledge is instituting control in such a way that more is learned about demographics, ecology, and economics of the system being managed. In its simplest form, ARM involves control and experimental units to assess the efficacy of the control effort. In its most complex form, ARM addresses specific hypotheses about ecosystem function and structure that are fundamental underpinnings of the control effort. The ARM is a management process of gaining reliable knowledge, while simultaneously addressing management objectives.

EUROPEAN WILD BOAR AT GREAT SMOKY MOUNTAINS NATIONAL PARK: A CASE STUDY

PARK SETTING

The southern Appalachian highlands are internationally known for their rich biological diversity. This region harbors a considerable number of endemic and disjunct plant species that date back to the Tertiary forests (Braun 1950). Great Smoky Mountains National Park is a refugium within this region of richly diversified flora and fauna. In 1976 the park was designated as one of the world's first International Biosphere Reserves and in 1987 it received the distinction of becoming a World Heritage Site. Both of these designations reflect the global significance of the resources of the park and region.

The Park lies on the borders of eastern Tennessee and western North Carolina at latitude 84°30″ and long. 35°35″ at the southernmost extension of the Canadian zone in eastern North America. Within its 1950 km² area are large tracts of virgin forest that Whittaker (1956) classified into the following 15 major forest types: cove, eastern hemlock, gray beech, red oak–pignut hickory, chestnut, chestnut–oak, chestnut–heath, red oak–chestnut, white oak–chestnut, Virginia pine, pitch pine–heath, Table Mountain pine–heath, grassy balds, red spruce, Fraser fir, and heath balds. Since the time of the Whittaker designation, the chestnuts have mostly died out, but these types are still recognized by their other components.

POLICY MANDATES

National Park Service management policies direct the control or eradication of nonnative animal species that have a negative impact on native ecosystems. The management objectives of the Park reiterate national policy in the statement from the General Management Plan: "on the basis of research and experimentation, direct management measures will be taken to reduce as much as feasible the impact of European wild boar on the park, particularly in areas of special scientific value, fragility, or aesthetic appeal" (NPS 1975).

ORIGIN AND INVASION

There are no true swine native to North America and the only related natives are of the peccary family (*Tayasuidae* spp.), represented by the collared peccary (*Eicotyles tajacu*) of Arizona, New Mexico, and Texas. Domestic pigs (*Sus scrofa domesticus*) first came to North America with the colonizing Spaniards in the early 16th century (Wood and Brenneman 1977; Towne and Wentworth 1950). By the 1900s, feral pig populations had spread throughout the southeastern U.S. (Hanson and Karstad 1959; Wood and Lynn 1977). In California, domestic swine were introduced in 1769 and wild boar were introduced in 1925 (Barrett 1977). Feral pigs brought by Polynesian settlers were already present on the Hawaiian Islands when the first Europeans arrived in 1778 (Baker 1979). The European wild hog (*Sus scrofa*) populations were first introduced in western North Carolina and eastern Tennessee in 1912 and later accidentally escaped from a hunting enclosure near Hoopers Bald, NC (Stegeman 1938). Hoopers Bald is now part of the Nantahala National Forest, located approximately 15 miles south of the Park. In the early 1920s, about 100 of these

European wild hogs escaped from their enclosure and dispersed throughout the surrounding area, interbreeding with feral domestic swine that roamed freely (Bratton 1977; Conley et al. 1972). They entered the southwestern quadrant of the park near Calderwood in the late 1940s (Jones 1959). Since then the invasion has steadily spread from west to east (Fox and Pelton 1978), averaging roughly 2.75 km/year (Singer 1981) until currently they occupy the entire park.

Pen-reared animals from the Hoopers Bald population were subsequently transported to various places around the country, including Monterey County, CA, in 1923 (Barrett 1977); Texas in the 1930s (Ramsey 1968); central Tennessee in 1971 (Conley 1977); West Virginia in 1975 (Decker 1978); and west Tennessee in 1979. These wild hog populations prospered in California and Texas and have freely interbred with already existing feral pig populations throughout their expanding ranges. Currently, wild pigs inhabit 13 National Park Service areas located in the States of Tennessee, North Carolina, Texas, Florida, South Carolina, Mississippi, Georgia, Hawaii, and the Virgin Islands. In spite of the cross breeding with domestic feral hogs over the years, wild hogs have retained typical traits of long guard hairs, e.g., middorsal (mane) of hair, split gray (brown hair) tips, fewer teats, agouti color, and longitudinally striped piglets (Springer 1977, Barrett 1978).

HOME RANGE AND ACTIVITY PATTERNS

Distribution, Density, and Migration

Even with research and active management spanning 3 decades in the park, the best estimates of distribution, density, and movement end up being mostly speculative. In the 1960s, the population estimate was around 500 for the park, and Singer and Ackerman (1981) estimated the population to be approximately 1500 in 1980. The current European wild hog management plan (GSMNP 1993) estimates the population at 1000 animals. Because of the lack of access to the park's rough terrain and because of the habits and markings of the animal, it is likely that a total population size for the park may never be accurately estimated.

Wild hogs concentrate in the northern hardwood forests from April to July and around grassy areas at old home sites and maintained meadows in lower elevations. Hogs are found in lower elevation forests in the winter months. Several authors have alluded to the seasonal migration of a large percentage of wild hogs up to the higher elevation northern hardwood forests (Howe et al. 1981; Singer and Ackerman 1981; Tipton and Otto 1979; Singer et al. 1979). In 1979, densities for this northern hardwood area were estimated to be 79 animals per square kilometer, compared to 2 animals per square kilometer in lower elevation pasture (oak/pine and oak/pine habitat) (Singer 1981). Highest densities have been reported 7 years after the initial occupation of an area, with stabilization occurring after 20 to 27 years (Tate 1984).

In late summer, wild hogs begin their migration down slope; a move that is correlated with the drop of acorns, which is their principal food during this time (Tipton and Otto 1979; Scott and Pelton 1975; Conley et al. 1972). During fall and winter months, hogs prefer the warm xeric slopes of low elevations, with oak/pine overstories and heath understories (Tipton and Otto 1979).

Mast failure and, to a lesser extent, severe winter weather conditions are two factors that can affect seasonal migration movement. During these periods, migrations are more erratic, movements are more extreme, and home range size increases as much as three fold. Hogs occupy a greater variety of forest types as well as low pastures and old home sites (Singer et al. 1979; Bratton et al. 1982). Another reason for these movements is probably related to behavioral thermal regulations. The sparse bristly hair, which provides inadequate protection from both temperature extremes, and the lack of any apparent sweat glands necessitate that European wild hogs seek cooler areas in warm months and vice versa (Belden and Pelton 1975). Another critical factor is mast abundance. During mast failure, winter home ranges at lower elevations increased in size three-fold (Singer et al. 1979). After a hard mast failure in 1984, hogs were found everywhere in the park, including the grassy bald of Spence Field (Bill Cook, pers. commun.).

While the European counterparts travel in large groups called sounders, hogs in the Park appear to be either solitary males or small units often consisting of a mother and piglets. Average group size for the Park is two to three individuals (Singer and Coleman n.d.).

Hogs have been frequently observed to become more nocturnal due to control efforts. Casual observation has suggested that coyotes may play a role in pig movement. In the past few years, park biologists have noticed a few areas where two to three sows with piglets are running together and have speculated that such grouping provides more protection from canids (Kim DeLosier, pers. commun.).

Other Factors Affecting Movements

Nonbreeding females occupied home ranges that were 20% smaller than those of males, but the difference was significant (Singer et al. 1981). One radio-collared female reduced her home range 94% while suckling piglets but resumed normal larger movements when the piglets were only 11 weeks old.

Hog movements vary with time, food supply, and disturbance. In all seasons, wild hogs are more significantly active during crepuscular and nocturnal periods than they are during the day (Singer et al. 1979). In winters of abundant mast, hogs move over 11% of their home range during a 24-hour period, compared to 22% during summers or the winter after a mast failure (Singer et al. 1979). The reduction in winter activity during good mast years certainly helps the animals maintain a positive energy balance. However, Singer and Ackerman (1981) reported starvation, reduced fat reserves, and reduced blood condition, parameters resulting from lack of food and their increased movement following mast failure.

Food Habits

Wild hogs are omnivorous and may eat fish, snakes, frogs, salamanders, crayfish, mussels, snails, small mammals, carrion, earthworms, immature and adult insects, and the young and eggs of ground-nesting birds. However, plants usually constitute the majority of the food items taken (Table 1). Much of the nourishment comes from the underground parts of the plants and from animals which inhabit the soil or leaf litter. Of 80 European food plants listed by Sablina (1955), the hogs eat the roots or rhizomes of 49 species, the leaves or shoots of 24, and fruit of 16. Rooting by wild hogs in northern hardwoods is concentrated in mesic herb communities in the flat ridge tops of north-facing slopes; ferns or sedge dominated communities were less affected (Bratton 1975). During the spring and early summer in the northern hardwood forest type (*Fagus grandifolia*, *Acer rubrum*, and *Betula lutea*), wild hogs feed on a diet of 58% corms of *Claytonia virginica*, 28% leaves and stems from mesic herbs, 11% other roots, 2% macroinvertebrates, and 1% leaves of shrubs; that is, their diet is 71% subterranean in origin (Singer 1981).

In late summer, hard mast (*Quercus* spp., *Carya* spp.) comprises 60 to 85% by volume of the wild hog diet (Conley et al. 1972; Scott and Pelton 1975). When the mast crop fails, alternative foods, particularly wild yam tubers (*Dioscorea batatas*) and other roots, comprise 83 to 91% of the diet, and hard mast drops to less than 10% (Conley et al. 1972; Scott and Pelton 1975; Howe and Bratton 1976).

While animal matter was found to make up only a small percentage of the diet, it was present in 94% of the stomachs examined. Items included flies, diptera larvae, ground beetles (*Caradidae*), land snails (*Polygidae*), and salamanders (*Plethodontidae*). Spring beauty corms were the most important food during the spring season, occurring in 98% of the stomachs and accounting for an average volume of 33% (Howe et al. 1981). Howe and Bratton (1976) reported that, in winter, rooting activity was concentrated in low-elevation successional tulip poplar forests and silverbell forests, and food items included mainly starchy tubers, bulbs, or rhizomes. However, these data were collected following an abundant mast crop and therefore the proportion of disturbance in

TABLE 1
Frequency of Occurrence of Food Items Identified in 128 European Wild Hog Stomachs Collected in Great Smoky Mountains National Park, 1971–1973

	Percent of frequency				
Items identified	Spring (30)[a]	Summer (14)	Fall (48)	Winter (36)	Total (128)
Plant Matter					
Roots	26.7	64.3	79.2	75.0	64.1
Leaves and stems	86.7	71.4	54.2	55.6	64.1
Fruits and seeds	63.3	50.0	39.6	44.4	47.6
Total plants	100.0	100.0	100.0	100. 0	100.0
Animal Matter					
Invertebrates	30.0	64.3	52.1	72.2	52.3
Other	16.7	14.3	20.8	25.0	20.3
Miscellaneous					
Garbage	—	7.1	—	2.8	1.6
Other (gravel, debris)	10.0	—	4.2	2.8	4.7

[a] Number in parentheses represents number of stomachs examined.

Data from Scott and Pelton 1975.

different forest types may vary in lean mast years. Except in hard mast failure situations, food is probably rarely a limiting factor, as hogs' adaptability enables them to utilize alternative food sources in all but the most extreme situations.

REPRODUCTIVE CHARACTERISTICS

The wild hog has an extremely high reproductive potential. It is perhaps the primary reason why control of the species is so difficult. The single most important reason is that there is no distinct rutting season and hogs can and do breed year-round. As a point of illustration, 41% of the piglets are born between March and May in the Park, compared to 100% in the Soviet Union (Sludskii 1974; Singer 1981). The earliest age of sexual maturity for both sexes in the Park is 7 to 8 months. Sexual maturity is delayed in years of food scarcity (Duncan 1974). The number of piglets produced averaged 4.36 per litter. Production of two successful litters in 1 year was limited to 5% of the sows in the Park. Although there were slightly fewer fetuses in the reproductive tracts, the litter size was slightly greater at the Park than in other areas reviewed by Singer (1981).

Rates of Increase

A population's observed rate of increase at a given time is determined by age specific survival, age specific fecundity, sex ratio, age distribution and place, immigration, and emigration (Caughley and Birch 1971). Hog populations vary tremendously from year to year in relation to food availability (Oloff 1951). In the Park, it was estimated that the wild hog population on an undisturbed study area of 11.6 km^2 increased 46% after an abundant mast crop year and declined 4% the following year (Singer et al. 1981). This statistic alone suggests that natural population fluctuations might be most directly related to mast productivity. As long as the hog control program remains active in the Park, the direct reduction of animals will hopefully have the greatest impact on the population size.

Resource Impacts

The growing number of hogs not only competes for space and food with virtually all types of wildlife, but also creates impacts on other resources.

Flora

Bratton et al. (1982), using vegetation survey plots from the western end of the Park, found that rooting was present at all elevations but was concentrated in mesic sites, except those having *Rhododendron maximum* understories. Hogs had also disrupted wet areas in Cades Cove known for their rare herbs. Singer et al. (1982) found that rooting by wild hogs mixed the A-1 and A-2 soil horizons and reduced ground vegetation cover and leaf litter.

Hog rooting in gray beech forests can reduce cover of herbaceous understory to less than 5% of its expected value (Bratton 1974a). Over 50 nonwoody species are known to be eaten, uprooted, or trampled. These disturbed species exhibit changes in population structure, including reduction in percentage of mature and flowering individuals. Changes in species composition favor plants with deep or poisonous roots (Bratton 1974b). Huff (1977) reported that rooting stimulates vegetative reproduction of gray beech (*Fagus grandiflora*), with root suckers being 4 to 44 times greater than in undisturbed plots. It is also suspected that a fungus infecting the beech forests (*Armellaria mellea*) proliferates because of aeration of soil from rooting. Hog rooting may indirectly represent a potential threat to other vegetation, especially reproduction of sugar maple (*Acer saccharum*) and the suppressing of foliar height of blackberry (*Rubus canadensis*).

Hog exclosures have been established in the Park to evaluate impacts of hog rooting on vegetation and the high elevation gray beech forests. Three sample locations have been established for these exclosures, representing a range of hog activity over time and a history of rooting impact. Quadrants were established both inside and outside the exclosures to evaluate hog effects. Total herbaceous cover within the exclosures has been measured to be approximately 70% while the range outside the exclosures ranged from 20 to 50% cover. Following hog rooting, total cover returned quickly to previous levels, but the species composition was slow to return to preimpact levels. A few plants which were nonfood items increased once the area was protected. Thus, there was a rather fast recovery in terms of cover but not in terms of species composition. By policing the exclosures in rooted areas, the recovery could be measured. When contrasting food and nonfood items at Spence Field, violets (*Viola* spp.), a food item, composed 21% of the herbaceous understory in the exclosure and less than 1% in the control area outside the exclosure. On another site, *Angelica triguinata*, a food item, made up 8% of the cover in the exclosure but only 3 to 4% outside (White 1984). An earlier set of exclosures established by Bratton and Singer revealed a reduction in herbaceous cover outside the exclosures, sometimes to less than 5% of the expected undisturbed value, and probably caused local extinction of individual wildflower species.

About one third of the herbaceous species listed in the wildflower guides specifically for the park (Campbell et al. 1962) are already known to be damaged by hogs. In addition to spring beauty (*Claytonia virginica*), fawn lily (*Erythronium americanum*) and wake robin (*Trillium erectum*) have been greatly reduced in both cover and number of blooming plants (Bratton 1974a). On grassy balds, hogs rake and roll the turf under the mountain oat grass (*Danthonia compressa*), causing soil erosion and a change in the local successional pattern. Weedy forbs such as *Potentilla rumex* usually invade the scars because the grasses are slow to recolonize. Hogs also root under shrubs and in patches of *Rubus*. These sites are reoccupied by *Potentilla* and semishade plants such as *Stachys clingmannii*.

Fauna

Singer et al. (1982) reported that red-back voles (*Coethrionomys gapperi*) and short-tailed shrews (*Blarina brevicauda*), that depend largely on leaf litter for habitat, were nearly eliminated from intensively rooted stands as their habitat was lost. Other species sampled which were more arboreal or subterranean seemed unaffected by hog activities. Two potentially threatened species that are in the diet of the hog include the red-cheeked salamander (*Plethodon jordani*), which is endemic to the Park, and the Jones middle-tooth snail (*Mesodon jonesianus*). An estimated 80% reduction in micro invertebrates in the soil in some areas could be attributed to habitat destruction as well as direct predation. Siltation or contamination of streams in the vicinity of rooting or wallow areas have had unknown effects on the aquatic environment that could be detrimental to native brook trout (*Salvelinus fontinalis*) (Howe et al. 1976; Ackerman et al. 1978). Wild hogs may compete for available food sources with other species such as deer (*Odocoileus virginianus*), wild turkeys (*Meleagris gallopava*), black bears (*Ursus americanus*), squirrels (*Sciurus carolinensis*, *Sciurus niger*, *Tamiasciurus hudsonicus*), and chipmunks (*Tamias striatus*). Matschke (1965) documented predation on nests of ruffed grouse (*Bonasa umbelous*) and wild turkey in highly populated areas of the Tellico Wildlife Management Area of Cherokee National Forest. Hogs may also have an impact on ground-nesting songbirds (Wilcove 1983).

Soil

Soil erosion was accelerated with the incidence of rooting, along with leaching of calcium, potassium, zinc, copper, and magnesium from leaf litter and soil. Nitrate concentrations, however, were higher in soil water and stream water from rooted stands, suggesting alterations in ecosystem nitrogen transportation processes with a potential loss of nitrogen from the soils. Rooting does not appear to increase the sediment load, apparently because of the high infiltration rate of loamy soils involved and because rooting decreased soil bulk density, therefore further promoting infiltration by rainfall (Singer et al. 1982).

Disease

Wild hogs may serve as cause, with other wildlife and livestock, for infectious and parasitic diseases and may serve as reservoirs for diseases which can spread to domestic livestock, such as hog cholera, brucellosis, trichinosis, hoof and mouth disease, African swine fever, giardia, and pseudorabies. A wild hog preserve in South Carolina has been quarantined for brucellosis and pseudorabies (Singer 1981). Blood from 108 hogs removed from the Park was tested in 1990 for selected viral and bacterial diseases. No antibodies were detected for swine brucellosis, pseudorabies, bovine virus, diarrhea virus or porcine rotavirus infection (New et al. 1994).

CONTROL PROGRAM

Early History

The hog control program began in 1960, in reaction to significant hog rooting of Gregory Bald and Spence Field (Singer 1981). The very early attempts at hog removal in the 1960s were focused primarily on trapping and direct reduction by shooting, which remains the mainstay of the hog control program in the park today. The total number of hogs removed over the life of this 36-year program is over 8000. Hog removal data from 1960 through 1995 is depicted in Figure 2. Over 55% of these animals were taken during the course of the last 9+ years. This is a result, in part, from 6 years of funds (1986 to 1988 and 1990 to 1992) associated with a special National Park Service initiative entitled the Natural Resource Preservation Program, dealing with critical problems in nation parks.

Setbacks

Although the control of the European wild hog and its mitigation impacts are well grounded in the legislative mandate and policy guidance for the National Park Service, the hog control program

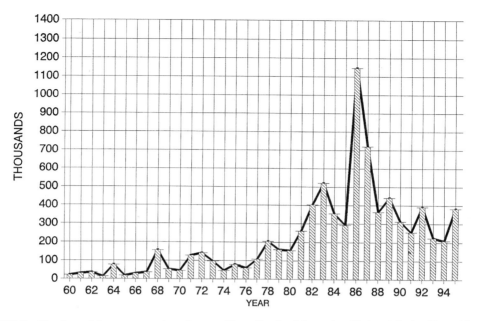

FIGURE 2 Number of hogs removed each year, Great Smoky Mountains National Park. (From National Park Service Records.)

has had its share of controversy. In August 1977, a controversy evolved following a field test of the use of hunters and dogs for the purpose of killing wild hogs in the Park. The handlers and dogs were not from nearby the Park, and hostility toward the program developed immediately and lingered for several years. In hindsight, the timing, choice of control, and personnel on the target area were all, unfortunately, ill-advised. As a result, a moratorium was established on further direct reduction by shooting in the Park until a management plan was written and approved. Simultaneously, a cooperative agreement on hog management with the North Carolina Wildlife Resource Commission was signed.

In April of 1978, the moratorium was lifted, and direct reduction was again used in concert with the uninterrupted trapping procedures. During this time, four additional part-time people were allocated to the hog control program of the Park. With this additional staff support, the number of hogs removed in the period from 1978 to 1983 surpassed the total number removed in the previous 19 years. In the summer of 1978 and subsequent summers, control emphasis was placed on high-elevation beech gaps and along northern hardwood forested ridge tops in the Park. Although direct reduction has been carried out on the North Carolina side of the Park, it has never been fully supported by the North Carolina Wildlife Resources Commission and has been low key at best.

In the summer of 1981 a second flare-up of hog controversy began with the publication of an article in the *Wall Street Journal,* which described government hog hunting in the Park. As the political temperature rose, park officials placed a moratorium on the killing of hogs in North Carolina. Several stipulations were place on the moratorium. Animals could be shot if they directly threatened an endangered species or if they were in the Cataloochee area, which at that time was the last remaining hog-free area in the Park. Other than that, they were not to be taken by shooting. The tradeoff was the formation of the Graham County Volunteer Citizens Group to trap hogs in the Park. They were permitted, under the joint supervision of Park personnel and the North Carolina Wildlife Resource Commission, to trap and remove as many hogs as possible. Resource support in the form of traps, bait, and sometimes boats was provided by the Park. Animals were transported to acceptable release sites in national forests in North Carolina. A core group of 10 to 12 individuals maintained traps in the Hazel Creek portion of the Park. The volunteer program successfully diffused much of the hostility toward the Park regarding the hog issue. It more than quadrupled the manpower that the Park has had to devote to the program. However, the project has had several drawbacks, including sporadic trapping efforts, the capture and release of nontarget species, less effective baiting techniques and trapping efforts, trapping restricted to areas of easy access, and the potential for illegal activity (Stuart Coleman, pers. commun.). The group is no longer active in the Park, largely due to a reduced population of hogs resulting in a greatly reduced trapping success rate; i.e., more effort for less reward. It is important to note that there has been very little controversy over the last 5 years with the wild hog control program.

Trapping Techniques

Early trapping success rates were relatively low in the Park, ranging from 0.0062 to 0.0328 captures per night (Duncan 1974). Traps are generally placed in areas of relatively easy access, due to the problems of transportation. This means that they tended to be placed in areas of open grassland along road edges and in agricultural districts such as Cades Cove. There has been an evolution in the design of the traps, and for a time they were fabricated by students of the Job Corps at Oconaluftee.

Trapping efforts in the 1960s and 1970s utilized stationary type traps or semiportable box traps. The box traps (cyclone fence with guillotine type door) were heavy, usually weighing around 300 pounds. The weight of the trap restricted movement, therefore limiting the trapping effort primarily to areas with vehicle access. In the 1980s, trap design changed, from dimensions 7' × 4' × 3' to 5' × 3' × 3' and in material utilized (chain link fence and aluminum instead of wooden trap doors) to reduce the weight to approximately 125 pounds, thus greatly increasing the mobility of the trapping efforts. The trapping program quickly expanded into park areas that were inaccessible to vehicles. Park officials have hand-carried these traps into areas up to 3 miles from roads. In remote back country areas, helicopters have been used to drop traps into grassy balds; the traps were then carried along the Appalachian Trail to desired trapping sites.

Shooting Techniques

Several weapons have proven effective in the shooting activities. The .357 magnum is the preferred sidearm which is used only for humanely euthanizing animals in traps. In the summer months, the vast majority of the hunting activity is at night. In the winter season, day hunting is the norm. The 12-gauge shotgun with an artificial light source directly fixed to the gun using .00 buckshot is preferred as the long gun for night hunting. The 30-06, the daytime rifle of choice in the control program, has proven effective. Shooting is the most effective method in terms of expenditure of man-hours. In 1980, 6.6 man-hours per hog were used for shooting as opposed to 9.4 man-hours per hog for trapping (Stuart Coleman, pers. commun.). The number of man-hours per capture/kill for both methods is highly variable due to varying hog densities, food availability and work performance of control personnel.

Hog Bait Study

As a result of a workshop among scientists and wildlife biologists in 1983 (Tate 1984) to assess research needs for the hog control program, a research initiative was launched in the Park to test for more effective attractants and baits for hogs. Attractants and baits tested are displayed in Table 2. Substantial enclosures were built and stocked with wild hogs, which were tested with several types of bait. After certain baits were designated as preferred, they were field-tested on free-ranging wild hogs. The study found that, overall, hogs significantly preferred fermented corn mash over other baits. Several olfactory attractants were also tested and generally there were no significant preferences (Wathen et al. 1988). The bait currently used most often in the trapping program is shelled corn.

TABLE 2
Attractant and Bait Combinations Used in Wild Hog Bait Enhancement Study, GSMNP, 1985 to 1988

Substance	Type of Testing [a]	Substance	Type of Testing [a]
Acorn scent	1	Corn with walnut extract	3, 4
Anise oil	1	Creosote	1, 2
Applesauce	1	Fish oil	1, 2
Apricot nectar	1	Molasses	1
Blood meal	1, 2	Motor oil (new)	1
Boar Mate™	1, 2	Motor oil (old)	1, 2
Boiled wheat	1	Peppermint extract	1
Chinese yams	1	Pig feed	3
Coal	1	Pine scent	1
Cod liver oil	1	Pine Sol™	2
Coffee beans	1	Snake feces	2
Corn	3	Snake scent	1
Corn mash	1, 2, 3, 4	Soybean oil	1
Corn with acorn scent	3	Spoiled milk	1
Corn with anise oil	3	Spoiled tofu	1
Corn with apple sauce	3	Strawberry flavoring	1, 2
Corn with beer	3, 4	Sweet potatoes	1
Corn with cinnamon	3	Truffles	1
Corn with Chinese yams	3	Violet scent (imitation)	1
Corn with cod liver oil	3	Walnut extract	1, 2
Corn with coffee beans	3	Wheat (boiled)	3
Corn with fish oil	3	Wheat (soaked in water)	3
Corn with molasses	3	Wheat with apricot nectar	3
Corn with peppermint extract	3	Wheat with molasses	3
Corn with spoiled milk	3, 4	Wheat with strawberry flavoring	3
Corn with strawberry flavoring	3, 4		

[a] 1 = Attractant preference testing — pen trials; 2 = Scent station surveys; 3 = Bait preference testing — pen trials; 4 = Bait preference testing — field trials.

Source: Wathen et al. 1988

Review of Control Activities

A review of the control program over a 4-year period is very instructive in assessing the potential for the park managers to control the population of hogs and to evaluate the overall effectiveness of the techniques of control that have evolved over the 36-year history of the program. Figure 3 shows management units that have evolved in the program. The letters refer to place names identifying the units. Comparing the place names with the removal data from 1986 to 1989 shown in Table 3 allows identification of where most of the hogs have been removed from the park. The distribution of the successive removal varies by year but, by and large, the greatest success lies in the area labeled BKY (backcountry) along the Appalachian Trail, where hogs are shot at night; and the areas labeled TWM (Twenty Mile), which borders Fontana Lake, and DCK (Deep Creek), near Deep Creek campground, where hogs are trapped and shot.

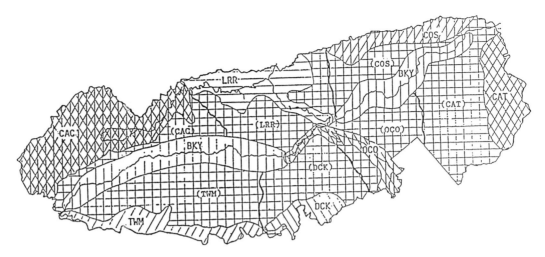

FIGURE 3 Great Smoky Mountains National Park: current management areas of concentrated hog removal and proposed extensions.

An interesting comparison may be seen in the ratio of adults and subadults by management unit. In those areas where trapping is the primary removal method (DCK, TWM), generally there are more juveniles removed than adults. Where hunting is the primary method (BKY), there are usually more adults taken. This may relate to the fact that lower elevation sites are preferred for farrowing.

There is an overall tendency for a change in the adult/juvenile ratio following a period of aggressive population reduction from an area. The Deep Creek management unit (DCK), for instance, showed an increase in the subadult group taken between 1986 and 1987. The high country area labeled BKY showed a tremendous decline in the taking of subadults vs. adults after the first year. The contrast from 1986 to 1987 is particularly striking. An additional interesting trend in control activity in the high elevation forests (BKY) is displayed in Table 4. In this situation shooting was the preferred method of control vs. trapping. As indicated in Table 4, there was a marked decline in the success rate of kill per hours hunted over the 3-year period, suggesting that the population was, in fact, being significantly reduced, assuming the vulnerability of the hogs to hunting remained constant.

MONITORING

Devising a means to monitor the hog population in the park has been extremely difficult. The most effective means utilized to date is to base the estimate on the level of effort required to remove the animals. The modeling activity in this regard is described below.

It is also important to monitor trends in the adverse effects of the animals on the environment. The most obvious evidence of adverse effects of hogs in the park is rooting. During the summer of 1983, trail and cross-country transects were run in two watersheds of the park to try to get a perspective on the overall distribution of rooting throughout watersheds (NPS 1983). Trails and cross-country routes were treated as 20-meter-wide belt transects. Results of the study tend to confirm the perspective that during early to midsummer the hog rooting is more intensive in the high elevations and is concentrated along trails, ridge lines, and grassy areas. The results were intended to be used in designing a park-wide and/or management unit monitoring program for hog rooting which would be utilized to evaluate the effectiveness of the overall hog control program. The hog rooting transects were intended to be utilized in conjunction with the system of hog exclosures to monitor long-term adverse impacts of hog activity. Unfortunately, monitoring activities have been deferred due to budget constraints.

TABLE 3
Hog Removals, 1986 to 1989,
Great Smoky Mountains National Park

Management unit	Total[a] + percent of total	Male	Female	Piglet[b]	Juv[c]	Adult[d]
1986						
BKY	285 (26)	93	160	23	136	115
CAC	72 (6.5)	40	43	13	44	27
CAT	3 (.3)	0	1	0	2	1
COS	6 (.5)	3	3	0	1	5
DCK	345 (31)	153	185	30	167	143
LRR	67 (6)	30	36	14	17	36
OCO	53 (5)	21	26	0	9	37
TWM	269 (24)	114	147	31	150	80
1987						
BKY	294 (41)	124	135	17	50	223
CAC	35 (5)	19	12	6	15	14
CAT	0					
COS	30 (4)	14	15	7	8	10
DCK	159 (22)	68	84	46	32	76
LRR	40 (6)	22	15	9	10	21
OCO	32 (5)	18	14	3	7	22
TWM	123 (17)	54	65	13	29	81
1988						
BKY	97 (27)	35	56	10	14	73
CAC	30 (8)	13	11	1	8	21
CAT	5 (1)	1	4	0	3	2
COS	8 (2)	5	2	0	3	5
DCK	55 (15)	24	29	0	4	50
LRR	29 (8)	13	16	4	6	19
OCO	41 (11)	19	21	0	4	37
TWM	96 (27)	27	67	21	10	65
1989						
BKY	54 (12)	23	30	4	21	29
CAC	45 (10)	15	30	1	31	13
CAT	1 (2)	1	0	0	1	0
COS	37 (7)	14	17	0	22	10
DCK	70 (16)	36	34	16	26	28

[a] The total may not equal the sum of the parameters because of unknowns.
[b] Animals weighing 15 lbs. or less.
[c] Animals weighing more than 15 but less than 60 lbs.
[d] Animals weighing 60 lbs. or more.

Source: Records at Great Smoky Mountains National Park.

TABLE 4
Removals from high elevations (BKY)

Pigs Shot			Pigs Trapped			Hours/Pig Shot		
1986	1987	1988	1986	1987	1988	1986	1987	1988
244	172	83	47	115	12	5.5	9.68	21.1

Source: Records at Great Smoky Mountains National Park.

Hog Population Models

Hog Population Model

As management techniques become more sophisticated, research needs become more and more specific. Research has progressed from the initial gathering of biological data to describing hog home ranges and movement patterns and the study of adverse impacts. For a time, research focused on the development of a computerized hog population model. A version of that model applied to the population of park hogs consisted of a user-friendly computer program that was adaptable to ongoing changes in the Park. This model could be modified to reflect the effects of changing conditions of weather, mast crop abundance, or other factors that might create dramatic effects on the hog population (Van Der Beast et al. 1990). The model was intended to be refined, based on procedures for validation, a process that was to continue indefinitely to mold the program as knowledge about population dynamics increased. Annual data input on population size estimates, population age/sex structure and reproductive rates gathered from captured/killed animals and sightings would be used to refine the model over time. The model was intended to become a key tool to resource management personnel in determining the most effective use of their control resources. The ultimate intention for the model was to be able to predict the minimum number of hogs that need to be killed in certain management areas in order to keep the population under control. This model would have been able to predict fluctuations in the hog population reacting to removal efforts and natural ecosystem fluctuations. Unfortunately, the project was abandoned in 1988 when the person responsible for maintaining the model left the Park employ.

Catch Per Unit Effort Model

Population estimates based on catch-per-unit-effort (CPUE) (Bishir and Lancia 1996; Lancia et al. 1996) can be used to evaluate the efficacy of removal efforts. CPUE is applicable in any situation where (1) the rate of detection of individuals, in this case the number of pigs removed per unit of hunting effort, is a function of the effort applied; and (2) some known change in population size takes place. The known change in this case is the numbers of pigs removed.

The CPUE model was applied to hog removals and effort expended data to estimate the size of the pig population in the high elevation, backcountry management unit (BKY). In spring and early summer a large proportion of the pig population migrates to these high elevations. The estimates applied only to the BKY, and therefore represent some unknown portion of the total pig population in the Park. It was assumed that patterns in BYK estimates were indicative of changes in the pig population throughout the Park. To make the BKY estimates, it was assumed that removal by shooting was the only source of change in population size during the 15- to 20-week summer hunting period, and effort used to remove them was recorded accurately.

In this case, estimates probably were accurate enough to assess efficacy because accuracy is positively correlated with proportion of the population that was removed: the larger the removal, the more accurate the population estimate. Obviously, the intent of control efforts is to remove as large a portion of the population as possible.

TABLE 5
Catch-Effort Estimates of the High Elevation Pig Population,
Great Smoky Mountains National Park, 1986 to 1988

Year	N	Se[a]	K	Total removed
1986	537[b]	121	0.000452	291
1987	329[c]	20.0	0.000740	287
1988	115[d]	15.2	0.000829	95

[a] SE's are from Seber's (1982) delta method and are only approximate (Bishir and Lancia 1995).

[b] CPUE estimate based on weeks 1 through 19.

[c] CPUE estimate based on weeks 4 through 19 is 308, which is the estimated population prior to the 4th week. Therefore, the total population estimate prior to the 1st week is 308 + 8 trapped + 13 shot = 329.

[d] CPUE estimate based on weeks 4 through 19 is 112, which is the estimated population prior to the 4th week. Therefore, the total population estimate prior to the 1st week is 112 + 0 trapped + 3 shot = 115.

From Lancia and Bishir. In press.

Weekly summaries (Bill Cook and Kim Delozier, pers. commun.) of man-hours spent hunting and total removals (both by hunting and livetrapping) were used to estimate the BKY population (Table 5). Yearly estimates for 1986 to 1988 were based on weekly effort and removal data obtained beginning the first week pig migration to high elevations appeared to be complete and ending in early August.

The pig population appeared to have been greatest in 1986, when intensive hunting began and then decreased by about 200 individuals per year (Table 5). Each year from about 50% to 90% of the estimated population was removed, based on the population estimates and number removed. Model assumptions might have been violated to some degree, which would probably lead to a underestimate of the true population by some unknown amount. However, if conditions were constant between years such that the population estimates represented a constant proportion of the true population, then the relative values between years and the decreasing trend in population would be accurate.

RECOMMENDATIONS FOR THE FUTURE

Adequate and Stable Funding

The hog control program in the park has emerged over a 36-year period with an invaluable working knowledge of control techniques. Until recently, there had been no base funding dedicated to hog control. In the early years, hog control duties were incorporated into the work assignments of park personnel and the limits of available staff time and resources significantly restricted activity. During federal fiscal years 1986 through 1988 and 1990 through 1992, a total of $1,184,000 from the National Park Service's Natural Resource Protection Program was applied to the hog control program in the Park. This funding level allowed demonstration that park personnel could make a difference and that the hog population could be controlled.

Unfortunately, all the progress made can be quickly lost if pressure is taken off the population, even for a short time, particularly during periods of favorable mast crop and weather conditions. The hogs are capable of very rapid population growth. An adequate level of base funding for the program is needed to ensure operation of an effective and efficient control program for the long term. Constant pressure must be placed on the population to avoid a rebound in population size. Base funding would provide an opportunity to stabilize personnel positions dedicated to the control program and thereby foster consistency in performance.

Staff Requirements

Ideally, the control program would be directed by a wildlife biologist dedicated full time to the task and a team of part- and full-time biological technicians appropriately distributed around the Park. The size of the team would vary in response to the size of the hog population. In addition, a research program should be facilitated to explore more efficient means to remove animals, monitor impacts, and design and verify a hog population model which would be a key tool of a highly efficient management strategy. The research program should be regionally based to serve the entire southern Appalachian highlands.

A key element to increase the success of both trapping and shooting is to know where the hogs are and how to get in position to either trap or shoot them quickly. There needs to be a wider geographical distribution of trained personnel stationed throughout the Park responsible for monitoring hog rooting and participating in hog removal.

Fencing

In some instances there may be a future need to protect rare and endangered plant and animal communities from hogs, which might require constructing extensive fencing to exclude the hogs from the sensitive areas. Additionally, there might be a need to consider fencing a portion of the Park to eliminate the immigration of hogs into the Park, particularly at the southwest corner. There has been some discussion that a power line right-of-way maintained by the Tennessee Valley Authority in the Park could be used for such fencing to minimize the potential for immigration of hogs from the Tellico Wildlife Area. Other possible invasion routes might be Deep Creek and Oconaluftee. As has been demonstrated in Hawaii Volcanoes National Park, fencing for hog control is a very expensive proposition.

Monitoring

In order to maintain an adequate monitoring program of the adverse impacts of the hogs in the Park, the hog exclosure system that has been established should be maintained, and the vegetation inside and outside the exclosures should be periodically evaluated. Vegetation monitoring should be staggered to distribute the work load evenly over the cycle period. There is a need to build larger exclosures so that evaluations beyond vegetation can be measured, such as small mammals, amphibians, and insects that use the forest litter layer.

In addition, a series of transects needs to be maintained that are monitored for hog rooting annually during the height of the season when the hogs are most likely to be in the vicinity of the transect. Biological technicians could be used to take the measurements on these rooting transects in the various management units as they hike into their control areas during the peak of the hunting season. The reporting of rooting sign needs to be more formalized and routinely documented on the park geographic information system. Hog routing sign in accessible portions of the park are much more likely to be reported. As a result, less attention is likely to be focused on the remote regions of the park.

A third component for monitoring would include the usage of Lancia and Bishir's (1996) catch effort estimator. Verification of the accuracy of his estimator would be a valuable addition to the management program.

Data Management

Protocols for data management should be developed for all aspects of the control program. The park geographic information system should be used to depict spatial patterns in hog removal statistics and rooting sign for each management unit.

Model Verification

Model development and verification is an evolutionary process. Over the long term, the draft hog population model that has been generically developed for the Park can be calibrated for specific management units within the park based on the accumulation of long-term systematic data that measure the key parameters driving the model. Over time, it is likely that the model will be modified to incorporate additional important factors that, experience will show, may contribute to the dynamic population fluctuations.

Research Agenda

Even though there has been a logical development of research topics over the last 15 years in the Park, research needs have not necessarily been met. If the control program is to become more refined and efficient in the removal of animals, which is certainly important in today's atmosphere of downsizing government, research will need to be conducted to evaluate the effectiveness of the current reduction program and to explore means to increase that efficiency. For instance, as management plans emerge on specific management units, additional insight on hog movement might be important to understand. We know that hogs tend to range on high elevation ridge tops during late spring and early summer but we really do not fully understand how that is triggered in relation to temperature; moisture; and the phonology of herbaceous plants, their primary spring food. It might be important, for instance, to map the population distributions of preferred high elevation species such as spring beauty (*Claytonia virginica*). When the hogs disperse from the high to lower elevations, key parameters of the microhabitat they seek out is not well understood. Very little is known about their midelevation behavior. Hog reaction to hunting pressure is not well understood beyond limited field observation by hunters. Additionally, research would be useful to establish specific correlates between weather, mast crop production, juvenile and adult survival, and reproductive potential.

A long-term commitment to research by the National Park Service would provide the opportunity to facilitate collaboration with academic institutions and private industry to find better, more cost effective solutions to the problem. The most vital long-term research need is to develop a species specific biological control mechanism for feral hogs which would not adversely impact domestic animals outside the Park. There are several options to explore in addressing this issue, and the payoffs are potentially enormous in the more efficient protection of the Park resources and potential long-term savings of funds that could then be directed to numerous other exotic species that go uncontrolled in the Park. Benefits from long-term research designed to develop species-specific biological controls would also benefit other efforts nationwide to control feral pig populations.

As funding levels have declined for the hog control program in the Park, research and monitoring activity has ceased and with it the investment in knowledge which could potentially lead to a much more efficient and humane means of control. Must all resources be dedicated to direct reduction activities without documenting resource benefits through monitoring of hog sign and exclosures and investing in research and development activity? Can the ecosystem manager afford not to make this investment? Such is the dilemma of the ecosystem manager.

Strategic Planning for Management

Sufficient base funding for the hog control program would provide for the establishment of a professional staff devoted to the program with the key added benefit of lower personnel turnover. Attention could be given to the development of a more comprehensive database on hog populations, such as seasonal location and movement, effort per capture, reproductive rate, population density, and the presence of disease, hog rooting sign, and systematic monitoring of the hog exclosures. These data could then be applied to develop a sophisticated predictive model for management as to population dynamics and adverse impacts projected from various control scenarios.

With the availability of adequate funding, the Park would be in a position to define specific management units, such as those suggested in Figure 2 for controlling hogs. Goals for acceptable damage from the hogs and the size of the population that is acceptable for the season of the year in which they are concentrated could then be set for these management units.

Review Hog Management Policy

The Park should consider getting out of the business of exporting hogs to other locations in the southeastern U.S. This practice has been greatly reduced in recent years as the population of hogs in the lower elevations has declined. However, this policy still continues the bad precedent of sanctioning exportation of an exotic species that does severe damage to soils and native herbaceous plants of the eastern deciduous forest biome. This compromise was made because the wild hog is a prized game species by hunters, and the Smokies population has been regarded as a source for these animals. The Park agreed to export them in exchange for political acceptance of their control inside the Park. Unfortunately, the result is acceptance of an obvious double standard motivated by prior political compromise. The existing policy is effectively saying "We don't want this devastating pest here but in order to placate local concerns, we have no qualms providing it elsewhere to forests in Tennessee and North Carolina." Such is the world of compromise required by ecosystem managers.

The control program should be more *regionally* focused on acknowledging areas being actively managed to enhance hog populations and how those efforts might influence activities in areas being managed to exclude them. How pervasive are these animals on the regional landscape? SAMAB could be asked to conduct a regional review of the wild hog management issue and provide recommendations for cooperation in control and long-term research.

CONCLUSIONS

Policy review, stable funding, qualified field personnel, program efficiency, a functional research agenda, data management, adequate monitoring, and development of a model to predict resource impacts from alternative control efforts are the essentials of a complete exotic species control program such as that of the wild hogs in the Smokies. A merging of those elements into a single functional strategic plan provides insight necessary to maintain a well-balanced and efficient hog control program.

Analysis of the current management program indicates that the methods and efforts of removal are successful in removing large numbers of animals. What is not known is what percentage of the population is being removed each year in general or by management unit. The regional context of the Park-based control program has not been described. The benefits of the removal program are not being measured nor monitored and financial support for the program is uncertain.

It may be worthwhile to take stock of the program now, even though the prospects for increased funding are not likely, and to develop a strategic plan to set long-term operational goals for acceptable limits of hog damage to the environment by defined management units such as those depicted in Figure 2. Over time, data can be systematically collected to build a model to estimate population levels by management units, and most importantly, to predict the number of hogs that need to be removed to sustain an acceptable level of disturbance on the ecosystem.

So what's the bottom line? Just how effective have the extraordinary efforts been to control this population of hogs? Have the taxpayers gotten their money's worth from the $1.32+ million public investment in this program from 1986 through 1993? All this expenditure of public funds has been directed to just one exotic species on the landscape at the expense of attention to hundreds of others. How do you justify the seemingly much greater importance of this exotic over virtually all the others? Have you conducted a comparative risk assessment of all exotic species on the land-scape? If not, why not? Are you using the most cost effective control technology? Why not let the public hunt them out for a fee and make money from the problem? Why is your concern for this exotic species confined to the Park?

These are examples from the hog control program of the type of difficult questions facing ecosystem managers concerned with controlling an exotic species, a situation exacerbated by the current political agenda to down-size government.

RISK ASSESSMENT REPORT CARD FOR ECOSYSTEM MANAGERS

Application of the *Risk Assessment Report Card for Ecosystem Managers* as applied to hog control program provides a kind of generic accounting system as follows:

Vision: B–

Well-defined control procedures, but lack of commitment to monitoring, research, modeling, and concern about the context of the regional feral hog population. Lack of investment in exploring alternative control strategies.

Resource risk: D

The severity of risk has been well documented.

Socioeconomic conflicts: C+

Social conflict in the past has adversely affected the control program. There is an inherent risk of shooting in a national park.

Procedural protocols: B–

Good protocols for control strategies, fair for monitoring, poor for building a viable population model.

Scientific validity: A–

Species biology and adverse effects well researched.

Legal jeopardy: B

Not likely to be legally challenged, but Park Service employees are using fire arms in a national park.

Public support: C+

North Carolina support for the program now OK but could change.

Adequacy of funding: C+

Compared to other exotic species, program is well funded but lacks in ideal funding levels for control program. Funds nonexistent for monitoring or research.

Policy precedent: A

Strong policy justification.

Administrative support: A

NPS has strongly supported this program.

Transferability: B

Control techniques developed for temperate forests of the southeast U.S. May be applicable to other international biosphere reserves.

All in all, professionals in the NPS have made remarkable progress in controlling the wild boar in Great Smoky Mountains National Park.

ACKNOWLEDGMENTS

The authors wish to express appreciation to Kim DeLosier and Bill Cook, wildlife biologists with the National Park Service, who managed the hog control program in Great Smoky Mountains National Park, for their review of the manuscript and sharing of information.

REFERENCES

Ackerman, B.B., M.E. Harmon, and F.J. Singer. 1978. Seasonal food habits of the European wild hog. Part II. Studies of the European wild boar in Great Smoky Mountains National Park, A Report for the Superintendent. Gatlinburg, TN.

Baker, J.K. 1979. The feral pig in Hawaii Volcanoes National Park. Pp 365–367 in R.M. Linn, Ed. Proc. 1st Conf. on Sci. Res. in the National Parks. U.S. Dept. Interior, National Park Serv. Trans. and Proc. Ser. 5. Washington, D.C.

Barrett, R.H. 1970. Management of wild hogs on private lands. *Trans. California-Nevada Section Wildl. Soc.* 17-71-78.

Barrett, R.H. 1977. Wild pigs in California. Pages 111–113 in G.W. Wood, Ed. Research and management of wild hog populations. Belle W. Baruch Forest Science Institute, Georgetown, SC.

Barrett, R.H. 1978. The feral hog on the Dye Creek Ranch, California. *Hilgardia.* 46283–355.

Belden, R.C. and M.R. Pelton. 1975. European wild hog rooting in the mountains of East Tennessee. *Proc. Annu. Conf. SE Assoc. Game Fish Comm.* 29:665–671.

Bishir, J.W. and R.A. Lancia. 1996. On catch-effort methods of estimating animal abundance. *Biometrics.* 52:1457–1466.

Bratton, S.P. 1974a. The effect of European wild boar (*Sus scrofa*) on the high-elevation vernal flora in Great Smoky Mountains National Park. *Bull. Torrey Bot. Club* 101(4):198–206.

Bratton, S.P. 1974b. An integrated ecological approach to the management of European wild boar (*Sus scrofa*) in Great Smoky Mountains National Park. Manage. Rep. 3, Uplands Field Res. Lab., Great Smoky Mountains National Park. Gatlinburg, TN.

Bratton, S.P. 1975. The effect of the European wild boar (*Sus scrofa*) on gray beech forest in the Great Smoky Mountains. *Ecology* 56:1356–1366.

Bratton, S.P. 1977. Wild hogs in the United States — origin and nomenclature. Pp 1–4 in G.W. Wood, Ed. *Research and Management of Wild Hog Populations:* Proc. of a Symp., Belle W. Baruch Forest Service Institute of Clemson University Georgetown, SC.

Bratton, S.P., M.E. Harmon and P.S. White. 1982. Patterns of European wild boar rooting in the Western Great Smoky Mountains. *Castanea* 47:230–242.

Braun, E.L. 1950. *Deciduous Forest of Eastern North America*. Blakiston: Philadelphia, PA. 596 pp.

Campbell, C.C., W.F. Hutson, and A.J. Sharp. 1962. *Great Smoky Mountains Wildflowers*. University of Tennessee Press: Knoxville. 112 pp.

Caughley, G. and L.C. Birch. 1971. Rate of increase. *J. Wildl. Manage.* 35:658–663.

Conley, R.H. 1977. Management and research of the European wild hog in Tennessee, pp 67–70 in G.W. Wood, Ed. *Research and Management of Wild Hog Populations*. Belle W. Baruch Forest Service Institute of Clemson University Georgetown, SC.

Conley, R.H., V.G. Henry, and G. H. Matschke. 1972. European wild hog research project W-34. Tenn. Game and Fish Comm. Final Rep. 259 pp.

Decker, E. 1978. Exotics, pp 249–256 in J.L. Schmidt and D. Gilbert, Eds. *Big Game in North America*. Stackpole Books: Harrisburg, PA.

Duncan., R.W. 1974. Reproductive Biology of the European Wild Hog (*Sus scrofa*) in the Great Smoky Mountains National Park. Unpublished M.S. thesis, University of Tennessee: Knoxville. 95 pp.

European Wild Boar Management Plan. Great Smoky Mountains National park. Gatlinburg, TN. 20 pp. + app.

Fowler, C.W. and T.D. Smith. 1981. *Dynamics of Large Mammal Populations.* John Wiley & Sons: New York. 477 pp.

Fox, J.R. and M.R. Pelton. 1978. An Evaluation of Control Techniques for the European Wild Hog in the Great Smoky Mountains National Park. M.S. thesis, University of Tennessee: Knoxville.

Hanson, R.P., and L. Karstad. 1959. Feral swine in the southeastern United States. *J. Wildl. Manage.* 23:64–74.

Hone, J. 1995. Spatial and temporal aspects of vertebrate pest damage with emphasis on feral pigs. *J. Appl. Ecol.* 32:311–319.

Hone, J. and G.E. Robards. 1980. Feral pigs: ecology and control *Wool Technol. Sheep Breed.* 28(4):7–11.

Hone, J. and C.P. Stone. 1989. A comparison and evaluation of feral pig management in two national parks. *Wildl. Soc. Bull.* 21:269–274.

Howe, T.D., and S.P. Bratton. 1976. Winter rooting activity of the European wild boar in Great Smoky Mountains National Park. *Castanea* 41:256–264.

Howe, T.D., F.J. Singer, and B.B. Ackerman. 1981. Forage relationships of European wild boar invading northern hardwood forests. *J. Wildl. Manage.* 45(3):748–754.

Huff, M.H. 1977. The effect of European wild boar (*Sus scrofa*) on the woody vegetation of gray beech forest in the Great Smoky Mountains. Manage. Rep. No. 18. Uplands Field Research Laboratory. Southeast Region, NPS, Great Smoky Mountains National Park. Gatlinburg, TN. 63 pp.

Izac, A.M.N. and P. O'Brien. 1991. Conflict, uncertainty and risk in feral pig management: the Australian approach. *J. Environ. Manage.*, Vol. 32:1–18.

Jones, P. 1959. The European wild boar in North Carolina. Game Division, North Carolina Wildlife Resource Committee Raleigh, NC. 26 pp.

Katahira, L.K., P. Finnegan, and C.P. Stone. 1993. Eradicating feral pigs in montaine mesic habitat at Hawaii Volcanoes National Park. *Wildl. Soc. Bull.* 21:269–274.

Lacki, M.J. and R.A. Lancia. 1983. Changes in soil properties of forests rooted by wild boars. *Proc. Annu. Conf. Southeastern Assoc. Fish Wildl. Agencies* 37:228–236.

Lacki, M.J. and R.A. Lancia. 1986. Effects of wild pigs on beech growth in Great Smoky Mountains National Park. *J. Wildl. Manage.* 50:655–659.

Lancia, R.A., J.W.Bishir, M.C.Conner, and C.S. Rosenberry. 1996. Use of catch effort to estimate population-size. *Wildl. Soc. Bull.* 24(4):731–737.

Lancia, R.A., K.H. Pollock, J.W. Bishir, and M.C. Conner. 1988. A white-tailed deer harvesting strategy. *J. Wildl. Manage.* 52:589–595.

Matschke, G.H. 1965. The influence of oak mast on European wild hog reproduction. Contrib. from Federal Aid to Wildlife Restoration Program, Tennessee, Pittman-Robertson Proj. W-35-R.

National Park Service. 1975. General Management Plan: Great Smoky Mountains National Park. Gatlinburg, TN.

National Park Service. 1983. Mapping Hog Rooting Patterns in Great Smoky Mountains National Park. Unpublished report. Gatlinburg, TN.

New, J.C., K. Delozier, C.E. Barton, P.J. Morris and N.D. Potgieter. 1994. A serologic survey of selected viral and bacterial diseases of European wild hogs, Great Smoky Mountains National Park, U.S.A. *J. Wildl. Dis.*, 30(1), 103–106.

Oloff, H.B. 1951. *Zur Biologie Ökologie des Wildschweines. Beitrage zur Tierkunde und Tierzucht,* Vol. 2. Verlag Dr. Paul Schors: Frankfurt. 95 pp.

Peine, J.D. and J.A. Farmer. 1990. Wild hog management program at Great Smoky Mountains National Park. In Proc. 14th Vertebrate Pest Conf. California Vertebrate Council, Sacramento. 25 pp.

Ramsey, C. 1968. Texotic line up. *Texas Parks Wildl.* 25:3–7.

Ricker, W.R. 1958. Handbook of computations for biological statistics of fish populations. *Fish. Res. Board Can. Bull.* 119. 300 pp.

Romesburg, H.C. 1981. Wildlife science: gaining reliable knowledge. *J. Wildl. Manage.* 45:293–313.

Sablina, T.B. 1955. Kapytnye Belovezhskoi Pushchi. *Tr. Inst. Morfol. Zhivatn. Akad. Nauk, USSR* 15:1–191.

Scott, C.D., and M.R. Pelton. 1975. Seasonal food habits of the European wild hog in the Great Smoky Mountains National Park. *Proc. Southeast Assoc. Game Fish Comm.* 29:585–593.

Singer, F.J. 1981. Wild pig populations in the national parks. *Environ. Manage.* 5(3):263–270.

Singer, F.J. and B.B. Ackerman. 1981. Food availability, reproduction, and condition of European wild boar in Great Smoky Mountains National Park. Res./Resour. Manage. Rep. 43, USDI, National Park Service, Southeast Region. Atlanta, GA. 52 pp.

Singer, F.J. and S. Coleman. n.d. Ecology and management of European wild boar in Great Smoky Mountains National Park. 12 pp.

Singer, F.J. D.K. Otto, A.R. Tipton, and C.P. Hable. 1979. Home ranges, movements, and habitat use of European wild boar. A Report for the Superintendent, Great Smoky Mountains National Park. Gatlinburg, TN.

Singer, F.J. D.K. Otto, A.R. Tipton, and C.P. Hable. 1981. Home ranges, movements, and habitat use of European wild boar in Tennessee. *J. Wildl. Manage.* 45(2):1981.

Singer, F.J. W.T. Swank, and E.E.C Clebsch. 1982. Some ecosystem responses to European wild boar rooting in a deciduous forest. Res./Resour. Manage. Rep. 54, USDI, National Park Service, Southeast Region. Atlanta, GA. 31 pp.

Singer, F.J., W.T. Swank, and E.E.C. Clebsch. 1984. Effects of wild pig rooting in a deciduous forest. J. Wildl. Manage. 48:464–473.

Sludskii, A.A. 1974. *The Wild Boar: Its Ecology And Economic Importance. Izdatel'svo Akademii Nauk Kazakhskoi Ssr, Alma-Ata.* 219 pp. [CSIRO trans. from Russian].

Springer, M.D. 1977. Ecological and economic aspects of wild hogs in Texas. Pp 37–46 in G.W. Wood, Ed. *Research and Management of Wild Hog Populations.* Belle W. Baruch Forest Science Institute, Georgetown, SC.

Stegeman, L.C. 1938. The European wild boar in the Cherokee National Forest, Tennessee. *J. Mammal.* 19:279–290.

Tate, J., compiler. 1984. Techniques for controlling wild hogs in Great Smoky Mountains National Park: Proc. of a Workshop, Nov. 29 to 40, 1983. Uplands Field Research Laboratory, Great Smoky Mountains National Park. Gatlinburg, TN. 87 pp.

Tipton, A.R. and K.K. Otto. 1979. Evaluating wild boar movement activity and distribution in Great Smoky Mountains National Park. Final proj. rep., Contract CX500061138. Gatlinburg, TN. 87 pp.

Van Der Beast, E., P. Thomas, M. Mahato, L. Mitchell, J. Hamilton, D. Johnson, L. Neal, and J. Farmer. 1990. BOAR: A wild hog population simulation model. (Unpublished paper.)

Walters, C.J. 1986. Adaptive management of renewable resources. MacMillan; New York. 374 pp.

Walters, C.J. and C.S. Holling. 1990. Large-scale management experiments and learning by doing. *Ecology* 71:53–74.

Wathen, G., J. Thomas, and J. Farmer. European wild hog bait enhancement study — Final report, December 1988. Res./Resour. Manage. Rep. Ser., USDI, National Park Service, Southeast Region. Atlanta, GA.

White, P. S. 1984. The Great Smoky Mountains National Park hog exclosure study. Pp. 23–24 in J. Tate, Ed., Control Methods for the Wild Hog, a Workshop. USDI National Park Service, Southeast Regional Office, Res./Resource Manage. Rep. SER-72.

Whittaker, R.H. 1956. Vegetation of the Great Smoky Mountains. *Ecol. Monogr.* 26:1–80.

Wilcove, D.S. 1983. Population changes in the neotropical migrants of the Great Smoky Mountains: 1947–1982. A rep. to the World Wildlife Fund, U.S. 13 pp.

Wood, G.W., Ed. 1977. Research & Management of Wild Hog Populations: Proc. of Symp. Belle W. Baruch Forest Science Institute, Georgetown, SC.

14 Control of Pests and Pathogens

Scott E. Schlarbaum, Robert L. Anderson,
and Faith Thompson Campbell

CONTENTS

Introduction 291
Forest Ecosystem Change and Stress 292
Examples of Exotic Pest Damage in Southern Appalachian Esosystems 293
    American Chestnut Blight...293
    Balsam Woolly Adelgid 294
    Butternut Canker 294
    Pear Thrips 295
    Dogwood Anthracnose 295
    Hemlock Woolly Adelgid...296
    Gypsy Moth...296
    Beech Bark Disease Complex..297
Examples of Exotic Plant Pest Invasions in a Southern Appalachian National Park 297
Impacts of Exotic Pests and Plant Pests on Habitat 299
The Mechanics of Ecosystem Protection and Species Reintroduction........................300
    Ecosystem Protection 300
    Species Reintroduction...301
    Genetic Conservation 301
    Pest Control 302
    Reintroduction Protocols..303
The Future 303
Risk Assessment Report Card 303
References 304

INTRODUCTION

The ecosystems of the southern Appalachian region contain a rich diversity of plant and animal species. This rich biodiversity, spread across a varying topography, provides a natural beauty that is enjoyed by millions of people each year. All is not well, however, in these ecosystems. Insects, pathogens, and plants that are alien to the region have altered and are altering the compositions of the flora and fauna. This chapter examines ecosystem changes and stresses over time, describes a selection of introduced pests, provides examples of their interactions with the ecosystem, discusses approaches for addressing the introduction and spread of exotic pests and exotic plant pests and presents a strategy for subsequent reintroduction of decimated species to southern Appalachian ecosystems.

FOREST ECOSYSTEM CHANGE AND STRESS

Southern Appalachian forests are in a constant state of change. Fire, environmental conditions, people, and pests have helped shape the current landscape. Native Americans used forest resources to support their way of life and cleared areas for villages and farming. These areas were normally near streams, on the more fertile, flat lands. Native Americans also burned forests frequently, causing tree mortality, decay, and soil erosion. Overall, the impact on the forest ecosystem was minor, and the forest quickly recovered from the disturbance (Anderson 1994).

Very early colonists had little impact on Appalachian ecosystems, but pressure on forest ecosystems increased proportional to the number of people moving to the mountains. The colonists used wood for housing, fencing, barns, wagons, tools, and fuel. Forested land was frequently cleared for farming. The number of people moving to the Appalachian region increased in the mid-to-late 1800s, rapidly changing the landscape. The new farmers cleared large areas of the forest for agricultural use, quickly depleting the soil of nutrients and eroding the upper soil layer. The remaining forest, that was typically too steep to farm, was subjected to burning, grazing, and/or highgrading (removal of the best trees). Many fires were natural in origin, but a large number of fires were deliberately set to produce woodland pasture for their cattle. In other areas, forests were harvested for activities such as pig iron production. These activities caused some forested areas to sustain major site degradation.

By the turn of the century, the exploitation of the southern Appalachian forests was at its peak. Vast areas of the forest were cut, and the practice of taking the best trees and leaving low quality trees may have reduced the genetic base. Irresponsible logging often resulted in erosion and site degradation. Some areas on steeper slopes initially escaped cutting, only to be logged when the demand for wood increased during World War I. When farms were abandoned during the Great Depression, the forests reclaimed the landscape naturally.

Conversion (or reversion) of farmlands to forests accelerated during World War II, when the nation focused on supporting the war effort. This trend was reversed after the war, when people returned to farming and received agricultural training. Improved farming practices were used by many of these farmers, but some continued the practice of exploiting the land. The forests again were burned to increase pastureland acreage for livestock, and most woodlots were heavily grazed. Farmers commonly let their hogs run wild in the woods, where they severely damaged tree roots and reproduction. Many of these hogs became wild, and their descendants continue to damage the forest. (See Peine and Lancia, Chapter 13 of this volume.) Soil loss and degradation occurred in deforested areas, while decay organisms caused tree quality losses.

Another major force, climate, continually affects the resource. Trees are affected by adverse climatic conditions such as frost, freezing injury, drought, and flooding. Local and regional conditions work together to cause multiple stresses on trees. Frost may cause the trees to defoliate and subsequently deplete starch reserves, while drought may cause stress from which the trees cannot recover, e.g., the severe drought of the 1950s is still a major factor in the health of today's forests. Finally, air pollutants, such as ozone, have increased, and studies report visible injury and potential interactions with other forces in forest settings.

Fire, environmental conditions, people and natural pests have produced a relatively even-aged forest across the southern Appalachian region. It is estimated that 75% of the forests in the region consist of mature trees in an overstocked condition. As land use and environmental conditions increased stress, a corresponding increase in forest declines, bark beetle damage, and a number of other problems has been observed.

Introduced forest pests and invasive pest plants have further augmented ecosystem alterations and have compounded the above stress levels in southern Appalachian forests. Exotic pests can be devastating as there often is no natural resistance present in host species. Coevolution of pests and hosts that allows for the existence of both organisms in the same environment has not occurred. Destruction of the American chestnut by the introduced chestnut blight fungus is the prime example of how an exotic pest can remove a dominant species from Eastern forest ecosystems within 5 decades (cf. Burnham et al. 1986).

Forest disturbance provides an opportunity for exotic plant pests to become established, although many exotic plants do not require disturbance to become established. These species are frequently more aggressive in occupying a disturbed site than native plants. Disturbances in forest ecosystems through pests, natural disasters, succession, or harvesting can offer an increased opportunity for occupation by invasive exotic plant species. The southern Appalachians are being invaded by groups of exotic trees, shrubs, vines, herbs, and grasses. Examples of these plant pests include tree-of-heaven, privet, kudzu, musk thistle, and Japanese grass. The end result can be the elimination of some native plant populations.

The pests and the associated damage are often viewed individually. In reality, they actually work as a complex, often compounded by the effects of native pests. Once established in a native ecosystem, exotic pests and exotic plant pests can be difficult, if not impossible to control or eradicate.

EXAMPLES OF EXOTIC PEST DAMAGE IN SOUTHERN APPALACHIAN ECOSYSTEMS

Southern Appalachian ecosystems have been impacted by a variety of exotic pathogens and insects for over 150 years. Pests that attack prevalent species, e.g., chestnut blight on American chestnut, have had a major effect in altering forest ecosystems. Other exotic pests that attack species with a limited occurrence, e.g., butternut canker on butternut, have had a relatively small effect on forest ecosystems. Below are some examples of exotic pest damage that have had a variable impact on southern Appalachian ecosystems.

American Chestnut Blight

The American chestnut (*Castanea dentata* [Marsh.] Borkh) was once the most important hardwood species in the eastern U.S. The species was found in vast stands from Maine to Georgia, before the turn of the century. It accounted for one quarter of all the standing timber in eastern forests. By the 1950s, however, virtually all mature American chestnuts had succumbed to the chestnut blight fungus (*Cryphonectria parasitica* [Murr.] Barr). Introduced to North America in shipments of Asian chestnut nursery stock, the chestnut blight was first recognized in New York City in 1904. From the New England region, the blight spread throughout eastern hardwood forests at a rate of approximately 24 miles per year (National Academy of Sciences 1975). Efforts by Pennsylvanian foresters to quarantine outbreaks were abandoned as a failure only 3 years later. Chestnut blight fungus also attacked Allegheny chinquapin (*Castanea pumila* Mill.), which was once a prevalent understory tree.

The American chestnut provided timber, food, and tannin that were important to early European settlers. The role of the species in wildlife food chains is often overlooked (Schlarbaum 1989). Annually the tree would produce large crops of hard mast, unlike the oaks, hickories, and other trees that have replaced the chestnut. American wildlife biology was not well developed as a science in the late 1800s and early 1900s. No surveys were conducted before or during the demise of the chestnut that could document the impact on animal species. However, historical accounts and old photographs clearly indicate that wildlife was much more abundant before the blight decimated the species.

The chestnut blight fungus did not affect root systems, although the above-ground portion of the trees were killed. American chestnut is now an understory component, existing as sprouts from old stumps with viable root systems. The sprouts are relatively short-lived (ca. 14 years), but can grow to over 50 feet in height. Allegheny chinquapin now forms small bushes that are subject to repeated die-back from blight. Observations of chestnut sprouts and chinquapin bushes indicate that there is a gradual decrease in populations in the southern Appalachians. Chestnut sprouts produce limited amounts of fruit, while chinquapin bushes can be prolific producers.

Different approaches have been used in attempts to produce a American chestnut with blight resistance. These approaches include breeding within the species (Thor 1976) and with blight resistant Asian chestnut species (Diller and Clapper 1969; Hebard 1994), and use of hypovirulent strains of the fungus (Anagnostakis 1990; MacDonald and Fulbright 1991; Brewer. 1995). A recent development is the genetic engineering of the virus that induces hypovirulence (Choi and Nuss 1992). Regardless of the approach, the end result will be a tree that is resistant to chestnut blight, but not immune. Resistant trees might survive in a forest situation, but could be more susceptible to secondary attack from other pests such as the introduced chestnut gall wasp (*Dryocosmus kuriphilus* Yasumatsu). In addition, introduction of resistant trees will encounter a problem that is paramount among all species that have been decimated: reclamation of their former niche from other species.

Balsam Wooly Adelgid

True firs (*Abies*) in North America are attacked by the balsam woolly adelgid (*Adelges picea* [Ratzeburg]). The adelgid was introduced in 1908 on European nursery stock into Maine (Kotinsky 1916). Vulnerability of North American fir species varies, depending on their ability to withstand abnormal water stress caused by premature heartwood formation which results from adelgid feeding.

Fraser fir (*Abies fraseri*) populations have been severely affected by balsam woolly adelgid. This species is endemic to high elevations in the southern Appalachians. The adelgid has eliminated mature trees from many locations. Immature trees still exist, but are attacked with increasing severity as they age. Therefore, the reproductive potential of populations has been greatly diminished, although some reproduction does still occur (Nicholas et al. 1992).

The North Carolina State Park Service tried intensive insecticide spraying at Mount Mitchell shortly after the detection of the adelgid in 1957. This technique proved effective in protecting individual trees, but was ill suited for use on a forestwide scale. The National Park Service in the Great Smoky Mountains National Park deployed an environmentally safe detergent spray to control the insect in aesthetically sensitive areas. This spray also proved too costly and labor intensive to be employed on a large scale (Johnson 1980).

Research has been conducted to find a form of biological control for the adelgid. Unfortunately, this research has produced no environmentally safe control schemes. Over a 35-year period, all apparently suitable insects were field-tested in eastern Canada. No predator(s) was found to be effective in controlling or eradicating adelgid populations. Although certain fungal diseases are known to attack the adelgid, studies in Quebec did not developed a successful control (Schooley et al. 1984). Putative resistant trees have been observed (Johnson, pers. commun.), but there are no resources to explore the use of these trees in a breeding program.

Butternut Canker

Butternut (*Juglans cinerea* L.) populations are being devastated by the exotic fungus, *Sirococcus clavigigenti-juglandacearum* (Nair, Kostichka, and Kuntz). The disease causes multiple branch and stem cankers with a characteristic black color. Death is caused by multiple trunk cankers that will eventually girdle the tree. The disease was first discovered in 1967 in southwestern Wisconsin, but is believed to have originated from the eastern coast at least 60 years ago (Anderson and LaMadeleine 1978). Butternut canker disease has gradually infected much of the species' range, but has caused the greatest decimation in southern populations. Approximately 77% of the butternut has been destroyed in southern forests (U.S. Department of Agriculture, Forest Service 1995). Currently, surviving butternuts are now usually found proximal to streams, and most trees are heavily affected (Manchester, pers. commun.). In contrast to American chestnut, butternuts killed by the disease will not sprout, young trees are subject to mortality, and even seed husks can be infected. Correspondingly, when a population becomes infected, that particular gene pool is permanently lost unless there is resistance present.

The rapid decimation of butternut populations has been considered so severe that the species was designated as Federal Category 2 candidate for listing under the Endangered Species Act (as of July 1995, the U.S. Fish and Wildlife Service no longer maintains a list of "Candidate 2" species; butternut now is a "species of great concern"). In northeastern National Forests and the National forests in Mississippi, butternut has been listed as a sensitive species. Other National Forests in the South and southern Appalachians have recommended the tree for sensitive status.

To date, there has been no proven strategy for protecting butternut trees from the disease or returning resistant butternuts to eastern forests. The survival of large butternuts thought to have canker resistance in areas where the species has been destroyed suggest that a backcross breeding approach is feasible for producing resistant trees. Additionally, a selection of Japanese walnut (*Juglans sieboldii* var. *cordiformis* [Maxim.] Rehd.), heartnut, and naturalized butternut–heartnut hybrids exhibit resistance to the disease and are potential sources of resistance in a breeding program (Orchard et al. 1982).

Pear Thrips

Sugar maple (*Acer saccharum* Marsh.) populations in some states have recently suffered severe defoliation by an introduced insect, pear thrips (*Taeniothrips inconsequens* [Uzel]). Pennsylvania foresters first mapped defoliated trees in 1979, but restudy of earlier defoliations has led scientists to believe that the thrips was causing damage in the mid-70s and probably earlier (Quimby 1990). Damage is highly variable by year and geographic location. The worst damage occurred in Pennsylvania and Vermont in 1988. Black cherry (*Prunus serotina* Ehrh.) and other trees are also eaten by the pear thrips, but the impact on these other tree species' growth rate and survival is not known.

Pear thrips infestations are found in New England and are spreading west and south. It is unknown why pear thrips, long a pest on fruit trees, have adapted to feeding upon maple leaves. Damage by pear thrips can be sporadic. In Pennsylvania, 100,000 acres of heavy defoliation occurred in 1982, none in 1983 and 1984, and 110,000 acres were defoliated in 1985 (Laudermilch 1988).

Control methods have not been developed yet. All pear thrips identified in the U.S. are females; they are believed to reproduce by parthenogenesis.

Dogwood Anthracnose

The flowering dogwood, *Cornus florida* L., is a small tree occurring in the understory of eastern North American forests. Dogwoods are important trees to the forest ecosystem. The leaves contain a large amount of calcium and act as a major soil builder. The fruit is high in protein and is a valuable food source for many migratory birds. The leaves and twigs of the dogwood provide browse for many herbivores, including deer.

In the late 1970s, a fungal disease, dogwood anthracnose, began to destroy the species in the northern U.S. The disease was first observed in 1978 in New York and had spread to the southern Appalachians by 1987 (Britton 1993). Populations at high elevations, ca. 3000'+, and populations proximal to water are particularly susceptible to damage. Symptoms include lower branch dieback, foliage blight, and twig, branch, and trunk cankers. Dogwoods below 3000 feet in elevation and in full sun or on dryer south-facing slopes are surviving and, in many cases, have good health. The loss of flowering dogwood in the forest is a serious concern.

Dogwood species and trees resistant to fungus *Discula destructiva* Redlin have been observed. Different studies have independently reported resistance to dogwood anthracnose in the Asiatic *Cornus kousa* Buerger ex Hance. A resistance-screening experiment involving seven *Cornus* species found that *C. alternifolia* L., *C. amomum* Mill., *C. kousa*, and *C. mas* L. were resistant. The most promising source of resistance, however, has been found in several flowering dogwood trees from Catoctin Mountain National Park. Using these materials, a breeding program could be initiated to develop resistant flowering dogwood genotypes adapted to different environments.

HEMLOCK WOOLLY ADELGID

The hemlock woolly adelgid was introduced into North America from Asia approximately 75 years ago (Annand 1924). In eastern forests, the pest feeds upon eastern hemlock (*Tsuga canadensis* [L.] Carr.), an important species in mountain watersheds. Although not completely understood, salivary secretions released during feeding are thought to be the causal agent of tree mortality. Mature trees can die within 4 years of infestation. Juvenile trees also are attacked. In the southeastern Appalachians, the adelgid has been found throughout Virginia to the border of northeastern Tennessee and has spread into North Carolina forests.

No resistance to hemlock woolly adelgid has been found in eastern hemlock. Resistance appears to be present in some trees of *Tsuga chinensis* (Franchet) Pritzel (Chinese hemlock) and *Tsuga diversifolia* (Maxim.) Masters (Japanese hemlock), based upon the relative degree of infestations in different hemlock species growing in the same proximity (Lewendowski and McClure, pers. commun.). Western hemlock, *Tsuga heterophylla* (Raf.) Sarg., appears to have some resistance as the insect was originally introduced to the West Coast and damage to western hemlock populations has been relatively minor. The National Arboretum currently has the only hemlock breeding program directed toward incorporating resistance into eastern hemlock cultivars used for landscape purposes (Townsend and Bentz, pers. commun.). This program was recently initiated and is limited by the number of sexually mature Asian hemlocks in this country. In addition, breeding is limited by annual variation in production of reproductive structures (USDA Forest Service). Hybridization barriers, however, do not appear to be significant factors in interspecific crossing among *Tsuga* species.

Research on biological control of hemlock woolly adelgid has been promising. A predatory beetle has been imported from Japan and released onto hemlock woolly adelgid infested trees in Connecticut (USDA APHIS 1995). Feeding by the beetle has significantly reduced populations of the pests.

GYPSY MOTH

Since the demise of the American chestnut, upland forests in the southern Appalachian mountains have been dominated by an oak–hickory species assemblage. This species composition is particularly vulnerable to attack by the European gypsy moth (*Lymantria dispar* [L.]), the most destructive insect in eastern hardwood forests. The larval stage (caterpillar) defoliates a wide variety of woody plants, although it prefers hardwood trees. Oak species are an especially favored food source. When preferred food sources are not available, older larvae feed on hemlock, cottonwood, and native eastern pines and spruces. In 1991, gypsy moth had infested an estimated 125 million acres nationwide, of which 4.1 million acres (3%) were defoliated (Burkman et al. 1993).

The insect was deliberately imported in 1869, with the hope of establishing a domestic silk industry. Within 10 years of arrival, gypsy moth was first observed as a forest pest in Massachusetts. The pest has slowly spread throughout northeastern states and then into midwestern and southern states in subsequent years.

Gypsy moth defoliation causes trees to use energy reserves in attempting to produce new leaves. A healthy tree can usually withstand several consecutive defoliations, where over 50% of the leaves are lost. The defoliation of large acres of forests by gypsy moth can have numerous secondary effects on the ecosystem (Allen and Bowersox 1989). The effects include: higher water temperatures, lower water quality, less hard mast (acorn) production with a potential for complete failure in some years, a higher proportion of woody debris in streams, altered microclimate, more patchiness within forests resulting from site-related tree mortality, and loss of nesting sites. Although there are many alternatives for gypsy moth control, forest ecosystem damage can be still severe, particularly on the advancing front where populations are perpetually high.

The gypsy moth has encountered many endemic enemies, including wasps, flies, ground beetles, ants, and many small mammals and birds. However, the naturally occurring virus, Nucleopolyhedrosis (NPV), is specific to the gypsy moth and has proved devastating to its survival. An exotic fungus (*Entomophaga maimaiga* Humber, Shimazu, & Soper) caused high mortality in gypsy moths in New England in 1989 (Hajek et al. 1995). This fungus was originally imported to control gypsy moth in 1920, but had not had a significant impact until almost 60 years after introduction. Currently, *Entomophthora maimaiga* has seriously reduced the defoliation by the gypsy moth. In 1996, there was not enough defoliation in Virginia to be mapped. Research is under way to examine the efficacy of this fungus as a biological control agent and its long-term effects.

Silvicultural techniques also are effective in reducing damage by the gypsy moth infestations. Prescriptions have been developed to reduce the risk to the stand prior to gypsy moth attack, reduce damage during an attack and reduce the risk of future attacks in areas that have been affected by gypsy moth feeding (Gottschalk 1993). Successful ecosystem restoration of an oak component after gypsy moth attack will depend upon proper management, and, if artificial regeneration is used, the ability of the planted trees to withstand occasional defoliation. Once present, gypsy moth populations are extremely resilient to total eradication. Periodic outbreaks, therefore, will occur at different locations and forest must be properly managed to minimize damage.

BEECH BARK DISEASE COMPLEX

American beech (*Fagus grandifolia* Ehrh.) populations in the southern Appalachian mountains are presently at risk. In 1993, beech bark disease was observed in the Great Smoky Mountains National Park (Rhea, pers. commun.). This potent disease/insect complex affects the American beech throughout much of its range in Northern hardwood forests. It is caused by an exotic *Nectria* fungus that gains entry into the tree through tiny holes in the bark caused by the feeding of the exotic beech scale (*Cryptococcus fagisuga* Lindinger). Both the scale and the fungus are thought to be exotic, brought to Nova Scotia from Europe around 1890. Diseased trees were first observed at Halifax, NS in 1920. In many northeastern stands, the disease has killed over 50% of the beech. Spread by wind, birds, and other insects, the scale and the fungus are moving steadily south and west. Trees infected by the fungus, but not killed, are suitable only for wood chips. The decline of beech populations has also affect wildlife populations, particularly the black bear, that feed upon the beech nuts.

Trees resistant to beech bark disease have been identified, but only in New England, where the disease pressure has been intense for a long period of time (Shigo 1964; Cammermeyer 1993). Information is needed on beech bark disease presence in the southern Appalachian mountains, so that affected areas can be monitored for potentially resistant trees and so that the disease spread can be monitored.

EXAMPLES OF EXOTIC PLANT PEST INVASIONS IN A SOUTHERN APPALACHIAN NATIONAL PARK

The Great Smoky Mountains National Park (hereinafter referred to as "the Park") encompasses approximately 520,000 acres in the heart of the southern Appalachian region. The Park is considered to be one of the "crown jewels" in the National Park System, attracting more than 10 million visits each year. It is renowned for possessing a rich diversity of animal and plant species and contains areas of truly virgin forests that have never been harvested. In recent years, this Park has encountered threats to its renowned natural values, especially the long-term threats posed by the gamut of introduced organisms, including nonindigenous plants.

Of the more than 300 exotic plant species found in the Park, 25 are considered to pose significant threats to Park resources. Among these are kudzu (*Pueraria montana* [Willd.] Ohwi), Japanese honeysuckle (*Lonicera japonica* Thunb.), oriental bittersweet (*Celastrus orbiculatus* Thunb.), Japanese grass (*Microstegium vimineum* [Trin.] A. Camus), Japanese barberry (*Berberis thunbergii* D. C.), periwinkle (*Vinca minor* L.), Tree-of-Heaven (*Ailanthus altissima* [Mill.] Swingel), privit (*Ligustrun vulgare* L.) multi-flora rose (*Rosa multiflora* Thunb. ex Murray), musk thistle (*Cardaria nutans* L.), Johnsongrass (*Solanum halepense* [L.] Pers.) and mimosa (*Albizia julibrissin* Durazz.).

The problem-causing species in the Park are typical of the region. Six species that were identified as among the most prevalent exotic trees, shrubs, vines and grasses in forests of southern U.S. (Miller and Street 1995) cause problems in the Park. These include mimosa, privet, kudzu, oriental bittersweet, Japanese barberry and tree-of-heaven. Due to its great topographic relief, the Park is invaded by the latter three species, which are considered as typical invasive species of more northern ecosystems.

The Park has good information about the extent and location of exotic species infestations, due to the considerable ecological research that has been conducted within its' perimeter. For example, surveys indicate that the number of princess trees (*Paulownia tomentosa* [Thunb.] Sib. & Zucc. ex Steud.) on the western (Tennessee) side of the Park increased from 8 in 1975 to more than 1000 in 1993. The increase may be related to efforts to restore fire-dependent communities through prescribed burns. These events are thought to open up additional areas to invasion by such exotic species as princess tree or tree-of-heaven.

Eradication efforts have focused on known and newly discovered sites for 30 target species. Manual eradication was used on over 7700 plants of musk thistle. The Park staff is also experimenting with techniques for killing plant species for which no acceptable technique had been developed. Among the experimental control strategies are use of brush cutters, followed by herbicidal treatment of the stumps for large areas of privet, use of a new herbicidal soap on Japanese honeysuckle and vinca, and varying concentrations of herbicide for treatment of stumps of mimosa. Mimosa has proved the most difficult species to eradicate.

According to Tom Remaley (pers. commun.), privet is one of the most abundant exotic plants in Great Smoky Mountains National Park. The species invades many habitats, tolerating a wide range of environmental conditions. It is found throughout the Park at old home sites, fence rows, disturbed areas, and forest edges. Privet quickly forms dense stands that crowd out native shrubs. Individual plants can grow to 18 feet in height. It can persist for years on sites after canopy closure. Privet readily reproduces from vegetative root sprouts. Dissemination is further aided by a prolific production of fruits that are eaten by birds, ensuring widespread dissemination. Seeds in soil remain viable for at least 10 years. Without control measures, privet can form thickets encompassing several areas. As of 1995, Park staff were actively controlling 58 privet sites, totaling more than 38,000 square meters. Surveys continue to locate previously undiscovered privet thickets.

Privet and other invasive shrubs are extremely difficult to manage because of dispersal into natural areas and the need for a combination of manual and chemical eradication. Each stem must be cut, stump treated, and cleared. Park crews spent 512 hours in 1995 eradicating privet, a "limited" response due to the need to manage 30 other exotic plant species.

Another problem shrub in the Park is the well-known invader of old fields and edge habitats, multiflora rose. This thorny shrub is an aggressive competitor with natural vegetation and an unwelcome denizen to natural resource recreationists. A possible biological control for this invasive shrub is rose rosette disease that is spread by mites (Epstein and Hill 1995). This disease spreads to the roots, and causes mortality usually after 1 to 2 years (Amerine and Hindal 1988). Unfortunately, hybrid ornamental roses are also vulnerable to rose rosette disease. Another biological control, rose seed chalcid, is expected to eventually infest almost 90% of multiflora rose seeds in the midwest States (Amerine et al. 1993).

Honeysuckle is one of several problem exotic vines in the Park. Honeysuckle is frequent in occurrence and forms dense mats in open and disturbed areas and edge habitat. Fortunately, establishment of this species in mature forests is limited by shading (Robertson et al. 1994). Honeysuckle prevents growth of native herbaceous species and tree seedlings and can cover mature trees (MacDonald et al. 1989). Individual plants can recover quickly from herbicide treatments, making control difficult and expensive (Cain 1992). This exotic honeysuckle is either less preferred by native insects and mammals than is the native honeysuckle (*Lonicerus sempervirens*), or is more aggressive in growth after feeding damage (Schierenbeck et al. 1994).

Two other exotic species, oriental bittersweet and tree-of-heaven, cause other problems separate from invasion. Oriental bittersweet can hybridize with the native bittersweet (*Celastrus scandens*), thereby causing a loss of species identity. Tree-of-heaven produces chemicals that strongly inhibit seed germination and seedling growth of other species (Heisey 1990). These chemicals probably assist the tree in occupying and dominating sites.

The Park staff is currently developing an integrated pest management (IPM) plan that will specify control strategies for individual plant species. The plan is being developed in cooperation with the Blue Ridge Parkway, that has already developed an IPM plan and strategies for several invasive species. However, the current control efforts of manual eradication and/or herbicide spraying require attacking each individual plant, a slow and costly procedure. Development of biological controls might be applicable to exotic plant species, although there are some instances where biological control would not be generally acceptable. Some invasive, exotic species are distributed by the ornamental nursery industry, and other species, e.g., oriental bittersweet and honeysuckle, are closely related to native species. Biological controls are unlikely to be sufficiently narrowly targeted to be acceptable for eradication of these species.

IMPACTS OF EXOTIC PESTS AND PLANT PESTS ON HABITAT

Exotic pests can be species specific in host selection, e.g., butternut canker disease, or affect a broad spectrum of species, e.g., gypsy moth. The loss of a species or devastation of a particular habitat, e.g., hemlock glade, can have dramatic ramifications on flora and fauna associated with the species or dependent upon the species for food or habitat.

For example, the demise of the American chestnut left large areas of the southern Appalachian forests without the dominant species. In some areas, the American chestnut and Allegheny chinquapin (also susceptible to chestnut blight) were extremely dense. Forest domination by chestnut in the southern Appalachians began to cease in the late 1920s, as the chestnut blight spread in the region. The demise left areas open to colonization, primarily by the upland oak–hickory assemblage. The loss of chestnut severely impacted the food chain in the region. Although records on black bear populations were not kept before or during the American chestnut demise, Pelton (pers. commun.) believes that the loss of this reliable source of mast dramatically reduced bear populations.

The impacts of the chestnut blight are still affecting present-day upland forests in the southern Appalachian region. Maintaining an oak component in eastern hardwood forests is currently a significant management problem. Within mixed hardwood forests of the southern Appalachians, a species such as northern red oak does not consistently regenerate after an even age harvest. Oak decline has been ascribed to a variety of causes ranging from air pollution to forest pests. Many of the oaks, however, colonized old chestnut sites, sites for which oaks may not be particularly adapted and are, correspondingly, entering into a state of decline.

Forest devastation can affect animal species in different ways. Animal species that are relatively mobile, e.g., birds and large mammals, may suffer population reductions, but can often migrate to more suitable environments. However, loss of habitat can dramatically affect animal species not capable of fast migration and plant species. In southern Appalachian spruce–fir forests, the spruce–fir moss spider (*Microbexura montivaga*) and rock gnome lichen (*Gymnoderma lineare*) were listed as Endangered Species (*Fed. Reg.*, January 18, 1995; February 6, 1995) due to the decline of Fraser fir and red spruce (*Picea rubens* Sarg.) forests. This forest type once cloaked southern Appalachian mountaintops, but the decline has exposed these formerly wet habitats to the drying effects of the sun. Correspondingly, unique species endemic to this forest type are declining toward extinction.

The above case histories involving American chestnut and the Appalachian spruce–fir forest could be written for each exotic pest, albeit on a smaller scale. The combined effects of species/genus specific exotic pests, e.g., dogwood anthracnose disease, hemlock woolly adelgid, or beech bark disease, on flora and fauna associated with the host species/genus will have a significant effect on biodiversity, food chains, visual appearance, etc. of southern Appalachian ecosystems, as the pests can affect different plant/animal communities. For example, dogwood anthracnose disease removes a prevalent understory species. Beech bark disease will heavily impact wildlife species (and food chains) that relay upon hard mast in their diets. Hemlocks grow in the riparian zones, i.e., streams and rivers, that harbor a number of plant and animal species with very specific environmental requirements (Lapin 1994). The loss of hemlock from southern Appalachian ecosystems also will directly affect water quality through increased debris and lack of shade and flow characteristics through downed logs in streams.

Exotic plant pests can be habitat specific or are able to occupy a wide range of sites. Exotic plant pests have not yet caused obvious alterations in southern Appalachian ecosystems, except in localized areas. Correspondingly, there has not been significant reductions in the economic value of these ecosystems from a timber harvesting, recreation, or tourism perspective. Consequently, relatively few studies have examined alternative methods to manual eradication or herbicide spraying of individual plants to control these invasive species. The potential does exist for eventual problems, however, as demonstrated by the invasion and dominance of *Melaleuca quinquenervia* in the Florida everglades.

THE MECHANICS OF ECOSYSTEM PROTECTION AND SPECIES REINTRODUCTION

Southern Appalachian ecosystems will always exist; native and exotic species will fill empty niches of species that have been decimated or species that cannot compete. The degree of ecosystem change due to exotic species will be related to efforts that prevent introduction or efforts to eradicate/control these species, coupled with subsequent reintroduction of affected, native species.

There is a need for long-term programs committed to ecosystem protection and reintroduction of species that have been decimated or displaced by exotic pests or exotic plants. The USDA Animal and Plant Health Inspection Service (APHIS) is charged with preventing the introduction into the U.S. of such pests and of invasive plant species (called "noxious weeds"). The USDA Forest Service is the primary federal agency involved in eradication and suppression of harmful exotic forest pests. State departments of agriculture work with APHIS and the USDA Forest Service to address exotic pest and plant pest problems. The reintroduction of a species destroyed by exotic pests or reestablishment of a species to a site dominated by exotic plants may involve state, federal, and private efforts, singly or in combination.

Ecosystem Protection

APHIS has traditionally concentrated on the prevention of new agronomic and horticultural crop-related pests. The first general plant health (phytosanitary) regulations that specifically govern timber imports were adopted in 1995. Perhaps the greatest weakness, however, is the inadequate numbers of APHIS inspectors to conduct thorough examinations of imported materials, vessels, and packing materials. Consequently, it is questionable whether the regulations will prove adequate to protect against additional introductions of foreign pests.

Prevention of exotic plant pest establishment is problematic under existing regulations. Unlike exotic insects or pathogens, APHIS can restrict the import of plant species only after they have been listed as a noxious weed under the provisions of the federal Noxious Weed Act. At present, that list includes only one species threatening natural ecosystems, *Melaleuca quinquenervia*.

The responsibility for curtailing the spread of weeds already in the country is shared among APHIS, state departments of agriculture, such land-managing agencies as the USDA Forest Service and National Park Service, and private landowners. APHIS lacks authority to eradicate outbreaks of foreign weeds when they are first identified in the country. This responsibility is usually addressed by a combination of state and federal efforts. Recently increased concern about the impact of invasive plant species by citizen groups has led to proposals to strengthen APHIS' authority and development of a strategy that aims at a more aggressive implementation of the law with respect to exotic plants that invade natural areas.

The USDA Forest Service has assumed principal responsibility from APHIS for containing the spread of forest-based exotic pests and exotic plant pests already in the country. The USDA Forest Service may conduct research and experiments to obtain, analyze, develop, demonstrate, and disseminate scientific information about protecting and managing forests for a multitude of purposes, under the auspices of the Forest and Rangeland Renewable Resources Research Act. Forest protection specifically includes addressing insect and disease problems. A second statute, the Cooperative Forestry Assistance Act, authorizes the USDA Forest Service to protect from insects and diseases trees and wood products in use on National Forests or, in cooperation with others, on other lands in the U.S. Such assistance may include surveys and determination and organization of control methods. The USDA Forest Service is further authorized to provide assistance to state foresters in the eradication and suppression of forest pests.

SPECIES REINTRODUCTION

Despite the best prevention and control efforts of state and federal agencies, it is probable that exotic pests and plant pests will continue to be introduced to this country and become established. Correspondingly, strategies for introduction of host species (in the case of exotic pests) or displaced species (by exotic plant pest species) eventually will need to be developed. These strategies can include three components: genetic conservation, pest control, and restoration protocols.

GENETIC CONSERVATION

Genetic conservation strategies can be elaborate and expensive to enact or restricted to a simple collection of seed followed by storage in an adequate facility. Conservation of a species' genetic diversity is related to the severity of decimation or displacement. It is essential when a species threatened with extinction by a exotic pest or plant pest. Unfortunately, genetic conservation is often neglected when a pest problem is recognized. Resources are usually diverted to pest control or monitoring the problem. Genetic conservation is a low-profile but necessary activity to insure the maintenance of genetic diversity in a threatened or endangered species.

The relative emphasis of genetic conservation activities will be dictated by priority level and available resources. Seed collection and storage is relatively inexpensive and would insure preservation of endangered populations. A systematic sampling of surviving populations of the affected species will help insure preservation of the bulk of genetic diversity. Seed collections from isolated or unique populations of the species should receive a high priority. These populations may contain rare alleles, and preservation would be highly desirable. Seed can be stored at the National Seed Storage Laboratory in Ft. Collins, CO (U.S. National Germplasm System 1991).

Additional conservation activities require greater commitment of resources. Biochemical studies to insure the bulk of diversity has been conserved, *ex situ* plantings in protected areas for seed production to restore devastated populations, and/or management of *in situ* populations to insure continuity are components of more sophisticated genetic conservation programs.

Unfortunately, gene conservation of southern Appalachian species threatened by exotic pests has been minimal. For example, the first genetic conservation planting of Fraser fir by the Great Smoky Mountains National Park was established in 1995, 37 years after the balsam woolly adelgid was first discovered in the region. Seed collections of flowering dogwoods and American beech have not been made despite the impending loss of whole populations at high elevations. Butternut has been nearly eliminated from the region, yet no systematic effort has been to locate surviving trees for possible seed collection. Unfortunately, tree species as above provide the most visible examples of genetic loss. Flora and fauna associated with the trees will also be impacted and should be considered for conservation efforts.

PEST CONTROL

Species restoration without pest control is futile. There are three major approaches to controlling pests: chemical, cultural, and biological, including incorporation of genetic resistance. Control measures are usually directed toward local suppression or eradication. Once a pest has been established, it is difficult to accomplish total eradication.

Chemical controls can be very effective, but often affect nontarget species. For example, some insecticides provide the best chemical control for gypsy moth infestations with high population levels. Unfortunately, these insecticides can affect both terrestrial and aquatic fauna. Proper use of chemical control should balance the damage to the environment by the chemical against the damage wrought by the pest.

Cultural control can be very effective at minimizing damage by certain pests. Reducing the population levels of the host species, altering species composition, planting more vigorous stock, e.g., genetically improved stock, and manual removal are cultural options. Gypsy moth damage can be minimized through silvicultural treatments prior to, during, and after attack (Gottschalk 1993). The Forest Service now has computer programs to aid in management decision for gypsy moth-related problems.

Biological control encompasses the use of a pathogen, virus, arthropod, or genetic resistance to allow the host species to live and reproduce under reduced levels or absence of the pest. The biocontrol agent may be native or exotic in origin. There are a number of examples of biocontrol of exotic pests through the use of insects or pathogens.

Breeding for genetic resistance is often not considered a biocontrol option. Nevertheless, in the instances where there is natural resistance to a pest or resistance in a close relative, genetic resistance breeding is a viable option. In addition, this approach has an advantage over other options in that it effects biocontrol while it is restoring propagules of the host species to the landscape. Other biocontrol options may solve the pest problem, but do nothing toward restoration of populations in devastated areas.

Resistance breeding begins with an accumulation of susceptible and putative or proven resistant genotypes. Simultaneous construction of breeding orchards and progeny testing should occur. Delay of breeding orchard construction until the progeny tests identify resistant genotypes would further delay reforestation with hemlock woolly adelgid resistant genotypes. The breeding orchards should be designed to include clones of putative resistant materials. The orchards can be thinned if resistance is not found in all materials. Inclusion of all materials at the initiation of the breeding orchard will help insure that sexually mature materials will be available for breeding when the progeny tests conclusively identify resistant genotypes.

The breeding scheme to incorporate resistance depends on source of resistance, e.g., biochemical or mechanical, and mode of inheritance. Resistance characterization depends upon reliable and appropriate evaluation procedures that consider pest pressure and host response.

Traditional tree improvement programs, i.e., breeding and field testing, are structured to provide materials for artificial regeneration. Development of resistant species or conservation of a specific population are routine activities of these programs. Tree improvement research and development requires a long-term commitment in personnel, resources, land base, and operating funds. Unfortunately, these programs are being seriously reduced from federal, state, and university research agendas, due to the associated expenses and reduced opportunities for extramural funding (Schlarbaum 1995). In addition, tree improvement activities generally are not encompassed in the infrastructure of a number of agencies, including the National Park Service.

REINTRODUCTION PROTOCOLS

Successful control of the pest allows for the existence or return of a species to a location. Species reintroduction may be as simple as letting existing genotypes multiply and spread or as complicated as reclaiming the site from successional vegetation. For populations devastated by a pest, i.e., reproductive potential is loss, the importance of previous genetic conservation activities is germane to species reintroduction. Without *a priori* genetic conservation, whole populations are lost forever. It may be possible to restore a species to a specific location; however, the seed source may not be ideal for successful survival and reproduction.

THE FUTURE

Unless proper strategies of prevention, eradication, and control of exotic pests are initiated, southern Appalachian ecosystems will continue to change. The progressively increasing ease of intercontinental movement requires a constant examination and revision of regulations and policies that govern importation of goods into this country that might harbor exotic pests. Importation of exotic species for the ornamental industry also needs to receive greater scrutiny to prevent "escapes" of invasive, exotic plant species. As indicated above, addressing the pest problem is only part of the solution to restoring a healthy, natural ecosystem. Reintroduction of decimated or displaced species that are locally adapted requires extensive planning and commitment. Unfortunately, the concentration of state and federal agencies has been on solving immediate pest problems, with little emphasis on repairing the damage to the ecosystem.

Exotic pests could dramatically increase if timber is imported from other countries to meet future demands. Lumber or fiber shortages in this country could promote the importation of timber from other countries. Risk assessments for importations of logs from Siberia (USDA Forest Service 1991), New Zealand (USDA Forest Service 1992) and Chile (USDA Forest Service 1993) stress the potential for introducing large numbers of exotic pests that could further devastate forest ecosystems. Solutions must be found to meet the wood and fiber needs of the U.S. without further risking our forest ecosystems, including the southern Appalachian region, to addition invasions from exotic pests and exotic plant pests.

RISK ASSESSMENT REPORT CARD

The control of pests and pathogens is one of the most difficult tasks facing the ecosystem manager as reflected by the low grades associated with the *Risk Assessment Report Card for Managers* assigned for this topic.

Vision: D

Not much systematic effort is going into documenting the genetic diversity of threatened species. As is documented above, genetic material of species threatened by the invasion of pests and pathogens is not being systematically protected. No public agency is taking responsibility for this most critical step. Genetic diversity of Frasier fir, dogwood, butternut and eastern hemlock is likely being lost in the southern Appalachians.

Resource risk: F

Several native species at the individual and/or community level are being displaced and their genetic diversity is likely being lost.

Resource conflicts: A

Control of exotic pests and pathogens is invariably good for the ecosystem and native species.

Socioeconomic conflicts: B to D

For gypsy moth control, there is a conflict whether or not to spray insecticides to control infestation due to the threats to nontargeted species such as other butterflies and moths. Generally, however, there is little conflict associated with control programs.

Procedural protocols: B to F

Again a mixed bag. Good for gypsy moth but not developed for most.

Scientific validity: B to F

Extreme variability in understanding the nature of the pest or pathogen, the impacts on the environment and methods of control.

Legal jeopardy: A

There is not much legal challenge to efforts to control pests and pathogens.

Public support: A

There is generally strong public support for control programs, especially for high profile problems such as gypsy moth and dogwood anthracnose.

Adequacy of funding: F–

There are known solutions through genetic resistance breeding for many of the exotic forest insects and pathogens. However, there is no real funding for such applied work. Nor is there much of an improvement infrastructure, i.e., state and federal programs left to execute long-term breeding programs on most of the species affected.

Policy precedent: B to F

Action by public agencies to control exotic pests and pathogens are supported by public policy.

Administrative support: B to D

Again a mixed bag here. The efforts to control exotic species has been very aggressively pursued for some pests and not so for others. The list is too long and the resources too limited to cover all the bases.

Transferability: A

Research into pests and pathogens and control procedures for them tend to be readily transferable.

REFERENCES

Allen, D. and T. W. Bowersox. 1989. Regeneration in oak stands following gypsy moth defoliations. In Proc. 7th Central Hardwood Conf., G. Rink and C. A. Budelsky, Eds., Gen. Tech. Rep. NC-132. U.S. Department of Agriculture, Forest Service, North Central Experiment Station. pp. 67–73.

Amerine, J. W., Jr. and D. F. Hindal. 1988. Rose rosette: a fatal disease of multiflora rose. Agricultural and Forestry Experiment Station, West Virginia University, 1988, Circular No. 147, 4 pp.

Amerine, J. W. Jr., T. A. Stansny, and B. N. McKnight, 1993. In Biocontrol of multiflora rose. Biological Pollution: the Control and Impact of Invasive Exotic Species. Proc. symp. held at Indianapolis, IN, October 25 to 26, 1991. pp. 9–21.

Anagnostakis, S. L. 1990. Improved chestnut tree condition maintained in two Connecticut plots after treatments with hypovirulent strains of the chestnut blight fungus. *For. Sci.* 36: 113–124.

Anagnostakis, S. L. and B. Hillman. 1992. Evolution of the chestnut tree and its blight. *Arnoldia* 52: 2–10. s. 24: 1058–1062.

Anderson, R. L. and L. A. LaMadeleine. 1978. The distribution of butternut decline in the eastern United States. USDA-Forest Service, Northeastern Area, State and Private Forestry, Rep. S-3-78. 4 p.

Anderson, R. L. 1994. How people, pests, and the environment have changed and continue to change, the southern Appalachian forest landscape. *In* Threats to Forest Health in the Southern Appalachians. C. Ferguson and P. Bowman, Eds., Southern Appalachian Man and the Biosphere Cooperative Workshop, Chattanooga, TN, February, 8, 1994.

Annand, P. N. 1924. A new species of *Adelges* (Hemiptera, Phylloxeridae). *Pan-Pac. Entomol.* 1: 79–82.

Boyce, J. S. 1961. *Forest Pathology,* 3rd ed., McGraw-Hill: New York.

Brewer, L. G. 1995. Ecology of survival and recovery from blight in American chestnut trees (*Castanea dentata* [Marsh.] Borkh.) in Michigan. *Bull. Torrey Bot. Club.* 122: 40–57.

Britton, K. O. 1993. Anthracnose infection of dogwood seedlings exposed to natural inoculum in western North Carolina. *Plant Dis.* 77: 34–37.

Burkman, W. Q., R. Chavez, S. Cooke, S. Cox, S. DeLost, T. Luther, M. Mielke, M. Miller-Weeks, F. Peterson, M. Roberts, P. Seve, and D. Trawdus. 1993. Northeastern Area Forest Health Report. NA-TP-03-93. U.S. Department of Agriculture, Forest Service Northeastern Area State and Private Forestry, Radnor, PA.

Burnham, C. R., P. A. Rutter, and D. W. French. 1986. Breeding blight-resistant chestnuts. *Plant Breed. Rev.* 4: 347–397.

Cain, M. O. 1992. Japanese honeysuckle in uneven-aged pine stands: problems with natural regeneration. *Proc. 45th Annu. Meeting of the Southern Weed Science Society.* 1992. pp. 264–269.

Cammermeyer, J. 1993. Life's a beech — & then you die. *Am. For.* July/August 1993, pp. 20–21, 46.

Campbell, F. T. and S. E. Schlarbaum. 1994. Fading forests — North American trees and the threat of exotic species. Natural Resources Defense Council, Washington, D.C., 47 pp.

Choi, G. H. and D. L. Nuss. 1992. Hypovirulence of chestnut blight fungus conferred by an infectious viral cDNA. *Science* 257: 800–803.

Diller, J. D. and R. E. Clapper. 1969. Asiatic and hybrid chestnut trees in the eastern United States. *J. For.* 67: 328–331.

Epstein, A. H. and J. H. Hill. 1995. The biology of rose rosette disease: a mite-associated disease of uncertain aetiology. *J. Phytopath.* 143: 350–360.

Ferguson, C. and P. Bowman, Eds., Southern Appalachian man and the Biosphere Cooperative Workshop, Chattanooga, TN, February, 8, 1994.

Gottschalk, K. W. 1993. Silvicultural guidelines for forest stands threatened by gypsy moth. USDA Forest Service, Northeastern Forest Experimental Station, Gen. Tech. REp. NE-171. 50 pp.

Hajek, A. E., L. Butler, and M. M. Wheeler. 1995. Laboratory bioassays testing the host range of the gypsy moth fungal pathogen *Entomophaga maimaiga*. *Biol. Control* 5: 520–544.

Hebard, F. 1994. The American Chestnut Foundation breeding plan: beginning and intermediate steps. Proc. Intl. Chestnut Symp., Morgantown, WV, July 10 to 14, 1992, pp. 70–73.

Heisey, R. M. 1990. Evidence for allelopathy by tree-of-heaven (Ailanthus altissima). *J. Chem. Ecol.* 16: 2039–2055.

Johnson, K. D. Personal communication to S. E. Schlarbaum.

Johnson, K. D. 1980. Fraser fir and woolly balsam adelgid: a summary of information. Southern Appalachian Research/Resource Management Cooperative, Western Carolina University, Cullowhee, NC. 62 p.

Kotinsky, J. 1916. The European fir trunk louse, *Chermes* (*Dreyfusia*) *piceae* (Ratz.). *Entomol. Proc. Soc. Washington* 18: 14–16.

Lapin, B. 1994. The impact of hemlock woolly adelgid on resources in the lower Connecticut River Valley. Report to USDA-Forest Service, Northeastern Center for Forest Health Research, Hamden, CT. 45 pp.

Laudermilch, G. In Parker, B. L., M. Skinner, and H. B. Teillon, Eds. Proc., Regional Meeting: The 1988 Thrips Infestation of Sugar Maple. 1988. Bennington, VT June 23, 1988. Vt. Agric. Exp. Stn. Bull. 696.

Lewandoski, R. D., personal communication to S. E. Schlarbaum.

MacDonald, I. A. W., L. L. Loope, M. B. Usher, and O. Haemin. 1989. Wildlife conservation and invasion of nature reserves by introduced species: a global perspective. In Biological Invasions: A Global Perspective, Drake, J.A., H.A. Mooney, F. diCastri, R.H. Groves, F.J. Kruger, M. Rejmanek, and M. Williamson, comps. SCOPE 37 (Scientific Committee on Problems of the Environment).

MacDonald, W. L. and D. W. Fulbright. 1991. Biological control of chestnut blight: use and limitations of transmissible hypovirulence. *Plant Dis.* 75: 656–661.

McClure, M. S., personal communication to S. E. Schlarbaum.

Miller, J. H. and J. E. Street. 1995. Exotic plants in southern forests: their nature and control. Proc. 48th Annu. Meeting of the Southern Weed Science Society, Memphis, TN, January 16 to 18, 1995, pp. 120–126.

National Academy of Sciences. 1975. *Forest Pest Control.* Washington, D.C.

Nicholas, N. S., S. M. Zedaker, C. Eagar, and F. T. Bonner. 1992. Seedling recruitment and stand regeneration in spruce–fir forests of the Great Smoky Mountains. *Bull. Torrey Bot. Club* 119: 289–299.

Orchard, L. P., J. E. Kuntz, and K. F. Kessler. 1982. Reactions of *Juglans* species to butternut canker and implications for disease resistance. USDA Forest Service Gen. Tech. Report NC-74 p. 27–31.

Payne, J. A., A. S. Menke, and P. M. Schroeder. 1975. *Dryocosmus kuriphilis* Yasumatsu, (Hymenoptera: Cynipidae), an oriental chestnut gall wasp in North America. *U.S. Dept. Agric. Coop. Econ. Insect Rep.* 25(49–52): 903–905.

Quimby, J. 1990. Historical summary of pear thrips in Pennsylvania. In Parker B. L., M. Skinner, S. H. Wilmot, and D. Souto, Eds. Pear Thrips Research and Management: Current Methods and Future Plans. *Vermont Agric Exp. Stn. Bull. 697,* p. 27.

Remaley, Tom, Great Smoky Mountains National Park, personal communication to Faith Campbell.

Rhea, J. R., USDA Forest Service, personal communication to S. E. Schlarbaum.

Robertson, D. J., M. C. Robertson, and T. Tague. 1994. Colonization dynamics of four exotic plants in a northern Piedmont natural area. *Bull. Torrey Bot. Club,* 121:2, 107–118.

Schierenbeck, K. A., R. N. Maci, and R. R. Shariat. 1994. Effects of herbivory on growth and biomass allocation in native and introduced species of *Lonicera. Ecology* 75: 1661–1672.

Schlarbaum, S. E. 1989. Returning the American chestnut to eastern North America. Proc. Southern Appalachian Mast Management Workshop. pp. 66–70. Knoxville, TN, August 14–16, 1989.

Schlarbaum, S. E. 1995. Exotic pests in North America: what has been done and what can be done. Proc. Northeastern Forest Pest Council and 27th Annu. Northeastern Forest Insect Work Conference, W. D. Ostrofsky, comp. pp. 12–15.

Schooley, H. O., J. W. E. Harris, and B. Pendrel. 1984. *Adelges piceae* (Ratz.), Balsam woolly adeldgid (Homoptera: Adelgidae). In J. S. Kelleher and M. A. Hulme, Eds. Biological Control Programmes against Insects and Weeds in Canada 1969–1980, Commonwealth Agricultural Bureaux, England, 1984.

Shigo, A. L. 1964. Organism interactions in the beech bark disease. *Phytopathology* 54: 263–269.

Thor, E. 1976. Tree Breeding at the University of Tennessee 1959–1975. *Univ. Tenn. Agric. Exp. Stn. Bull. 554.* 48 pp.

U.S. Department of Agriculture — Animal and Plant Health Inspection Service. 1995. Filed release of a nonindigenous lady beetle, *Pseudoscymnus* sp. (Coleoptera: Coccinellidae), for biological control of hemlock woolly adelgid, *Adelges tsugae* (Homoptera: Adelgidae). Environmental Assessment, April 1995. 7 pp.

U.S. Department of Agriculture, Forest Service. 1991. Pest risk assessment of the importation of larch from Siberia and the Soviet Far East. Misc. Publ. No. 1495.

U.S. Department of Agriculture, Forest Service. 1992. Pest risk assessment of the importation of *Pinus radiata* and Douglas-fir logs from New Zealand. Misc. Publ. No. 1508.

U.S. Department of Agriculture, Forest Service. 1993. Pest risk assessment of the importation of *Pinus radiata*, *Nothofagus dombeyi*, and *Laurelia philippiana* logs from Chile. Misc. Publ. No. 1517.

U.S. Department of Agriculture, Forest Service. 1995. Forest insect and disease conditions in the United States 1994. USDA Forest Service, Forest Pest Management, Washington, D.C., October, 1995. 74 p.

U.S. National Plant Germplasm System, Committee on Managing Global Genetic Resources: Agricultural Imperatives. 1991. Managing global genetic resources. National Academy Press: Washington, D.C. 171 pp.

15 Air Quality Management: A Policy Perspective

John D. Peine, Leslie Cox Montgomery, Barry R. Stephens,
William R. Miller III, Brian J. Morton, and Karen A. Malkin

CONTENTS

Air As a Natural Resource..307
    The Role of Air Quality in Ecosystem Management...307
Principles of Air Quality Management..308
    Responsibilities of the Ecosystem Manager...308
    Regulating Sources of Air Pollution...308
    Inventorying Sources of Air Pollution..309
    Modeling the Transport of Airborne Pollutants..310
    Monitoring Air Quality 310
    Defining Air Quality-Related Values 310
    Supporting Research to Define Adverse Effects from Air Pollution 311
    Integrating Functions in the Management Process 311
Science and the Federal Land Manager 312
Southern Appalachian Mountains Initiative 313
    The Organizational Structure and Function..313
    Effectiveness of the Decision-Making Process 317
        Perspectives of a SAMI Member Representing a State Government Regulator 317
        Perspectives of a SAMI Member Representing a Federal Land Management Agency 318
        Perspectives of a SAMI Member Representing a Voluntary Association Dedicated
        to Environmental Conservation..319
        Perspective of a SAMI Member Representing Industry 321
New Source Permit Application Reviews for Great Smoky Mountains Park 322
    Number and Nature of PSD Permit Applications 322
    Eastman Chemical Company Permit Application 323
    Tenn Luttrell Permit Application 323
Conclusions 324
    Risk Assessment Report Card for Ecosystem Managers 324
References 326

AIR AS A NATURAL RESOURCE

THE ROLE OF AIR QUALITY IN ECOSYSTEM MANAGEMENT

Not until air pollution was determined to be a serious problem in natural areas was the tropospheric atmosphere recognized as a natural resource of equal importance to terrestrial and aquatic natural resources. Ecosystem managers are beginning to recognize the inextricable connection between good air quality and a healthy natural environment. In fact, air pollution has been ranked as a primary problem concerning natural resources and the quality of the visitor experience at many national parks in the U.S., ranging from Yosemite and Grand Canyon in the west to Shenandoah and Great Smokies in the east.

For the sake of this discussion, "air" refers to the mixture of odorless and tasteless gasses of nitrogen and oxygen that surrounds the earth, its movement patterns, and the plethora of other gasses, liquids, and particulate matter that are carried in it from natural and anthropogenic sources. The sun, via the atmosphere and associated climate conditions, provides the primary source of energy for ecosystem processes. The effect of this energy driver for ecosystem dynamics and sensitive species can be greatly impacted by the condition of the atmosphere. Thus, pollutants in the atmosphere can have significant adverse indirect effects on ecosystem processes and direct effects on vulnerable species, most of which are commonly found on the landscape.

PRINCIPLES OF AIR QUALITY MANAGEMENT

RESPONSIBILITIES OF THE ECOSYSTEM MANAGER

As illustrated elsewhere in this volume, the myriad of challenges facing the enlightened manager of natural ecosystems is daunting indeed. Air pollution is oftentimes off the chart of assumed responsibilities. The topic is foreign to the typical disciplines associated with natural resources management. The lack of a sense of control over the issue and its complexity tend to discourage aggressive involvement by land managers. However, the ecosystem manager should provide a balanced stewardship for the triad of ecosystem elements; e.g., land, water, *and* air-associated natural resources.

REGULATING SOURCES OF AIR POLLUTION

The 1970 Clean Air Act (CAA), as amended in 1977 and 1990, provides the framework for air pollution control policy in the U.S. [42 U.S.C. § 7401 (b)(1)]. The primary mechanism to "protect and enhance" the nation's air is the National Ambient Air Quality Standards (NAAQS), which set pollutant concentration levels for six atmospheric pollutants: sulfur dioxide, ozone, nitrogen dioxide, particulate matter less than 10 $\mu g/m^3$, lead, and carbon monoxide. Each NAAQS is divided into primary standards related to human health and secondary standards meant to protect the public welfare, which includes natural resources. Both primary and secondary standards for criteria pollutants ozone and particulate matter are currently under review as to their adequacy to protect the natural environment. If an accredited air quality monitor records concentrations of pollutant(s) that violate the NAAQS, the area (which may be at the county, metropolitan statistical area or other geographic level) falls into some classification of non-attainment. An area can be in non-attainment for one or more pollutants. The designation may result in various mandatory control measures, depending on the severity and frequency of exceedence (Peine et al. 1995).

The Prevention of Significant Deterioration Program (PSD) is applied to the construction or modification of major new sources of pollution or modification of existing major stationary sources with "major" being defined as emitting a critical pollutant at a level of 100 tons per year (TPY) or 250 TPY, the applicability depending on source category. A PSD permit is required before these sources can be constructed or modified. Ideally, these sources are supposed to install the best available control technology (BACT) and must meet an allowable PSD increment before the permit is issued [42 U.S.C. § 7475 (a)(1)(2)]. The application of BACT requirements is applied on a case-by-case basis, taking into account economic impact considerations, as well as efficiency of the pollution controls (Peine et al. 1995). The cost of application of a given BACT weighs heavily in the decisions. Interpretation of what constitutes BACT varies among states, greatly complicating and weakening the regulatory process of this interstate commodity of air pollution.

Special protection was afforded by Congress to national parks at least 6000 acres in size and wilderness areas at least 5000 acres in size established as of August 7, 1977 by defining them as "Class I areas." The PSD program requires that the federal land manager (FLM) takes on the "affirmative responsibility to protect the air quality related values of [Class I areas] and to consider, in consultation with [U.S. Environmental Protection Agency (EPA) and state officials], whether a major [stationary] source will have an adverse impact on such values" [42 U.S.C. § 7475 (d)(2)(B)]. The PSD program requires the FLM to quantify any adverse effects that an incremental increase in emissions contributed by the new source will have on the air quality related values (AQRV). These values include but are not limited to the effects of air quality on visibility, water, soils, and vegetation. Such cause–effect relationships are extremely difficult to quantify, except in the case of the largest stationary sources (Peine et al. 1995).

Unfortunately, the limitations of the PSD program and other Clean Air Act provisions restrict the ability of the FLM to adequately protect the air quality in national parks and wilderness areas for a multitude of reasons.

- Authority of the FLM is strictly an advisory role of reviewing PSD permit applications.
- Air quality management of mobile sources, most existing and "minor" point sources and toxic emission sources is not addressed in the PSD program and therefore does not involve the FLM.
- State regulators of air pollution have significant discretion in interpreting federal regulations associated with this interstate commodity of air pollution.
- Interpretation of BACT varies significantly among the states.
- There are currently no NAAQS secondary standards for most of the criteria pollutants to protect against adverse air pollution effects on water and soil chemistry and sensitive plants and animals.
- The concern to hold air quality standards higher in the nation's national parks and wilderness areas is laudable, but the designation of Class I airsheds is unfortunately impractical, considering the continuous fluidity of air currents and the long-range transport of pollutants. In most cases, there is little correlation between a watershed and an airshed.
- The cause/effect relationship between pollutant levels and corresponding adverse effects is invariably complicated by a myriad of extenuating circumstances, making generalizations about the relationship difficult.
- Models used to predict adverse effects that an incremental increase in air pollution emissions will have on AQRVs tend to be imprecise, allowing for a wide range of interpretation in the permitting process.
- It is much easier to define the direct costs associated with air pollution controls that can be readily quantified than the considerable and wide ranging benefits from a clean air environment, which are more likely to be measured in qualitative terms.

As a result, the ecosystem manager is left to address the serious problems of air pollution with little influence over a complex regulatory process and an incomplete information base from which to make informed decisions.

INVENTORYING SOURCES OF AIR POLLUTION

The concept is becoming more accepted that ecosystem managers should have a regional perspective (Berish et al., Chapter 7 in this volume). This is particularly true in the context of air quality. An inventory of stationary and mobile sources of air pollution should be compiled within the regional context of the ecosystem of concern. The size and configuration of the region of influence from air pollution sources is dependent on the prevailing weather patterns, the topographic characteristics of the protected landscape of concern, and the location of major sources that contribute significant pollution levels, particularly during periods of air stagnation. In addition, it is important to include an inventory of major point sources within recognized long range transport corridors of pollutants associated with prevailing weather patterns. The toxic release inventory compiled by the EPA is a convenient and readily available source of data on point sources compiled by EPA is a convenient and readily available source into the atmosphere (EPA 1993).

MODELING THE TRANSPORT OF AIRBORNE POLLUTANTS

In order to make the connection between sources of pollution, as defined by the source inventories, and air quality conditions in the ecosystem of concern, transport models are devised. These models are extremely complex and tend to lack precision. Because air current velocity and direction is so variable and the transport of pollutants can be for such a long distance, it is difficult to model the cause/effect relationship between specific sources of pollution and their contribution to ambient levels of various pollutants occurring at a specific location. This conundrum is exacerbated further by the fact that chemical transformations take place during long-range transport due to exposure to the sun and/or association with other elements/chemical compounds. Models for the eastern U.S. tend not to be as reliable as those constructed in the west, due to the multiplicity of sources, the complexity of the weather patterns, and the greater degree of moisture in the atmosphere.

Both the federal land managers as part of their responsibility to protect air quality and industries seeking permits for new sources of pollution sometimes construct pollution transport and chemistry models to estimate the incremental influence of point sources of pollution and to defend their respective positions concerning permit applications.

MONITORING AIR QUALITY

Resource monitoring is becoming an accepted cornerstone of ecosystem management (Smith et al., Chapter 8 of this volume.). Trends in ecosystem processes and functions; and the distribution, abundance, and productivity of indicator species provide key information to the ecosystem manager. Monitoring air quality is an important component of a resource monitoring program. Instrumentation to monitor air quality ranges from the recording of meteorological conditions to measuring the acidity and chemical composition of wet, dry, and cloud/fog deposition, and gaseous pollutants. Visibility is often monitored as well.

Operation of quality controlled monitoring stations is complex and the chemical analysis of samples expensive. Many of the instruments must be operated within stringent protocols, such as periodic calibration and operation in climate-controlled conditions, as defined by the EPA in order for the data to be accepted as valid for the regulatory process.

DEFINING AIR QUALITY-RELATED VALUES

Ecosystem managers have a responsibility to define the values of resources and visitor experiences that potentially can be adversely impacted by air pollution. The AQRVs are defined by the FLM. These AQRVs provide the basis for the rationale to protect the resources from the adverse effects of air pollution. Ideally, they should be defined in specific geophysical, chemical, and biological terms as well as more generally defined social, economic, political, and spiritual terms. Too often, the debate concerning air pollution controls is very specific in its economic definition of cost for control and very vague in its description of benefits associated with the anticipated improvement in air quality. This is a major concern of the Southern Appalachian Mountains Initiative discussed later.

AQRVs associated with natural ecosystems typically focus on potential adverse effects to ecosystem processes reflected by soil and water chemistry which can influence stream acidity and the availability of soil nutrients such as nitrogen, calcium, magnesium, and nitrogen. Adverse impacts on native plants and animals are of primary concern, as well. Plants can be adversely impacted directly by gaseous pollutants, such as ozone, and indirectly by wet and dry deposition of pollutants, such as sulfates and nitrates affecting soil conditions. A wide range of organisms are being demonstrated by science to be sensitive to the wide range of pollutants transported by air. Many of the lesser known adverse effects are secondary to the impact on sensitive species which are in some way related to the life cycle of a myriad of other co-dependent species occurring in the natural environment.

Another major group of air quality related values are those related to visibility. Air pollution can result in an extreme degradation in visibility, which can adversely impact a multitude of social and economic values associated with the environment.

The social, economic, and spiritual values associated with a clean atmosphere are generally not well understood nor universally accepted by those engaged in the business of managing air quality. The most precisely measured economic adverse impacts are those associated with threats to agricultural row crops whose sensitivity to air pollution can be precisely measured in dose–response chamber experiments that can be readily extrapolated to estimate crop damage (EPA 1996). Much attention has also been given to potential adverse impacts on timber production related to soil nutrient loss due to acid deposition and foliar injury from tropospheric ozone, which may result in the loss of growth and/or genetic diversity within species. Visibility loss due to air pollution can potentially impact land values and the quality of the recreational/tourist experience. The considerable social value placed on air quality as an integral part of the nation's great natural landscapes protected as national parks and wilderness areas is acknowledged in the 1977 Amendments to the Clean Air Act through the designation of such areas as Class I. As reflected by the political process, parks provide a focal point for expressing society's moral and ethical sense of obligation to protect nature for the benefit of present and future generations, reflecting stewardship and bequest values. Examples of visibility-related values are the long distance views of mountains, a highlight of a park experience whether observed on foot or by horseback or automobile. The communal and spiritual values of mountain peaks is closely related to air quality. Ironically, these values so strongly held by society are the ones first ignored in the PSD regulatory process.

SUPPORTING RESEARCH TO DEFINE ADVERSE EFFECTS FROM AIR POLLUTION

It is one thing for a manager to recognize the broad array of AQRVs. It is quite another to contribute resources to conduct research on the degree to which they are being adversely impacted by ambient air pollution levels in the ecosystem of concern. The nature of effects research ranges from the effects of ambient ozone levels on foliar injury, growth, biomass, reproduction, and the potential synergistic effects with other stressors to the effects of acid rain on soil nutrients and stream chemistry and their impacts on aquatic organisms. The nature and degree of visibility impairment from air pollution is another topic of extensive research. Such research is very expensive, requiring considerable investment from the federal land management agencies, the EPA, and state air quality regulatory agencies. Industry has invested heavily into this type of research as well. The nature of the effects research is discussed in some detail by Berish et al. in Chapter 7 of this volume, reporting results of the Southern Appalachian Assessment.

INTEGRATING FUNCTIONS IN THE MANAGEMENT PROCESS

Needless to say, air quality management is an extremely complex issue. The circumstances require extensive collaboration among the major stakeholders: state governors, industries producing the primary pollution sources, state regulators, federal regulators, federal land managers, public sector economic development facilitators, scientists modeling pollution transport and dispersion and conducting effects research, and nongovernment organizations representing various interests, such as economic development and environmental conservation. The stakes are perceived to be quite high by many of the stakeholders. As described later in this chapter, a heroic effort is underway in the southern Appalachians to create and facilitate an integrated strategy for the management of air pollution emissions.

SCIENCE AND THE FEDERAL LAND MANAGER

Awareness and understanding of the nature, cause, severity, and impact of air pollution in the Class I area of Great Smoky Mountains National Park has steadily evolved over the last 18 years as a result of a series of research and monitoring projects that collectively provided key pieces of the puzzle. The insight gained from air quality research and monitoring serves as a valuable illustration of the indispensable role played by science in the process of ecosystem management. A very brief synopsis of key research findings related to visibility impairment in the Smokies is used to illustrate the point as follows.

An initial step was to define long-term trends in visibility, utilizing the visual range data collected at airports by the National Weather Service across the nation. Analysis of the data compiled seasonally documents the steady decline of visibility in the eastern U.S. from the late 1940s through the early 1990s, followed by some improvement except in the vicinity of the southern Appalachians. The loss of visibility over the years has been much greater in the summer months (Husar 1996). A compilation of the total number of extreme air stagnation days from 1936 through 1965 in the eastern U.S. revealed that the most frequent stagnation events (400+ days) occurred over the southern Appalachian highlands (University Corporation for Atmospheric Research 1991), as do the greatest number of days of severely degraded air quality.

Other vital research findings have resulted from work conducted at the Look Rock visibility research and monitoring station. Located along the Foothills Parkway, the station affords an unobstructed view into the northwest portion of the Park. Look Rock is the longest continuously operated visibility monitoring station in the eastern U.S. Numerous research projects have been conducted there. By the mid 1980s, scientists from the Tennessee Valley Authority (TVA) had documented the degree and seasonal fluctuation in visual range and conducted elemental analysis of the fine particulate matter demonstrating the high percentage of sulfur dioxide derivatives therein (Reisinger and Valente 1985).

As an aside, the Integrated Forest Study, directed by scientists at Oak Ridge National Laboratory in the late 1980s and early 1990s, measured the deposition rates of sulfates and nitrates from rain, fog, and dry particulates. The study was conducted at 11 sites in the U.S., including a high elevation site at Noland Divide in the Great Smokies. Results indicated that, among all sites (including two in New England), the greatest cumulative deposition rate of sulfates was in the high elevation Smokies site, almost twice the next highest loading site. Additionally, measurements showed that the media delivering the pollution in greatest concentrations were cloud/fog (45%) and dry particulates (29%), as opposed to rain (26%) (Johnson and Lindberg 1992).

Another key analysis demonstrated the extremely high correlation between the levels of sulfur dioxide gaseous emissions and the increase in haziness during summer months in the southeast region of the U.S. (Malm and Pitchford 1994). Sulfur dioxide emissions increased by a factor of five, and visual range has been reduced to one fourth of what was determined to be natural levels. The current annual average visibility at the Park is 20 miles (Sisler et al. 1993), well below the estimated natural level of 93 ± 30 miles (Trijonis et al. 1991). Data from the Interagency Monitoring of Protected Visual Environments monitoring network (including the Look Rock site) corroborated findings previously determined at the Look Rock station by TVA scientists — as the visual range declines, the concentrations and percentage of sulfates in the particulate matter in the atmosphere significantly increases. Sulfate particles are very efficient at scattering light which causes the haze.

Near term analysis of trends has recently revealed that a slight decline in concentrations of sulfates occurred at Look Rock in 1991. Coincidentally, this decline followed a major reduction in sulfur dioxide emissions at some of the power plants of the TVA system. However, by 1994, there had been a resumption of the trend of inclination (Cahill et al. 1996). In fact, by far the highest levels of sulfur and ammonium sulfate concentrations in the atmosphere ever recorded occurred in the summer of 1996, representing a 23% increase over the previous summer season (Renfro pers. commun.). The reason for the resumption of this trend of increased sulfate levels may, in part, be related to an increase in sulfur emissions near the Park. Data recently released by TVA reveal that for coal-fired power plants near the Smokies, the release of sulfur dioxide has recently become greatest in the summer months, whereas, previously they were highest in the winter (Pyles 1996). This trend is presumably due to changing patterns of peak use demand (e.g., demands on power in the summer have greatly increased near the Park).

For instance, the TVA coal fired power plants at Kingston (39 air miles from the Park) emitted 96,000 tons, John Sevier (44 air miles from the Park) emitted 58,000 tons, and Bull Run (31 air miles from the Park) emitted 51,000 tons of sulfur dioxide in 1995 (Vincent 1996). Nearby pollution sources of this magnitude (combined 205,000 tons of sulfur dioxide annually) are much greater contributors to regional haze during periods of air stagnation in the summer months when visibility degradation is most severe, which coincidentally occurs during the peak tourist season at Great Smoky Mountains National Park (GSMNP), the nation's most visited national park.

The most popular recreation activities pursued in the Park — viewing scenery and driving for pleasure — are extremely influenced by visual range (Peine and Renfro 1988). In a survey conducted in 1985, park visitors ranked "clean, clear air" among the top four of 23 features ranked on a Likert importance scale. The average visitor ranked "viewing scenery" as "very important" to their overall experience (Morse 1988).

Last, a study sponsored by member institutions of the Electric Power Research Institute, the National Park Service, and others as part of the Southeastern Aerosol and Visibility Study was conducted at Look Rock from July 15 through August 26, 1995. The study was, in part, designed to quantify the contributions to visibility degradation attributable to ambient moisture conditions and naturally occurring volatile organic compounds in the atmosphere (Day et al. 1996). The sampling period began with a period of extremely clean air and high visibility with normal humidity fluctuations. This phenomenon was followed later in the season by an extensive period of air stagnation, extreme pollution levels, and a high degree of visibility impairment. Although the results of the study are not yet fully published, preliminary findings suggest that the dramatic range of atmospheric conditions occurring during the experiment period clearly demonstrated that the phenomenon of visibility degradation is overwhelmingly due to anthropogenic sources of air pollution (primarily sulfate particles), the effects of which are magnified by high moisture content in the atmosphere.

The cumulative insight gained from these studies and the application of various pollution transport models strongly suggests that during periods of air stagnation, the coal-fired electric power plants in close proximity to the Smokies are the most significant contributors to the conditions of visibility degradation in the Class I area of GSMNP.

For an overview of the condition of air quality in the southern Appalachians, see Berish et al. in Chapter 7 concerning the southern Appalachian Assessment. Bill Jackson prepared the section of the chapter concerning air quality.

SOUTHERN APPALACHIAN MOUNTAINS INITIATIVE

THE ORGANIZATIONAL STRUCTURE AND FUNCTION.*

Research and monitoring in national parks and national forest wilderness areas of the southern Appalachian mountains have documented adverse air pollution effects on visibility, streams, soils, and vegetation. Beginning in 1990, the federal land managers (FLM) for Shenandoah National

---

* Leslie Cox Montgomery, then SAMI Coordinator.

Park, Great Smoky Mountains National Park, and Jefferson National Forest/James River Face Wilderness Area made several adverse impact determinations in the review of proposed air permits for major new sources of air pollution. These data and subsequent actions led to the voluntary formation of a community-based environmental protection project called the Southern Appalachian Mountains Initiative (SAMI) in 1992. Now a nonprofit organization, SAMI's goal is to provide a regional strategy for assessing and improving air quality, based on credible data and peer-reviewed science, to protect this unique and sensitive ecosystem.

Expansion of human activity has significantly contributed to the decline in air quality in the southern Appalachians. In addition, many of the pollutants reaching the southern Appalachian mountains are transported by the wind from other parts of the country, as well as from growing urban areas in the Southeast. Emissions from motor vehicles, power plants and industries are the primary sources of human-induced air pollution. Some of the highest deposition rates in the entire country for sulfates and nitrates have been recorded in the Great Smoky Mountains National Park (Johnson and Lindberg 1992). This phenomenon may be attributed largely to the topographic position of the high elevation spruce–fir forests intercepting air currents laden with high concentrations of pollutants in wet, dry, and gaseous form.

SAMI is a partnership of more than 100 organizations, including eight state environmental regulatory agencies (AL, GA, KY, NC, SC, TN, VA, and WV), several federal agencies, industries, academia, environmental organizations, and other stakeholders across the region. SAMI addresses the public, policy, and technical aspects of air quality issues through the consensus-building efforts of three main advisory committees comprised of leading scientific experts, as well as corporate, citizen, and government stakeholders. SAMI gives affected states, federal agencies, regulated industry, and the public an opportunity to broadly debate environmental issues and to propose reasonable solutions to identified problems, based on available science. This group recognizes the urgent need for better management of the vast ecosystem of southern Appalachia — one that will provide for continued economic growth of local communities while improving the region's air quality.

In 1990, Congress strengthened the Clean Air Act by requiring major reductions in air pollutants from a wide variety of sources, making it the most expansive and expensive effort ever undertaken to clean our nation's air. Affected sources will be implementing changes needed to meet these new requirements over the coming years, but the majority of changes should be in place by the turn of the century. Although this effort is expected to improve air quality in general, it is not targeted specifically at the southern Appalachian highlands region. Additionally, since much of the region is in attainment with the national ambient air quality standards, individual stakeholders are currently not mandated to resolve documented adverse effects on the natural resources of concern. Unfortunately, the current national regulatory standards in place are designed to protect human health, rather than the more sensitive organisms associated with the natural environment. The EPA is required to review the national ambient air quality standards every 5 years. The current primary standards (set to protect human health) for ozone and particulate matter are now under the EPA's review. Consequently, these standards could be revised to become more stringent, or a secondary standard (aimed in part to protect the environment) could be added.

At this writing, the EPA has proposed that the particulate standard currently requiring regulation of particles 10 μ or smaller (PM-10) in concentrations of 50 mg/m^3 annually and 150 mg/m^3 daily be reduced to particles of 2.5 μ or smaller (PM-2.5) in concentrations of 15 mg/m^3 annually and 50 mg/m^3 daily. In addition, EPA also proposed maintaining the current standard for PM-10 so that larger, coarse particles would continue to be regulated. The current ozone standard of 0.12 ppm measured over 1 hour is recommended to be changed to 0.08 ppm measured over 8 hours or some optional variation thereof (EPA 1996). These are major changes in regulatory standards, designed to alleviate the perceived health risk that will undoubtedly have significant environmental benefit as well, assuming they survive the currently ongoing review process.

Using current data, SAMI is in the process of building an integrated assessment framework to evaluate how changes in levels of air emissions will result in changes in air pollutant exposures, as well as their environmental and socioeconomic impacts in the southern Appalachian mountains. Although well documented, the adverse effects of air pollution are not well understood in this region. The air transport chemistry, background geology, complex topography, and meteorological characteristics of the mountainous terrain add to the complexity of the problems. A comprehensive assessment of how fluctuations in air emissions cause changes in exposures, and subsequently visibility and the ecological health of streams, soils, and vegetation, has not been conducted for the southern Appalachian mountains ecosystem. Likewise, a socioeconomic analysis corresponding to the resultant emission, exposure, and effects changes has not been performed. Policy decisions on emission management scenarios exceeding reductions mandated by the Clean Air Act Amendments await recommendations founded on scientifically validated and peer-reviewed databases and analyses. The technical assessment, monitoring, and subsequent actions SAMI plans to implement in the southern Appalachian mountains will address the adverse effect human-induced air pollution has on visibility, streams, soils, and vegetation and will provide policy makers with validated and peer-reviewed information.

Internally, SAMI utilizes a hierarchical committee structure, much like the one employed by the Grand Canyon Visibility Transport Commission. SAMI is directed by a Governing Body — a 14-member group comprised of the eight participating states' primary environmental officials, the Regional Administrators of EPA Regions III and IV, the Director or designee for the Southeast Region of the National Park Service, the Forester for the Southern Region of the USDA Forest Service, or designees of these officials. This body also has representatives of industry and public interest groups. The SAMI Coordinator and two other paid staff members, guided by an Operations Committee appointed by the Governing Body, manage the day-to-day activities of SAMI. Three advisory committees — the Policy Committee, Technical Oversight Committee, and Public Advisory Committee — review, analyze, and evaluate air pollution research and control strategies, recommend policy positions, and develop outreach materials to build public support. The Socioeconomic Work Group is a standing committee made up of appointees from the three primary committees. The advisory committees and their subcommittees are made up of a balanced and diverse group of volunteers who work together through a consensus-building process. Additional ad hoc groups are formed as needed by these committees to carry out their tasks. Although decision-making by committee consensus can be slow, the decisions reflecting stakeholder buy-in at the beginning of the process run less risk of being disputed at a later date.

The Technical Oversight Committee (TOC) is working to ensure that SAMI's technical assessments are based on sound scientific data. Assisted by experts on four subcommittees — monitoring, modeling, emissions inventory, and effects — this committee is building an integrated assessment framework (IAF) that will project the environmental and socioeconomic responses of compliance with the Clean Air Act Amendments of 1990 and with additional changes in air emissions resulting from various emission management options (EMOs). The IAF is divided into six linked modules: (1) base year emission inventory, emissions projections, and control costs; (2) atmospheric transport and air chemistry; (3) effects — acid deposition (aquatic and terrestrial); (4) effects — ozone (vegetation); (5) effects — visibility; and (6) socioeconomic consequences.

The most heated debate in designing the IAF has been the issue of what to include or exclude in the socioeconomic analyses. Socioeconomic consequences include anything from the cost of installing and maintaining pollution control technology and foregone economic development on the debit side to improved environmental and human health and related social values, such as increased visibility, important to the tourism industry, on the credit side. In order to address all concerns expressed, it will be necessary to attempt to strike a balance between full disclosure of the expression of EMOs' costs and benefits with improved air quality.

This integrated assessment is projected to cost a total of $2 to 3 million, barring unforeseen delays and complications. The TOC has worked to identify, gather, and evaluate all existing data, models, and studies to establish a foundation of current knowledge and identify critical information gaps. During this phase, SAMI collaborated with other organizations with similar regional concerns to avoid duplication of efforts. SAMI peer-reviewed reports have been compiled on the following topics which describe the current state of knowledge as it pertains to air quality related values of the southern Appalachian region: (1) emission inventories, (2) atmospheric transport and air chemistry, (3) acid deposition effects to aquatic resources, (4) acid deposition effects to terrestrial resources, (5) ozone effects to terrestrial resources, (6) visibility degradation, and (7) IAF design. The information gathering phase is complete.

In order to evaluate how changes in emissions will affect natural resources, the TOC will establish an emission-response relationship for the entire SAMI region by a series of computer model runs. By first characterizing an emission-response "surface," the TOC hopes to produce an analytical tool that can be used by decision makers to estimate the benefits and costs of custom "what if" emission management scenarios. Currently, the TOC is attempting to determine what pollutants and magnitude of emissions reductions will be necessary to detect a change at the resource (receptor) of concern.

For instance, work in the acid deposition area is occurring in two phases. The first phase focuses on understanding how selected sensitive receptors might respond to changes in deposition levels of sulfate and nitrate using indicators, such as soil solution chemistry, stream water quality, vegetation nutrient content, or forest productivity. The second phase will take a more regional approach to assessing resource responses to changes in deposition and will use indicators that are more meaningful to the general public, such as acres of forests that are healthy or miles of streams that support fish. Work in the other IAF modules is proceeding concurrently and in phases, as appropriate. Once the IAF is built and tested, SAMI will move from a development phase to one of application.

The Policy Committee (PC) members focus on EMOs that will be applied to the IAF tool. This committee is responsible for developing, analyzing, and evaluating the environmental and socio-economic implications of both regulatory and nonregulatory strategies for improving air quality. The committee developed criteria for evaluating EMOs and prioritized pollutants of concern. They also were engaged in the development of some of the initial EMOs. Twelve EMOs not requiring further technical evaluation were approved for implementation by the Governing Body in June 1995. The approved EMOs focus primarily on energy efficiency and conservation measures and environmental education efforts with the potential to reduce emissions at little or no adverse economic cost to the public, but for which the overall benefits may be difficult to quantify.

A myriad of EMOs, ranging from mobile to stationary source controls to energy conservation and environmental education to incentives for reducing pollution, is under review by the PC. A process to refine and group the slate of suggestions for use with the IAF is also under development. This process is much like that used by the Grand Canyon Visibility Transport Commission. Additional strategies will likely be developed in light of the results of the technical work being performed.

Once the EMOs are applied to the IAF and the results are analyzed, SAMI will make recommendations to the public. If measures beyond current mandates are recommended, then an implementation phase will follow. The scope of the recommendations (i.e., source-specific measures, regional cap and controls, federal standards, voluntary changes, etc.) will depend on what source categories and geographic locations are targeted for emission reductions.

The Public Advisory Committee (PAC) is charged with educating the public about air quality in the southern Appalachians and about SAMI, facilitating public involvement in the SAMI process and providing a conduit for public concerns that also provides feedback to the public about their concerns. Initially, the PAC developed a brochure and several fact sheets about SAMI and air quality issues. The PAC developed strategies for implementing the 12 approved EMOs and is now in the process of implimenting them. This committee is working on ways to enhance public education without duplicating efforts. Currently, the PAC is developing an infrastructure to handle information undergoing development by the other committees as it becomes ready for public consumption. As SAMI moves into the application and implementation phases of its technical work efforts, the PAC will assume a more central role in SAMI's activities.

With so much focus and care having been spent on process and assessment inputs, tangible progress with SAMI has been relatively slow over the last several years, causing a portion of the membership to become impatient. It is important to recognize that SAMI has undertaken a task of monumental proportions with enormous implications for future economic development and environmental sustainability. The most extraordinary aspect of SAMI is that it is a voluntary effort not required by federal nor state statutes. This is truly the first attempt to define an equitable and objective process for addressing complex environmental issues fraught with uncertainties. It is hoped that this process will stimulate efforts to develop cost-effective, innovative, and flexible solutions to balance future economic growth with environmental protection. Whether or not these issues can, in fact, be adequately addressed without the hammer of regulatory reform is yet to be answered. SAMI's success and shortcomings in developing a regional air quality management strategy will serve as a model for the whole country, but the success of SAMI cannot be judged until the process is complete in 1998.

Until then, SAMI provides a unique opportunity to educate the public about important air quality issues, dispel misperceptions, and help people to understand the challenges faced in trying to sustain the environment while promoting economic growth. Perhaps even more important, the SAMI process allows stakeholders to have a greater voice in determining how environmental problems will be handled in the future. One thing is certain, slow progress or not, SAMI's greatest achievement has been its ability to bring together groups with very diverse interests to constructively discuss complex issues. Without SAMI, this dialogue might never have effectively occurred.

EFFECTIVENESS OF THE DECISION-MAKING PROCESS

The SAMI process has been ongoing for over 4 years. A reported lack of progress is a major source of frustration for the federal land managers and environmental advocacy participants. The admittedly slow progress is largely due to a lack of adequate funding, but also can be attributed, in part, to the divergent expectations of the various stakeholders engaged in the process. In order to illustrate these different perspectives, representatives of various stakeholder interests have prepared the following brief statements in response to the question:

> *"What are your expectations for SAMI and how might it best be achieved, given the history of the organization?"*

Their statements of opinion which follow illustrate the diversity of perspectives on a complex resource issue of which the ecosystem manager must be cognizant. Understanding key stakeholder perspectives on a controversial issue is just as critically important as compiling the technical information defining the environmental aspects of an issue. These statements represent the opinions of the authors and do not necessarily represent those of the organizations with which they are affiliated.

Perspectives of a SAMI member representing a state government regulator.*

It is always difficult to require costly additional reductions in air pollution emissions when a problem that needs to be addressed is not obvious. From this perspective, the National Park Service (NPS) has a difficult job, as do the state government environmental programs. The NPS is charged with protecting the AQRVs of the Class I areas, such as Great Smoky Mountains National Park (GSMNP). Often these AQRVs are not readily observable, and the NPS must provide scientific documentation to justify the need for more stringent air pollution limits and standards in order to prevent adverse impacts on a Class I area. What lies at the heart of the occasional permitting controversy between the NPS and the various state environmental regulatory agencies is the real world definition of

---

* Barry R. Stephens, Air Pollution Control, State of Tennessee.

"adverse impact" on the fauna and flora of the various Class I areas. The NPS is of the opinion that it has provided all the evidence that is needed to show that AQRVs are threatened and that the impacts of new industrial sources must be held in check even below those "significant impact" increment levels specified in the PSD rules.

Most government agencies consider the more stringent emission limitations and requirements proposed to be imposed on prospective new industrial sources allegedly impacting a Class I area to have an "adverse impact" on economic growth in a state. For this reason, the states want indisputable evidence that the AQRVs are in jeopardy before requiring controversial, very stringent requirements, such as securing emission offsets, before allowing a new source to build.

In Tennessee, the Air Quality Act, Section 68-201-103, specifically states that "it is the intent and purpose of the part to maintain purity of the air resources of the state consistent with the protection of normal health, general welfare and physical property of the people, maximum employment and the full industrial development of the state." Because of this, the State must maintain a balance between development and the protection of the environment — a difficult job at best. The NPS is charged with the protection of many of the country's most pristine natural resources. In Tennessee, the Great Smoky Mountains are an invaluable natural resource for the state, and the park's protection is a primary concern of the state.

It is recognized that the only legal mechanism to limit the impacts of air pollution is provided to the NPS in EPA's PSD regulations and the Federal Clean Air Act. The regulations only apply to new source growth and provide the only avenue available to the NPS to try and minimize any adverse impacts on Class I areas. However, new sources may not be the main problem. This is not to say that there cannot be an impact from new sources. For GSMNP, it may be that the contribution of existing sources of air pollution emissions may be the real problem — those emissions emitted by existing point, area, and mobile sources.

With the NPS stating that no additional increases in air emission impacts is acceptable in GSMNP, an organization known as the Southern Appalachian Mountains Initiative (SAMI) was created to study and find a solution to any adverse air quality impacts in GSMNP. The no growth strategy by the NPS is not acceptable to the states involved. SAMI must develop a strategy that will still allow for new source development while providing for protection of the AQRVs set by the FLM. The NPS must be willing to have its data subjected to a critical review providing for input by all involved parties.

Perspectives of a SAMI member representing a federal land management agency*

My original expectations for SAMI seem wildly optimistic today. I thought SAMI would develop emission management options (EMO) to address the regional air quality problems affecting the southern Appalachian Class I areas *and* be the catalyst for implementation of these options through EPA and state legal processes. That is, once SAMI adopted an EMO, EPA as well as each SAMI state, would implement that EMO into law, perhaps seeking advice on specific aspects of rule making from qualified SAMI participants. More than 4 years have passed since SAMI was initiated in 1992. I would have expected SAMI to have adopted at least one EMO and seen it through to implementation by now.

Based on my experience in SAMI, my expectations are now somewhat different, although I hope my original expectations concerning SAMI's basic "action-oriented," problem-solving purpose will become a reality. Currently, I see SAMI as predominantly assessment-oriented. That is, SAMI will develop and perform (or oversee) a massive, semi-integrated assessment of the costs and benefits associated with various methods for reducing emissions or EMOs. At this time, given the political situation and fact that most of the eight SAMI states have a prohibition by statute or executive order against requiring any action beyond what federal law requires, I do not expect the

---

* Karen A. Malkin, then with the Natural Resource Stewardship and Science Directorate, Air Resources Division, National Park Service.

assessment by statute or executive order to necessarily trigger any new legal requirements. I do expect the assessment to heighten industry awareness of measures, such as pollution prevention measures, which may be both cost-effective from a relatively short-term business perspective and effectively maintain the pollution status quo; i.e., keep the situation from worsening.

In an attempt to merge my present expectations for assessment with my more action-oriented original expectations and, based on discussion with others involved with SAMI, the following plan is suggested:

Interagency work-group. Form a small interagency work group to discuss funding, oversight, strategies to implement previously approved "near term" EMOs, and time frames (not to exceed 18 months). The group should designate responsible parties for the various actions recommended. The group would also review the feasibility of additional incentive based EMOs and make recommendations to the governing body for possible implementation on a trial basis. Additional meetings of the proposed interagency work group are anticipated to discuss staffing, progress and issues as they arise.

SAMI buy-in. After the interagency work group meets and comes up with proposed funding as well as "strings," that group would invite key SAMI participants in for a meeting to discuss the proposal and take comment. Once the states have bought into the general concepts, the next step would involve memorializing the understanding and procedures through a memorandum of understanding or some other means. The understanding would probably be among federal agency SAMI members, funders, and the members of the full SAMI governing body.

Strategies to evaluate EMOs in the Integrated Assessment Framework. There is a need to involve additional respected government technical and academic experts in the IAF process to avoid needless duplication of other studies, to monitor developments in contractor work, and to help frame details of the assessment.

Consideration should be given to establishing and supporting a "blue ribbon" peer-review of the IAF process and modeling runs associated with the process.

It is important to structure the IAF to allow full analysis of one pollutant effects category as soon as possible, beginning with visibility, followed by acid deposition on aquatic and terrestrial resources. Before beginning the SAMI ozone assessment, the Technical Oversight Committee and others as appropriate should review efforts of the Ozone Transport Assessment Group, the Southern Oxidant Study, and EPA's proposed revision of the ozone NAAQS.

The IAF should comprehensively address socioeconomic implications, including human health, affected property values, and the social values directly and indirectly associated with environmental health.

Overarching issues. There is a need to consider the pros and cons of anointing SAMI as a Visibility Transport Commission (VTC), to help support getting the visibility part of SAMI accomplished sooner. Maybe this model could be applied to ozone and acid deposition as well. Finally, SAMI should be in a position to support and adjust to any regulatory initiatives on the horizon such as the recently proposed NAAQS for ozone and fine particulates.

Perspectives of a SAMI member representing a voluntary association dedicated to environmental conservation*

Proven ideas and tools for economically reducing air pollution in the southern Appalachians are in hand. Some are decades old but easily updated. Others will still seem new a decade from now, but even today one can be sure of their potential. SAMI should spend less effort on the integrated assessment and most of its effort on assisting communities with reengineering, using the means of sustainable development.

---

* Brian J. Morton, Ph.D., then with the North Carolina office of the Environmental Defense Fund.

The overall strategy which I propose for SAMI is founded on this design principle articulated by the architect Christopher Alexander: "when you build a thing you cannot merely build that thing in isolation, but must also repair the world around it, and within it, so that the larger world at that one place becomes more coherent, and more whole; and the thing which you make takes its place in the web of nature (Alexander 1977)." The strategy for protecting Great Smoky Mountains National Park and the other special places in the southern Appalachians comprises new approaches to the organization of production and urbanized areas — sustainable communities and eco-industrial parks — and an overarching ecological constraint on the use of the atmosphere as a receptacle for unwanted byproducts of the combustion of fossil fuels — a cap on aggregate NO_x emissions implemented with an emissions trading program.

Each component addresses a progressively larger field of action: industrial park, community, airshed. A smaller project facilitates creation of a larger, more encompassing project, and a larger project integrates smaller ones, creating otherwise unachievable environmental and economic gains. The enterprises in an eco-industrial park reduce impact on the environment, conserve energy and materials, and reduce the expense of residuals management through pollution prevention, energy-efficient building design, and coordination of on-site flows of energy and materials. Because discharges to the environment have been reduced, the environs of the industrial park may be sufficiently attractive to allow development of residential and commercial buildings in the neighborhood of the park. Co-location of places to work, reside, and shop increases travel by foot, bicycle, and public transit, further reducing emissions of NO_x. Thus, redeveloped, our communities will be more healthful and intrude less into the countryside; they may even be able to save urban forests instead of pave them.

An eco-industrial park helps to make a sustainable community. Conversely, the planners of a sustainable community facilitate development of eco-industrial parks by encouraging pedestrian- and transit-oriented development, providing incentives for energy-efficient buildings and surface transportation, encouraging cooperation among stakeholders, and publicly expressing commitment to sustainable development and evaluating progress. Finally, reduced NO_x emissions in the airshed and implementation of an emissions cap and trading program ensure that regional emissions are limited by an ecological constraint, provide polluters with flexibility consistent with emission reduction responsibility, and reward innovators of less costly ways to prevent or abate NO_x emissions.

At its core, this strategy for protecting southern Appalachian forests from ground-level ozone is a strategy for restructuring — by making mutually supportive — the connection between industry, urbanized areas, and wilderness. This idea has a long history, rooted in the Appalachians. Benton MacKaye, cofounder of the Regional Planning Association of America and of the Wilderness Society, proposed in 1921 the Appalachian Trail as the backbone of a system of linked wild areas and parks (MacKaye called them levees) which would protect all the primeval and rural landscapes of the East — and sustain the cities as well.

> Here is the barrier of barriers within this world-empire of industrial and metropolitan upheaval. We have here already laid, both on the ground and in the public mind, the thread on which to weave this basic barrier. This is the projected mountain footway known as the Appalachian Trail… Here is marked the main open way across the metropolitan deluge issuing from the ports of the Atlantic seaboard. This open way, when once it really opens, would form the base throughout eastern populous America for controlling the metropolitan invasion.

MacKaye believed that balance between the elemental environments, the primeval, rural, and urban, produced the highest quality of life (MacKaye 1990). Tony Hiss expresses MacKaye's thesis in terms of the moral obligation of the current generation to subsequent generations: "people properly grounded in a complete and rounded environment could begin to get a better feeling for the day-to-day aspect of the many multi-generational decisions in modern life — those actions we initiate that pile up assets or debts for our children or grandchildren (Hiss 1990)." The danger of alienation from any of these landscapes by careless development is not just damage to ecosystems, but putting "our own safety and health in peril, by cutting ourselves off from settings that — undefaced, as MacKaye said — could let us live as one "unit of humanity" with the next two or three generations after our own time." (MacKaye 1990)

MacKaye's insights continue to be valid and relevant to protection of the southern Appalachians. Now, though, when long-distance transport of air pollution threatens ecosystem health, we must build levees with the tools of industrial ecology, sustainable communities, and market-based incentives for emission reductions.

Perspectives of a SAMI member representing industry*

It is safe to say that no piece of environmental legislation impacts industry more than the Clean Air Act Amendments (CAAA) of 1990. Not only does it impact industry at its manufacturing sites but it frequently affects the products made by that industry. The generation of electric power, the production of raw and finished chemical goods, and the production of automobiles are all manufacturing sectors whose stationary sources and finished products are directly affected by the CAAA. Add to this the complication that the control of air pollution is perhaps the most challenging and complex environmental issue facing our country and you can begin to appreciate the scope of this issue.

The CAAA mandates protection of our most cherished natural resources, i.e., Class I areas. Frequently, industry and the federal land managers, who are charged with protecting air quality in Class I areas, are at odds over air quality issues. Industry, on one hand has economic goals it is striving to meet; e.g., increased production, increased jobs, greater market share, operational flexibility, and profit margins — all of which in one way or another, translates into increased emissions of air pollutants. On the other hand, FLMs have documented damages to air quality related values (AQRVs) caused by air pollution from the existing mix of air pollution sources and are charged with protecting these values.

Take, for example, the issue of ozone. Ozone is generated by a complex mix of pollutants (i.e., VOCs and oxides of nitrogen) emitted from a variety of stationary sources (e.g., industry), mobile sources (e.g., automobiles), naturally occurring sources (e.g., vegetation), and consumer products (e.g., paints). When an industry that emits ozone precursors proposes to build a major new source near a Class I areas it must demonstrate that these additional emissions will not adversely impact AQRVs identified by the FLM. What is to be done if the Class I area's air quality has already deteriorated to the point that no additional emissions are acceptable? The conflict then evolves into an environment vs. jobs debate that, more often than not, comes down on the side of continued degradation of air quality in Class I areas; i.e., the economic interests of the state or region end up taking precedence over protecting the Class I area. Forgotten in this thinking are the real economic advantages a state or region enjoys due to the Class I area or the fact that an industry might have located in the area for the same "quality of life" type of values that were there because of the Class I area.

This is but one example of the many complexities surrounding the protection of AQRVs in Class I areas. The point is that these air quality problems are extremely complex and require regional solutions developed through a cooperative effort of *all* stakeholders. This is where the Southern Appalachian Mountains Initiative comes into the picture. SAMI is taking this regional approach with input from a number of different areas. Representatives from our national parks and forests, environmental organizations, organic chemical manufacturers, regulators, power companies, citizen action groups, and academia all have come together to address the issues surrounding the identification and implementation of the best approaches to solving the air quality problems experienced by Class I areas in the SAMI region.

---

* William R. Miller III, The Saturn Corporation. Spring Hill, TN.

Many have faulted SAMI for not moving fast enough or not having produced tangible results. Many have faulted industry with intentionally slowing SAMI's progress in the hopes that "paralysis by analysis" will set in and derail the attainment of SAMI's mission. This criticism is valid in many cases. However, in a larger context, the real benefit of SAMI is the bringing together of diverse views and differing environmental agendas in a forum that promotes a free exchange of ideas. Only once an industry understands the frustrations of the FLM or a FLM understands the economic constraints of an industry will there be an opportunity for making real progress. To a large extent this has occurred within SAMI.

From an industrial perspective, there are a number of issues under consideration by SAMI that are of both environmental and economic importance. Companies are in business to make money. Consequently, when faced with the considerable expenses that sometimes accompany pollution control projects (e.g., SO_2 scrubbers on power plants or VOC control on a coating operation) it is reasonable for a company to ask:

1. What is the benefit to the environment of the expenditure?
2. How is air quality in the near-field and far-field affected by the pollution reduction?
3. What is the best mix of pollution control options to affect a given degree of change in an air quality characteristic?
4. Will the SAMI effort be consistent with other national programs and initiatives (e.g., visibility transport regions and commissions, expected changes in the ozone and fine particle standards)?
5. Will the voluntary efforts that evolve out of SAMI wind up being regulatory mandates at some point and, if so, will industries in the SAMI region be put at a competitive disadvantage because of the mandates?

It is extremely difficult to answer such questions, given the current state of the science in such disciplines as long range transport, complex air models, and valuing the societal benefits of a visitor's experience in a Class I area, but the framework SAMI is working on could help in this regard.

The real test of SAMI's effectiveness will come when the agreed upon emission management options are assessed, prioritized, and implemented. SAMI will be deemed a success if this happens or a failure if the process stops short of implementing strategies that produce real improvements in the region's Class I areas' air quality.

NEW SOURCE PERMIT APPLICATION REVIEWS FOR GREAT SMOKY MOUNTAINS NATIONAL PARK

This section provides insight concerning the PSD permitting process from the perspective of the U.S. Department of the Interior in it's role as the federal land manager (FLM). The following statement was prepared and presented at a public hearing on July 19, 1996 in Knoxville, Tennessee by Molly Ross, U.S. Department of the Interior, representing the U.S. Secretary of the Interior. Mr. Barry Stephens of the Tennessee Air Pollution Control Division edited the piece for accuracy. His contribution in no way implies his concurrence with the points made by Ms. Ross.

NUMBER AND NATURE OF PSD PERMIT APPLICATIONS

Since 1978, the National Park Service (NPS) has reviewed over 500 major new source permits proposing to locate within approximately 100 miles of Class I designated national parks. Of these permits, 32 were for sources near the Smokies, more than any other park except Shenandoah, thus illustrating the growth of industry in the southern Appalachian region. Most of those applications were processed quickly without problems. Only one in ten raised concerns. Of these, the National

Park Service reached an adverse impact determination, calling for permit denial unless mitigation measures were taken, in only a dozen cases, three of which were for sources in Tennessee. The adverse impact determinations affecting sources in other states were resolved primarily through settlement agreements. In Tennessee, none of the three adverse impact determinations were resolved through settlement agreements. The State awarded the permit agreements over the objection of the FLM. These Tennessee permits illustrate the difficulty in protecting air quality by the FLM.

Eastman Chemical Company Permit Application

The Eastman Chemical Company of Kingsport, Tennessee is one of the largest sources of toxic air pollution in the entire southern Appalachian region. The company reports emitting over 29 million pounds of toxic chemicals per year, accounting for one quarter of the entire amount emitted within the state of Tennessee, which in turn is ranked second in the nation for toxic chemical emissions (EPA 1993). In November 1990, the NPS filed comments on three proposed gas-fired boilers for Eastman Chemical, requesting better pollution control technology, lower nitrogen oxide limits, and emission offsets. The State rejected the requests and issued the permit over the objection of the FLM. The NPS appealed the permit to no avail. In denying the request for emission offsets, the State indicated that it lacked authority to require offsets in so-called attainment areas. Both the FLM and the EPA believed the State does have such authority to require offsets in order to permit a new source that would otherwise adversely affect a Class I area. The State currently does not have regulations providing for such action. Nevertheless, at the State's suggestion, in April 1991, the NPS filed a formal rule-making petition with Tennessee to provide explicit authority for emission offsets to protect Class I areas. In August 1991, the NPS appeared before the Tennessee Air Pollution Control Board to address questions about the petition. The State never acted on the petition, providing no direction concerning this matter to the Tennessee Air Pollution Control Division. Such a provision for offsets in emission sources is extremely sensitive politically since it can be mis-construed to imply, from the NPS perspective and from the state's perspective, placing a cap on economic growth in the region. Since much of the country's political capital is based on fostering unrestrained opportunity toward the pursuit of wealth, the conflict of interest presented by air quality management poses a particularly difficult conundrum.

In November 1991, the NPS filed comments on a proposed new coal-fired boiler for Eastman Chemical, requesting better pollution control technology and emission offsets. The neighboring state of Virginia, as well as the EPA, supported the technology recommendations. The EPA specifically recommended that synthetic catalytic reduction should be evaluated. In March of 1992, the NPS reached a final adverse impact determination with respect to this source, and requested permit denial unless the mitigation was required of the permittee. The State neither required the advanced pollution control technology nor accepted the NPS adverse impact determination. The State's evaluation was that the new source was predicted to be the equivalent of less than one automobile driving through the park per day and they therefore felt that this source was not going to significantly contribute to any adverse impact.

In February 1992, the NPS published a preliminary "adverse impact determination" notice in the *Federal Register,* to alert all interested parties of the air pollution impacts at Great Smoky Mountains National Park. In March 1992, with the sponsorship of the Southern Appalachian Man And Biosphere Cooperative, a large symposium was held to discuss the Park air pollution issues (Peine et al. 1995).

Tenn Luttrell Permit Application

After 31 permit reviews, multiple meetings, a rule-making petition, a large symposium, two adverse impact determinations, and the establishment of the SAMI, the new source permitting process in Tennessee had not yielded any concessions to the FLM. In 1995, the State gave early notice of a proposal by the Tenn Luttrell Company to build two new lime kilns 35 miles from the boundary of the Park, and invited the Park Service to a preapplication meeting. The NPS felt that the company had not performed the analysis needed to evaluate impacts on park resources. Facing a short deadline, the NPS performed a conservative screening analysis, which determined that the company's nitrogen oxide emissions would cause up to an 8% increase in nitrogen oxides at the Park, whose resources were already experiencing adverse impacts from nitrogen oxides. Based on the available information, NPS requested the State to either perform a "more refined" analysis which might well exculpate Tenn Luttrell, or require emission offsets to mitigate the identified adverse impact. The State refused both requests and issued the permit.

The NPS appealed issuance of the permit along with the National Parks and Conservation Association. As a result, Luttrell was required to contribute $40,000 to the National Parks and Conservation Association to be used to purchase emission offset options from industry in the future. In addition, the State and NPS were able to agree on a Memorandum Of Understanding (MOU) concerning future review and litigation of new source permit reviews in Tennessee that was designed to facilitate cooperation and collaboration. The MOU was not tied to the Tenn Luttrell permit. It did not require emission offsets before Tenn Luttrell began operations. In fact, project construction was underway before the memorandum was executed. The MOU agreement did not make new law. It was not a radical nor revolutionary document. It simply explained clearly how the permit review process was handled. From the NPS perspective, the agreement did not make Tennessee different from other states in the region, since it simply articulated NPS procedures followed throughout the nation. Representatives of Tennessee industry did not agree with this perspective.

As a result, in March 1996 under political pressure, the Division Director of Tennessee Air Pollution Control rescinded the MOU, negating the agreement. The Division Director who had signed the MOU with the NPS rescinded it after much debate before the Tennessee Air Pollution Control Board (TAPCB) and upon input from the Tennessee Department of Economic Development. TAPCB recommended rescinding the MOU because of some language believed to put Tennessee industry in a noncompetitive situation. There was an immediate negative reaction to the decision by the conservation community that was widely reported by the media. The Director's decision was very unpopular among those interest groups. Because of the concerns expressed by various parties, the Governor appointed an advisory committee that held two hearings in Knoxville and quickly agreed to recommend that the MOU be renegotiated and the old one be honored until the new one is in place. So goes the rocky road of air quality management.

CONCLUSIONS

RISK ASSESSMENT REPORT CARD FOR ECOSYSTEM MANAGERS

Next to climate change, probably the most complex threat commonly facing natural ecosystems comes from air pollution. The myriad of pollution sources; the complexity of their transport, transformation, and deposition; the nature of their adverse effects; and the labyrinth of regulations to control emissions is perplexing, to say the least. In addition, the ecosystem manager has little leverage in the regulatory process for air quality. The Southern Appalachian Mountains Initiative is an ambitious attempt to tame this beast in the region. The SAMI work in progress has had mixed reviews. It should be noted that the SAMI process has never before been tried, so the uncertainty is to be expected. Application of the *Risk Assessment Report Card for Ecosystem Managers* to the SAMI experience, at the time of this writing, is judged to be as follows:

Vision: B–

The vision that led to the creation of SAMI is to be applauded. However, the vision became and remains blurred by the SAMI process, while the debate as to the nature and significance of the adverse impacts continues.

Resource risk: D

According to the NPS as the FLM, air pollution is one of the greatest problems facing the natural resources in Great Smoky Mountains National Park. This concern is central to SAMI's charge.

Resource conflicts: F–

The lack of resolution of PSD permit applications and the slow rate of progress of the SAMI process is testimony to the level of conflicting issues associated with air quality.

Procedural protocols: B+ and C–

The B+ is in recognition of the quality of the science applied to measure pollutant levels and deposition rates, and associated adverse impacts. The C- is in recognition of the inevitable shortcomings of the framework and data to conduct the integrated assessment of the proposed emissionmanagement options.

Scientific validity: B+ to C–

Some SAMI elements are well defined via scientific research, such as pollutant deposition rates and adverse effects. Other elements, such as the contribution to deposition rates by long distant sources, the relative contribution of precursor pollutants to the formation of tropospheric ozone, the potential adverse effects of pollution on hardwood timber production and genetic diversity, and the socioeconomic benefits of clean air, have not received enough attention by the scientific community to provide adequate tools for the SAMI IAF process.

Legal jeopardy: C+

If adverse impacts from air pollution continue to escalate unabated, the voluntary spirit of SAMI may some day be replaced by litigation.

Public support: B

Generally speaking, the public is very supportive of clean air. However, the SAMI effort has not generated much attention by the general public. Some of the members of the environmental conservation community have dropped out of the SAMI process in frustration.

Adequacy of funding: C

Research and monitoring in the southern Appalachians has been well supported by various federal agencies. SAMI has received considerable funding from federal and state agencies. Over $1 million was recently allocated to conduct the first phase of the integrated assessment. Twice that amount is needed to complete the process.

Policy precedent: C

There is no policy precedent for the organization or mission of SAMI within the legal construct of the Clean Air Act and associated amendments, but if the process is successful, it will undoubtedly set a policy precedence for the future.

Administrative support: B

Representatives of member institutions attend SAMI meetings and have become more active since the $1 million funding support became available.

Transferability: B

If the SAMI experiment in voluntary conflict resolution works, it could become a model for the management of air quality in other regions of the country.

The air quality issue is a classic example of a bedevilment to the ecosystem manager that will not go away and will most likely only get worse over the long term, given the inevitable increase in human presence on the landscape. The primary role of the ecosystem steward is to raise the issue, facilitate documentation as to its severity and adverse impacts and to continue to participate in the never ending process of controlling emissions. Never give up on the hope for solution!

REFERENCES

Cahill, T.A., R.A. Eldred, and P.H. Wakabayashi. 1996. Trends in Fine Particle Concentrations at Great Smoky Mountains National Park. Presented at the annual meeting of the Air and Waste Management Association, Nashville, TN, 1996.11 pp.

Alexander, C. et al. 1977. *A Pattern Language: Towns, Buildings, Construction.* (Oxford University Press: New York. 1977), p. xiii.

Cole, D.W. 1992. Nitrogen, chemistry, deposition and cycling in forests. In *Atmospheric Deposition and Forest Nutrient Cycling: A Synthesis of the Integrated Forest Study.*, Eds. D.W. Johnson and S.E. Lindberg. Springer-Verlag: New York, p. 158.

Day, D., W.C. Malm, S.M. Kreidenweis. 1996. Aerosol light scattering as a function of relative humidity at Great Smoky Mountains National Park. Unpublished paper of the National Park Service, Denver, CO. 10 pp.

Johnson, D.W. and S.E. Lindberg. 1992. *Atmospheric Deposition and Nutrient Cycling: A Synthesis of the Integrated Forest Study.* Springer-Verlag: New York.

Hiss, T., 1990. *The Experience of Place*, Vintage Books: New York. p. 190.

Husar, R. 1996. *U.S. Visibility Trends — Haze Trend Summary.* Washington University, St. Louis, MO.

Malm B. and M. Pitchford. 1994. Interagency Monitoring of Protected Visual Environments. National Park Service, Ft. Collins, CO. 8 pp.

MacKaye, B. 1990. *The New Exploration: A Philosophy of Regional Planning*, introduction by Lewis Mumford, forward by David N. Startzell University of Illinois Press for the Appalachian Trail Conference: Urbana-Champaign, IL. p. 200.

Mitchell, M.J. and S.E. Lindberg. 1992. Chemistry deposition and recycling in forests. In *Atmospheric Deposition and Forest Nutrient Cycling: A Synthesis of the Integrated Forest Study.*, Eds. D.W. Johnson and S.E. Lindberg. Springer-Verlag: New York p. 79.

Morse, D. 1988. Air Quality in National Parks. USDI National Park Service, Air Quality Division, Natural Resources Programs, Nat. Res. Rep. 88-1. Denver, CO, pp. 3.1–13.

Peine, J.D., J.C. Randolph, and J.J. Presswood. July 1995. Evaluating the effectiveness of air quality management within the Class I Area of Great Smoky Mountains National Park. *Environmental Management*, University of Massachusetts, Amherst, MA. 40 pp.

Peine, J.D. and J.R. Renfro. 1988. Visitor use patterns at Great Smoky Montains National Park. USDI, National Park Service Southeast Region Res./Resour. Manage. Rep. 90. Atlanta, GA. 93 pp.

Pyles, D. 1996. Local and regional influences on air quality at eastern national parks and wilderness areas: a report to the Environmental Defense Fund. Department of Land and Water Resources, University of California, Davis.

Reisinger, L.M. and R.J. Valente. 1985. Visibility and Other Air Quality Measurements Made at the Smoky Mountains National Park 1980–1983. Office of Natural Resources and Economic Development, Air Quality Branch. Research Station, Muscle Shoals, AL. 47 pp.

Sisler, J.F., D. Huffman, and D.A.Latimer. 1993. Spatial and temporal patterns and the chemical composition of haze in the U.S. and analysis of data from the IMPROVE network, 1988–1991. Report submitted to the National Park Service (W. C. Malm, Principal Investigator) and the U.S. Environmental Protection Agency (M.L. Pitchford, Principal Investigator). ISSN No. 0737-5352-26. 141 pp.

Trijonis, J.C., W.C. Malm, M. Pitchford (and others). 1991. Visibility: existing and historical conditions–causes and effects. In Acidic Precipitation Assessment Program, Acidic Deposition: State of Science and Technology. Vol. III, Rep. 24. 129 pp.

University Corporation for Atmospheric Research. 1991. In Southern Oxidant Study document dated August, 1991.

Vincent, G.M. 1996. Historical records of emissions from TVA power plants. Fuel Supply and Engineering Division, TVA. Chattanooga, TN.

U.S. Environmental Protection Agency (EPA). 1993. Toxic Release Inventory. Region IV Atlanta, GA.

U.S. Environmental Protection Agency (EPA). 1996. Air quality criteria for ozone and other related photochemical oxidants Volume II. EPA/P93/004BF National Center for Environmental Assessment. Research Triangle Park, NC.

16 Fire Management

Edward R. Buckner and Nicole L. Turrill

CONTENTS

Introduction 329
Fire and the Southern Appalachian Region...330
    Prehistoric Setting 330
    Historic Setting...333
    Federal Land Acquisition 334
Fire and Ecosystem Management...336
    Fire Behavior..336
    Fire Effects on Soil 337
    Vegetative Response...337
    Biodiversity 339
Fire Management Options 340
Conclusions 342
    Risk Assessment Report Card for Ecosystem Managers 343
Acknowledgments..345
References 345

INTRODUCTION

Fire is an evolutionary force that has helped shape many terrestrial ecosystems. With the exception of the coldest and wettest regions of the earth, great tracts of land have been subject to periodic fires for millennia. This is true for the southern Appalachian region where several fire-associated and/or fire-dependent pine and oak species require fire to maintain their "natural" community structures. Most of these fire-dependent communities are located on federally owned land.

Seventy to 90 years of highly effective fire prevention and suppression has placed several of these pine–oak ecosystems in jeopardy. In the absence of fire, regeneration niches needed to establish fire-dependent species are lost. Today, extensive areas that once supported these communities are without a seed source for their fire-dependent components. On many sites the remaining seed source is from old, decadent trees that are highly vulnerable to insects, especially the southern pine bark beetle — *Dendroctonous frontalis* — and diseases. Where a viable seed source remains, prompt action to regenerate these communities is essential if they are to remain as ecosystem components in the southern Appalachian region.

Increasing and convincing evidence shows that cultural fires maintained these fire-dependent communities over prehistoric millennia. This poses a perceptual problem in that some people argue that forest communities exibiting cultural influences are not "natural." Accepting this hypothesis, the policy of many public agencies has been to protect forests from fire allowing them to succeed to their "natural" (e.g., no cultural influence) endpoints. However, since landscape and species characteristics evolved under cultural burning, protecting them from such fires would not restore the perceived "natural" condition. Removing cultural burning from landscape management practices would create a condition previously unknown and unnatural to the landscape.

MacCleery (1992) proposes that maintaining "naturalness" in public land management, with man in the environment, requires accepting as "natural" all events and conditions that occurred and existed prior to 1492. Management guidelines can be developed based on a knowledge of those conditions provided by archaeologists, palynologists, geographers, and ethnobotanists. Unfortunately, in current literture the use of "natural" as an opposite of "cultural" is too deeply ingrained to be discarded.

Managing fire is an extremely complicated and dynamic task. Ecosystem managers face the consequences of using a high-risk tool backed by uncertain science. In addition, prescribed burning is a costly practice. With these constraints, will a public that often has a predetermined, value-laden, preferred ecosystem condition support widespread prescribed burning?

This chapter discusses the degree to which fire molded the prehistoric southern Appalachian landscape and suggests management strategies for reintroducing fire as a vector in ecosystem dynamics. When developing a fire management program, land managers need to consider the following principles and questions:

1. *Fire History* — How prevalent were natural and/or cultural fires during prehistoric and historic times? How has that fire regime changed following federal land acquisition?
2. *Regional Vegetation* — How has fire influenced the vegetation of the region? Where are fire-dependent communities, or remnants of these communities, located on the landscape? Can fire be reintroduced into these areas?
3. *Validity of Science and Procedures* — How does fire behave on the landscape? What is known about the effects of fire on regional soils and vegetation? Considering the topography of the region, are current prescribed burning guidelines truly applicable and effective for maintaining fire-dependent communities?
4. *Biodiversity* — Which vegetative communites will be gained and/or lost if fire is used as a management tool? Will regional biodiversity increase? Will landscape heterogeneity increase?
5. *Desired Future Condition* — What is the desired future condition of the landscape? Do land managers want to maintain fire-dependent communities or allow forest succession to remove these communities from the region?
6. *Public Support* — Are residents and visitors acceptive to using fire in forest management practices? Do they understand the role of fire in ecosystem dynamics?
7. *Policy and Economics* — Do current federal policies allow for effective prescribed burning in the regional terrain? Will operating budgets support operations to maintain or restore fire-dependent communities?

FIRE AND THE SOUTHERN APPALACHIAN REGION

PREHISTORIC SETTING

Fire played an important role in shaping the species rich landscape of the southeastern U.S. Fires of both natural and cultural origin were common on the landscape when the present arborescent flora migrated into the region after the last ice age, 8000 to 10,000 years ago (Delcourt and Delcourt, 1996). Natural fires, originating from lightning strikes, shaped vegetation of the southeastern Piedmont and Coastal Plain for thousands of years before humans arrived in North America (Komarek 1974). Lightning-caused fires in the southern Appalachian Mountains were not as frequent as in the pine–grasslands of the adjacent Piedmont and Coastal Plain (Van Lear and Waldrop 1989). The prehistoric landscape most likely contained open pine and oak forests with herbaceous understories (Delcourt and Delcourt, 1996). These fuels would have supported light surface fires that maintained open, park-like stands.

Today, considerable lightning is present in these mountainous areas but the opportunity for catastrophic fires under present fuel conditions is small (Barden and Woods 1974). The mesic broad-leaved forest that dominates as a consequence of high and evenly distributed rainfall and high humidity is not conducive to large, widespread fires of natural origin (Komarek 1974). Lightning strikes and subsequent fires are more common on ridgetops and at higher elevations. When extremely dry conditions prevailed, large fires could occur. Generally, however, lightning strikes result in small, low intensity, down-slope burns (Komarek 1974). Prior to human settlement, the fire mosaic of the southern Appalachian region was likely a pattern of light burns interspersed over the landscape at irregular intervals (Komarek 1974) with occasional large fires.

Modern man (*Homo sapiens sapiens*) became a part of the North American landscape between 12,000 to 15,000 years ago (Chapman 1985). These original human inhabitants entered North America via Beringia, a land bridge connecting North America with Siberia during much of the last ice age. These first Americans brought with them controlled use of fire, a trait characteristic of various *Homo* species for over a million years (Bass, pers. commun.). It was likely this ability that allowed Neanderthal man to be the first *Homo* species to move northward into the colder reaches of Europe during the Pleistocene ice ages. Likewise, control of fire allowed *Homo sapiens* to spread eastward across Berengia in pursuit of large, grazing megafauna (e.g., woolly mammoths, mastodons, etc.). Archaeological evidence suggests rapid spread of these peoples across the Americas as living conditions (both climate and food sources) improved as they moved southward.

When nomadic Paleo-Indians first arrived in the southern Appalachian region, the landscape was dominated by boreal forests (tundra or taiga), conducive habitat for the megafauna that was their primary food-source (Table 1). Over thousands of years, upland hardwood forests became predominant over much of the southeast in response to global warming culminating in their dominance over the past 7000 years (Delcourt and Delcourt 1991). Indian tribes of the Archaic Period were well dispersed throughout the eastern U.S. A rise in pines coincided with a rapid temperature rise during this period, suggesting the widespread occurrence of fire in the landscape. During the Woodland Period, eastern deciduous forests migrated into the southern Appalachian region, forming closed forests with two to three canopy layers. Without disturbance, deciduous forests develop a deep forest floor which, along with closed canopy conditions, essentially eliminates grasses from the landscape. The loss of grasses decreased the carrying capacity for grazing animals. This threat to their primary food source likely encouraged early cultures to initiate widespread burning for preserving grazing habitat.

As pre-Columbian Indian tribes became more sedentary, fire was used to clear fertile floodplains for cultivation (Chapman 1985). In the southern Appalachian region, these alluvial sites serve as nutrient and moisture "sinks" for surrounding areas. While these areas provide very favorable conditions for agriculture, they also support aggressive growth of native vegetation. Clearing agricultural sites and subsequent suppression of vegetative regrowth by prehistoric cultures was accomplished by burning. However, when the sheltered alluvial sites that were generally selected for farming were sufficiently dry for successful burning, fuels in the surrounding uplands were much drier. Under such conditions it is not likely that agricultural fires were restricted to agricultural sites. More important, fires escaping into the surrounding landscape provided numerous other benefits to prehistoric cultures (grazing habitat for wildlife, exposing nuts, encouraging blueberries and blackberries, etc.) (Williams 1994).

Pertinent to the degree to which cultural fires influenced the landscape is the question of human population densities in prehistoric America? This is a subject of much debate among geographers, anthropologists, and ethnologists. Until the 1980s, Mooney's estimate of less than one million people in North America (north of Mexico) was the accepted answer (Dobyns 1983). Today, new

TABLE 1
Relationships between Major Cultural Periods and Fire-Related Events, Conditions, and Attitudes in the Southern Applachians

Cultural period	Date	YBP[a]	Fire-related conditions and events
Present day	1990–1996	0	Most unintentional fires suppressed; the role of natural fire being evaluated
Technological Age	1900–1950	45	All unintended fires suppressed; Smoky Bear reigns supreme
Birth and implementation of a conservation ethic and movement	1900–1950	75	Widespread exploitation of resources; massive fires burning, largely in logging slash; destructive wildfires stimulate interest in, and action toward, a conservation movement
Post-Civil War	1865–1900	125	Discovery and exploitation of forest resources in the southern Appalachians; forest fires commonly followed logging; lower slopes cleared for agriculture.
The Settlement Period	1800–1865	160	Land "ownership" changes from Native Americans to Euro-Americans; Indian impacts much reduced due to disease/displacement.
The Post-Columbian Period	1500–1800	350	Landscape initially a mosaic of many forest/grassland types, seral stages, and vegetative conditions, from open grasslands to closed forests due to frequent landscape fires. During this period forests closed due to pandemics that largely eliminated cultural burning
The Mississippian Period	900–1500	800	Mound-building with highly structured Indian societies; landscape character largely controlled by Indian use of fire for agricultural purposes. Fire was also used in hunting and to maintain an open landscape for ease-of-travel and defense
The Woodland Period	1000 B.C.–900 A.D.	2000	Cultures increasingly agrarian; fire used to maintain grasses in the landscape for grazing animals and, later, to clear land for agriculture. Archaeological evidence of marked population increases that were widely dispersed over the southern Appalachian landscape
The Archaic Period	8000–1000 B.C.	6500	The beginning of agriculture, with fire used to clear suitable sites, usually along streams; high population numbers widely dispersed throughout the southern Appalachians. Fire used to expose nuts in grasses and forest litter and to favor berries
The Paleo-Indian Period	12,000–8000 B.C.	10000	Primary food source was the megafauna (mastodons, woolly mammoth, etc.), most of which became extinct at the end of this period. Landscape conditions were tundra/boreal forest. Fire was used in hunting these large animals, which maintained the open conditions needed for their survival. All North American cultures used fire for cooking and heating

[a] YBP = Years Before Present

evidence and new methods of calculating population numbers give much higher figures. Around 1200, the Indian city of Cahokia, located near present-day Saint Louis, MO, is estimated to have had a population in excess of London, England (Kennedy 1994). Some estimates of native American populations in 1492 go as high as 100 million. While this estimate seems unrealistically high, the more widely accepted range of 18 to 20 million native Americans dispersed over the North American landscape in 1492 (Dobyns 1983) would have made most areas subject to frequent cultural fires.

HISTORIC SETTING

Of the pre-Columbian Indian tribes encountered by early European settlers, the Cherokee Indians were the largest and occupied the greatest proportion of the southern Appalachian region (primarily western North Carolina, east Tennessee, north Georgia, and northwestern South Carolina with additional land claims in West Virginia and Kentucky) (DeVivo 1991). Their use of fire in hunting, gathering, and agriculture produced a "shifting mosaic of open grasslands, woodlands, and closed forests with widely scattered Indian villages" Buckner (1989). Early accounts by Europeans in eastern North America emphasize the open character of the forest (Guffey 1977) as well as people's misconceptions of fire's effects. Maxwell (1910) quotes the Discoveries of John Lederer (1891) concerning the burning practices of Virginia Indian tribes as follows:

> Virginia, …, was passing through its firey ordeal, and was approaching a crisis, at the time the colonists snatched the fagot from the Indian's hand. The tribes were burning everything that would burn, …, if the discovery of America had been postponed 500 years, Virginia would have been a pasture land or desert.

Other evidence of a more open landscape previous to the early 16th century is found both in pollen profiles and remnant plant populations that suggest parts of the region once supported a plains-type flora. DeSoto's chronicles of 1540 describe traveling for days through abandoned agricultural fields in the piedmont area in what is now South Carolina. DeVivo (1991) claims these same chronicles allude to a treeless French Broad Valley in western North Carolina up to the 3000 foot contour (2000 feet is the approximate elevation of the valley floor). Maxwell (1910) states that the Shennandoah Valley of Virginia, when first seen by Europeans, was treeless its entire length.

The greatest impact of European contact on the pre-Columbian fire regime was not the further addition of a cultural influence, it was the removal of one. Pandemics caused by diseases introduced after Cortes' incursion into Mexico began around 1519. Assuming that there were 20 million Native Americans in North America in 1492 and that pandemics killed approximately 90% of the native population, then only two million Native American inhabitants remained. This reduction in cultural pressure resulted in development of closed forests over vast areas that had been kept open by frequent cultural fires for thousands of years. Historical accounts of the Appalachian region, dating to approximately the Revolutionary War (1770s) often described a "wilderness." However, this wilderness was likely composed of the 200 to 250-year-old forests that established after the decline of the Native American population resulted in greatly reduced cultural burning.

Reduced by disease and warfare with settlers, Native Americans of the 16th, 17th, and 18th centuries had a more localized influence on landscape conditions. European settlers adopted the firing practices of the Indians (Pyne 1982) and augmented land-clearing practices with steel plows pulled by domestic livestock. Since there were no improved pastures, livestock "management" was accomplished by "firing the woods" to open the forest canopy and encourage grasses.

European settlement increased rapidly after the Revolutionary War. The rich bottomlands along the major streams were "developed" first. Further settlement moved farming and logging (and associated burning) practices up-slope into the mountains (Van Lear and Waldrop 1989). The southeastern landscape evolved in response to a mixture of European and Indian fire practices, one for farming and the other for hunting and ranging, respectively (Pyne 1982). As Indian influence in the region lessened, European settlers expanded burning throughout the mountains, primarily to clear land and enhance grazing in open forests (Van Lear and Waldrop 1989). Settlement activity in the Appalachian uplands increased markedly after the Civil War when land grants were given for military service.

Around 1880, railroads were constructed along both sides of the Appalachian uplands making their rich timber resources accessible to national and world markets. This resulted in rapid exploitation of the forests. Utilization standards of the time were such that most of each tree was left in the woods. Upon drying, the logging slash became highly flammable fuel for the inevitable spark that would escape from a locomotive's tinder box or from the clearing of garden sites.

FEDERAL LAND ACQUISITION

Localized fires continued to be common throughout the Appalachian uplands during late 1800s and early 1900s. As recently as the 1950s, "old-timers" claim that "over a 5-year period, every acre in Swain County, NC, would burn at least once." The fledgling conservation movement that developed in the late 19th and early 20th centuries did little to change this trend. Although the Weeks Act of 1911 and the Clark McNairy Act of 1924 authorized purchase of headwater areas in the Appalachian Mountains, it was not until manpower was made available by New Deal Programs (especially the Civilian Conservation Corps [CCC]) in the early 1930s that serious efforts were made to control fires.

Eliminating all fire from public forests was the official stance of federal agencies in the early 1900s when national forest acquisitions began and the U.S. Forest Service was established. In the 1920s, the USDA Forest Service was opposed to any use of fire in forests (Pyne 1982). At this time, even light burning was prohibited in the newly established National Forests. Foresters of that time did not realize the benefits of fire or that fire had played a major ecological role in the development and maintenance of the ecosystems they were trying to protect (Van Lear and Waldrop 1989). Their highly effective prevention and suppression practices essentially removed fire as a vector for shaping landscapes in the southern Appalachian region. A similar pattern occurred on private lands as state agencies improved their capabilities for both fire detection and fire suppression.

The creation of the Great Smoky Mountains National Park (hereafter referred to as the Park) in Tennessee and North Carolina, just as manpower became available through the CCC and similar programs, enabled the abrupt removal of fire as a vector shaping landscapes on this land. This was documented in a study of the fire history of the westernmost portion of the Park by Harmon (1982). During the period of Indian and European settlement (1856 to 1940), fires in pine forests of the Park were more frequent at lower elevations where they occurred once every 10 to 40 years (Harmon 1982). With the establishment of the National Park in the 1930s, the fire rotation increased to over 2000 years (Harmon 1982).

In the 1940s, Smokey Bear became the symbol of a highly effective campaign carried out by public agencies to eliminate all fire from North American forests. The need for this approach was more than justified by the conflagrations that followed the "cut-out and get-out" policies that characterized logging in the Great Lakes and Southern regions in the late 19th and early 20th centuries. Clearfelling was the standard practice in most even-aged stands. Low utilization standards of this era resulted in deep accumulations of logging slash which, upon drying, provided fuel that resulted in some of the largest and most destructive forest fires in recorded history. Individual fires consumed as much as 3 million acres and the Peshtigo fire, MI, killed 1500 people. As a result of Smokey's campaign, two to three generations have been taught that fire in the forest is bad. The success of this program has made research on the ecological role of fire in North Amereican forests difficult.

Although fire frequencies have been greatly reduced, fire has not been eliminated from the southern Appalachian landscape. Data from the Southern Appalachian Assessment (see Berish et al., Chapter 7 of this volume) documented that fires of anthropogenic origin are common on National Forest lands (Figure 1). Contrary to most historic burning practices that used fire for maintaining agricultural areas and wildlife habitat, the majority of human caused fires today are due to carelessness and arson. Records show that a few fires are started by lightning. However, most of these fires are restricted to ridge tops and are small in size. Due to effective suppression efforts, very few fires exceed 100 acres (40 hectares) in size.

Historically, appropriations to fire prevention and suppression programs have dominated forest management budgets, often to the exclusion of forest management activities. The continued need

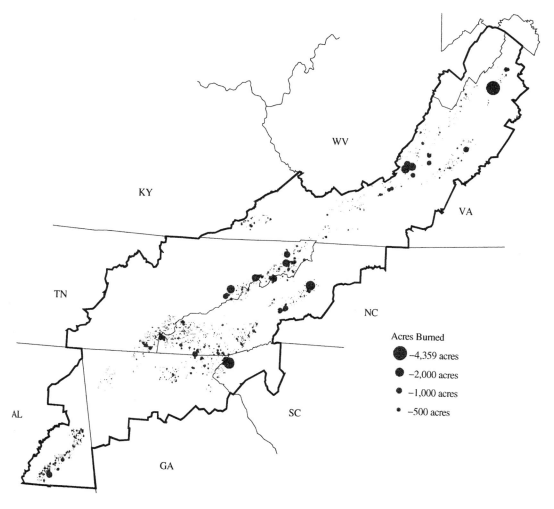

FIGURE 1 Fire occurrence on National Forest lands. (From National Interagency Fire Management Integrated Database.)

for these programs is apparent. However, the beneficial role of fire both in shaping and maintaining forest communities and in reducing fuels to prevent disasterous wildfires has not, until recently, been widely recognized.

Today public land managers are aware of the beneficial role of fire in forested ecosystems. Prescribed burning was first used in the "flatwoods" of the deep South for hardwood control, fuel reduction and to enhance wildlife habitat. Most of the operational guidelines for using prescribed fire in pine stands have been developed in this region where topography (slope, aspect, etc.) is not an important factor in fire behavior amd effects. For decades, fire has been used throughout the deep South as a site preparation technique to regenerate stands. Prescribed fire has not been widely used in the southern Appalachians because fire behavior is more erratic and the benefits are still questionable since hardwoods predominate in these systems (Van Lear and Waldrop 1989). Establishing parameters (both fuel and weather conditions) and firing techniques for using fire for ecological and other management benefits in the southern Appalachians should be a major concern of public land managers in this region.

TABLE 2
Microclimatic and Vegetation Conditions of Topographic Positions in the Southern Appalachian Mountains

Slope position	Aspect			
	SW	SE	NW	NE
Upper slope	Highly xeric Yellow pines	Xeric Pine–hardwood	Xeric Pine–hardwood	Somewhat mesic Hardwood
Mid-slope	Xeric Pine–hardwood	Somewhat xeric Pine–hardwood	Somewhat mesic Hardwood–pine	Mesic hardwood
Lower slope	Somewhat mesic Hardwood	Mesic Hardwood	Mesic Hardwood	Highly mesic Mixed mesophytic

FIRE AND ECOSYSTEM MANAGEMENT

Recognition of the role of fire in ecosystem processes has come from studies of past vegetation, identification of soil charcoal layers, fire scars on trees, the even-aged (pioneer) character of many forests, and records of explorers (Ahlgren 1974). Modern prescribed burning research has provided quantitative information about fire effects on forested ecosystems. Fires of both natural and anthropogenic origin have known abiotic and biotic effects on ecosystem properties and predictable behavior on the landscape. Abiotic effects include changing localized light and temperature conditions as well as altering properties of regional soils (Viro 1974; Christensen 1987; Boring et al. 1991; Groeschl et al. 1991). The primary biotic effect is the role of fire in controlling regional vegetation (Whittaker 1956; Quarterman and Keever 1962; Zobel 1969; Komarek 1974; Barden 1977; Greller 1988; Hartnett and Krofta 1989; Sanders 1992). The nature of these effects depends upon whether the fire is a fuel-reducing surface fire, a stand-replacing crown fire, or a destructive ground fire (Barbour et al. 1987). In turn, local environmental factors such as fuel type and arrangement, topography, and microclimate conditions influence fire behavior and severity.

FIRE BEHAVIOR

Fire behavior and frequency in the southern Appalachians is strongly influenced by topography. The extent to which fire carries and persists in the landscape depends upon factors such as fuel levels, time since the last burn, relative humidity, fuel moisture, wind, slope position and steepness, and aspect. Two abiotic variables of great importance are aspect and slope. These variables along with fire history largely control a third, biotic variable, vegetation. The interaction of fire with these variables produces a highly diverse landscape mosaic. Represented in this mosaic are essentially all of the site types that accomodate the full complement of the native flora (Table 2).

Morning sunlight is more effective than afternoon sunlight at driving transpiration and drying sites. As a result, the most xeric sites are found on southwest slopes and the most mesic sites on northeast slopes (Table 2). On all aspects, sites tend to become more xeric as one moves upslope.

On most lower-slope positions and throughout north- and east-facing slopes, forest cover generally consists of hardwood species whose fuels have relatively low combustability. Also, these positions generally have higher fuel moisture and relative humidity. Consequently, lower slopes burn less frequently and with less intensity than upper-slope positions. In general, lower-slope positions and north- and east-facing areas remain in hardwoods and hemlock and represent later seral stages. Surface fires in these areas typically kill only small hardwood stems. These fires also create basal wounds in some larger trees, enabling the entrance of saprophytic fungi, causing heart rot and resulting in hollow trees.

In contrast to north- to east-facing slopes, south- to west-facing slopes have higher transpiration demands that keep sites much drier with xeric conditions that increase upslope. Fires burn more frequently and, as permitted by fuel-load-recovery, with greater intensity in these areas. Such open canopy conditions allow shade intolerant, drought-resistant yellow pines to become canopy dominants. Under these conditions there is generally a transition from pure hardwoods on foot-slopes to mixed pine–hardwoods (most commonly oak–pine) on midslopes to pure yellow pine stands on upper slope positions.

Variations in fire frequency can maintain two distinct vegetative types on upper southwest-facing slopes. Intense fires occurring at wide intervals (decades) maintain pure yellow pine communities that grade downslope into pine–hardwood and more mesic hardwood stands, respectively. Stand replacing fires promoting regeneration of pure yellow pine stands are common on upper slopes with such burning. In contrast, where fires are more frequent, and less intense due to decreased fuel loading, fire stabilizes the community and open stands with abundant grasses develop. Where this condition persists, the eventual death of scattered trees would result in open, grassy balds.

FIRE EFFECTS ON SOIL

Regardless of ecosystem type, fire has generalized effects on soil physical and chemical properties. Differences between pre- and post-burn measurements typically show increases in soil temperature due to both the direct heating of fire and to the opening of the canopy. Soil warming improves conditions for microbial activity and nutrient uptake (Barbour et al. 1987; Viro 1974). Water holding capacity and wetability of mineral soil is generally enhanced by volatilization of hydrophobic substances such as monoterpenes (Barbour et al. 1987; Groeschl et al. 1991).

Following a fire, soil pH is higher due to the release of mineral bases in the soluble ash (Barbour et al. 1987). Changes in soil nutrients result from ash deposition, increased mineralization rates, and alteration of soil ion exchange properties, as well as the loss of nutrients through leaching and volatilization (Christensen 1987). Significant nutrient loss by leaching or volatilization only occurs with intense fires (Christensen 1987). Fire typically has positive effects on nutrient availability. Groeschl et al. (1991) reported increased total carbon and nitrogen levels of the surface 10 centimeters of mineral soil following low intensity prescribed burns in a fire-dependent table mountain pine (*Pinus pungens* Lam.)–pitch pine (*Pinus rigida* Mill.) community of the Shenandoah National Park, VA. These authors also reported increased mineral soil inorganic nitrogen levels in burned areas, conducive to increased plant uptake.

VEGETATIVE RESPONSE

Perhaps the most profound ecosystem effect of fire is its influence in altering and/or maintaining successional stages. Forest succession may be viewed as a directional, cumulative change in the species that occupy a given area through time. Specifically, forest succession is true secondary succession whereby species invade previously vegetated land following removal of the preexisting vegetation by natural or human-caused disturbance (Barbour et al. 1987). Fire drives forest succession by stabilizing, replacing, and/or directing community composition.

The duration of a seral stage depends upon the conditions that maintain it. As an exogenous disturbance, fire disrupts community and ecosystem structure by changing resource availability and the physical environment. These changes in turn produce heterogeneous patches across the landscape (Pickett and White 1985). The degree of heterogeneity that results (as related to fire intensity) affects the survival of residual organisms in the patch and the rate of invasion and success of establishment of new species (Pickett and White 1985). Therefore, forest vegetation following fire is the result of "the fluctuating and fortuitous immigration of plants and an equally fluctuating and variable environment" (Gleason 1926). In such situations, fire acts as an environmental filter which removes species lacking traits for survival in fire maintained communities from the species pool (Keddy 1992). Different species utilize different regeneration niches. The absence of appropriate regeneration niches is an important factor limiting species success (Clark 1991).

In the absence of disturbance, forests of the southeastern U.S. generally proceed from pine, to pine–hardwood, and finally to hardwood dominated communities (Quarterman and Keever 1962; Komarek 1974; Hartnett and Krofta 1989). When fire is present in these systems, forest succession is maintained in the pine (high fire frequency) or pine–hardwood (moderate fire frequency) seral stages (Quarterman and Keever 1962; Komarek 1974; Hartnett and Krofta 1989) with fire-dependent or fire-associated species dominating the community. When fire is excluded or suppressed in the ecosystem, regeneration niches for fire-dependent species are lost. Strong competitors (e.g., many hardwood species) can insure local extinction of weaker competitors (e.g., pine species) when both compete for the same regeneration niche (Clark 1991). Competition then influences the probability of surviving until the next recruitment opportunity, or patch-creating fire (Clark 1991).

Table mountain pine (*Pinus pungens* Lamb.) is the most obvious fire-dependent tree indigenous to the southern Appalachian Mountains. It is a shade intolerant, Appalachian endemic that occurs on thin soils, on southern and western slopes from central Pennsylvania to northeastern Georgia at elevations 300 to 1220 m (Zobel 1969). Table mountain pine is one of four yellow pines native to the southern Appalachian Mountains (the others being Virginia pine [*P. virginiana* Mill.], shortleaf pine [*P. echinata* Mill.], and pitch pine [USDA Forest Service 1965]). These are all pioneer species that only establish stands following disturbance (USDA Forest Service 1965). For successful yellow pine regeneration disturbances must create sites with exposed mineral soil in full sunlight. Table mountain pine also requires hot fires to open their serotinous cones (Zobel 1969; Barden 1978; Sanders 1992). Cone production begins when saplings are 5 to 7 years old. Thus the period during which there is no seed source is short. Reduced fuel levels following the hot fires that establish table mountain pine stands make it unlikely that another fire would occur during this 5- to 7-year period.

The success of table mountain pine regeneration also depends upon the type and intensity of fire as well as the understory vegetation. A post-fire study of the 1986 Bote Mountain, TN, fire in the Great Smoky Mountains National Park demonstrated that moderate to high intensity combined surface and crown fires are most effective in clearing organic matter to expose mineral soil, opening serotinous cones, and eliminating competition from hardwoods (Sanders 1992). The greatest proportion of table mountain pine seedlings (96%) was located in high and moderate intensity burns. Low intensity fires, however, were also beneficial as they killed back smaller competing hardwoods and occasionally created openings large enough to allow table mountain pine regeneration (Sanders 1992).

In the absence of disturbance, individual table mountain pine trees may persist for up to 200 years (Zobel 1969). Although Barden (1977) reported self-maintaining populations of table mountain pine on steep, xeric southwest slopes in North Carolina, regeneration is most often not successful in the absence of fire (Williams and Johnson 1990, 1992). Table mountain pine stands in the southern Appalachians have both other yellow pines (pitch, shortleaf, and Virginia) and hardwoods (primarily scarlet [*Quercus coccinea* Mueschh.] and chestnut [*Q. prinus* L.] oak) as codominants. The understory commonly contains mountain laurel (*Kalmia latifolia* L.) and scattered oaks, as well as other hardwoods (Zobel 1969; Barden 1977; Williams and Johnson 1992).

Because table mountain pine is very shade intolerant, populations are usually replaced by more shade-tolerant hardwoods in later seral stages in the absence of fire (Sanders 1992). Williams and Johnson (1990, 1992) noted the decline in recruitment and population maintenance of table mountain pine in southwestern Virginia pine–oak forests. In the absence of fire, the number of stems per hectare was greatly skewed toward larger diameter classes (15 to 35 cm diameter breast height) for table mountain pine, indicating little regeneration of this species. Chestnut and scarlet oak, on the other hand, were successfully regenerating in the understory and were predicted to become canopy dominants under continued fire suppression practices.

BIODIVERSITY

As an agent of disturbance, fire is important in maintaining high levels of species richness and diversity. Of particular interest to maintaining enhanced species diversity are the frequency, intensity, and size of fires (Malanson 1987; Pickett and White 1985). According to Connell's (1978) intermediate disturbance hypothesis, moderate levels of disturbance permit coexistence of species adapted to more extremes, thus maintaining richness and diversity. Fires of moderate intensity, size, and frequency should maintain the highest species diversity for a given community. In reality, it is not this simple. The level of intermediacy varies greatly from one community to the next. Huston (1979) suggests that a high level of richness and diversity should be maintained when disturbance recurs more frequently than the time required for competitive exclusion. In other words, the greatest packing of species niches along a resource gradient (i.e., the highest species richness and diversity) would be maintained as long as interspecific competition for limiting resources does not exceed intraspecific competition.

Malanson (1987) suggests that species diversity would be greatest where a mixture of "inertial species" (those tolerant of disturbance [e.g., fire-adapted species]) and "resilient species" (those species quick to respond following disturbance in order to reestablish ecosystem structure and function [e.g., herbaceous vegetation]) can coexist. Denslow (1980), however, emphasizes that regardless of the species or mechanisms involved, there is a strong relationship between site history and species diversity. She concluded that the relationship between the disturbance regime that characterized the evolutionary history of the community and the current disturbance regime was the key factor in determining competitive viability of species and, thus, diversity. This conclusion perhaps explains why "intermediate" levels of disturbance differ among communities and why deviations from these levels would actually decrease species richness and diversity. In these regards, knowledge of an area's fire history is crucial to maintaining the full compliment of native plant species.

Fire is important in maintaining diversity on landscape scales in addition to regional scales. Fire is a physical disturbance that alters landscape dynamics creating internal patch boundaries that differ from those created by edaphic and topographic factors alone (Wiens et al. 1985). Fire can be very local or widespread, depending upon burning conditions. Immediately following a fire, patchiness tends to increase. Abiotic and biotic changes induced by fire alter ecosystem processes (e.g., nutrient cycling, stand composition) and, in turn, landscape pattern (Turner 1989). If fires are suppressed, regional edaphic factors eventually level out landscape heterogeneity (Wiens et al. 1985) and the landscape appears more homogeneous.

Most systems in which fire is prevalent can be described as nonequilibrium systems, or those that never reach a stable end point (Sprugel 1991; Turner et al. 1994). The frequency and severity of fire determines the vegetation of particular regions and the various life history strategies of individual plants. Variations in fire frequency alter the mixture of species that dominate a landscape (Shugart 1984). It is these variations in fire frequency and severity that maintain high levels of heterogeneity and prevent fire-dependent systems from reaching equilibrium.

Perhaps the best example of a fire-dependent, nonequilibrium system is the lodgepole pine (*Pinus contort* Dougl.) forests of Yellowstone National Park. Historic fire records of Yellowstone National Park show that regenerating lodgepole pine forests do not burn readily until they are at least 200–250 years old (Sprugel 1991). Prior to the fires of 1988, Romme and Despain (1989) showed that these forests were dominated by decadent, 250- to 300-year old stands of lodgepole pine. Forests less than 250 years old were uncommon. Following the fires of 1988, young regenerating forests covered wide areas of the Park (Sprugel 1991).

As Christensen et al. (1989) reported, one of the most striking features of the 1988 Yellowstone fires was the resulting heterogeneity of the burned landscape and the variability of fire severity within burned areas. Fires previous to 1988 typically produced a mosaic of burned and unburned patches across the landscape. Within burned areas, varying degrees of burn intensities resulted in a variety of stand ages and, thus, a great degree of landscape heterogeneity (Turner et al. 1994). In 1988, however, fires in some areas were so extensive and so severe that the burned landscape actually appeared more homogeneous than the mosaic of forest age classes that had been present before the fires. She showed that post-fire landscape heterogeneity is a function of the size of fire-created patches. Specifically, fires burning at smaller scales are sensitive to environmental variables such as fuel moisture, fuel types, atmospheric humidity, wind, temperature, and topography. These smaller patches are very complex and irregular in shape and have a mixture of burn severity classes. In contrast, large fires respond primarily to wind velocity and direction. These fires are more uniformly severe and had more rectilinear (simple) shapes with a large proportion of the patch being severely burned (Turner et al. 1994). In other words, smaller scale fires of varying intensity increase landscape heterogeneity, whereas large-scale, severe fires most often decrease landscape heterogeneity.

FIRE MANAGEMENT OPTIONS

The importance of cultural fires in the prehistoric, southern Appalachian landscape poses serious problems for resource managers. Management guidelines commonly accept "prescribed natural fires," which allow fires of lightning origin to burn in areas where this "prescription" has been approved. Other policy directives call for "maintaining natural or pre-Columbian conditions" or those "untrammeled by man," both of which assume there was little or no cultural impact on the land. However, previous sections of this chapter demonstrate that most fires in the prehistoric southern Appalachian landscape were of cultural origin. Furthermore, it was largely these fires that set morphological traits making some regional species, such as table mountain pine, fire-dependent.

Habitat for such fire-dependent species has rapidly disappeared on protected lands throughout the southern Appalachian region. Whittaker's (1956) widely used 1952 vegetation diagram of the Great Smoky Mountains National Park showed a vast area in open oak–pine forests, relics of frequent fires that burned the area in pre-Park days. This condition is essentially nonexistent in the Park today, due to over 60 years of highly effective fire prevention and suppression. In addition, declines in fire-dependent communities have decreased biodiversity and landscape heterogeneity in the region.

Failure to take action to restore some of the fire-dependent communities will likely result in their permanent loss as ecosystem components. During the senior author's 40-year tenure as a dendrology teacher in the region, he has witnessed the loss of yellow pine communities over broad areas. The march of forest succession, combined with aggressive action to eliminate fire disturbances, has made this loss constant and pervasive. It is periodically speeded by outbreaks of the southern pine bark beetle as has recently occurred over much of the southern Appalachian region. On many sites that once supported yellow pine communities, the seed source has been completely eliminated. These changes often go unnoticed by federal land managers, due to their usually short tenure at any one location.

Ecosystem managers in the souhern Appalachian region should consider the following management options (and combinations thereof) in regard to the role of fire in ecosystem management policy (Figure 2):

Option #1:   Support a policy of suppression of fire with few or no management-ignited fires. This "safe" option will result in continued homogenization of the landscape. It will also result in rapid loss of biodiversity.

Option #2:   Maintain selected areas of fire-associated and fire-dependent species. This would be accomplished by choosing an area meeting management objectives. The area would be surrounded by artificial boundaries (e.g., roads, fire lines) to contain the

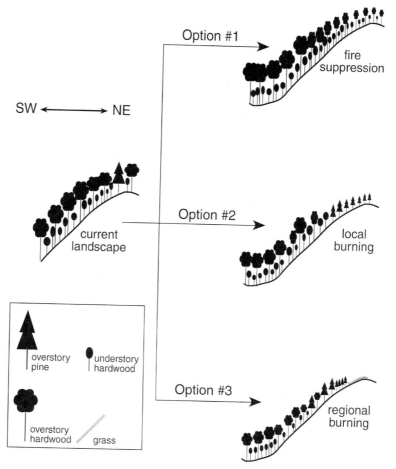

FIGURE 2 Resulting landscapes of fire management options available to land managers of the southern Appalachian region.

prescribed burn. The result will be a dissected, patchy landscape with sharp boundaries between treated and untreated areas. Lacking will be the normal ecotones between ecotypes. This option would not maintain the full range of seral conditions required to maximize biodiversity.

Option #3: Burn an entire watershed(s) containing a wide range of aspect and topographic conditions. This watershed would be fired in a stochastic manner at random intervals representing a wide range of burning conditions. While the fires will be confined to the watershed by natural boundaries (e.g., streams, ridge lines), within the watershed no atttempt will be made to control them. This treatment will maintain the widest range in seral stages resulting in maximum biodiversity. It will, in the long run, maintain an asthetically pleasing landscape containing the full range of ecotones between diverse landscape conditions (open meadows, grassy balds, large trees in old forests, etc.). Ecological, wildlife and asthetic benefits will be gained at the expense of productivity for forest products. In the long run there would be valuable timber only in valley bottoms, lower north and east slopes and in yellow pine stands on some upper southwest slopes.

Much of an ecosystem manager's decision as to which of these paths to follow should be determined after defining the "desired future condition" for the landscape. If increased hardwood dominance is preferred, then Option #1 would be best. Most of the public is not aware of the loss of fire-dependent species. Since these communities disappear by virtue of replacement by other vegetation there is a continuous vegetative cover on the land. Without aggressive public education as to the ecological role of fire in the life cycle of these species, the continuous forest cover provided by Option #1 would likely be more accepted compared to the periodic "scorched earth" condition needed to restore fire-dependent communities. However, Option #1 does not meet the guidelines of ecosystem management. Leaving it alone will result in continued loss of biodiversity in the landscape. It will not restore the "original" or "natural" landscape condition, but will tend toward a homogeneous condition that has not previously existed in the southern Appalachians. Option #2 would benefit an ecosystem manager who simply wished to maintain small, token "vignettes" of fire-dependent communities. This approach will also be used by the research community to establish guidelines for fire use and behavior.

To truly maintain fire-dependent communities and to maximize biodiversity and landscape heterogeneity, Option #3 would be necessary. Using table mountain pine as an example, one can conceptualize the steps needed to return this species to its prehistoric and historic dominance in ridge-top communities. Declines in table mountain pine populations over the years of fire suppression have greatly reduced the seed source for this species. Oaks have invaded these communities and, in many cases, now dominate the canopy. In addition, there is a tremendous fuel build-up on these areas. First, intense, small scale prescribed burning (Option #2) would be required to clear the forest floor down to mineral soil, eliminate hardwood competition, and allow table mountain pine regeneration. If the previous table mountain pine stand was much depleted, this process might have to be repeated on a 5- to 7-year cycle. Such burning practices would restore table mountain pine populations on a localized scale.

To restore a landscape-scale mosaic that includes the site-dependent gradient from oak to pine—oak to pine, watershed scale burning, as described in Option #3, will be necessary. Firing the watershed from the bottom and allowing it to run upslope would best restore the desired gradient. Then, using the stochastic firing method described in Option #3 would maintain the gradient and maximize biodiversity and landscape heterogeneity. Correctly determining the firing rotation necessary to maintain these conditions would require the establishment of permanent vegetation plots. Monitoring species composition of the overstory, understory, and regeneration will give clues as to the needed firing regime.

CONCLUSIONS

Land managers have much evidence that justifies the reintroduction of fire to the southern Appalachian landscape. Cultural fires were prevalent in the region during prehistoric and historic times. However, federal land acquisition in the early 1900s resulted in a marked reduction in fire frequency. Fire-dependent species that evolved during times of frequent burning have declined. Forest succession has established hardwood species on areas previously dominated by yellow pines.

Gaps remain in the knowledge of how fire should be used to maintain and restore fire-dependent communities in the southern Appalachian region. While prescribed burning has been used widely in "flatwood" regions of the southeastern U.S., it is a relatively new management tool in the eastern mountains. Although the effects of fire on regional soils and vegetation are well documented, knowledge of how a fire will behave on the steep topography of the southern Appalachian landscape is scant. Research should be conducted at localized and watershed scales, respectively, to determine how fire can be reintroduced effectively into these areas.

From the authors' viewpoints, the desired future condition for selected areas on some public forests in the southern Appalachian Mountains is the pre-Columbian condition. That landscape condition would include the following forest types: (1) mixed mesophytic communities on lower slope positions, (2) various hardwood–hemlock communities on north and east slopes (3) pine–oak mixtures on mid-south-to-west slopes, (4) yellow pine or grass-dominated communities on upper, south-to-west slopes and ridge tops. This gradation of communities would maintain the majority of southern Appalachian seral stages, thus maximizing biodiversity and landscape heterogeneity.

Lightning fires alone will not maintain these conditions. Management-ignited burns designed to mimic prehistoric cultural burning practices will be necessary. Specifically, both Option #2 and Option #3 will be needed to develop guidelines for reestablishing the desired condition.

Although most public land managers are well aware that such burning practices must be initiated to maintain healthy ecosystem dynamics, there are obstacles impeding implementation of effective fire management programs. Negative public opinion is one such obstacle. Most residents and visitors in the southern Appalachian region are not aware of the benefits of prescribed burning. Following research at a localized scale, programs and campaigns promoting the role of fire in ecosystem dynamics and forest health are needed before landscape-scale burning will be accepted.

Differences between the USDA National Forest Service and USDI National Park Service approaches to land management also limit regional fire management programs. The Forest Service focuses heavily upon ecosystem management. In contrast, accommodating the needs of tourists in an aesthetically pleasing, wilderness setting is more the focus of Park Service land management policy. While tourism is very important on Forest Service lands, maintaining ecosystem dynamics takes priority. Timber operations and prescribed burning are common. The Park Service appears to take a more protective approach to land management. Ecological processes are of great concern, but invasive actions such as management-ignited fires are rarely used for fear of adverse public perception.

Currently, the Great Smoky Mountains National Park fire policy is being revised to allow prescribed natural fire and some management-ignited fires. To date, however, no fires have been used to purposely restore fire-dependent communities. In contrast, over the past 2 years, six prescribed fires have been set on four National Forests specifically to study yellow pine regeneration as part of a research project being conducted by Region 8 of the USDA Forest Service and the authors.

RISK ASSESSMENT REPORT CARD FOR ECOSYSTEM MANAGERS

With these differences in the use of prescribed burning in mind, application of the *Risk Assessment Report Card for Ecosystem Managers* as applied to the role of fire in southern Appalachian ecosystem management is as follows:

Vision: A

Most federal land managers understand the importance of fire in regional vegetation dynamics.

Resource risk: D

In any prescribed buring project there is always the risk of fire escaping the control area. Land (both public and private) outside the burn area could contain marketable timber or other valuable resources or campgrounds. Furthermore, development near federal lands could mean that an escaped fire would enter residential or business districts.

Resource conflicts: B–

Debate continues on the desired future condition of the southern Appalachian landscape. Ecosystem managers must decide between a minimum biodiversity, homogeneous landscape with hardwood dominance or a maximum biodiversity, heterogeneous landscape with a mosaic of hardwood, pine–hardwood, pine, and herbaceous communities.

Socioeconomic conflicts: B–

Both residential and business encroachment near federal lands complicates prescribed burning procedures. Smoke management problems are of great concern during a prescribed burn. Insuring that wind is blowing smoke away from developed areas restricts burning conditions. Furthermore, tourism demands will likely limit burning to remote locations and to a time of the year when crowds are few.

Procedural protocols: B–

Prescribed burning protocols are well established. Long term monitoring programs for oberving the effects of the burn are not. Also missing is the knowledge of when to burn. How far should pine stands succeed to hardwoods before burning is necessary?

Scientific validity: B

The rugged topography, wide range in elevation and diverse cover types that characterize the southern Appalachian region will require concentrated research programs before safe, effective fire management programs can be initiated. In addition, further knowledge is needed on the season and intensity of fire best for regenerating fire-dependent communities.

Legal jeopardy: B–

The public may sue federal agencies in opposition to planned burns. While Environmental Assessments and Environmental Impact Statements prepared before the burn provide a legal blanket for the agency, lawsuits slow progress. In addition, a fire escaping the control area and possibly escaping boundaries of National Forest or National Park land could expose the respective agency to many costly lawsuits.

Public support: C+

For over 60 years, Smokey Bear has told the public that all fire in the forest is bad. Only since the Yellowstone fires of 1988 has there been a widespread effort to educate the public on the beneficial role of fire in ecosystem dynamics. Many residents of the southern Appalachian region view the mountains and their vegetation as majestic and sacred. Most do not understand that fire played a vital role in the evolution of the forests they hold so dear.

Adequacy of funding: C

The types of fires needed for landscape restoration are large in size and intensity and carry with them a risk for escape. Establishing control lines around proposed burn areas involves many hours of labor. In addition, a large crew would be needed to carry out the burn. Furthermore, helicopters are generally used both to ignite prescribed burns as well as to contain escaped spotfires. These factors combined make fire management programs very expensive. Current budgets will not permit burning on a landscape scale.

Policy precedent: C+

Federal policy philosophically supports the role of fire in ecosystem management. However, rigid guidelines stating when prescribed fire can be used often prevent technological support of the fire policy. These guidelines involve complicated restrictions based on weather, wildlife habitat, and proximity of residential or business areas. The windows of opportunity allowed by these guidelines are so narrow that significant prescribed fire programs cannot be implemented. For safety reasons, most of these guidelines are so conservative that landscape restoration burns are not possible. Furthermore, when conditions are right for large, intense prescribed burns, necessary support and standby crews are often fighting wildfires and are not available.

Administrative support: B–

Administrations of both the Forest Service and the Park Service support use of fire in ecosystem management under current policy. Congressional support and funding for expanded use of fire in the southern Appalachian region is poor.

Transferability: A–

Landscape restoration burning techniques learned in one region of the southern Appalachian Mountains should be applicable to other montane areas within the region. Cooperative studies involving multiple National Forests and National Parks would increase the efficiency in which knowledge is shared.

ACKNOWLEDGMENTS

The authors' involvement with public land managers in the southern Appalachian region has given them an appreciation for the benefits and difficulties that come with using fire as a managment tool. Many thanks go to Dr. David Van Lear, Clemson University, for his suggestions and review of this manuscript. Gary Brunk, Ginger Brudevold, Brian Spears (all of the USDA Forest Service), Roger Tankersley, and Carl Herman (of the Tennessee Valley Authority and the National Biological Service, respectively) graciously provided all maps. Saundra Campbell, The University of Tennessee, generated Figure 1.

REFERENCES

Ahlgren, C.E. 1974. Introduction, pp. 1–5. In T.T. Kozlowski and C.E. Ahlgren, Eds., *Fire and Ecosystems.* Academic Press: New York.

Barbour, M.G., J.H. Burk, and W.D. Pitts. 1987. *Terrestrial Plant Ecology.* 2nd ed. Benjamin/Cummings: Menlo Park, CA. 634 pp.

Barden, L.S. 1977. Self-maintaining populations of *Pinus pungens* Lam. in the southern Appalachian Mountains. *Castanea* 42:316–323.

Barden, L.S. 1978. Serotiny and seed viability of *Pinus pungens* in the southern Appalachians. *Castanea* 44:44–47.

Barden, L.S. and F.W. Woods. 1974. Characteristics of lightning fires in southern Appalachian forests. *Proc. Annu. Tall Timbers Fire Ecol. Conf.,* No. 13:345–361.

Bass, W. 1996. Department of Anthropology, University of Tennessee, Knoxville, TN. Personal communication.

Boring, L.R., J.J. Hendricks, and M.B. Edwards. 1991. Loss, retention, and replacement of nitrogen associated with site preparation burning in southern pine-hardwood forests. In S.C. Nodvin and T.A. Waldrop, Eds., Fire and the environment: ecological and cultural perspectives. *USDA For. Serv. SE For. Exp. Stn. Gen. Tech. Rep. SE-69.*

Buckner, E.R. 1989. The changing landscape of eastern North America. Pp. 55–59. In: 100 Years of Professional Forestry, Proc. Appalachian Soc. Am. Foresters, 71st Annu. Meeting, Asheville, N.C.

Chapman, J. 1985. Tellico Archaeology: 12,000 Years of Native American History. Rep. No. 43. Department of Anthropology, University of Tennessee, Knoxville, TN. 135 pp.

Christensen, N.L. 1987. The biogeochemical consequences of fire and their effects of the vegetation of the Coastal Plain of the southeastern United States. Pp. 1–23. In: L. Trabaud, Ed., *The Role of Fire in Ecological Systems.* Academic Publishing: The Hague, The Netherlands.

Christensen, N.L., J.K. Agee, P.F. Brussard, J. Hughes, D.H. Knight, G.W. Minshall, M. Peek, S.J. Pyne, F.J. Swanson, J.W. Thomas, S. Wells, S.E. Williams, and H.A. Wright. 1989. Interpreting the Yellowstone fires of 1988. *BioScience* 39:678–685.

Clark, J.S. 1991. Disturbance and population structure on the shifting mosaic landscape. *Ecology* 72:1119–1137.

Connell, J.H. 1978. Diversity in tropical rainforests and coral reefs. Science 199:1302–1310.

Delcourt, P.A. and H.R. Delcourt. 1996. Holocene vegetation history of the northern Chattooga Basin, North Carolina. *Conserv. Biol.* 11:1010–1014.

Delcourt, H.R. and P.A. Delcourt. 1991. *Quaternary Ecology, a Paleoecological Perspective.* Chapman & Hall: New York.

Denslow, J.S. 1980. Patterns of plant species: diversity during succession under different disturbance regimes. *Oecologia* 46:18–21.

DeVivo, M.S. 1991. Indian use of fire and land clearance in the southern Appalachians. Pp. 306–312. In S.C. Nodvin and T.A. Waldrop, Eds., Fire and the environment: ecological and cultural perspectives. *USDA For. Serv. SE For. Exp. Stn. Gen. Tech. Rep.* SE-69.

Dobyns, H.F. 1983. *Their Numbers Become Thinned.* University of Tennessee Press: Knoxville. 378 pp.

Gleason, H.A. 1926. The individualistic concept of the plant association. *Bull. Torrey Bot. Club* 53:1–20.

Greller, A.M. 1988. Deciduous forest. Pp. 287–316. In M. Barbour and W.D. Billings, Eds., *The Terrestrial Vegetation of North America.* Cambridge University Press: London.

Groeschl, D.A., J.E. Johnson, and D.W. Smith. 1991. Forest soil characteristics following wildfire in the Shenandoah National Park, Virginia. Pp. 129–137. In S.C. Nodvin and T.A. Waldrop, Eds., Fire and the environment:ecological and cultural perspectives. *USDA For. Serv. SE For. Exp. Stn. Gen. Tech. Rep.* SE-69.

Guffey, S.Z. 1977. A review and analysis of the effects of pre-Columbian man on the eastern North American forests. *Tenn. Anthropol.* 2:121–137.

Harmon, M. 1982. Fire history of the westernmost portion of the Great Smoky Mountains National Park. *Bull. Torrey Bot. Club* 109:74–79.

Hartnett, D.C. and D.M. Krofta 1989. Fifty-five years of post-fire succession in a southern mixed hardwood forest. *Bull. Torrey Bot. Club* 116:107–113.

Huston, M. 1979. A general hypothesis of species diversity. *Am. Nat.* 113:81–101.

Keddy, P.A. 1992. Assembly and response rules: two goals for predictive community ecology. *J. Veg. Sci.* 3:157–164.

Kennedy, R.G. 1994. *Hidden Cities: the Discovery and Loss of Ancient North American Civilization.* Free Press: New York. 372 pp.

Komarek, E.V. 1974. Effects of fire on temperate forests and related ecosystems: southeastern United States. Pp. 251–278. In T.T. Kozlowski and C.E. Ahlgren, Eds., *Fire and Ecosystems.* Academic Press; New York.

MacCleery, D.W. 1996. American forests: a history of resiliency and recovery. USDA Forest Service, FS 540. Forest History Society, Durham, NC.

Malanson, G.P. 1987. Diversity, stability, and resilience: effects of fire regime. Pp. 49–63. In L. Trabaud, Ed., *The Role of Fire in Ecological Systems.* Academic Publishing: The Hague, The Netherlands.

Maxwell, H. 1910. The use and abuse of forests by the Virginia Indians. *William and Mary College Quarterly Historical Magazine* 19:73–103.

Pickett, S.T.A. and P.S. White. 1985. Patch dynamics: a synthesis. Pp. 371–384. In S.T.A. Pickett and P.S. White, Eds., *The Ecology of Natural Disturbance and Patch Dynamics.* Academic Press; New York.

Pyne, S.J. 1982. *Fire in America: a Cultural History of Wildland and Rural Fire.* Princeton University Press; Princeton, New Jersey. 654 pp.

Quarterman, E. and C. Keever. 1962. Southern mixed hardwood forest: climax in the southeastern Coastal Plain, U.S.A. *Ecol. Monogr.* 32:167–185.

Romme, W.H. and D.G. Despain. 1989. Historical perspective on the Yellowstone fires of 1988. *BioScience* 39:695–699.

Sanders, G.L. 1992. The Role of Fire in the Regeneration of Table Mountain Pine in the Southern Appalachian Mountains. Master's thesis, University of Tennessee, Knoxville, TN. 125 pp.

Shugart, H.H. 1984. Categories of dynamic landscapes. Pp. 158–180. In H.H. Shugart, Ed., *A Theory of Forest Dynamics.* Springer Verlag: New York.

Sprugel, D.G. 1991. Disturbance, equilibrium, and environmental variability: what is 'natural' vegetation in a changing environment? *Biol. Conserv.* 58:1–18.

Trabaud, L. 1987. Fire and survival traits of plants. Pp. 65–89. In L. Trabaud, Ed., *The Role of Fire in Ecological Systems.* Academic Publishing; The Hague, The Netherlands.

Turner, M.G. 1989. Landscape ecology: the effect of pattern on process. *Annu. Rev. Ecol. System.* 20:171–197.

Turner, M.G., W.W. Hargrove, R.H. Gardner, and W.H. Romme. 1994. Effects of fire on landscape heterogeneity in Yellowstone National Park, Wyoming. *J. Veg. Sci.* 5:731–742.

USDA Forest Service. 1965. Silvics of forest trees of the United States. *Agriculture Handbook,* No. 271. 762 pp.

Van Lear, D.H. and T.A. Waldrop. 1989. History, use, and effects of fire in the Appalachians. *USDA For. Serv. SE For. Exp. Stn. Gen. Tech. Rep.* SE-54. 20 pp.

Viro, P.J. 1974. Effects of forest fire on soil. Pp. 7–46. In T.T. Kozlowski and C.E. Ahlgren, Eds., *Fire and Ecosystems.* Academic Press; New York.

Weins, J.A., C.S. Crawford, and J.R. Gosz. 1985. Boundary dynamics: a conceptual framework for studying landscape ecosystems. *Oikos* 45:421–427.

Whittaker, R.H. 1956. Vegetation of the Great Smoky Mountains. *Ecol. Monogr.* 26:1–80.

Williams, C.E. and W.C. Johnson. 1992. Factors affecting recruitment of *Pinus pungens* in the southern Appalachian Mountains. *Can. J. For. Res.* 22:878–887.

Williams, C.E. and W.C. Johnson. 1990. Age structure and the maintenance of *Pinus pungens* in pine-oak forests of southwestern Virginia. *Am. Midl. Nat.* 124:130–141.

Williams, M. *Americans and Their Forests: a Historical Geography.* Cambridge University Press; Cambridge, MA. 599 pp.

Zobel, D.B. 1969. Factors affecting the distribution of *Pinus pungens*, an Appalachian endemic. *Ecol. Monogr.* 39:303–333.

17 Land Use Planning: Sustainable Tourism

John D. Peine

CONTENTS

The Growth of the Tourism Industry..350
The Relationship of Tourism to the Environment..351
Principles of Sustainable Development Applied to Tourism 351
    Introduction 351
    Ecosystem Sustainability..352
        Habitat Degradation and Fragmentation, the Island Effect............................352
        Invasion of Alien Species...352
        Intrusion by Domestic Animals 352
        Conflicts with Wildlife 353
        Introduction of Pollutants..353
        Light and Noise Pollution..353
        Ecotourism Principles of Low Impact Services and Facilities........................353
    Cultural Sustainability...353
        Ecotourism Principles of Sustaining Local Culture 354
    Social and Economic Sustainability 354
        Ecotourism Principles Applied to Building a Sustainable Economy..............354
Principles of Land Use Planning for Sustainable Development.......................................355
    Steps in Land-Use Planning...355
        Visioning..355
        The Inventory Process 356
        Opportunities and Constraints Map 356
        Constructing the Plan 357
        Implementation Strategies 357
Tourism in the Southern Appalachians...357
    The Regional Setting for Recreation and Tourism...357
    Gateway Communities 358
    Sevier County...360
The Pittman Center Case Study 362
    History of the Community..362
    Proximity to National Park Service Properties 362
    Community Planning Process 362
        The Visioning Process..363
        Resources Inventory 364
    Conceptual Plan...365
        Community Vision Statement 366
        Ridge Top Ordinance...366
        Commercial Sign Ordinance 366

0-57444-053-5/99/$0.00+$.50
© 1999 by CRC Press LLC

    Architectural Guidelines for Commercial Structures 367
    River Corridor Ordinance 368
    Primary Street Design Standards 368
    Performance Zoning 369
    Sustainable Development Commission 370
    Relation to Greenbriar Park Entrance 371
    Implementation Strategies 371
Conclusion 371
    Risk Assessment Report Card for Ecosystem Managers 372
References 373

Tourism can be a mixed blessing. Many new tourism jobs require little training of rural residents because of the menial nature of the jobs, such as food servers, maids and retail clerks. These jobs are often seasonal, with low wages, few benefits and little chance for advancement. The better paying jobs, such as those in management, could require the costly training of rural residents and, as a result, often go to outsiders.

Fredrick 1992

...Then the side effects hit home: the sparkling, clear air begins to turn gray; traffic slows and snarls; parking gets more difficult; doors must be locked; taxes go up; and the quaint old buildings that made the town feel like home are replaced by massive blocks of cement and glass.What had been a close knit community begins to feel like an amusement park to which locals drive each day to operate the rides and sell trinkets.

Kinsley 1994

THE GROWTH OF THE TOURISM INDUSTRY

According to the World Tourism Council, tourism is now the world's largest revenue-producing industry. More money was spent on worldwide travel in 1993 than on world armaments. The industry employs 204 million people worldwide, or one in nine workers. Tourism is the world's leading economic contributor, producing 10.2% of the world gross national product. It is the leading producer of tax revenues at $655 billion and is the world's largest industry in terms of gross output, approaching $3.4 trillion. Tourism accounts for 10.9% of all consumer spending, 10.7% of all capital investment, and 6.9% of all government spending (Nasbitt 1994).

The travel industry is second largest in terms of employment behind health services in the U.S. Travel and tourism in the U.S. in 1993 generated $397 billion in expenditures — more than the furniture, clothing, or drug industries. In 1993, travel and tourism was once again the nation's leading export. More and more rural communities in the U.S. are looking to tourism for economic growth. Of the 2400 rural counties, only approximately 400 are considered to have economies centered on agriculture (Weaver 1993). In seeking alternatives to agriculture, mining, and forestry for sustainable economic growth, many communities are exploring tourism as a viable option because it tends to capitalize on the area's cultural, historic, ethnic, geographic, and natural resource uniqueness. Such strategies are increasingly being viewed as opportunities for keeping rural communities economically viable.

In the Southeast there were over 296 million tourist trips in 1994, accounting for $50.3 billion (Southeast Tourism Society Newsletter 1995). With major convention destinations like Atlanta, New Orleans, and Orlando; and with major attractions like Disney World, Historic Savannah and Charleston, the Florida beaches, the Gulf coast, and the Great Smoky Mountains in the southern Appalachians, the Southeast is a major tourist destination region for vacationers throughout North and South America. Tourism is a good investment for states in the Southeast where every dollar spent on tourism promotion helps generate $388 in travel related expenditures, a healthy return on the investment (Southeast Tourism Society Newsletter 1995).

As a $7.3 billion industry providing 136,700 jobs, tourism is the second largest employer in Tennessee behind health services. By the year 2000, it is projected to be the number one revenue-producing industry (TN Department of Tourism Development 1995). In the three most eastern counties in Tennessee, which encompass the Tennessee side of Great Smoky Mountains National Park, tourism is a leading industry (Economic Impacts of Tourism in TN 1994). The population of Sevier County increased by 23.2% (1990 U.S. Census; Bureau of the Census 1993) and its labor force by 44% (Tennessee Statistical Abstract 1992/1993) during the 1980s, making it one of the fastest growing counties in the state, due primarily to the explosion in investment in the tourism industry. Expenditures by tourists in Sevier County are estimated by the U.S. Travel Data Center to include $426.8 million in travel and tourism expenditures, $23.6 million in state tax receipts, and $11.5 million in local taxes. These expenditures are estimated to provide 10,660 jobs in the county. These estimates are quite low considering that the combined total gross receipts from sales reported for Pigeon Forge and Gatlinburg, TN was $833,360,205 in 1995 (Betty Goode, Gatlinburg City Manager's Office and John Jaggers, Director of Tourism Marketing, Pigeon Forge, pers. commun.). The receipts figure for Pigeon Forge ($529,320,594) represents an increase of 9% from 1994. Pigeon Forge has a population of 3619 and Gatlinburg 3400 permanent residents.

THE RELATIONSHIP OF TOURISM TO THE ENVIRONMENT

Like the circumstances in Sevier County, many tourist areas are anchored by natural resource attractions such as lakes, beaches, waterfalls, canyons/gorges, and mountains. Forests, prairies, deserts, and wetlands support wildlife that attract people in their search for renewal. A significant segment of the U.S. population prefers to spend a portion of their leisure time getting away from the day to day urban/suburban lifestyle and reconnecting with the natural environment. There is a huge latent demand for this type of leisure experience since 75% of the U.S. population lives in a nonrural environment (Bureau of the Census 1993). One of the fastest growing sectors in the tourism industry is ecotourism, where people are particularly seeking a more intimate experience with nature and/or indigenous culture closely tied to the natural environment.

As the use of natural landscapes for leisure and tourism escalates, a pattern of related tourism facilities and services can quickly evolve as well. All too often, the tourism industry becomes more than just support facilities and services for those interested in environmentally oriented recreation. As the tourism industry evolves, it can become a compilation of primary attractions totally unrelated to the natural environment and risks the potential to ultimately overwhelm the local culture and natural landscape. As the industry expands and develops, new clients representing different tourism markets are attracted to the area. These people may have less interest in the local environment and culture. In the southern Appalachians, this dynamic is best demonstrated in Pigeon Forge, TN, home of the nation's first outlet mall, which has become a haven for a plethora of motels, restaurants, amusements, music theaters, gift and souvenir shops, and additional outlet malls. This community is located only 3 miles from the main entrance to Great Smoky Mountains National Park.

PRINCIPLES OF SUSTAINABLE DEVELOPMENT APPLIED TO TOURISM

INTRODUCTION

As described by McCormick in Chapter 1 of this volume, sustainable development integrates economic, environmental, and social values during planning; distributes benefits equitably across socioeconomic strata and gender upon implementation, and ensures that opportunities for continuing development remain undiminished to future generations. This concept is extremely complex and suggests a dramatic shift in philosophy in the capitalist society that is so ingrained in the U.S. Can economic forces be refocused to facilitate *sustaining over the long term* the integrity of cultural and natural resources when they generally have not previously been so oriented even in the short term? In a corporate world directed by chief executive officers whose primary reward for maximizing short-term return to their shareholders is to receive millions of dollars of stock options themselves, the disincentives in society toward such a philosophy are numerous. Federal entitlement programs such as Social Security, Medicaid, and Medicare are certainly not geared to long-term financial sustainability either. This situation is recognized by the President's Council on Sustainable Development. One of their principles is that sustainable development requires fundamental changes in the conduct of government, private institutions and individuals (President's Council 1994). These principles are a very tough sell in the corporate–political landscape of the U.S., the epicenter of instant gratification, but they are critical to the future long-term social and economic vitality of the country and the world.

Application of these principles to the tourism industry is being most actively adopted by the rapidly emerging ecotourism phenomenon, the fastest growing sector of the tourism industry. The Ecotourism Society has related these principles to the environmental tourism industry and is promoting their adaptation as described below (Ecotourism Society 1993).

ECOSYSTEM SUSTAINABILITY

Human intervention should not impact ecosystem sustainability by destroying or significantly degrading components that effect ecosystem capability (Kaufmann et al. 1994). There are a wide range of potential adverse impacts associated with tourist-oriented communities which tend to be exacerbated adjacent to protected natural areas such as national parks and forests. Categories of impacts are as follows.

Habitat Degradation and Fragmentation, the Island Effect

Of particular concern is the loss of habitat for rare, threatened or endangered species. Land use conversion from a natural to a built environment creates physical barriers which can adversely impact ecosystem dynamics such as nutrient cycling and species propagation and foraging. As communities adjacent to protected natural areas expand, their boundaries extend along an ever-increasing segment of their borders with the protected area. There is potential to eventually establish such a formidable barrier as to create an island effect, isolating the protected area from the surrounding natural habitats occurring on the broader scale landscape. This is discussed by Berish et al. in Chapter 7 in this volume.

Invasion of Alien Species

The built environment can provide a venue for the invasion of alien (nonnative) species of plants and animals, many of which can be devastating in their impact on native species. Many national parks including the Great Smokies, list alien species among the greatest threats to their biological resources. In most cases, control of exotic species is at best problematic. An example of how tourism can accelerate dispersion is the transport of gypsy moth on recreational vehicles from locations of infestation to uninfested areas (see Schlarbaum et al., Chapter 14, this volume). This topic is also discussed in the chapter on European wild boar (Peine et al., Chapter 13 in this volume).

Intrusion by Domestic Animals

Feral dogs and cats hunt and run wildlife along the boundaries of protected areas. They can disturb ground-nesting birds. Some are used to illegally hunt wildlife such as black bear (Clark and Pelton, Chapter 10 in this volume).

Conflicts with Wildlife

Artificial food sources in the built environment, such as unprotected garbage, can create problems of altering the natural behavior patterns and diet of wildlife. This is alluded to in Chapter 10 on black bear management. People can come in contact with wild animals carrying infectious disease such as rabies, when lured by artificial foods.

Introduction of Pollutants

There are a multitude of pollutants routinely introduced into the air, water, and soil. The levels of toxicity and longevity in the environment vary dramatically and can be devastating to entire surface and/or ground water systems and airsheds. Chemical, power generation, and agriculture industries are all primary sources. This is addressed in the chapter on regional assessments (Berish et al. in Chapter 7 and Peine et al. in Chapter 15 in this volume).

Light and Noise Pollution

The natural ambiance in the protected area can be seriously invaded by an intrusive adjacent presence of noise from traffic and/or sightseeing helicopters, lights from adjacent communities, and visual blight of the built environment. In Sevier County, TN for instance, all the communities except Pittman Center have adapted various versions of "festivals of light." Hundreds of thousands of dollars are collectively being invested by communities every year to increase their displays creating ribbons of light as much as 23 miles long throughout the county from Interstate 40 in the north to the national park boundary in Gatlinburg. Highways are being widened to increase traffic to the primary national park entrance. The nation's first outlet mall, called the Red Roof Mall, is located in Pigeon Forge. On a clear day, those red roofs can be seen from numerous trails in the park and from Mount LeConte and Clingmans Dome, two of the Park's most famous vista points. At night the view of these communities from the Park becomes even more obtrusive.

The greatest annoyance to community residents beyond traffic congestion is the noise from helicopters giving sightseeing tours. The town government of Pigeon Forge has conducted a long legal battle to control their use in their community, but there are no controls elsewhere around the Park.

Ecotourism Principles of Low Impact Services and Facilities

Application of these principles can help mitigate some of the aforementioned adverse impacts among geophysical, biological, and social components of an ecosystem. Facilities and services associated with the ecotourism industry should ideally utilize energy conservation practices at every phase of the operation. Utilization of alternative energy sources to fossil fuels provides a particularly poignant demonstration of adapting sustainable life styles. The tours should have minimal impact on the environment. Care should be given to minimally disturb native plants and animals during the environmental tour. Native plants should be used in landscaping. Waste should be minimized, composting practiced, and byproducts recycled to the extent practical. Organic materials should replace the use of toxic chemicals wherever practical. The interconnectedness and cumulative effects of all aspects of the operation should be conveyed to the tour clientele. The educational value of the operation will probably be its most significant benefit. In the southern Appalachians, the Smoky Mountain Field School is a good example of using a natural area for tourism purposes with minimal impact on natural resources.

Cultural Sustainability

Cultural resources take many forms, ranging from historic structures to ethnic communities and neighborhoods, festivals, crafts, traditions, cuisine, and dialect. These cultural elements constitute the social fabric of the landscape and oftentimes reflect a community's collective orientation to the natural environment. These poignant dimensions of society can be neglected in community development planning. The tourism industry that adversely effects the natural environment invariably adversely impacts indigenous cultures as well. The influx of tourists changes the scale of the community. Local culture is in effect sold to the tourist in terms of community cultural events, local cuisine, tours of historic structures, etc. Even the hospitality and goodwill of the local population is marketed. Commercialization of local culture can at best become disingenuous and at worst become highly distorted, accentuating negative stereotypes. In the tourism industry, this practice takes the form of performing and visual art, souvenirs, and architecture. In Sevier County, TN for instance, numerous commercial tourist establishments play on the negative stereotyping of the indigenous southern Appalachian culture. Reference to "hillbillies" is very pervasive.

Ecotourism Principles of Sustaining Local Culture

A key theme is to maintain the integrity of the indigenous culture. Ecotours should strive to demonstrate by experience the interconnectedness of indigenous cultures and the natural environment. These isolated cultures are likely to have inherently practiced sustainable living for centuries, opportunistically taking advantage of natural food sources without exhausting them. The relationship of culture to nature is invariably manifest in ceremony, art, cuisine, architecture, spiritual doctrine, and life style. Self discovery of these dynamics by the tourist can be most introspective and exhilarating; yet an isolated distinctive culture can be jeopardized when tourists are interjected. Care must be taken to judge how much exposure is allowable without destroying the integrity of the local culture.

Examples of sustainable community heritage associated with tourism in the U.S. can be found in the Mennonite and Amish communities and on Native American reservations.

Social and Economic Sustainability

The tourism industry can be a key economic sector, particularly in small, rural communities with minimal economic diversity. The tourism industry is labor intensive. However, many of the jobs are seasonal in nature, occurring only during the peak of the tourist season. The majority tend to be minimum wage, with few or no benefits. Many motel, gift shop, and restaurant operators utilize employees on an as-needed basis. In areas where the tourist industry is expanding rapidly, the work force willing to work for low wages and minimal benefits can become fully employed, creating opportunities for employment from adjacent communities. Sevier County businesses employ numerous individuals from neighboring Cocke County, which has much higher unemployment. The social equity of these circumstances has been debated over the years (Smith 1989; Fredrick 1992).

Another social ramification of a rapidly growing tourist-based economy is the strain on the infrastructure of the community, such as schools, roads, water supply, solid waste management, police, and emergency services. At the same time, taxes collected from tourist expenditures are often the primary funding source that support those facilities and services.

Michael Kinsley challenges the notion that growth is the solution of economic problems (1994). He points out some of the fallacies of this almost universally held axiom. The trappings that come with rapid growth can soon overwhelm the character of the community. Clean air and water and a distinctive sense-of-place are obscured by traffic jams, polluted streams, and generic architecture reflecting the franchising of the American commercial landscape (Kinsley 1994). The greatest tragedy is the potential loss of sense-of-place that provides the cultural roots of a community (Hiss 1990).

Ecotourism Principles Applied to Building a Sustainable Economy

The term "sustainable development" has become the catch phrase of the 1990s in community planning circles. The question is, to what is the term being referred? In most cases, it is suspected to be referring simply to sustaining economic growth over the longer term and not particularly reflecting the social values of sustaining cultural and natural resources of the community for future generations.

Key economic tenets of sustainable ecotourism are to turn over tourist expenditures in the local economy, provide for an equitable distribution of benefits, diversify the local economy, alter consumer patterns to support conservation practices, actively market to attract premium value markets, and always focus on long-term consequences.

These principles may be manifest by the following strategies to capture and circulate the tourist expenditures at the local level:

Promote local ownership and/or operation of ecotour businesses.
Use local raw materials.
Use local people for laborers, service providers, guides, managers and contractors.
Expose tourists to the local cuisine.
Provide an opportunity for them to purchase locally made arts and crafts.
Provide a quality experience for which clients are willing to pay a premium, thereby providing
    more cash flow per tourist into the local economy.

PRINCIPLES OF LAND-USE PLANNING FOR SUSTAINABLE DEVELOPMENT

Frederic Sargent and his co-authors in their book, *Rural Environmental Planning for Sustainable Communities* (1991), listed the following values associated with the concept of sustainable development in rural communities:

- "Rural people place a high value on self reliance and self-determination. They have experience with techniques for cultural and economic survival. They can make decisions regarding their long-term interests, design and carry out programs, evaluate the results of their work, and make necessary adjustments.
- Rural people value cooperation as a guide to problem solving. This attitude has evolved from generations of experience in rural living, where cooperation is a major tool of survival and community maintenance.
- Long-term sustainability of a rural environment is achieved when citizens guide economic development according to the "physical carrying capacity" of the ecosystem. Land ownership is valued not just for its market value but also for sustaining ways of life. Consideration of the ecosystem's physical carrying capacity assumes that, although efficiency of use can vary, physical and natural resources are finite and can bear only so much use.
- Increasing the self-reliance of citizens in rural communities can be the basis for sustainability. A self-reliant community possesses the knowledge, skills, resources, and vision to identify changing conditions, locate appropriate technical assistance and initiate actions in a manner that conserves the rural environment and distributes benefits in an equitable manner."

Most of the communities with tourism based economics in the southern Appalachians are of small scale and occur in a rural setting. Examples include Rugby, TN; Helen, GA, and Beech Mountain, NC.

Steps in Land-Use Planning

Visioning

The process of building a consensus among stakeholders to define a vision for the future is time-consuming, tedious, and occasionally controversial, yet of fundamental importance. Any strategy for managing growth must be well grounded in community support as defined by the visioning process. Without it, the land use plan quickly becomes delegated to a shelf and is soon forgotten. This should be the first phase of the planning process and should continue throughout. Visioning is a continuous process that invariably occurs as decisions concerning development are made incrementally over time.

Community stakeholders need to provide input through the planning process. They can define the cultural and natural resource significance for the community. They can become involved in the analysis of the inventory database. In order to communicate adequately and cajole broad based cooperation, a variety of techniques are suggested: newsletter; community meetings; guest speakers describing other community efforts in land use planning and growth management; focus groups; questionnaires; school educational materials; planning charettes to collectively compile desired features in designs for commercial signs and buildings, roadways, and parks; and special events to dramatize community cultural and natural heritage. Also, citizens can be requested to send in pictures defining the people and places of importance. See the case example of Pittman Center, TN which is described later for more specific examples (Peine and Welsh 1990).

The Inventory Process

The natural resources inventory should include vegetation patterns; wildlife and wildflower resources; critical habitats such as wetlands and riparian zones; existing and potential greenway linkages among natural areas; aquatic resources such as rivers, streams, and lakes; topographic related features such as steep slopes, ridge or mountain tops, overlooks, and gorges; and agricultural and forest lands. Population distribution and habitat for species of concern should be identified as well. Cultural resources should include archeological sites, historic structures, neighborhoods, and landmarks that distinguish the local heritage.

The built environment inventory should include the infrastructure of public facilities and services such as roads and pedestrian pathways, sewer lines and treatment facilities, water plants and distribution systems, power plants and distribution systems, storm water systems, and school and recreation facilities. The public facilities that have the greatest impact on development are transportation systems, public water supply, and sewer systems. Point sources of pollution into the air, surface waters, and ground should be identified.

Socioeconomic information from the U.S. Census can be applied to characterize housing, income, employment, ethnicity, age and gender distribution, and other parameters that help to define the population. Information on types of employment and the location of major employers and employment centers is important as well. Defining the patterns of commuting to work is also useful. Community population and economic forecasts are also extremely helpful.

Other information that should be obtained in the inventory phase includes land ownership, current zoning regulations, and any federal, state, or local laws, regulations, and ordinances that may be applicable to land use. This information is very useful in determining the likelihood of development in various areas of the community.

Opportunities and Constraints Map

The next step is to analyze and interpret information collected during the inventory phase. This process will demonstrate the adequacy of the information compiled. Based on information collected during the inventory phase, an opportunities and constraints map can be prepared to show where new development is most suitable and where it is the least or totally unsuitable. This map can be created using a series of prioritized themes of constraints and resource attributes. The principles for such an analysis were introduced by Ian McHarg in his book *Design with Nature* in 1969. Such a process can be greatly enhanced by applying computerized mapping, a geographic information system, to create the envisioned overlays to screen for prioritized hazards and protection of resources of concern.

Resources desirable to protect might include ecologically sensitive areas such as wetlands, beaches and sand dune systems, wildlife habitat and wildflower areas; economically productive resources such as mineral deposits, productive soils for agriculture; aquifers for drinking water supply; recreational opportunity areas such as streams, rivers, lakes, greenways or meadows, and designated parks and publicly owned open space; visual resources such as overlooks, ridgetops, and steep slopes, unusual land forms and shorelines; and historically and archeologically significant sites and structures that help to preserve an area's heritage and create a sense of place (Hendler 1977).

Hazardous areas to avoid might include steep slopes subject to erosion, flood prone areas and major water drainage paths, seismic zones, soils that do not percolate well or support heavy structures, sink holes or cavernous areas that are subject to collapse, and solid waste and point source pollution sites that could endanger human health.

Constructing the Plan

Using the building blocks expressed above, a conceptual land use plan can be devised. This is an extension of the visioning process where one transitions from expression of general principles to actual resource implications on the ground. As expressed by Godschalk et al. 1979, communities and regions that desire to manage their growth "seek to influence the location, amount, type, timing, quality, and/or cost of development in accordance with public goals."

Implementation Strategies

This is the reality check part of the process. As expressed by Kaiser et al.(1995), communities can apply a wide variety of tools to implement their land-use plan. These range from impact fees, urban line limits, design guidelines, development agreements, capital improvement programs and public facility ordinances to traditional zoning and subdivision controls. The innovative performance zoning concept is illustrated in the following Pittman Center case study. Realistic implementation strategies are an integral part of the planning process. Any land use or community plan without specific implementation strategies is ultimately of little use.

TOURISM IN THE SOUTHERN APPALACHIANS

THE REGIONAL SETTING FOR RECREATION AND TOURISM

The southern Appalachians are well known for scenery, recreation opportunities, and the traditions and culture of its people. A 1994 report by the Appalachian Regional Commission notes that:

> The well-known positive images of the region — beautiful mountainous scenery and unlimited outdoor recreation possibilities — are why most tourists who go to Appalachia do so. However, the unknown attractions and activities in the region represent an untapped potential for the region to attract the types of travelers who are looking for a wide range of different activities in a vacation destination.

Of the 37 million acres of land in the 135-county, 6-state region defined in the Southern Appalachian Assessment (Berish et al., Chapter 7 in this volume), 16% is publicly owned. Approximately 845,000 acres are in national parks and an additional 580,000 acres are in state parks and other state land classifications. There are six national forests in the region, totaling 4.4 million acres. In 1990, over 15 million recreation visits occurred in those national forests. The 470-mile-long Blue Ridge Parkway connects with the Shenandoah National Park in the north and Great Smoky Mountains National Park in the south. In 1995, the Blue Ridge Parkway had 17.5 million recreational visits, Shenandoah National Park had 1.8 million visits, and the Great Smokies National Park entertained 9.8 million visits from an estimated 1.8 million individuals. The Smokies is not visited by 9+ million people per year as is frequently reported. This wide disparity between the number of visits vs. visitors is much greater in the Smokies than in most other national parks because the visitors make multiple visits into the park per day (1.3), per trip to the area (3.9) and per year (5.3) (Peine and Renfro 1988).

There are a total of 3500 miles of scenic byways and 10,000 miles of trails, mostly through the mountainous highlands in the region. Approximately 45% of the 37-million-acre region is in a rural setting. Approximately 18% is developed into urban, suburban or transitional settings. Nearly 8% of the region (2,980,000 acres) is considered remote or has the potential for remoteness. Large rivers and lakes account for 2% of the region.

There are 36,000 camp sites in 218 campgrounds. Some 70% are privately owned. There are 33,000 miles of trout streams. There are over 17,000 day use picnic and/or swimming areas — and there are over 4 million acres of game lands suitable for hunting in the region (SAMAB 1996).

GATEWAY COMMUNITIES

Communities adjacent to national parks and forests in the southern Appalachians have particularly strong vested interests in the natural resources in the region. By their proximity, these communities also have the potential for greater adverse effects on natural resources than those located in long-since disturbed landscapes. Examples of gateway communities include Cumberland Gap, TN; Dahlinega, GA; Maggie Valley, NC; Gatlinburg, TN, and Stearns, KY. First and foremost, residents of these communities are bound to identify with the natural resources within the townscape and regional setting. This intimate relationship to parks and forests impacts the sense of place that citizens hold for their communities. Segments of the local economy invariably will utilize natural resources directly via agriculture and timber harvest or indirectly via recreation, tourism, and second/retirement home development. The social and economic association with the environment very much reflects the public attitudes concerning related values. People living in these communities with a strong natural resource-related sense of place tend to have stronger opinions concerning the environment in which they live than those residing in a highly urbanized environment.

As the term implies, gateway communities can have a significant impact on the visitor experience associated with adjacent national parks or forests. The community sets the regional context of the access corridor. The first impression of those entering the area sets levels of expectation from which stereotypes are formed. This impression is not limited to visual stimuli, but includes the quality and variety of services provided to the tourists as well. A four-star gateway community might enjoy visitor comments like "Wow, what a beautiful setting. The food was great! Everyone was so helpful and friendly to us. We loved our room in the lodge."

As discussed above, activities within these communities have the potential to adversely impact resources within protected areas. Concern for the impact of rapid tourist development in gateway communities in the southern Appalachians led to a study funded by the Economic Development Administration for the Southern Appalachian Man And Biosphere Cooperative (SAMAB) which explored attitudes of residents concerning the impact of tourism on their communities. The study was designed by the SAMAB Sustainable Development Committee and executed by Dr. Paul Jackus and Dr. Paul Segal of the Department of Economics and Rural Sociology, the University of Tennessee at Knoxville. In the summer of 1993, mail back questionnaires were distributed to public officials, business owners/operators and conservationists living in the six counties surrounding Great Smoky Mountains National Park. As displayed in Table 1, respondents felt that tourism brought many positive economic and social benefits to their communities. Most believed that the standard of living is greatly increased due to the tourism industry. Over 80% however, did not feel that economic gains were more important than environmental protection. As displayed in Table 2, there

TABLE 1
Response to Question:

"Tourism development can result in trade-offs between the positive and negative aspects of the tourism industry. Given existing tourism opportunities in the community, please indicate whether you agree or disagree with the following statements:"

	Strongly agree	Agree	Uncertain	Disagree	Strongly disagree
Tourist attractions and facilities improve the appearance of this community (N=570)	13.2	34.7	26.3	19.6	6.1
Tourism increases the quality of life in this community (N-574)	25.4	45.3	15.0	10.1	4.2
Tourism reduces the quality of outdoor recreation in this area (N-572)	7.0	17.7	19.2	42.1	14.0
Increasing the number of tourists visiting this community would improve the local economy (N-578)	51.9	36.7	7.3	2.9	1.2
Tourism brings better jobs to this community (N=576)	27.4	28.6	22.9	16.5	4.5
Our standard of living increases considerably because of the money that tourists spend and the tax revenue they generate (N=578)	35.6	37.9	14.2	9.9	2.4
The negative social impacts of tourism outweigh the positive economic contribution of tourism (N=570)	9.8	14.4	20.9	35.4	19.5
The economic gains of tourism are more important than environmental protection (N=576)	2.3	5.2	11.8	40.6	40.1
Local residents have suffered from living in a tourist area (N=576)	4.0	13.0	20.1	45.3	17.5

Source: Jackus and Segal 1993.

TABLE 2
Response to Question:

"Please think about your vision for the future of the community. Now…, give your vision of the future"

	Strongly agree	Agree	Uncertain	Disagree	Strongly disagree
Tourism should play a major role in the community's future (N=583)	56.4	29.8	5.5	5.1	3.1
The character of the community should be preserved (N=583)	63.8	27.4	7.2	1.0	0.5
This community should control tourism development (N=580)	54.8	34.3	6.9	2.4	1.6
Long-term planning by public agencies could moderate the impact of tourism on the environment (N=577)	54.1	34.7	8.3	1.7	1.2
Because of tourism development there will be more recreational activities for local residents (N=582)	35.2	40.4	16.3	5.5	2.4

Source: Jackus and Segal 1993.

was strong consensus that tourism should play a major role in the future but that the community's character should be preserved. Over 85% felt that the community should control development and that long-term planning by public agencies could moderate the impact of tourism on the environment.

Responses to the question, "Please indicate the degree to which the following issues associated with tourism development in the community are of concern to you?" are displayed in Table 3. The primary concerns expressed were with traffic congestion and the seasonality and low pay of the

TABLE 3
Response to Question:

"Please indicate the degree to which the following issues associated with tourism development in the community are of concern to you."

	Major concern	Minor concern	Not a concern	Don't know
Jobs in the tourism industry tend to be low paying with minimum benefits (N=572)	42.7	41.4	13.5	2.4
Jobs in the tourism industry tend to be seasonal part-time jobs (N=575)	48.5	37.7	12.9	0.9
Jobs in the tourism industry are service oriented (N=565)	18.1	31.5	48.3	2.1
Growth in the tourism industry may stifle economic diversity in the community (N=566)	14.0	30.0	50.2	5.8
Growth in the tourist industry will...				
Cause traffic congestion (N=578)	48.4	39.8	11.2	0.5
Strain the community's ability to maintain roads (N=573)	31.1	37.7	29.0	2.3
Strain the community's police and fire protection services (N=576)	30.4	40.5	26.9	2.3
Strain the community's water supply (N=571)	32.9	35.4	28.7	3.0
Increase housing costs (N=568)	24.3	42.8	30.3	2.6
Strain the community's sewage treatment facilities (N=570)	40.4	33.0	22.3	4.4
Strain the community's solid waste disposal capacity (N=571)	45.4	32.9	17.7	4.0
Strain the community's primary and/or secondary school system (N=572)	18.7	34.1	43.7	3.5
Strain the community's emergency medical services (N=572)	21.7	43.0	33.0	2.3

Source: Jackus and Segal 1993.

labor force. Strains on sewage and solid waste were also expressed as a concern by approximately three out of four respondents. Table 4 displays responses to the question "Which of the following factors associated with tourism adversely affect the community?" Signage was the most frequently mentioned distraction, followed by vehicular traffic and architectural style.

The most significant finding from the study was the response to the question, "Which of the following factors associated with tourism adversely affect the community?" At the top of the list of factors ranked as "very adversely affected" was a lack of adequate zoning and land use planning.

SEVIER COUNTY

The tourism industry is exploding in Sevier County, TN. As graphic evidence, the following is a list of projects that are being seriously considered, planned, or where construction is well underway in Sevier County, as reported in an article from the Sunday, February 25, 1996 issue of the Knoxville "News Sentinel." This is a remarkable number of projects for just 1 year. All of the projects listed are along the primary tourist transportation corridor from Interstate 40 south through Sevierville, Pigeon Forge, and Gatlinburg to the main entrance to Great Smoky Mountains National Park. It is interesting to note that none of the announced development has any relationship to the nearby presence of the national park.

- $200 million development located near the boundary between Sevierville and Pigeon Forge consisting of music theaters, restaurants, and a shopping mall consisting of factory outlet stores and gift shops
- $50 to 60 million multiple tourist development by the Bass Angler's Sportsman's Society Headquarters (in planning stages)
- $10 million Lee Greenwood Theater (construction completed)

TABLE 4
Response to Question:

"Which of the following factors associated with tourism adversely affect the community?"

	Very adversely affected	Slightly affected	Not affected	Don't know
Architectural distractions				
Large scale (N=480)	29.0	29.9	33.0	8.1
Garish colors (N=483)	42.0	23.8	28.6	5.6
Inappropriate style (N=485)	42.9	26.2	26.2	4.7
Lack of building codes (N=488)	50.8	15.4	24.4	9.4
Positioning on steep slopes (N=488)	34.8	25.4	32.4	7.4
Land-use conversion				
Loss of agricultural lands (N=486)	28.2	37.2	31.3	3.3
Building on ridgetops (N=486)	36.8	30.0	29.0	4.1
Building on steep slopes (N=488)	35.7	30.7	28.5	5.1
Building in river floodplain (N=485)	33.8	32.6	26.6	6.8
Lack of land-use planning (N=487)	56.6	19.3	17.9	6.2
Lack of adequate zoning (N=489)	57.8	16.7	18.8	6.7
Commercial signs				
Presence of billboards (N=497)	54.3	30.2	14.5	1.0
Overwhelming numbers (N=491)	48.5	27.5	22.8	1.2
Obstruction of scenic views (N=493)	51.9	23.9	23.3	0.8
Changes in population				
Ethnic conflicts (N=488)	7.2	22.7	64.3	5.7
Rapid population growth (N=491)	16.9	36.0	43.8	3.3
Conflicts between new and long-term residents (N=497)	17.9	48.9	28.6	4.6
Vehicular traffic				
Congestion (N=504)	56.3	30.4	12.7	0.6
Parking (N=501)	44.7	33.7	20.2	1.4
Air pollution (N=493)	34.5	34.3	27.0	4.3
Difficult to get around during tourist season (N=505)	47.9	36.6	14.3	1.2

- $30 million aquarium on Lee Greenwood site plus restaurant and 250 unit motel (motel and restaurant completed and aquarium in planning stages)
- $8 million Dollywood addition (cinema theme development construction completed)
- $9 million Yesteryear Music Palace in Pigeon Forge (under construction)
- Smoky Mountain Broadway on 100-acre site including (in planning stages)
- Gospel Music Hall of Fame, theater, crafts village and lodging on a 300-acre River Bluff Landing site on Highway 66 on south side of French Broad River (in planning stages)
- An additional 10 other hotels including 700 units, 4 music theaters and at least three other substantial amusement/attractions projects (under construction)
- $70 million dollar road building program for Sevier County for eight projects, including among others, improving the I-40 interchange, widening highways to six lanes from I-40 to Pigeon Forge, and improving access through Sevierville (funding announced, interchange near completion, other projects in planning stage)

The state government is investing heavily along with local communities to alleviate some of the traffic problems, but the fundamental problem of restricted traffic flow through the county has not yet been comprehensively addressed. Other than transportation, the implications of all of this development on services and facilities provided by the communities in Sevier County is not well understood.

THE PITTMAN CENTER CASE STUDY

History of the Community

The community of Pittman Center, TN is a prototypical rural community of 6000 acres, including 325 households. The roots of this typical southern Appalachian community can be traced to the first European family settling in the area that came to be known as Emerts Cove. Legend has it that the river dominating the landscape in the area, the Middle Prong of the Little Pigeon River was so named for the preponderance of now extinct passenger pigeons (*Ectopistes migratorious*) that foraged on the oak hickory forests which to this day dominate the landscape. Following World War I, educator Dr. Eli Pittman, the community's namesake, and John Sevier convinced the Methodist Church to establish a missionary Appalachian school in the then-isolated community. From 1921 to 1963, the school was central to the area. Drawing teachers from around the U.S. and as far away as Scotland, Pittman School may have given its students a broader, more universal view of the world than was available to students in the much larger communities of the region (FutureScape 1995). Perhaps Pittman's legacy provided the intellectual foundation for the enlightened community planning efforts which followed decades later.

In the mid 1930s, numerous families moved to the area who were being displaced from the nearby Greenbrier watershed, which was being purchased by the federal government as part of efforts to create Great Smoky Mountains National Park. In 1974, Conley Huskey organized the citizenry to incorporate the community. This action was taken in large part to block attempts by neighboring Gatlinburg to extend its boundaries and extract water from the local river for municipal use. The community of Pittman Center remains today largely undeveloped, representing an opportunity for the citizenry to get in front of the development curve and manage growth to their vision for the community's future.

Proximity to National Park Service Properties

The setting of the community is literally sandwiched between two land parcels managed by the National Park Service. The Great Smoky Mountains National Park forms the southern boundary of the community, and the northern boundary roughly follows the right-of-way for the Foothills Parkway connecting the four-lane highway (locally referred to as the Spur) between Gatlinburg and Pigeon Forge and that portion of the parkway that has been constructed from I-40 to Cosby in neighboring Cocke County (Figure 1). The Middle Prong of the Little Pigeon River drains the Greenbrier watershed in the park through the community. The Greenbrier entrance to the park onto State Highway 321 is right next to the river. This section of the park remains relatively undeveloped. The National Park Service has not yet established a Development Concept Plan for that portion of the park so *the opportunity remains for collaborative planning between the park and the community.*

Community Planning Process

In 1978, a city planning commission was formed in Pittman Center. In the late 1980s, planning assistance from the state of Tennessee was solicited to begin a community-wide land-use planning process. Community planners from the Tennessee Economic and Community Development Department, Local Planning Office in Knoxville facilitated the planning process. The community began a visioning process that was carefully crafted over the course of 5 years involving three distinct planning efforts.

In 1990, SAMAB began participating in the planning process. As part of SAMAB, the Economic Development Administration provided funding for the effort. Technical support staff and information databases were provided by other member institutions. As a result of those efforts, a planning document was developed, entitled "Sustainable Development Strategies for Communities with Tourism-Based Economies in the Southern Appalachian Highlands" (Peine and Welch 1990). This

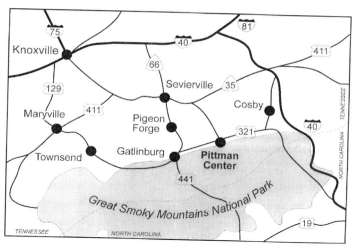

FIGURE 1 Location map for the community of Pittman Center adjacent to Great Smoky Mountains National Park.

publication documents the planning process applied and key growth management strategies devised. There has been a wide distribution of the publication throughout the region and the country.

Following completion of that project, Pittman Center applied for and was selected for a pilot project known as FutureScapes to demonstrate sustainable development strategies for rural communities in the southern Appalachians. The community was chosen primarily due to the comprehensive visioning process that had previously taken place. Initiated in 1990 by the East Tennessee Community Design Center in Knoxville, the FutureScapes process creates innovative plans to promote economic development, while at the same time preserving key cultural and environmental resources.

The notion that communities could bring economic and resource protection interests together found a champion in Robert Yaro, a University of Massachusetts planner who proposed a new style of rural development that was aesthetically, environmentally, and economically harmonious (Yaro et al. 1990). His ideas received national acclaim. "Yaro says, 'First figure out what you want to save — then figure out what you want to build,'" according to Annette Anderson, then Director of the Community Design Center (FutureScape 1995). This straightforward strategy became the central focus of the FutureScapes activity in Pittman Center, which culminated in the production of a planning report in July 1995.

The above mentioned three phases of planning were somewhat redundant, but each additional phase more explicitly defined the vision and described the tools necessary to appropriately manage growth in the community. The critical job still left undone in spite of all the planning effort is the construction of innovative regulations that explicitly define a regulatory process that is flexible, rational and enforceable, and thereby provides the tools for the community to manage growth to realize its vision for the future. Until this work is accomplished, the land use planning will remain incomplete and the community's envisioned future is less than assured.

The Visioning Process

All three planning efforts in Pittman Center focused on the visioning process. This phase of planning in the community utilized a variety of methods to conduct the visioning process. Speakers were invited to present ideas applied in other communities around the country. Public meetings were held to solicit public comment. A newsletter was prepared by local citizens and distributed to every household and absentee property owner. Workshops were held to discuss specific design preferences.

Pictures and personal items were used to define important places and things and from that exercise construct a sense of place. Committees were formed to study special issues such as ridge top protection, water supply and sewage treatment, flood plain protection, and the design of State Highway 321.

Perhaps the most important visioning exercise was the administration of mail back questionnaires to all residents and property owners. The results provided a credible census of the array of opinions on visions expressed for the community's future. The vast majority (91%) supported the draft goal statement of the Pittman Center Planning Commission. Strong support was expressed for the following management options as well (Peine and Welsh 1990):

92% protection of flow rate and water quality of the Little Pigeon River
89% control of billboards
80% monitor ground water quality
75% limit location of commercial development
68% encourage low density development
63% protect ridge tops from development
55% increase public access to the river

Resources Inventory

A geographic information system was developed for the community by the SAMAB team. Existing data sources were utilized to construct the data themes describing natural and cultural resources. No new resource inventories were created for any of the three planning efforts.

- Data themes for the natural resources included:
    Topography
        Slope steepness
        Ridgetops
    Vegetation
        Type distribution
    Wildlife
        Species habitat
        Migration/movement corridor
    Water resources
        Quantity
            Rivers and streams
            Runoff flow rates
        Quality
        Aquatic biology
    Soils
    Flood plains
    Agricultural lands
    Rare and endangered species
        Abundance and distribution
        Critical habitat
    Pollution emissions
- Data themes for cultural resources included:
    Archaeological resources
    Historic structures
    Historic events tied to the landscape

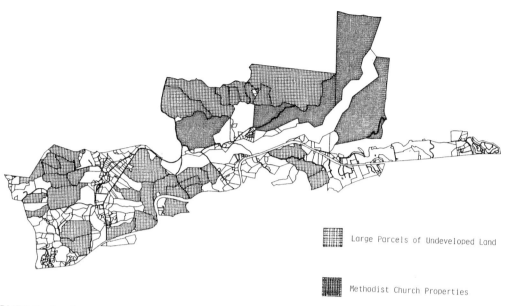

FIGURE 2 Land ownership map for the community of Pittman Center. (From Peine, J. and H. Welsh, Sustainable Development Strategies for Communities with Tourism-Based Economies in the Southern Appalachian Highlands, Southern Appalachian Man and the Biosphere Cooperative, Gatlinburg, TN, 1990.)

Recreation resources
Viewsheds
Social foci
Town center
Neighborhoods
Property boundaries
• Economic resources
Commercial businesses
Industrial sites
• Land ownership

For a rural, sparsely developed community like Pittman Center, the most important database was the land ownership map. See Figure 2. This data theme was used to predict the pattern of future growth. The most alarming statistic generated in the inventory analysis was that over 60% of the land in Pittman Center is owned by absentee owners or known land speculators, suggesting that a considerable percentage is being held for speculative purposes (Annette Anderson, pers. commun.).

CONCEPTUAL PLAN

Included in this section is a series of outlines of components of the Pittman Center conceptual plan as they appear in the report by Peine and Welsh (1990), which provide key tools to manage growth and shape the future in a community with a tourism based economy. The information is by no means a complete statement concerning all aspects of drafting ordinances. The level of detail associated with such ordinances ranges widely from community to community. The more detailed the ordinances, the less potential for ambiguity but the less opportunity for flexibility, as well.

Community Vision Statement

In the first phase of the planning process, the following goal statement was determined for the Pittman Center Planning Commission:

> To create and perpetuate a quality living environment and to encourage quality development that supports that end. To encourage development that supports a tourist-oriented economic base that relates to and magnifies our unique relation to and with the Great Smoky Mountains.

The town motto was determined to be "A community dedicated to our mountain heritage."

Ridge-Top Ordinance

The landscape of Pittman Center is mountainous. Steep slopes and sharply defined ridge lines dominate view points around the community. Protection of ridge lines is of critical importance in achieving the goal of maintaining a close association in the community with the natural landscape. Flexibility of application of ridgetop ordinances is suggested by the conceptual plan, since some ridgetops are more critical to the overall visual impact in the community than others. Suggested key components follow.

- Discourage the building of structures on ridgetops, using the performance zoning ordinance
- Strictly prohibit building structures on ridgetops that are prominent foreground in priority viewsheds of the community
- Compensate landowners of ridgetops targeted for priority protection by direct purchase of development rights by the community or by transferring them to another parcel via purchase by a developer
- Incorporate the following restrictions for structures that are allowed to be built:
  Set back 50 feet from the brow of hill or set back so not visible from a public roadway, whichever is more restrictive
  Use nonreflective materials for roofs and walls
  Use dark earth-tone paints
  Restrict height to 25 feet above the low point of the natural terrain slope occupied by the structure
- Minimize removal of native herbaceous plants, shrubs and trees.

Commercial Sign Ordinance

The citizens of Pittman Center who have expressed concerns about the adverse effect of billboards and inappropriate signage on the community are not alone. Over the last few years, more than 500 U.S. cities have enacted new regulations to control the proliferation of billboards. As a result, there are many very fine examples of sign ordinances that have been tested in the courts and can serve as models for Pittman Center. Lubbock, TX, for instance, has a very detailed sign ordinance that is described in Mantell et al.'s 1990 publication, *Creating Successful Communities*. Lake Wales, FL has taken great pains to clearly define the difference between an on-premise sign and a billboard. Hilton Head, SC, restricts sign lighting, size, and design. All of the on-premise sign controls suggested below are taken from existing sign ordinances.

- Enforce state sign regulations associated with scenic highways
- Ban off-premise signs
- Amortize nonconforming signs and billboards
- Adopt a style of street sign which reflects the cultural themes of the community

- Control on-premise signs by:
    Limit free-standing signs to one per business, with a maximum height of 8 feet and area of 100 square feet along roadways, with speed limits of 45 to 55 miles per hour, and 50 square feet for roadways with speed limits under 45 miles per hour
    Allow size bonuses for monument style signs as compared to freestanding pole signs
    Limit lighting to one 150-watt light per side per 40 feet^2 and two per side for signs larger than 40 feet^2
    Prohibit problem signs such as portable signs, pennants, banners, streamers, and flashing or intermittent lights
    Limit shopping centers, malls, office parking lots and similar large developments to one group identity sign for individual businesses
- Develop a Tourist Information Program associated with the sign ordinance
    Utilize interpretive signs and markers to enhance appreciation of natural and cultural heritage
    Utilize scenic turnouts at key viewsheds
    Develop a uniform system of symbols and signs to orient visitors to community services
    Devise a series of logo signs to direct visitors to tourism enterprises
    Operate a tourist information center and provide maps to commercial enterprises

Christopher Duerksen (1986) provides a word of caution on sign ordinances. He indicates it is important, in case the ordinance is contested in court, to define the distinctions portraying temporary and portable signs differently from free-standing commercial signs. Communities would be well advised to have adequate documentation to support the more stringent regulations on these types of signs. The Southern Environmental Law Center has prepared a very useful publication entitled, "Visual Pollution and Sign Control: A Legal Handbook on Billboard Reform."

Architectural Guidelines for Commercial Structures

Michael Mantell et al. (1990) made the following suggestions on how to improve the effectiveness of an environmental review program. These suggestions are relevant to an architectural review function.

- Build community consensus about assets that are distinctive, desirable, and worthy of conserving
- Integrate the design review with the comprehensive plans for the community
- Focus design standards and review on important aspects of design rather than minute details
- Include specific design standards or guidelines. Provide as much detail and direction as possible
- Include both design professionals and lay people on the review boards
- Provide an opportunity for developers to consult with the board or its staff in order to resolve questions and ambiguities early in the development process
- Document the review process to demonstrate the fair exercise of design judgment. Provide specific reasons for denial and indicate what revisions could be made to make the proposal acceptable

An architectural review committee should be formed to evaluate the appropriateness of planned new commercial structures in the community. It is of critical importance that the decisions by the architectural review committee do not appear arbitrary. Frequent judgment of this sort would quickly undermine the intent of the program. Again, flexibility without being arbitrary is the name of the game. The architectural review function is designed more to encourage appropriate development than to outright forbid certain kinds of development. There must be a commitment between the community officials involved and the developers to work toward solutions that preserve the vision of Pittman Center for the future. Key components follow.

- Encourage designs to showcase community architectural and cultural traditions
- Encourage use of native materials in structure construction and landscape design
- Use earth-tone exterior colors
- Minimize the use of reflective material
- Allow for a maximum of 10 units per motel/hotel/condominium structure
- Allow for a maximum building height of 34 feet
- Allow for a maximum building height for necessary buildings in critical viewsheds of 16 feet

River Corridor Ordinance

The setting for virtually every community adjacent to Great Smoky Mountains National Park includes a scenic river. Unfortunately, in virtually every case the layout of these communities is mistakenly directed to the primary roadways. The rivers are invariably relegated to the back end of the property lots. Architectural design and building entrances are oriented to vehicular parking or sidewalk access. Parks occasionally are located along the river. For the most part, the opportunity to use the river as an aesthetic focal point of the community is lost. Communities that have adopted river themes in other parts of the country, such as San Antonio, TX, have gotten significant recognition and enormous economic benefit for their forethought. The citizens of Pittman Center have indicated that the river is their most important resource. The challenge is to devise a means to focus future community development on this centerpiece resource without destroying it.

- Orient townscape focus on the river
- Obtain public access to the corridor as the opportunity arises
- Utilize performance zoning to encourage commercial structural placement on a property to focus on the river
- Prohibit structures that divert flow
- Prohibit the direct taking of water from the river
- Prohibit construction of structures in the floodplain zone so designated by the community
- Prohibit the removal of native plants for a 50-feet setback from the bank
- Develop recreational facilities along the river at selected locations
- Develop lineal pedestrian pathways along the waterway linking community neighbor-hoods and commercial districts with recreation areas

Primary Street Design Standards

More than anything else, primary street design in conjunction with commercial development provides a first and lasting impression for tourists. It sets the tone of the townscape and is a reflection of the people living in the community. Hilton Head, SC, is a community with carefully considered designs of their primary streets. Their townscape is one of the major drawing cards to the region. Such a townscape in Pittman Center would likewise be a positive tourist attraction for the community and a stark contrast to the other tourist communities in Sevier County. The primary streets of Pittman Center to focus on first are the designated State Scenic Highways U.S. 321 and State Route 416 that run adjacent to areas designated for commercial development on the Pittman Center land-use plan. There is growing pressure to widen U.S. Highway 321 to four lanes from Gatlinburg to I-40. It is suggested that federal and state funds to do this work should incorporate the design interests of the community of Pittman Center as follows:

- Limit access from parking lots and shopping areas
- Provide for pedestrians and bicyclists on one side of street
- Use a 15-foot landscaped setback from each side of roadway to screen parking lots and commercial structures
- Use only high pressure sodium street lights to conserve energy (best lumens per watt performance) and minimize glare

Performance Zoning

Standard zoning regulations do not address the issue of cluster development designed to position structures, roads, and parking lots in such a way on a property so as to protect flood plains, natural areas, riparian habitats, agricultural lands and open space. In fact, standard low density zoning levels of 5 to 10 acres per residence actually promote fragmentation of open space. The performance zoning provides an incentive to reward those individuals proposing to subdivide land and/or build new commercial structures who adhere to the priorities of the community in planning development of the landscape. The reward could be to grant the permittee permission to construct at a higher density than otherwise allowed in appropriate areas. This is predicated on the setting of a conservative development density standard. The lower the standard, the greater the leverage to influence the actions of the developers. This strategy holds the potential to actually increase the unit density in a community beyond that commonly held via standard zoning ordinances. An example of suggested cluster development applied in Pittman Center townscape is displayed in Figure 3.

CLUSTERED RESIDENTIAL / MIXED USE

FIGURE 3 Illustration of cluster development to protect flood plain and open space in the community of Pittman Center. (From FutureScape of Pittman Center, East Tennessee Community Design Center, Knoxville, 1995.)

MODEL PERFORMANCE ZONING ASSESSMENT FORM	Score (0-10)	Weight factor	Points earned
Percent of parcel left as open space		3.1	
Percent of ridge top protected		2.5	
Percent of steep slopes protected		2.1	
Unobtrusive architecture design		2.0	
Percent of structures clustered		1.8	
Utilize community cultural theme in architecture design		1.7	
Low density development adjacent to national park and parkway right of way		1.5	
Provide for recreational corridors		0.8	
Scale of structure appropriate to community; i.e., motel 50 units-score high, over 10 units-score low		0.6	
Property oriented to river or other key community natural or cultural resource		0.5	
Utilize native plant material in landscaping		0.5	
Pedestrian trails provided		0.5	
Percent of trees protected		0.5	
			TOTAL

FIGURE 4   Model performance zoning assessment form, including a list of attributes relevant to the community of Pittman Center.

The implementation of such a system is obviously more involved than standard zoning but does not have to be promulgated in a complex manner. At the present time, the amount of development activity in Pittman Center is such that existing resources and citizen participation is sufficient to administer the program with the exception of the enforcement component. Some of the costs of permit review should be borne by the developer.

The prototype assessment form appearing in Figure 4 provides a mechanism to prioritize the relative merits of various contributions to the community's future.

Sustainable Development Commission

The type of tourist and other commercial enterprises envisioned in the FutureScape Plan for Pittman Center will not likely come to the community without a very aggressive recruitment campaign by property owners and community leaders. A commission is proposed to be formulated with the explicit purpose of matching landowner interests with an active campaign to recruit businesses that reflect the community vision. Example appropriate enterprises to recruit would include bed and breakfast inns, organic farming, arts and crafts, ecotourism outfitters, and specialty restaurants. This activity can now be pursued at a small scale before water and sewer service comes to the community and bigger scale projects rapidly emerge.

Relation to Greenbrier Park Entrance

Possibly the most intriguing opportunity for Pittman Center as a gateway community is to collaborate with the National Park Service to provide facilities and services relevant to residents and visitors to the park and the community. Such collaboration is rarely accomplished on anything more than a somewhat superficial basis. Some specific areas of common interest are as follows:

- Wildlife-proof solid waste management would reduce the potential for human–black bear confrontations
- Linking greenway and accompanying trail system between the community and the park with focus on the river would enhance both and would be an important marketing feature for bed and breakfast inns in the community
- A shared visitor center for orientation for community attractions and park trails and natural features. This precedence has been set with neighboring Gatlinburg.
- Public parking in the vicinity of the visitor center and shuttle bus service into the park making Greenbrier the *only* park entrance that is free of private vehicular traffic.
- Buffer zone ordinances such as pet leash laws and lighting restriction to minimize the obtrusiveness of the built environment in the community.

Implementation Strategies

Conceptual plans are not worth much it they are never implemented, if the vision is not realized. Small communities like Pittman Center have few resources to facilitate an implementation strategy. Enormous energy is required to maintain the dialogue within the community to draft managed growth regulations negotiated with developers, and to police regulatory compliance. What the community has accomplished is truly extraordinary but to date it is only a road map. Some suggestions to aid in the difficult task of implementation are as follows:

- Prepare pamphlets to distribute to land owners concerning key regulations likely to be violated, particularly from a bulldozer seat
- Establish a voluntary technical advisory team to support the planning commission
- Establish a not-for-profit community development corporation
- Articulate and recruit needed ongoing support from SAMAB member institutions
- Maintain a close relationship with the citizenry via continuation of the citizen news letter
- Establish early dialogue with the developers and landowners
- Complete project review in a timely fashion
- Recognize those developments and signs representing the community values
- Keep focusing on the community vision
- Maintain consistency in supporting the conceptual plan

CONCLUSION

Ecosystem management is not usually associated with community affairs such as land use planning. However, federal land managers are more and more recognizing their inextricable link to the built environment. As a result, several agencies are engaged in community based technical assistance programs. The National Park Service, concerned with gateway communities, is involved in selective community planning efforts. The U.S. Forest Service has a program to help rural communities whose economies are dependent on the timber industry to strive to diversify their economies.

RISK ASSESSMENT REPORT CARD FOR ECOSYSTEM MANAGERS

Applying the **Risk Assessment Report Card for Ecosystem Managers,** as referenced in the other chapters in Section IV of this volume, to the Pittman Center case study provides the following generic accounting system of risk to managers as follows.

Vision: A+

The federal land manager actively supported and participated in this process.

Resource risk: C+

The nature of the risks to natural resources in the community were not thoroughly documented but remain high as development escalates.

Socioeconomic conflicts: C+

The potential conflicts have yet to emerge since the community remains largely undeveloped but the situation is ominous.

Procedural protocols: B–

The explicit regulations necessary to effectively manage growth are still under development, but many key statutes are in place.

Scientific validity: B

The scientific value of resource protection are documented in adjacent communities with tourist economies.

Legal jeopardy: C–

Some of the innovative regulations proposed here to manage growth have not been tested much in the courts.

Public support: A+

The community has worked diligently to involve the citizenry in the process of managing growth.

Adequate funding: D–

The community is extremely small with a limited tax base from which to promulgate the innovative tools to manage growth.

Policy president: B

Though there is not much formal policy for supporting local community planning, the partnership role demonstrated in Pittman Center is being encouraged by the current federal administration in Washington, D.C.

Transferability: B+

Many of the planning procedures and tools for managed growth explored by the community are applicable elsewhere in the southern Appalachians. On the other hand, each community is unique as to its natural and cultural resource orientation, socioeconomic makeup, and collective vision by the citizenry.

Ecosystem managers of the future will routinely work within the context of the built environment. The tentative steps being taken now in enlightened communities such as Pittman Center will have to become the norm in the future if there is any hope for sustainable ecosystem management.

REFERENCES

Appalachian Regional Commission. 1994. Annual Report. 1666 Connecticut Avenue., N.W., Washington, D.C. 160 pp.

Bureau of the Census. 1993. 1990 Census of the Population and Housing. Washington, D.C.: U.S. Department of Commerce. CPH-2-1.

Duerksen, C.J. 1986. Aesthetics and Land Use Controls — Beyond Ecology and Economics. Washington, D.C., American Planning Association. 45 pp.

Ecotourism Society. 1993. Ecotourism Guidelines for Nature Tour Operators. P.O. Box 755, Bennington, VT.

Fredrick, M. 1992. Tourism as a Rural Economic Development Tool: An Exploration of the Literature. USDA. Washington, D.C.: Economic Research Service, Bibliographies and Literature of Agriculture N. 122 pp.

FutureScape of Pittman Center. 1995. A Partnership Project of the East Tennessee Community Design Center and Tennessee Valley Authority. Knoxville: East Tennessee Community Design Center. 36 pp.

Godschalk, D. 1979. *Constitutional Issues of Growth Management*. Chicago: Planners Press.

Hendler, B. 1977. *Caring for the Land*. Chicago: American Society of Planning Officials.

Hiss, T. 1991. *The Experience of Place*. New York: Vintage Books. 223 pp.

Jackus, P. and P. Siegel. 1993. Attitudes Toward Tourism Development in the Appalachian Highlands of Tennessee and North Carolina. Report Submitted to H. Hinote, Tennessee Valley Authority and J. Peine, Gatlinburg, TN: National Park Service. 164 pp.

Kaiser, E.J., D.R. Godschalk, and F.S. Chaplin, Jr. 1995. *Urban Land Use Planning*. Urbana and Chicago, IL: University of Illinois Press. 493 pp.

Kaufmann, M.R., R.T. Graham, Boyce, B.A. Jr., D.A. Moir, W.H. Perry, R.T. Reynolds, R.L. Bassett, P. Mehlhop, C.B. Edminster, W. Block, and P.S. Corn., 1994. An ecological basis for ecosystem management. Fort Collins, CO.: USDA U.S. Forest Service, Rocky Mountain Forest and Range Exp. Stn. GTR-RM-24. 22 pp.

Kinsley, M. 1994. Sustainability: Prosperity Without Growth. *Public Manage*. V76:10:6–9.

Mantell,M., S. Harper and L. Propst. 1990. *Creating Successful Communities — A Guidebook to Growth Management Strategies*. Washington, D.C.: Conservation Foundation. Island Press. 233 pp.

McHarg, I. 1969. *Design with Nature*. Garden City, NY: Natural History Press.

Nasbitt, J. 1994. *Global Paradox*. New York: William Morrow. 304 pp.

Peine, J. and J. Renfro. 1988. Visitor Use Patterns at Great Smoky Mountains National Park. Research/Resources Management Report SER-90. Atlanta, GA: National Park Service. 93 pp.

Peine, J. and H.Welsh. 1990. Sustainable Development Strategies for Communities with Tourism-Based Economies in the Southern Appalachian Highlands. Southern Appalachian Man and the Biosphere Cooperative. Gatlinburg, TN: Great Smoky Mountains National Park, Uplands Field Research Laboratory. 54 pp.

President's Council on Sustainable Development. 1994. Draft Vision and Principles of Sustainable Development. June 9, 1994. Washington, D.C: White House. 8 pp.

SAMAB (Southern Appalachian Man and the Biosphere). 1996. The Southern Appalachian Assessment Social-Cultural-Economic Technical Report. Report 4 of U.S. Department of Agriculture, Forest Service, Southern Region. Atlanta. In Press.

Sargent,F., P. Lusk, J. Rivrea, M. Varela. 1991. Rural Environmental Planning for Sustainable Communities. Washington, D.C.: Island Press. 254 pp.

Smith, M. 1989. Behind the Glitter: The Impact of Tourism on Rural Women in the Southeast. Lexington, KY: Southeast Women's Coalition.

Southeast Tourism Society Newsletter. March 1995. Atlanta, GA: Bill Hardman [publisher]. 4 pp.

Southern Environmental Law Center. Visual Pollution and Sign Control: A Legal Handbook on Billboard Reform. 1988. Charlottesville, NC: Papercraft Printing and Design Co. 39 pp.

Tennessee Statistical Abstract. 1992/1993. Nashville, TN: Tennessee Department of Economic and Community Development. 54 pp.

Tennessee Department of Tourism Development. Nashville, TN. 1994. Economic Impacts of Tourism in Tennessee. 42 pp.

Tennessee Department of Tourism Development. Nashville, TN:Tourism News. Spring 1995. 8 pp.

U.S. Travel Data Center. 1992. Impact of Travel on State Economies. Bureau of Labor Statistics. Washington, D.C.

U.S. Travel Data Center. 1994. Economic Impacts of Tourism in Tennessee Counties. Nashville, TN: Tennessee
    Department of Tourist Development. 40 pp.
Weaver, G.D. 1993. Tourism Development: A Guideline for Rural Communities. Columbia, MO: University
    Extension, Department of Parks, Recreation and Tourism, University of Missouri.
Yaro, R.D., R.G. Arendt, H.L. Dodson, and E.A. Brabec. 1990. Dealing with Change in the Connecticut River
    Valley: A Design Manual for Conservation and Development. Amherst, MA: Center for Rural Massa-
    chusetts, University of Massachusetts. 182 pp.

18 Managing Biodiversity in Historic Habitats: Case History of the Southern Appalachian Grassy Balds

Peter S. White and Robert D. Sutter

CONTENTS

Introduction 375
Distribution and Floristics of the Grassy Balds 377
Origin, Maintenance, and Succession..380
    Climate and Topography 382
    Geology and Soils..382
    The Ecotonal Hypothesis 382
    Native Grazers..383
    Other Natural Factors: Windthrows, Ice, Frost Damage, and Insects 383
    Fire...383
    Native American Activities 384
    Clearing by European Settlers and Subsequent Grazing 384
    What We Know about Bald Origin and History: a Summary 385
    The Rate of Succession...386
Management of Grassy Balds..387
    Two Examples of Management 388
    Gregory and Andrews Balds in Great Smoky Mountains National Park 388
    Roan Mountain...389
Distribution of the Grassy Balds Flora and Lessons for Regional Conservation 390
Conclusions 391
    Risk Assessment Report Card for Ecosystem Managers 392
References 393

INTRODUCTION

Southern Appalachian grassy balds represent a situation which has become a frequent challenge to ecosystem managers: the presence of discrete areas managed for historic, scenic, or biodiversity purposes within the context of larger natural areas. In these situations, ecosystem managers will require information in several key areas: surveys of the biota present, documentation of the relationship of the biota and the environment, special documentation of rare or otherwise endangered species, synthesis of information on historic and prehistoric influences on the habitat, an assessment of social values of the habitat (e.g., visual, historic, or spiritual values), design of cost effective

TABLE 1
Generic Research Objectives for Management of Discrete Historic, Aesthetic, and
Biological Values in a Larger Natural Area Context, with Examples from Recent Grassy
Balds Research in Great Smoky Mountains National Park

Objective	Examples from grassy balds
Conduct a botanical survey	Stratton and White 1982
Conduct a survey of animal use	
Conduct a review of prehistoric and and historic human influence	Lindsay and Bratton 1979
Conduct a field survey of cultural artifacts	
Assess populations of endangered species	Bratton and White 1980
Assess current habitat condition, including composition, structure, natural processes, and successional trends	Lindsay and Bratton 1980
Conduct an assessment of social values (visual, historic, spiritual, uniqueness of setting)	
Determine the most cost effective and environmentally safe techniques for management	Lindsay 1977
Conduct a cost–benefit analysis of management	

and environmentally safe management techniques, and cost/benefit analysis of management alternatives. Grassy balds restoration and management was made possible largely by an accumulation
of answers to these questions (Table 1). The earliest questions that were raised were ones having
to do with history, environment, and naturalness: are grassy balds natural or historic communities?
What natural or historic factors created or maintained grassy balds? If these factors no longer
operate, how can human management restore and maintain balds?

Management of grassy balds raises the two paradoxes in conservation: we seek to conserve a
natural world that must change and we often interject human management into ecosystems that we
would ideally like to see human-free (White and Bratton 1980). These two paradoxes were not as
apparent in North American conservation 2 or 3 decades ago as they are today. The implicit
assumption of conservation management in the past was that nature's balance produced a dynamic
equilibrium in habitats and populations. It was also assumed that significant human influence
occurred only in postsettlement times and that natural processes would reestablish original ecosystems once human influence was removed. These ideas lead to a decidedly passive or "hands-off"
approach to management.

Conservation of grassy balds in the southern Appalachians presented an early challenge to passive
management. Succession from open grassy habitat to shrub thickets and forest is universal on the
balds — balds will not remain an element of landscape diversity without active management.
Further, balds may not represent the remnants of a naturally occurring or previously human-free
habitat (see the fuller discussion below).

As land managers began to deal with grassy balds, the issue of naturalness and past human
influence became critical and contributed to a wide variety of opinions about balds management.
Some conservationists felt that grassy balds should be managed only if they were natural communities or remnants of natural communities. Research, however, tended to support the importance
of a human role in origin or at least persistence of balds. Given this evidence, some conservationists
believed that natural succession should be allowed to take place, whereas others supported restoration
and management of the balds to preserve human history (White 1984). Among those who supported
conservation of grassy balds, some supported balds restoration and maintenance primarily because they
considered them to be of Native American origin and therefore relatively old, whereas others were
indifferent as to whether the balds were created by Native Americans or European settlers.

Naturalness and human history, however, are not the only reasons for balds conservation.
Populations of some rare plants are found on grassy balds (Bratton and White 1981; Wiser and

White, in press, and below), and balds thus contribute to local and regional biodiversity. Further, the southern Appalachians are not high enough for a climatic treeline (Cogbill et al. in press, Gersmehl 1973), and the mountains are heavily forested to their summits. Grassy balds are one of the few habitats with panoramic views of the mountains and are thus treasured because of experiences they provide (e.g., a feeling of being "in" a dramatic landscape or being "one" with the mountains). In addition, early succession on grassy balds is characterized by diverse wildflowers and abundant berry crops, aspects of the balds present in the period 1940 to 1970 when management of these habitats was first discussed.

Grassy balds, thus, raised the issue of management, not only for natural or historic values, but also for aesthetic values and produced an early example of the conflicts that can arise when there are multiple management goals. Some people supported retaining balds of biological diversity, landscape diversity, recreation, or aesthetics, regardless of whether they were human-caused or not.

In sum, the dilemma of grassy balds management challenged land managers in the southern Appalachians to think in new ways about conservation (for an early summary, see Saunders 1981). In addition to presenting issues of naturalness, past human effects, and the need to actively manage against succession, grassy balds are an important regional case study because these rare and island-like habitats vary in species composition, environment, and history. Because no one grassy bald can represent the whole, we will show that a regional approach to conservation of this habitat is desirable. Because balds must be actively restored and maintained (i.e., there will be a cost to management), we must also address the questions: How many balds must we manage to maintain the community and species-level variation of this habitat on a regional basis? How do we determine this number? What are the costs and benefits of management alternatives?

This chapter will address: (1) distribution and floristics of the grassy balds, (2) the origin, successional status, and maintenance of grassy balds, (3) examples of current management to restore and/or maintain balds, and (4) the lessons that grassy balds can teach us about approaches to regional conservation. We will conclude by summarizing regional management of this distinctive habitat.

DISTRIBUTION AND FLORISTICS OF THE GRASSY BALDS

Grassy balds are high elevation, open communities dominated by herbaceous grasses and sedges, with a mix of herbaceous forbs, shrubs, and scattered trees. They are distributed throughout the southern Blue Ridge, from southern Virginia to southern North Carolina and from the Blue Ridge escarpment to the Tennessee state line (the best regional summary is that of Gersmehl 1971). Occurrences are clustered within this range, with balds being most common in the Great Smoky Mountain National Park (GSMNP) (Stratton and White 1982) and the Roan Mountain Massif (Mark 1958a).

The scientific and aesthetic interest in southern Appalachian grassy balds has existed for well over a century. Much of the early interest centered on Roan Mountain, along the northern border of North Carolina and Tennessee. Elisha Mitchell stated in 1835 that Roan Mountain was "… the elysium of a southern botanist as a number of plants are found growing in this cold and humid atmosphere, which are not seen again until we have gone hundreds of miles further north" (in Gray 1842; see also Smathers 1981). He also said:

> Roan is … the easiest of access and the most beautiful of all the high mountains of that region. With the exception of a body of granitic rocks looking like the ruins of a castle near its southwestern extremity, the top of Roan may be described as a vast meadow about nine miles in length … without a tree to obstruct the prospect, where a person may gallop his horse for a mile or two with Carolina at his feet on one side and Tennessee on the other, and a green ocean of mountains raised in tremendous billows immediately about him.

Roan Mountain's ease of access, spectacular views from rock outcrops and ridgetop grassy meadows, and unique plants have attracted numerous botanists such that an outing in the late 1800s was described as a Who's Who of the botanical community (Wilson 1991). In the 1900s, Roan Mountain became a site of much botanical research, some of which were designed to test hypotheses concerning the origin of balds (Brown 1938, 1941, and 1953; Billings and Mark 1957; Mark 1958). Balds farther south, however, were not being ignored. Stratton and White (1982) state that the bald communities in GSMNP are the "most throughly studied" of any community in the park (e.g., Wells 1936a; Gilbert 1954; Bruhn 1964; Radford 1968; Lindsay and Bratton 1979a, 1980).

Grassy balds occur in a matrix of high elevation forest communities including spruce–fir, northern hardwood, and oak-dominated forests. They sometimes occur with several other high elevation treeless communities, including heath balds, cliff faces (Wiser 1994), seepage springs, burn scars, fields, and other human-created clearings (e.g., fire tower sites, campsites). We follow Ramseur (1960) in distinguishing shrub balds from heath balds. Heath balds are evergreen shrub-dominated, low diversity communities of dry upper slopes and ridges. Shrub balds are more species rich and are dominated by herbaceous plants and shrubs. In fact, shrub balds are a successional community that may develop from grassy balds, as well as other high elevation clearings. Lindsay and Bratton (1979b) described compositional differences between grassy balds and other high elevation open communities such as burn scars in the Great Smoky Mountains.

Grassy balds are anomalous in that they are treeless communities in a mountain range that is not high enough for a climatic treeline (Gersmehl 1973). Cogbill et al. (1997) used several methods to suggest that the elevation of the expected climatic treeline in the southern Appalachians would be 7000 to 8000 feet (2150 to 2500 m), at least 400 feet (120 m) above the highest summit elevations and 1600 ft (480 m) above average elevations of the grassy balds. Few grassy balds species have alpine affinities, and those that do are found in adjacent habitats, such as rock outcrops, as well. Grassy balds also are very rare in the southern Blue Ridge, occurring in a very small percentage of even the most appropriate topographic positions in the region. Stratton and White (1982) found that balds occupied only 0.015% of the landscape of GSMNP.

The number of grassy balds has been estimated differently by different investigators, ranging from 43 (a list that combines the 27 balds of Wells (1937), with the partially overlapping set of 34 in Marks 1958; Pittillo 1981) to 119 (Gersmehl 1971; this total includes 56 that Gersmehl considered as historic, that is, no longer in the open condition, but does not include "artificial balds" or fields that Gersmehl determined were created by people relatively recently). Each study used different methods for locating balds (maps, surveys, place names, pers. commun.), different criteria for separating adjacent balds, and different definitions of a grassy bald to differentiate them from high elevation old fields. These estimates illustrate the difficulty in differentiating grassy balds from old fields and other human clearings occurring at high elevations without compositional or historical data. The lack of a definitive list of the status and location of balds, as well as the array of public and private land owners involved, suggests the importance of a regional assessment and conservation plan for this habitat.

Gersmehl's tally is the most complete and illustrates some of the problems in accounting for all regional occurrences of this habitat. Gersmehl investigated a total of 240 sites, 44 of these within GSMNP. Gersmehl listed 34 true balds (sites with characteristics of balds plus historic reference to an open condition; 9 of these were within GSMNP), 56 historic balds (sites inferred to have been true balds, but now dominated by woody plants; 7 of these were within GSMNP), 29 apparent balds (sites with characteristics of balds but no historic reference to an open condition; 5 of these were within GSMNP), 71 nominal balds (sites with neither characteristics of balds nor historic reference to ever having been open; 17 of these were within GSMNP), 49 fields and other "artificial" balds (5 of these were within GSMNP), and 1 unclassified. Lindsay and Bratton (1979a) confirmed that some forested areas with good understory forage in GSMNP were called "balds" by tradition (the nominal balds of Gersmehl). If we accept Gersmehl's apparent balds, his tally suggests that 47% of the balds (56 of 119) had been lost by 1970 because of succession. Even balds that are

TABLE 2
Community Diversity of Grassy Balds as Described by the Nature Conservancy's Draft National Community Classification

Rubus alleghaniensis–Rubus canadensis shrubland alliance
   Rubus alleghaniensis–Rubus canadensis/Carex pensylvanica shrubland
Danthonia compressa shrub herbaceous alliance
   Rhododendron calendulaceum/Danthonia compressa shrub herbaceous vegetation
Danthonia (compressa, spicata) dwarf–shrub herbaceous alliance
   Sibbaldiopsis tridentata/Danthonia compressa dwarf–shrub herbaceous vegetation
Carex pensylvanica herbaceous alliance
   Carex pensylvanica herbaceous vegetation
Phleum pratense herbaceous alliance
   Phleum pratense–Bromus pubescens–Helenium autumnale herbaceous vegetation

Source: Karen Patterson, unpublished report.

still extant but unmanaged lose surface area due to invasion by shrubs and trees at a rate that has been estimated at approximately 1 to 2% per year. Lindsay and Bratton (1980) found that Andrews and Gregory Balds in the Great Smoky Mountains lost 33 and 50% of their areas, respectively, between 1940 and 1975.

The physical parameters that characterize southern Appalachian grassy balds are not distinctive in that several forest communities occupy the same range of sites that grassy balds do. Their elevations range from 4600 to 6100 feet (1415 to 1875 m), with the majority between 5200 to 5800 (1600 to 1780 m), and a peak at 5400 (1660 m) (Mark 1958). Their topography includes domes, ridgetops, gentle slopes, and gaps (Mark 1958), often facing to the south or southwest. The geology of grassy balds is not unique and the soils do not differ from forest soils, except slight differences that can be attributed to the grass turf itself (Mark 1958). As these findings suggest, grassy balds occupy only a small fraction of the sites that seem appropriate for their development based on the topographic and substrate variables that most often correlate with vegetation (Stratton and White 1982). Most grassy balds are small in size, ranging primarily from 1 to 8 ha in GSMNP (Stratton and White 1982), to the series of larger balds on the Roan Mountain Massif that range to over 200 ha in size.

The Nature Conservancy's draft national community classification (Weakley et al. 1996) classifies grassy balds in five alliances, each with a single community association (Table 2). Common vascular plant species are listed in Table 3. Species dominance varies both between and within sites. Balds that are currently or recently grazed have a higher cover of *Danthonia* (mountain oat grass) and *Potentilla* (cinquefoil), while ungrazed sites have a higher cover of sedges (*Carex* spp.). Woody cover varies with aspect, disturbance, grazing history, and seed source.

Grassy balds are characterized by high species richness when compared with other high elevation communities in the southern Appalachians. Over 150 species are recorded for a number of sites, with nearly 400 species listed for all balds (Mark 1958; see also Stratton and White 1982 who list nearly 300 species for balds in the Great Smoky Mountains). Exotics are present on all balds and range from less than 5% to 30% of the flora.

Grassy balds are distinctive in appearance and species dominance, but not unique in the species that are present. Almost all the species found on the balds are found in other high elevation communities as well, occurring on rock outcrops, in the forest understory, and in high elevation seeps (Wiser and White, in press; Wiser 1994). This is also true for the rare species, the exception being Robbin's ragwort (*Senecio schweinitzii*), a plant disjunct from northern New York. The diversity of rare plant species on grassy balds, however, is among the highest of any similar sized habitat. More than a quarter of all rare plant species occurring in less than 20 locations range wide

TABLE 3
Common Herbaceous Vascular Plants of Southern Appalachian Grassy Balds

Grasses	Other herbaceous plants (forbs)	
Danthonia compressa	*Achillea millefolium*[a]	*Lysimachia quadrifolia*
Danthonia spicata	*Angelica triquinata*	*Oxalis stricta*[a]
Agrostis perennans	*Aster* spp.	*Potentilla canadensis*
Phleum pratense[a]	*Chrysanthemum leucanthemum*[a]	*Prenanthes* spp.
Poa spp., including exotic spp.[a]	*Fragaria virginiana*	*Prunella vulgaris*
Sedges	*Gentiana decora*	*Rumex acetosella*[a]
Carex brunnescens	*Hieracium* spp. incl. *exotic spp.*[a]	*Solidago* spp.
Carex debilis	*Houstonia purpurea*	*Stachys clingmanii*
Carex normalis	*Houstonia serpyllifolia*	*Stenanthium gramineum*
Carex pensylvanica	*Hypericum* spp. incl. *exotic spp.*[a]	*Trifolium spp.*[a]
Rushes and relatives	*Lechea racemulosa*	*Viola* spp.
Juncus tenuis	*Lilium superbum*	
Luzula spp.		
Ferns		
Athyrium asplenoides		
Botrychium dissectum		
Dennstaedtia punctilobula		

[a] Exotic species.

Adapted from Stratton and White 1982 and Mark 1958.

in the southern Blue Ridge have at least one occurrence on grassy balds (Table 4). While harboring important rare plants, grassy balds do not possess a flora related to that of the northern Appalachians alpine flora or other alpine floras (e.g., Stratton and White 1982, Wiser 1994). The seven vascular plant species that do show a relation to the alpine flora of New England are characteristic of rock outcrops rather than grassy balds (Wiser and White, in press). In fact, we postulate that these species occur on grassy balds only if other habitats (cliffs, seepage areas) that harbor these species are nearby, thus allowing expansion of the populations onto the bald itself.

The hybrid swarm of azaleas (*Rhododendron* spp.) on Gregory Bald in Great Smoky Mountains National Park is a unique floristic element. Other balds both within and outside the park also have azalea populations that may include unique genotypes. However, this is not necessarily evidence for the naturalness or antiquity of the balds — the open condition of balds and the transport of livestock each year from low elevations might have be the causal factors for the establishment and of parental species and success of the hybrids. Species from various other habitats may have been brought together in a small area and low competition from other woody plants may have promoted survival of the resulting hybrids.

Although vertebrates use the grassy bald habitat, it appears that none are dependent upon or specific to it. We do not know enough about invertebrates to speculate about the possible role that this habitat plays in their diversity.

ORIGIN, MAINTENANCE, AND SUCCESSION

Because of the enigmatic nature of treeless communities in a forested landscape, the first questions about grassy balds concerned their origin, successional status, and maintenance. Conservationists sought an explanation for the origin of grassy balds in order to grapple with the question of whether and how to manage them. For example, in the mid-1970s, the managers of GSMNP, while finalizing

TABLE 4
Rare and Endemic Vascular Plants Found on Grassy Balds

Taxon	Distribution	Rarity	Other habitats
Agrostis mertensii	Northern	S	Outcrops, streamsides, heath balds
Alnus viridis ssp. *crispa*	Northern	S	Outcrops
Calamagrostis canadensis	Northern	S	Seeps, openings
Carex aenea	Northern	S	Dry openings, rocky areas
Carex cristatella	Northern	S	Bogs
Carex misera	Endemic	G3	Outcrops
Carex ruthii	Endemic	S	Seeps, streamsides
Delphinium exaltatum	Northern	G3	Rich woods, rocky slopes
Geum geniculatum	Endemic	G1	Moist, rocky woods
Geum radiatum	Endemic	G3	Outcrops
Glyceria nubigena	Endemic	G2	Seeps, streamsides
Helianthemum bicknellii	Northern	S	Dry sandy soil
Houstonia purpurea var. *montana*	Endemic	G1	Rocky summits
Huperzia selago	Northern	S	Outcrops
Hypericum graveolens	Endemic	S	Seeps, outcrops
Hypericum mitchellianum	Endemic	S	Seeps, outcrops
Krigia montana	Endemic	S	Outcrops, streamsides
Lilium grayi	Endemic	G2	Forest openings, meadows, seeps
Lilium philadelphicum	Northern	S	Glades
Minuartia groenlandica	Northern	S	Rocky summits
Phlox subulata	Midwestern	S	Outcrops, glades
Platanthera grandiflora	Northern	S	Seeps
Platanthera psycodes	Northern	S	Seeps
Poa palustris	Northern	S	Spruce–fir forests
Polygonum cilinode	Northern	S	Thickets
Prenanthes roanensis	Endemic	G3	Seeps, woods
Rhododendron cumberlandense	Endemic	G2Q	Openings
Rhododendron vaseyi	Endemic	G3	Seeps, swamps
Rubus ideaeus ssp. *sachalinensis*	Northern	S	Thickets
Rugelia nudicaulis	Endemic	G3	Woods
Senecio schweinitzianus	Northern	S	Grassy openings
Sibaldiopsis tridentata	Northern	S	Outcrops
Spiranthes ochroleuca	Northern	S	Meadows, woods
Stachys clingmanii	Endemic	G3Q	Seeps, woods
Trisetum spicatum	Northern	S	Rocky summits

Note: Under rarity, "S" indicates rarity at the state level only (North Carolina or Tennessee), and G1, G2, and G3 are the three top levels for globally rare plants in the classificaton scheme of the Nature Conservancy (G1 is the most rare).

From Stratton and White (1982), Mark (1958), and unpublished data of the authors.

a management plan for the park, were faced with several questions: Should grassy balds be treated as historic resources or natural resources? Should they be considered natural sites and be included within designated wilderness areas? Should grassy balds be managed? How many and which ones should be conserved?

Numerous natural and anthropogenic mechanisms have been proposed, either singly or in combination, for the formation and original maintenance of these grasslands (Wells 1936a, 1936b, 1937, 1938, 1946, 1956, 1961; Marks 1958; Radford 1968; Gersmehl 1969, 1970a, 1970b; Saunders

1981). As Brown noted in 1941, origin and maintenance are really different questions and factors that are dismissed for bald origins might nonetheless have been important in maintenance. Marks (1958) also discussed possible mechanisms for bald extension, as a separate question from origin and maintenance.

The major dichotomy among the hypotheses for origin is, of course, between anthropogenic and natural factors. The anthropogenic hypotheses include both activities of European settlers and Native Americans. Below we will first discuss natural processes that have been proposed for bald origin and maintenance. We will then describe human influence. Fire can be either natural or human-caused, so it forms a transition in the following discussion from natural to human factors.

CLIMATE AND TOPOGRAPHY

The distribution of balds does not seem to be controlled by topography in a way that would implicate the physical environment as the ultimate causal factor. The elevations, aspects, slope steepnesses, slope positions, and slope shapes of balds broadly overlap with sites dominated by forests; these factors are just not distinctive for the balds. Further, balds occupy a very small percentage of the seemingly appropriate sites, as judged by the topographic distribution of known balds. Finally, the grassy balds are uniformly invaded by trees and shrubs — the physical environment seems unlikely in balds maintenance as well as origin. These factors do not exclude woody plants from the balds.

Marks (1958) compared the microclimate of the balds and nearby forests. He found that climatic conditions did not exclude woody plants, but did find that the physical environment (wind, a tendency for more fluctuation in temperature and moisture compared to forest understories) might slow woody plant invasion. A number of investigators have noted wind-swept tree shapes ("flagging") on balds. However, these microclimatic differences are the result of the open bald habitat, rather than its cause. Recent forest clearings near the balds do not become grassy balds and revert to forest despite the microclimatic conditions of treeless habitats. Again, it would appear that physical factors do not currently prevent woody plants from colonizing and dominating grassy bald sites.

GEOLOGY AND SOILS

The geology of grassy balds is not unique and the soils do not differ from nearby forest soils on the same parent material (Mark 1958), except for the influence in the upper horizon of the grass turf itself. Bald soils are not droughtier, thinner, or more acid than forest soils nearby (Mark 1958). Mark also found that fungi of balds soils were not different that those of nearby forested sites; they were not similar to typical grassland soil fungi.

THE ECOTONAL HYPOTHESIS

Whittaker (1956; Gersmehl 1969) hypothesized that during the warmest postglacial periods the lower boundary of spruce–fir was at 5700 feet (1750 m), basing this on the observation that peaks that only reach 5700 feet (1750 m) in elevation lack spruce–fir, while those that are higher than 5700 feet (1750 m) have well-developed spruce–fir forests that extend down to 5500 feet (1680 m). Whittaker's model was used by Billings and Mark (1957; Mark 1958) to explain the formation of grassy balds. Their model has spruce–fir forests migrating upward during warmer periods and being extirpated from mountains less than 5700 feet (1750 m) in elevation. With cooler climatic conditions, the surrounding hardwood forest migrated down slope. Peaks around 5700 feet (1750 m) in elevation, without a dispersal source of spruce and fir, developed into bald communities.

Several lines of evidence suggest that this scenario does not explain the presence of balds. First, while spruce and fir may be absent from some slopes that are high enough in elevation for these species, many species of hardwoods are capable of growing at the elevations and locations of current balds. Second, after grazing animals were removed, succession did begin to establish a woody cover on all balds. There does not appear to be a lack of appropriately adapted woody species for this environmental situation. Many "ecotonal" areas have been disturbed by logging and fire but these return to a forest cover.

Native Grazers

Currently there are no grazers that are actively maintaining the open nature of grassy balds. However, two extirpated large native grazers, elk and woodland bison, were present in the southern Blue Ridge until the 1800s. There were also numerous Pleistocene grazers that are now extinct. These could explain the presence of a grazing-adapted flora, but landscape-vegetation relations and climate have undergone so much change that these animals do not present us with an explanation of modern grassy balds.

Other Natural Factors: Windthrows, Ice, Frost Damage, and Insects

While these factors do kill trees, there is no evidence that they have operated or are currently operating to create or maintain balds. For example, windthrows in the forests surrounding balds undergo gap dynamics and eventual restoration of the canopy. Most insect attacks are isolated in time and space and are specific to particular species. The invasion of grassy balds involves diverse woody plants and though insects might limit particular species at particular times (Gates 1941), they do not prevent succession from taking place.

General support for the fact that natural factors do not create or maintain balds can be found in the results of transplant experiments (Brown 1953; Marks 1958). These studies show that woody plants are able to survive and establish populations on the balds. This reinforces the observational studies that show universal, if sometimes slow and patchy, succession on the balds.

Several natural factors, while ultimately responsible for neither grassy bald origin nor maintenance, might contribute to the observed slow rate of succession on grassy balds. An isolated sapling within an open habitat may very well be more susceptible to freezes, ice storms, and dessication from wind (in fact, trees and shrubs on the open parts of grassy balds often have a wind-influenced shape). However, isolated woody plants on the bald surface also speed succession by serving as a locus for woody plant establishment because of seed dispersal by birds using the woody plants as perches, and because the canopy overtops and shades the grasses, thereby reducing competition for seedlings or sprouts.

Fire

The evidence, however, does not suggest that fire is an active process in these communities. There is no evidence of charcoal in the soil; fire is difficult to ignite in such a moist environment; and the effect of fire has not enhanced an open grassy condition in several management trials (Barden 1978; Lindsay and Bratton 1979a; Murdock 1986). Natural fire frequencies are particularly low at these elevations (Harmon 1982). Studies have shown that fire does not eliminate woody species — in fact, it can stimulate the sprouting and spreading of *Rubus*, *Vaccinium*, and other woody plants (Lindsay and Bratton 1979a; Murdock 1986). Similar conclusions have been drawn from work in other parts of the southern Blue Ridge (USFS-Mount Rogers, pers. commun.). Very hot fires on logging debris did create open communities in the southern Appalachians, but these were different in composition (Lindsay and Bratton 1979), topography, other site factors, and size compared to grassy balds. Unless soil erosion accompanied these fires, they return to woody plant cover during succession.

It is possible, however, that fire, under the right conditions, could eliminate or slow the spread of trees into the balds. This would contribute to maintenance only if the bald were large enough and entirely grass dominated (so that clonal shrubs were not present and woody plant seed input was low) and the fire was hot enough, uniform enough, and frequent enough (so that seedlings did not survive to fire-resistant size classes) to eliminate seedlings. For example, Barden (1978) suggested that fires every 5 to 8 years might slow tree invasion. That such conditions have held historically seems unlikely, but this remains a management possibility for fully restored balds.

NATIVE AMERICAN ACTIVITIES

Native Americans are known to have burned the woods to improve hunting and the gathering of berries and nuts. They also maintained hunting camps and trail systems in these mountains. There is, however, no concentration of Native American artifacts that would suggest camping or active use for hunting near or on the balds (Mark 1958; Lindsay and Bratton 1979a). Wells (1937) cites traditional Native American myths that refer to open habitats as evidence of the antiquity of balds and labels these habitats "archaeological disclimaxes." One such story is that related by Gilbert (1954), in which Native Americans called Gregory Bald in the Great Smoky Mountains the "Great Rabbit Place" and infers that this bald predates European settlement. However, Gersmehl (1971) has argued that these stories may postdate, rather than predate, European influences in the mountains. Further, at least some of the myths may refer to open habitats other than balds.

Native American influence was not constant in time or space. Smathers (1981) argues that the balds are possibly the old camp sites or "nodes" along high elevation trails from a Native American culture that predates the most recent ones. Smathers (1981) suggests that European settlers took advantage of, and sometimes expanded, these already vanishing sites when they arrived. Unfortunately, our current evidence of Native American influence is too fragmentary to allow evaluation of this hypothesis. The fact that some grassy balds were open habitats when the first European settlers arrived indirectly suggests Native American influence or an interaction of Native American influence and the influence of elk or woodland bison.

CLEARING BY EUROPEAN SETTLERS AND SUBSEQUENT GRAZING

Early European settlers cleared some forests on ridge lines for summer pasture (Wells 1936). Historical photographs and literature provide evidence that settlers girdled trees to create balds for summer livestock pastures (Lindsay and Bratton 1979a). These activities were happening with the earliest settlement, certainly by the early to mid-1800s. As a result, some "balds" were called "fields" (Gersmehl 1971). For example, Spence and Russell Fields in the Great Smoky Mountains were cleared by 1880, and there are frequent observations of the use of high elevation summer pastures in the period 1860 to 1880 (Lindsay and Bratton 1979a). Settlers also used forests with open understories for upland summer forage; sometimes these were called "balds" despite being forested (some of these are the "nominal" balds of Gersmehl 1971). The earliest botanists did note that the balds and adjacent forest understories were being used for livestock grazing when these habitats were first visited.

The recent encroachment of woody vegetation on balds since grazing was eliminated provides additional evidence for their anthropogenic origin, but it is also possible that the invasion took place because of grazing's impact. It is certainly true that many open fields at high elevations were cleared by early settlers, but whether particular balds were the result of the expansion of preexisting balds or the creation of new ones is not known. Since the balds, regardless of history or origin, were used heavily for livestock for periods ranging from 50 to 100 years or more, all balds have a heavy overlay of historic land use practices.

Settlers took their cattle, sheep, and horses to high elevation summer pastures for a number of reasons. This freed the productive low elevation valley farmland for crop plants and simultaneously eliminated the need for the extensive fencing of these fields. Fences were not built in the high elevations, and large herds could be tended by a few individuals, and livestock also grazed in woods around the balds. Because of the high rainfall in these mountains, springs were easy to find for water supplies. The use of high elevation summer pastures was also a tradition transmitted from European to American cultures.

Some balds seem to have been open spots in this heavily forested landscape when the first explorers and surveyors arrived. Gregory and Parsons Balds in the Great Smoky Mountains were described as open during the survey of the state line between North Carolina and Tennessee in 1821 (Keith Langdon, unpublished file copies of surveyor notes), just 2 years after the first few settlers arrived in Cades Cove, the nearest agricultural area. Lyon observed the open bald on Roan Mountain in 1799 (from Peterson 1981). Several other reports date this bald to about 1750 (Keith Langdon, pers. commun.). Some of these early reports do not completely rule out creation by European settlers since they are contemporaneous with settlement, but they would certainly require a seemingly unrealistically rapid development of this high elevation habitat. As Smathers (1981) noted these early reports that suggest the presence of presettlement grassy balds are only available for a few balds.

WHAT WE KNOW ABOUT BALD ORIGIN AND HISTORY: A SUMMARY

We know that some balds were created by European settlers. However, some balds seem to predate that period. The origin of grassy balds is likely to remain controversial, since we know the origin of only a few balds with certainty and the balds themselves have possibly been altered by livestock grazing and further clearing. Nonetheless, we can state the following:

1. No combination of environmental factors (e.g., elevation, slope aspect, slope shape, or soil) explains the balds
2. Balds occupy a very low percentage ($\ll 1\%$) of the sites that seem appropriate for balds, as judged by the sites where balds occur today
3. No new balds are being created by natural mechanisms (i.e., the forces that created the balds are not currently active)
4. The balds are universally invaded by trees and shrubs and thus whatever forces may have once maintained balds are no longer active)
5. Transplant experiments show that woody plants are not excluded from the sites by the physical environment
6. There is no positive evidence that Native Americans created or maintained the balds, but in essence this consists of an absence of evidence rather than a rejection of the hypothesis outright
7. All grassy balds have a history of grazing in the last hundred years
8. After grazing animals were removed, primarily between 1920 to 1940 as a result of socioeconomic changes and land protection, balds were universally invaded by trees and shrubs
9. Where grazing persists or is used in management and/or where mowing is used, the balds can be maintained in their open condition
10. While they represent unique communities, they do not harbor a flora that is restricted to this habitat
11. Some grassy balds were created through girdling of trees in the 1800s, perhaps in forests that had a prior history of burning or grazing and thus an open, herbaceous dominated understory (e.g., we know that grassy balds can be created by human action).

It is tempting to conclude from this list that all balds were created by European settlers seeking to establish high elevation summer pastures for their livestock. Again, however, we note that some balds seem to have been open at the time that the first European settlers arrived, so we would have to argue for a rapid development of these high elevation pastures. It is also possible that grazing itself altered the original plant community beyond recognition, so that the original evidence for the origin and maintenance of these balds was destroyed.

One possible scenario for the occurrence of presettlement grassy balds includes three elements: native grazers (woodland bison and elk), Native American influence (i.e., frequent fires that would maintain grassy understories in forests), and past climates (i.e., climates that would be conducive to frequent fires in the high elevations). It is possible that these factors combined to convert forests with grassy understories to grassy balds. It is also possible that they maintained balds that were remnants of other conditions that are no longer extant (whether human, climatic, or biological). However, even if extirpated native grazers, Native American influence, and past climatic conditions worked together to explain some grassy balds, it is important to note that these influences operated in very few places in the landscape (there are literally thousands of seemingly appropriate sites that are unoccupied by grassy balds). The rarity of grassy balds suggests that human influence would likely be an important ingredient of any explanation. Thus, we hypothesize that human influence was crucial. Native American impacts changed through time, and references to Native American effects come most heavily from the period just prior to European settlement, so that our knowledge of Native American effects is fragmentary. We might even question whether we will ever be able to definitively prove how the grassy balds came to be. Whether caused by European settlers, Native Americans of more recent or more distant periods, or some combination of Native Americans, past climate, and extirpated native grazers, grassy balds are certainly relictual communities in the sense that no causal factor now operates in the landscape.

THE RATE OF SUCCESSION

While the balds are steadily invaded by trees and shrubs, the rate of succession is not as fast as occurs after fire or windstorm in the surrounding forests (Brown 1941; P. White, unpublished data). The dense herbaceous community slows invasion by seed, and the rate limiting step for succession is the establishment of new individuals rather than the growth rates of those individuals that are established (Table 5). Seedling densities on the grassy turf ranged from 0 to 0.24 and average $0.12/m^2$, which is more than an order of magnitude lower than surrounding forests. On the other hand, extension growth rates of the woody plants, once established, ranged from 10.1 to 23.0 cm and averaged 15.2 cm/year (Table 5), which is similar to growth rates in canopy gaps in the surrounding forests.

Establishment by seed is thus a slow mechanism for the invasion of grassy balds. In fact, many of the early successional woody plants on the balds spread clonally through the turf (e.g., blueberries). Invasion by seed often occurs more frequently where the grass is shaded (e.g., along the bald edge and under isolated trees within the bald itself, Brown 1953) or broken (e.g., areas rooted by hogs, exposed by soil erosion, or near boulders; Bruhn 1964; P. White, unpublished data). Thus, woody plant islands tend to spread outward from a starting point, either because the plants spread clonally or because an overarching tree shades the grass turf below (and also serves as a roosting place for birds that spread seeds to the bald).

The rate of succession probably also varies with the composition of the open, herbaceous community. Some balds are grass dominated, others have mixes of grasses and sedges, and Roan Mountain has areas that are dominated by sedges. Additional research is needed on the relationship between bald composition and invasion rate.

The slowness of the succession gives managers a window of opportunity for determining management goals and experimenting with management techniques. Where the grass turf is thick and continuous, it is remarkably resistant to seedling invasion, presumably because the seeds either do not establish roots that fully penetrate the grass turf or are densely shaded by the grass itself, or both. In such cases, succession proceeds from the edges (where the grass is shaded by an overarching canopy), around trees that have somehow managed to establish, or by vegetative spread of clonal species. However, some balds are already completely covered by woody plants and the

TABLE 5
Seedling Invasion and Extension Growth Rates on Parts of the Grassy Balds in Great Smoky Mountains National Park That Are Still Dominated by Mountain Oat Grass

Grassy bald	No. of observations	Density of seedlings per m²	Extension growth (cm)
Andrews Bald	100	.24	12.2
Blockhouse Bald	50	.04	13.8
Gregory Bald	70	.13	15.5
Hemphill Bald	50	.04	19.0
High Springs Bald	50	.02	14.0
Little Bald	50	0	10.1
Little Bald Knob	25	.6	23.0
Parsons Bald	100	.17	19.3
Russell Field	50	.08	14.6
Silers Bald	50	.24	12.2
Spence Field	100	0	13.3
Total	695	.12	15.2

Note: The number of observations is the number of 1 × 1 m quadrats for seedlings and the number of trees measured on each bald. Seedlings were defined as less than 1 m tall. Extension growth is the annual increment of the longest leader for each sampled stem.

From unpublished data of P. White.

window of opportunity will close for others within the coming decades. We noted above that Gersmehl's (1971) figures suggest that 47% of the balds were already lost by 1970. Lindsay and Bratton (1980) showed that one of the larger balds in the Great Smoky Mountains, Gregory Bald, became 87% covered with woody plants in the period ca. 1935 to 1975 and blueberry (*Vaccinium*) cover more than doubled in one 11-year period. Lindsay and Bratton (1980) reported rates of succession that would produce a 0.4 to 1.9% loss of bald area per year, values that would result in loss of herbaceous dominance in 50 to 100 years. Their data and those of others suggest that all remaining balds, without active management, would be woody plant covered by the mid-21st century. Restoration will also become more expensive as successional time proceeds.

MANAGEMENT OF GRASSY BALDS

Whatever the origin, balds are significant sites for rare species, species richness, aesthetics, recreational opportunities, and historic value. Observation and research has shown that to maintain these resources, all of which are dependent on the openness of the balds, management is needed. The question of management, however, is more complicated than removing woody vegetation. The issue of origin haunts these decisions, and the many possible reasons for keeping balds open and the number of stakeholders makes management decisions difficult. There are numerous questions that arise when addressing the management of grassy balds. First:

1. If grassy balds are not natural or if we cannot prove whether they are or not, should they be maintained or should we let natural processes result in the succession to forest? More specifically, what are the costs and benefits of management vs. abandonment of balds?

TABLE 6
Possible Goals for Grassy Balds Management

Management for rare species populations
Management for rare plant communities
Management for historic state, for example:
  ca. 1750: Presettlement condition of true balds (unknown)
  ca. 1880–1930: Livestock (sheep and cattle) grazing, closely cropped vegetation and grass dominance, probably low diversity
  ca. 1940–1970: Postgrazing succession
    Early succession with diverse wildflowers, azalea hybrid swarms, and berry crops
Management for aesthetics
  Wildflower displays, wild berry crops, and vistas
Management for recreation

A number of managers have answered that we should restore and manage at least some of the balds, leading to the next set of questions:

2. How many and which grassy balds should be managed? What criteria should be used to prioritize balds for management?
3. If these balds are managed, what should the management goal be? The range of goals is listed in Table 6.
4. What are the most cost effective and environmentally safe techniques for restoration and management (hand cutting, fire, herbicide, seeding, mowing, grazing)?

Two Examples of Management

Active management is taking place at two places in the southern Appalachians, Great Smoky Mountains National Park and Roan Mountain.

Gregory and Andrews Balds in Great Smoky Mountains National Park

The formulation of a management plan for GSMNP in the 1970s brought the grassy balds issue to the forefront. Eventually, the management plan called for the restoration and maintenance of two of the Park's 21 grassy balds (Stratton and White 1982). Gersmehl (1971) classified these 21 sites as follows: 9 true balds (relatively high quality extant balds), 7 historic balds (already lost), and 5 apparent balds (balds that lack a historic record, but otherwise fit the definition of true bald); an additional 5 balds are classified as fields or artificial balds, but several of these are high quality sites in terms of current vegetation. The two selected for management were Gregory Bald and Andrews Bald, which were two of the best known, most diverse, most studied, and most popular balds in the Park. Both are true balds in Gersmehl's sense. Gregory Bald has a renowned swarm of azalea hybrids that display a diversity of flower colors and has long been a popular hiking destination. Andrews Bald is on the edge of the spruce–fir zone near Clingmans Dome. Because of its higher elevation relative to Gregory Bald, it harbors some species not found on that bald. In addition, it possesses a seepage meadow below a spring that has one of the few Park locations of the round-leaved sundew (*Drosera rotundifolia*).

The management plan recognized that it was unknown whether the balds were natural communities. They were not designated as historic areas, however, because of the uncertainty of origin and because the difficulty of choosing a particular historic period for management. The most recent and best documented historic period, during which livestock grazed on the balds, may have negatively impacted plant diversity and blueberry (*Vaccinium*) abundance, both of which are popular with Park users. Management trials had shown that livestock grazing would be expensive and might not protect rare species. In addition, before maintenance was put in place, the balds would have to be restored to their earlier configuration. In the end, the balds were designated "experimental areas" with natural area zones.

Restoration of the two balds began in the late 1980s. A field survey of bald edges had been carried out in the 1940s by A. E. Bye, and his maps proved to be very valuable, along with field inspection of tree sizes and growth rates on the edges of the balds. The Park's Resource Management Staff, with Student Conservation Corps workers, successfully expanded the balds towards their original sizes and selectively cut trees and shrubs within the bald surface.

Mary Lindsay and Susan Bratton had carried out a series of management experiments in the 1970s (Lindsay and Bratton 1980). They had shown that fire would not restore the balds — it often stimulated vegetative sprouting of the established woody species. They also showed that livestock grazing by a number of different species would be difficult. In the confined areas on the balds as they now exist, the animals lost weight. They were not selective in terms of their effect on rare vs. common plants. There were other logistical problems associated with livestock, such as whether to fence the balds and what to do about the interaction between Park users and the livestock. The management technique of choice was therefore handcutting of woody plants, some herbicide use on cut stumps to eliminated resprouting, and seeding of opened areas with a mulch harvested from health grass patches that were in seed. Rare plant populations and the azaleas on Gregory Bald were carefully mapped.

The restoration of these balds has been a success. The grass surface of the balds will be monitored and mowed as needed. However, only two of some 20 to 30 balds in the Park were restored, representing a conservative commitment to this habitat type. No cost–benefit analysis has been carried out for other balds in the Park. Among the high quality sites remaining are several true balds (e.g., Parsons Bald, Thunderhead Mountain Bald and Rocky Top Bald, and Silers Bald) and several artificial balds (e.g., Spence Field and Russel Field).

ROAN MOUNTAIN

The Roan Mountain balds may be the most unusual in the southern Blue Ridge because of their relatively large size (to 200 ha) and apparent age (early reports suggest they were open at least by ca. 1750). The Roan Mountain Massif contains six distinct grassy balds. The U.S. Forest Service manages the majority of the balds, but two private conservation groups, the Nature Conservancy and the Southern Appalachian Highlands Conservancy, also own sections of balds. The management of these balds has been a team effort that includes the U.S. Forest Service, the U.S. Fish and Wildlife Service, the Nature Conservancy, Southern Appalachian Highlands Conservancy, Natural Heritage Programs in both Tennessee and North Carolina, the North Carolina Plant Conservation Program, and the Appalachian Trail Council, with the review of the public stakeholders.

Of all the balds in the southern Appalachians, some of the Roan Mountain balds are among those thought to predate European settlement (nonetheless, European settlement probably changed their configuration and composition). They were present during the earliest time of European settlement: their occurrence was first recorded in 1799 (Peterson 1981). While these balds have a long history of grazing and currently there is the invasion of woody species, the invasion is slow, selective in locations (lower slopes and heavily disturbed areas), and dominated by *Rubus* (blackberry). Few trees have established in the thick sedge mat, and planted spruce on one bald has survived but not expanded much beyond their original location. Exotic plant species are fewer than on many other balds.

The management plan recognizes that multiple management goals exist. All balds on the massif are being managed, since they all have significant characteristics. The plan adopts multiple and unique objectives for each bald, reflecting the interests of all stakeholders. For example, on Jane Bald, the bald closest to road access with the Appalachian Trail traversing its ridgetop, the management goals reflect a balance among recreational and aesthetic, rare species and communities,

TABLE 7
Total Vascular Plant Species, Rare Species (G1 to G3 species of the Nature Conservancy), and Noteworthy Southern Appalachian Endemics in Several Grassy Balds Surveys

Study	Total species	G1 to G3 species	Southern Appalachian endemics
Mark 1958	382	15	56
Round Bald	175 (46%)	10 (66%)	28 (50%)
Great Smoky Mts. NP	346 (91%)	11 (73%)	34 (61%)

Note: Mark (1958) presented a survey of important sites across the entire southern Appalachian region. Percent figures are based on Mark's list.

and historic values. In Grassy Ridge, which is further from the road access and does not have the Appalachian Trail crossing its expanse, rare species and communities are the dominant management goal. The management alternatives reflect these management goals, with the most conservation management techniques (hand cutting) being used on Grassy Ridge while different combinations of actions (mowing, grazing by goats, herbicides, and prescribed fire) are being used on less sensitive sites.

DISTRIBUTION OF THE GRASSY BALDS FLORA AND LESSONS FOR REGIONAL CONSERVATION

Which grassy balds need to be protected to conserve their suite of biological diversity? How does the protection of grassy balds relate to the protection of biological diversity in the southern Blue Ridge? To explore these questions, P. White compiled species lists from the region in order to explore the distribution of vascular plant species.

The data analyzed here comes from: (1) a species list for Round Bald (a relatively undisturbed and species rich bald in the Roan Mountain Massif) from unpublished surveys done by Smith and Heiman (unpublished data), (2) a species list for grassy balds that occur in the Great Smoky Mountain National Park from Stratton and White (1982), and (3) a species list for all southern Appalachian grassy balds compiled by Mark (1959).

A single grassy bald (Round), one that is relatively undisturbed and species rich, has only 46% of all the species recorded for grassy balds in the region (Table 7). This suggests that species richness is distributed among balds and is not contained within a single bald. In contrast, a cluster of balds occurring in Great Smoky Mountains National Park, has 91% of species recorded for grassy balds. GSMNP has approximately half of all extant grassy balds, and these balds differ in their environmental characteristics (elevation, aspect, successional stage). These data suggests that species distribution throughout the range of balds in the southern Appalachians is influenced more by environmental variability than geography.

The rare species found on grassy balds have a different pattern of distribution (Table 7). While many of these species are found on numerous balds (examples being Fraser fir (*Abies fraseri*), wretched sedge (*Carex misera*), mountain avens (*Geum radiatum*), and mountain rattlesnake root (*Prenanthes roanensis*), a significant number are specific to balds in certain subregions. These narrowly distributed species include for GSMNP: Smoky Mountain manna grass (*Glyceria nubigena*) and Rugel's ragwort (*Rugelia nudicaulis*), and for Roan Mountain: bent avens (*Geum geniculatum*), Gray's lily (*Lilium grayi*), and mountain bluets (*Houstonia purpurea* var. *montana*). Of the 15 rare species, only 7 species occur in both areas. Thus, rare species are influenced more by geography than by environmental variability (or geography based on environmental variability). When endemics are added to the rarer species, a similar pattern occurs.

There are two messages here. The first is that the patterns of distribution of rare and endemic species are different than that of the common species. While a number of grassy balds within an area can protect the occurrences of the majority of grassy bald species, to protect the rare and endemic species of these communities requires the protection of specific grassy balds throughout their total range. This in essence protects both environmental and geographic gradients. Second, if one is concerned about the long-term viability and genetic diversity of rare species, numerous grassy balds would need to be protected. This would provide the multiple populations to maintain the genetic diversity of species and act as a buffer against extinction from demographic and environmental stochasticity. The pattern of distribution of rare species results in the need to protect a large number of the grassy balds to ensure the long-term protection of those species.

These data suggest the need for an assessment and cost–benefit analysis of grassy balds restoration and management on a regional and cross-agency basis. The Southern Appalachian Man and the Biosphere Program is ideally suited to facilitating this task. Some of the necessary data are being assembled. The Southern Appalachian Assessment has resulted in a computerized database on rare taxa and their locations. Additional field work will be required on all remnants balds and the full suite of questions phrased in Table 1 must be addressed on a regional basis. Only when we combined the biological and social values of the balds into a regional perspective will we be able to write a conservation plan specifying the number and distribution of grassy balds that should be restored and maintained in this region.

CONCLUSIONS

We begin our summary of the grassy bald discourse with a story about the Badgeworth Nature Reserve in Great Britain (Frost 1981). The Reserve was purchased in 1932 to protect the rarest buttercup species in Great Britain. Just enough land to protect the population was acquired from the farmer who owned and used the marshy field — 290 m² — an amount that qualified the Reserve as the smallest nature reserve in the world, according to the *Guinness Book of World Records*. To protect the plant from cows, a fence was established in 1933. As biomass accumulated within the fenced area, the population of buttercups began immediately to decline. Between 1934 and 1962, there were 5 years in which no buttercups occurred within the Reserve, and yet hundreds flourished just outside the fence (Frost 1981). The buttercup maintains a seedbank in the soil and germinates profusely on open mud banks. Managers were able to restore the buttercup population after they realized that they had to combat a natural process in these fields — succession.

This situation is parallel to that of the grassy balds in many senses. We don't know the original nature of the wetland habitat in this field. Obviously, the site had been used by people, of changing technologies, for many generations. Given the history of human use, natural processes will not maintain this rare plant population. In these cases we are faced with a decision: become caretakers of biological diversity or stand passively by while "natural" processes (themselves, though, the result of other human changes in the environment, like the alteration of the original wetland habitat when it became part of a farmer's field) reduce diversity.

From a global perspective, human use of ecosystems has often resulted in the dependence of diversity on traditional patterns of human use. This includes other cases like the grassy balds, such as the high elevation summer pastures in Scandinavia and elsewhere in Europe (Olsson 1991). Within the southern Appalachians, there are other cases. For example, wetland species depend on the drainage ditches dug by farmers trying to drain wet meadows (White 1984), and pine forests are dependent on fires, some of which were set by Native Americans and then European settlers. Because the world has changed in so many ways (e.g., the loss of large native grazers and predators) and because some of those changes have erased the original conditions (e.g., the natural fire frequency, the original nature of wetland habitats or the grassy balds), we are faced with losing diversity or finding ways to incorporate the human presence (vs. the use of strictly "natural" criteria) into our management actions.

There is, of course, unfinished business for the grassy balds. For example, one possible scenario is that large native grazers are important in maintenance of the grassland areas. In addition, we should reexamine the question about how many balds to restore and maintain. This question demands a regional approach and will require additional inventory of existing sites and their biological diversity. It is not too late to restore many of the remaining sites, but the opportunity will likely pass within the next 20 to 50 years, with costs increasing through time.

RISK ASSESSMENT REPORT CARD FOR ECOSYSTEM MANAGERS

Application of the *Risk Assessment Report Card for Ecosystem Managers* to grassy balds management provides this accounting:

Vision: B

The vision is well-defined for specific sites and would merit an "A" on a local scale, but a regional vision has not been explored or articulated and the rationale for management of a limited number of sites (e.g., only 2 of some 21 sites in GSMNP) has been set by budget constraints and not biological/sociological criteria.

Resource risk: D

It is clear that the grassy balds are successional and will be gradually lost through time. The woody invasion rate is relatively slow, but succession has been occurring on sites for 50 to 70 years and many sites are already lost. Additional sites will be lost over the next 20 to 50 years. While this may not rank the problem as an immediate crisis, it is clear that the resource is at risk.

Socioeconomic conflicts: A–

The vistas, flower displays, and berries of grassy balds make them tourist attractions. Use of livestock grazing (including fencing of the balds) or fire could be rated a negative factor for tourism by local communities if these techniques are used.

Procedural protocols: A

Experiments have been and are being used to refine restoration and management protocols. Hand cutting and local herbicide treatment, reseeding, and mowing are effective. Grazing and fire are subjects of management experiments on some balds. Adequate monitoring and evaluation are in place to refine protocols. However, no one has attempted a wholesale restoration from an already forested site.

Scientific validity: A–

The literature is well-developed for the grassy balds. It is unlikely that new methods for tracing the vegetation history of sites will be found. Several areas might be addressed in further research: compilations of early descriptions of balds (e.g., by earlier travelers or surveys that list trees and habitats along survey routes); influence of elk and woodland bison grazers; and effects of Native Americans. The high grade assigned reflects the fact that this work is unlikely to be definitive and is unlikely to greatly change our understanding of the balds.

Legal jeopardy: A

Balds management is not likely to be legally challenged.

Public support: A

Public support is likely to be strong; some individuals traditionally supportive of conservation goals may object to active management in settings that are dominated by wilderness natural areas. Some individuals may feel that the costs are too high or that other areas have higher priority.

Adequacy of funding: C+

Support has been adequate only for a subset of the balds, but has been strong within those sites.

Policy precedent: C+

The policies have been contested, debated, and defined for over 2 decades. However, natural area managers may still favor passive, hands-off management and be reluctant to restore and manage additional balds or commit to a regional cross-agency plan.

Administrative support: C+

While administrative managers have strongly supported restoration and management of individual balds, there has been no ongoing evaluation of costs and benefits for expansion to additional sites.

Transferability: B+

While the specifics of grassy balds restoration and management are unlikely to be transferable, grassy balds can serve as an important model for approaching the issues of historic and aesthetic management in the context of large natural areas.

REFERENCES

Barden, L. S. 1978. Regrowth of shrubs in grassy balds of the southern Appalachians after prescribed burning. *Castanea* 43:238–246.

Billings, W. D. and A. F. Mark. 1957. Factors involved in the persistence of montane treeless balds. *Ecology* 38:140–142.

Bratton, S. P. and P. S. White. 1981. Grassy balds management in parks and nature preserves: issues and problems. Pp. 96–114 in P. R. Saunders, Ed., Status and management of southern Appalachian mountain balds. Southern Appalachian Research/Resources Management Cooperative, Western Carolina University, Cullowhee, NC.

Brown, D. M. 1938. The Vegetation of Roan Mountain: an Ecological Study. Ph.D. dissertation, Duke University. Durham. 152 pp.

Brown, D. M. 1941. Vegetation of Roan Mountain: a phytosociological and successional study. *Ecol. Monogr.* 11:61–97.

Brown, D. M. 1953. Conifer transplants to a grassy bald on Roan Mountain. *Ecology* 34:614–617.

Bruhn, M. E. 1964. Vegetation Succession on Three Grassy Balds of the Great Smoky Mountains. M.S. thesis, University of Tennessee, Knoxville. 84 pp.

Camp, W. H. 1931. The grass balds of the Great Smoky Mountains of Tennessee and North Carolina. *Ohio J. Sci.* 31:157–164.

Cogbill, C. V., P. S. White, and S. K. Wiser. 1997. Predicting treeline elevation in the southern Appalachians. *Castanea* 62:137–146.

Frost, L. C. 1981. The study of *Ranunculus ophioglossifolius* and its successful conservation at the Badgeworth Nature Reserve, Gloucestershire. Pp. 481–489 *in* H. Synge, Ed., The biological aspects of rare plant conservation. John Wiley: New York.

Gates, W. H. 1941. Observations on the possible origin of the balds of the southern Appalachians. *Contrib. Dept. Zool., Louisiana State Univ.* 53:1–16.

Gersmehl, P. J. 1969. A geographic evaluation of the ecotonal hypothesis of bald location in the southern Appalachians. *Assoc. Am. Geogr. Proc.* 3:56–61.

Gersmehl, P. J. 1970a. Factors involved in the persistence of southern Appalachian treeless balds: an experimental study. *Assoc. Am. Geogr. Proc.* 3:56–61.

Gersmehl, P. J. 1970b. Factors leading to mountaintop grazing in the southern Appalachians. *SE Geogr.* 10:67–72.

Gersmehl, P. J. 1971. A Geographic Approach to a Vegetation Problem: the Case of the Southern Appalachian Grassy Balds. Ph.D. dissertation, University of Georgia, Athens. 463 pp.

Gersmehl, P. J. 1973. Pseudo-timberline: the southern Appalachians grassy balds. *Arctic Alpine Res.* 5:137–138.

Gilbert, V. C., Jr. 1954. Vegetation of the Grassy Balds of the Great Smoky Mountains National Park. M.S. thesis, University of Tennessee, Knoxville. 73 pp.

Gray, A. 1842. Notes of a botanical excursion to the mountains of North Carolina, in a letter to Sir William J. Hooker. *Am. J. Sci. Arts* 42:1–49.

Harmon, M. E. 1982. The fire history of the westernmost portion of Great Smoky Mountains National Park. *Bull. Torrey Bot. Club* 109:74–79.

Lindsay, M.M. 1977. Management of grassy balds in Great Smoky Mountains National Park. U.S.D.I. National Park Service, Southeast Regional Office Res./Res. Manage. Rep., SER-17.

Lindsay, M. M. and S. P. Bratton. 1979a. Grassy balds of the Great Smoky Mountains: their history and flora in relation to potential management. *Environ. Manage.* 3:417–430.

Lindsay, M. M. and S. P. Bratton. 1979b. The vegetation of grassy balds and other high elevation disturbed areas in Great Smoky Mountains National Park. *Bull. Torrey Bot. Club* 106:264–275.

Lindsay, M. M. and S. P. Bratton. 1980. The rate of woody plant invasion on two grassy balds. *Castanea* 45:75–87.

Mark, A. F. 1958a. An Ecological Study of the Grass Balds of the Southern Appalachian Mountains. Ph.D. dissertation, Duke University, Durham, NC. 284 p.

Mark, A. F. 1958b. The ecology of the southern Appalachian grass balds. *Ecol. Monogr.* 28:293–336.

Mark, A. F. 1959. The flora of the grass balds and the fields of the southern Appalachian mountains. *Castanea* 24:1–21.

Murdock, N. 1986. Evaluation of Management Techniques on a Southern Appalachian Bald. M.S. thesis, Western Carolina University. Culowhee, NC.

Olsson, E. G. A. 1991. Agro-ecosystems from Neolithic time to the present. Pp. 293–314 in B. E. Berlund, Ed., The cultural landscape during 6000 years in southern Sweden — the Ystad Project. *Ecol. Bull.* 41.

Petersen, K. M. 1981. Natural origin and maintenance of southern Appalachian balds: a review of hypotheses. Pp. 7–17 in P. R. Saunders, Ed., Status and management of southern Appalachian mountain balds. Southern Appalachian Research/Resources Management Cooperative, Western Carolina University, Cullowhee, NC.

Pittillo, J. D. 1981. Status and dynamics of balds in southern Appalachian mountains. Pp. 39–51 in P. R. Saunders, Ed., Status and management of southern Appalachian mountain balds. Southern Appalachian Research/Resources Management Cooperative, Western Carolina University, Cullowhee, NC.

Radford, S. W. 1968. Factors Involved in the Maintenance of the Grassy Balds of Great Smoky Mountains National Park. M.S. thesis, University of Tennessee, Knoxville. 74 pp.

Ramseur, G. S. 1960. The vascular flora of the high mountain communities of the southern Appalachians. *J. Elisha Mitchell Sci. Soc.* 76:82–112.

Saunders, P. R., Ed. 1981. Status and management of southern Appalachian mountain balds. Southern Appalachian Research/Resources Management Cooperative, Western Carolina Univ., Cullowhee, NC.

Smathers, G. A. 1981. The anthropic factor in southern Appalachian bald formation. Pp. 18–38 in P. R. Saunders, Ed., Status and management of southern Appalachian mountain balds. Southern Appalachian Research/Resources Management Cooperative, Western Carolina University, Cullowhee, NC.

Stratton, D. A. and P. S. White. 1982. Grassy balds of Great Smoky Mountains National Park: vascular plant floristics, rare plant distributions, and an assessment of the floristic database. USDI, National Park Service, Southeast Regional Office, Res./Res. Manage. Rep. SER-58. Atlanta. 33 pp.

Weakley, A. S., K. D. Patterson, S. Landaal, and M. Gallyoun. 1996. International classification of ecological communities: terrestrial vegetation of the southeastern United States. The Nature Conservancy, Southeast Regional Office, Chapel Hill, NC.

Wells, B. W. 1936a. Andrews Bald: the problem of its origin. *Castanea* 1:59–62.

Wells, B. W. 1936b. Origin of southern Appalachian grass balds. *Science* 83:283.

Wells, B. W. 1937. Southern Appalachian grass balds. *J. Elisha Mitch. Sci.* Soc. 53:1–26.

Wells, B. W. 1938. Southern Appalachian grass balds as evidence of Indian occupation. *Bull. Arch. Soc. No. Car.* 5:2–7.

Wells, B. W. 1946. Archeological disclimaxes. *J. Elisha Mitch. Sci. Soc.* 62:51–53.

Wells, B. W. 1956. The origin of southern Appalachian grass balds. *Ecology* 37:592.

Wells, B. W. 1961. The southern Appalachian grass bald problem. *Castanea* 26:98–100.

White, P. S. 1984. Impacts of cultural and historic resources on natural diversity: lessons from Great Smoky Mountains National Park, North Carolina and Tennessee. Pp. 119–132 in J. L. Cooley and J. H. Cooley, Eds., *Natural Diversity in Forested Ecosystems*. Institute of Ecology, University of Georgia. Athens.

White, P. S., and S. P. Bratton. 1980. After preservation: the philosophical and practical problems of change. *Biol. Conserv.* 18:241–255.

Whittaker, R. H. 1956. The vegetation of the Great Smoky Mountains. *Ecol. Monogr.* 26:1–80.

Wilson, J. B. 1991. *Roan Mountain, a Passage of Time*. John F. Blair: Winston-Salem, NC 162 pp.

Wiser, S. K. 1994. High elevation cliffs and outcrops of the southern Appalachians: vascular plants and biogeography. *Castanea* 59:85–116.

Wiser, S. K. and P. S. White. In press. High elevation outcrops and barrens of the southern Appalachian mountains. In R. C. Anderson, J. S. Fralish, and J. Baskin, Eds., *The Savanna, Barren and Rock Outcrop Communties of North America*. Cambridge University Press: Boston.

19 Climate Change: Potential Effects in the Southern Appalachians

John D. Peine and Cory W. Berish

CONTENTS

Introduction 397
Ecological and Social Effects of Climate Change 399
Mountainous Areas: Appropriate Landscapes for Early Detection of the Ecological
and Social Effects of Climate Change 400
Steps Ecosystem Managers Should Take Now to Assess the Potential Impacts
of Climate Change 402
    Framework for Sensitivity Analysis...402
    Preliminary Natural Resource Sensitivity Analysis..403
    Preliminary Human Resource Sensitivity Analysis..404
    Strategic Plan for Species Conservation...405
    Monitoring for Climate Change and Its Effects 406
Case Example: Great Smoky Mountains National Park 406
    Goals and Measurement Endpoints 407
    Political and Ecological Boundaries...407
    Definition of Valued Resources 407
    Assessment of Data Needs..408
    Time Frame and Climate Scenarios...408
    Preliminary Natural Resource Sensitivity Analysis..409
    Preliminary Cultural Resource Sensitivity Analysis 410
    Strategic Plan for Species Conservation...410
    Monitoring for Climate Change and Its Effects 411
Conclusion...411
    Risk Assessment Report Card for Ecosystem Managers 412
References 413

INTRODUCTION

On June 23, 1988, Dr. James E. Hansen, a prominent scientist at the National Aeronautical and Space Administration, testified before the U.S. Senate Committee on Energy and Natural Resources that the so-called "greenhouse warming" had in fact begun (Climate Alert 1988). This public statement galvanized the interests of many governmental and conservation interests in the U.S. It contributed to the national debate on environmental issues that were a big part of the presidential election of that year. Since that time, the threat of global climate change has captured the imagination of the world. It has been the focal point of numerous debates on international environmental policy (Response Strategies Working Group 1990). In 1990, the then newly elected U.S. President, George Bush, established a U.S. global change research program that was developed by the Committee on Earth Sciences (1991). This commitment of over $1 billion per year represented as much money in the first year of the program as was expended during the entire 1980s to study the phenomenon of acid rain under the National Acid Deposition Assessment Program. The stakes to better understand this phenomenon are very high, as expressed by an editorial in the newspaper, the *Boston Globe*, on August 10, 1988:

> The threat that does not lessen, but grows stronger, is the one posed by the gradual warming of the earth's atmosphere — a climatological phenomenon known as the greenhouse effect. Scientists now warn that early in the 21st century — within the lifetimes of most people now alive — temperatures will rise significantly to trigger devastating effects... [Solutions] will require international cooperation of a kind never before undertaken or achieved. They also require a degree of political leadership that has rarely been displayed — except in time of national peril.

Eight years later in 1996, there is a general consensus in the scientific community that global climate will change in the coming decades at a rate unprecedented in human history (Houghton and Woodwell 1989; Schneider 1989; Hansen and Lebedeff 1988; IPCC 1995). Recently, the Intergovernmental Panel on Climate Change (IPCC) wrote that "Based on sensitivities of climate to increases in greenhouse gas concentrations reported by the IPCC Working Group I and plausible ranges of emissions, climate models, taking into account greenhouse gases and aerosols, project an increase in global mean surface temperature of about 1–3.5°C by 2100, and an associated increase in sea level of about 15–95 centimeters" (IPCC 1995). The frequency of numerous other extreme events, such as heat waves, droughts, floods, storms, and hurricanes is predicted to increase. Human mortality rates could increase from cardiorespiratory related heat failure and the spread of vectors of infectious diseases. Many other climate related events and impacts; e.g., agricultural, animal, and fisheries productivity are also predicted to change, and indeed, numerous articles in the popular press are already questioning whether the process of climate change has already started (e.g., *Newsweek* 1996). The predicted costs associated with potential climate change impacts are in the billions of dollars (Titus 1992). At a recent conference on the effects of climate change on the insurance industry, projections were that the entire industry could go bankrupt with the predicted increase in extreme weather/storm events (Leggette 1992)

An increasing concentration of carbon dioxide in the atmosphere is primarily responsible for global warming and the resultant global climate change. The concentration of carbon dioxide in the atmosphere is steadily increasing from the combustion of fossil fuels such as coal, oil, and natural gas and from the wanton destruction of forests, which release carbon dioxide into the atmosphere when they are burned or cut down. The increase in atmospheric carbon dioxide and other greenhouse gases such as nitrous oxide and methane (natural gas) over the last century has been documented in gas samples analyzed from ice caps and glaciers (Watson 1990). Other "greenhouse gases" which contribute to global climate change are fluorocarbons, which also disturb the protective ozone layer in the upper atmosphere and ozone in the lower atmosphere (World Resources Institute 1989). Approximately 6 billion tons of carbon dioxide are estimated to be emitted into the atmosphere each year, which equals approximately one half of the anthropogenic sources contributing to the greenhouse effect. It is estimated that in the U.S., 33% of the carbon dioxide emissions can be attributed to the burning of fossil fuels by electric utilities, 31% by transportation activity, 24% by industry, and 12% by residential use (World Resources Institute 1989). An American car driven 10,000 miles will release approximately its own weight, 1 to 2 tons, of carbon dioxide. Relative contribution of CO_2 by industrialized countries includes 26% by the

U.S. 21% by the Commonwealth of Republics that once made up the Soviet Union, 17% by the countries of western Europe, and 11% by China (World Resources Institute 1989).

ECOLOGICAL AND SOCIAL EFFECTS OF CLIMATE CHANGE

Regional changes in temperature will affect local rainfall, snowfall, and soil moisture conditions (Mitchell 1990). Abrupt change of established climate conditions often creates stress on ecosystems (Houghton and Woodwell 1989; Graham and Grim 1990; Overpeck et al. 1990). Climate is the most important factor influencing the relationships between soil, vegetation and site properties, such as primary productivity. Climate as a source of energy and moisture, acts as the primary control for ecosystems (Bailey 1990). An increase in global temperature will lead to an accelerated rate of sea-level rise; an increase in weather extremes such as hurricanes, tornadoes, floods, droughts, heavy snowfall and accompanying avalanches; variability in frost free days; temporal distribution of moisture accumulation seasonally; and the distribution of temperature extremes throughout the course of the year (IPCC 1992).

The reaction to temperature rise is also complicated by an accompanying biotic response to increased carbon dioxide which is anticipated to accompany global climate warming. In various fumigation studies, vascular plant responses measured due to carbon dioxide enrichment varied by species, but included changes in photosynthesis, respiration, water use efficiency, reproduction, growth rate, crown and nutritional qualities of grasses and ratios of root to shoot, and seed production to vegetation growth (Strain and Cure 1985).

It is extremely difficult to estimate the magnitude of potential impacts on ecological systems and human societies by the phenomenon of global climate change. First, it is very difficult to predict direct regional and landscape impacts, and second, there is not enough knowledge about how these various systems function and how they interrelate with climate to make broad-brush judgments on a global scale. The general circulation models used to model global climate change phenomenon are not precise enough to provide exact predictions for climate change on a regional scale (USGCRP 1995; Katz 1988). Complicating regional modeling is the wide degree of uncertainty associated with normal climatic variation. A number of general circulation models are used to predict global and regional changes in climate and related factors, such as temperature, precipitation, cloud cover, and sea level rise (Root and Schneider 1993). Annual average global temperatures are projected to increase between 1.5 and 4.5C (Strzepek and Smith 1995). With global warming, general predictions include a greater occurrence of extreme events, including high temperature, drought, floods, snow-storms and hurricanes.

Superimposed on the potential effects of climate change, such as changes in the range of annual and seasonal temperatures; alterations in the quantity and timing of precipitation; and reduced soil moisture are a series of events which could be triggered by climate change such as increased incidents of pests and pathogens, fire frequency and intensity, and extended periods of stagnent air resulting in the build up of air pollutants. The biological response to this litany of stressors could include a decline in forest productivity; shifts in the structure and/or function of plant and animal communities; changes in population distribution; and an overall reduction in biodiversity and nutrient availability. Indeed, Likens et al. (1996) recently demonstrated that acid rain can cause a decrease in forest productivity. It is difficult to speculate how the interaction of changing climatic conditions, increased atmospheric pollutants and increase temperatures will affect terrestrial ecosystems.

Complicating modeling predictions are factors such as population growth, resource use, and demand for more energy. For instance, the world population is expected to triple before reaching a plateau in the next century (Keyfitz 1991). How this dramatic growth in population and its utilization of resources and expulsion of waste into the atmosphere will affect the phenomenon of global climate change is next to impossible to estimate. One can only speculate that the global change phenomenon will accelerate exponentially. As the Third World countries become more industrialized, one can assume that the escalation of air pollutants into the atmosphere will continue to increase at a faster rate than population growth. The accompanying ever-accelerating loss of natural habitat will ultimately threaten the survival of an unknown percentage of the currently described 1.4 million species of the estimated 4 to 30 million organisms with which we share the planet. An extensive loss of species is predicted, even without the exacerbating influence of global climate change (Soule 1991).

There have been many efforts to model the effects of global climate change on forested regions (Pastor and Post 1986). Dramatic shifts in distribution of selected species have been predicted (EPA 1989). Today's assemblages of organisms will likely change due to changing climate conditions (Davis 1983). Vulnerable species include those on the edge of range, geographically localized, genetically impoverished, poor dispersers, slow producers, localized and annuals, highly special-ized, and migratory (McNeely 1990). The ability of plants to adapt to new habitat in response to climate change may be hindered by constraints associated with soil conditions. In temperate mountain ranges, soils have short evolution, so that even as seeds reach them suitable for their climate conditions, the soil conditions may not be appropriate and therefore may not be an adequate refugium from climate change (Retzer 1974). Insects have a generation time much shorter than their host plants and therefore can adapt more quickly to climate change. As a result, there will be more pressure on their hosts with the advent of climate change (Bale 1991). Pests and pathogens of all types are expected to increase. Disturbances can have a much greater influence on species association than gradual evolution of ecosystems (Pine 1981). All of these factors and many others associated with the complexity of the functions of ecosystems make it extremely difficult to adequately predict the effects on plants from global climate change.

Some landscapes and specific ecosystems are by their evolutionary history vulnerable to rapid change in climatic conditions. For example, many coastal ecosystems are only marginally above sea level and cannot retreat from a rapid rise in sea level (Titus et al. 1991). Many mountain ecosystems have, at their upper elevations, vegetation characteristically found in colder, more polar ecotones. A rapid climatic warming could easily put such systems at risk.

MOUNTAINOUS AREAS: APPROPRIATE LANDSCAPES FOR EARLY DETECTION OF THE ECOLOGICAL AND SOCIAL EFFECTS OF CLIMATE CHANGE

The diverse relief characteristics of mountainous landscapes vary in terms of aspect, slope, and elevation, providing dramatic temporal and spatial variability (Barry 1981). The climate of moun-tainous areas is also generally complex, particularly when compared to neighboring areas (Barry 1990). Complex landscapes yield gradients of temperature and moisture that are functions of slope, aspect, and elevation. Ecosystems associated with these diverse land forms occur in very complex patterns (Rowe 1984, 1991). Many mountain ranges tend to be rich in biological diversity, with many isolated plant and animal communities on mountain peaks, populations of rare and endangered species, and ecotones of transition between mesic and xeric vegetation assemblages.

Mountain peaks can provide a great deal of isolation, which would make it difficult for some species to locate refugia during periods of rapid climate change (Graham 1972; Davis 1989). On the other hand, the elevation gradient provides an invaluably convenient mechanism for biological migration as the climate gradually changes. Appropriate land form characteristics include features such as glacial substrate, surface slope, slope length, and aspect (Peterson et al. 1991). For example, the dynamics of subalpine pine forests are extremely complex. Temperature, precipitation, and storm frequency all affect the growth and productivity of these communities and any changes could alter the location of the subalpine, alpine, and mountain subalpine ecotones (Canaday and Fonda 1974; IPCC 1992). The timing, quantity, and distribution of precipitation (primarily snowfall) are particularly important in high elevations (Peterson et al. 1991). The altitudinal tree line is a very

---

TABLE 1
Mountain Protected Areas (MPAs) by Biogeographical Realm

Realm	Number of MPAs	Area (ha)
Indomalaya	53	8,811,898
Afrotropical	35	10,986,512
Western Palaearctic	36	2,313,370
Eastern Palaearctic	62	20,721,323
South/Central Palaearctic	48	8,027,363
Nearctic	93	153,707,666
Neotropical	82	30,393,615
Oceania	8	3,598,032
Antarctic	11	1,510,044
Australia	3	160,512,705

---

good place to detect potential effects of global climate change (Tranquillini 1979; Wells 1983; LaMarche et al. 1984; Graumlich 1991; Peterson et al. 1991).

Mountainous areas are more likely to include undisturbed landscapes than lowlands, which are more suitable for human habitation and utilization. Steep slopes do not lend themselves to intense development. A total of 432 mountain protected areas have been identified that fall within the World Conservation Union's (IUCN) hierarchy of designated protected areas under Categories I to IV, with a minimum size of 10,000 hectares and a minimum relief of 1500 meters. This network constitutes 42% of the world's area devoted to nature conservation, or 24% of the system if the huge Greenland National Park is excluded from the total (Thorsell and Harrison 1991). The large number of mountain protected areas is quite well represented in the biogeographic realms defined by Udvardi (1975). Table 1 lists the number and total of mountain protected areas by geographic realm distributed throughout the world (Thorsell and Harrison 1991). These mountain protected areas collectively represent an important global network of natural areas that can be used to detect the early effects of climate change.

The sociocultural aspects of mountainous landscapes provide just as compelling a rationale as the natural resources to focus study for early detection of the effects of climate change. Indigenous people residing in mountain areas who rely on subsistence agriculture and wood gathering often live on the edge of sustainability. They tolerate extreme weather conditions and marginal growing seasons to sustain their life style of independence. Their daily lives tend to be very closely intertwined between providing for the basic sustenance of life and the close social and spiritual relationships that they hold with the landscape. Although these people tend to be very resourceful in combating extreme climatic conditions, long-term changes in climate could very much affect the delicate balance they have established with nature to sustain their life style. Changes in the availability of woody plants, grasslands, and water supplies due to climate change could either enhance or detract from their capabilities to sustain their activities.

Human consumption of plants and animals in mountainous areas is of critical importance and should be considered in the context of impact from global climate change. Some local people rely on mountain ranges seasonally for summer grazing of livestock and hunting and trapping. New grazing patterns might be necessary due to changes in vegetation in high elevation meadows and the availability of water supplies. The mountain headwaters of watersheds provide important sources of water which, in some environments, could be much more restricted with the onset of global climate change. Not only the quantity but the quality of water made available from these headwater areas might become one of the most scarce resources, limiting the ability to sustain populations over the long term in many parts of the world.

Recreation and tourism is one of the fastest growing industries in the world (Peine, Chapter 17 in this volume), and mountainous landscapes are like a magnet attracting people interested in viewing magnificent scenery which is quite a counterpoint to their regular urban environs. Many mountain parks have become so popular that it has been necessary to develop mechanisms to establish social and natural resource carrying capacities so that the resources are not overwhelmed by people. As the world population continues to grow rapidly, the social demand for physically touching base with a natural landscape in a montane setting is expected to rise as well. It is speculated that the growth in popular interest in mountainous areas for recreation will increase faster than interest associated with many other landscapes. This phenomenon will likely be exacerbated when the climate warms and people seek refuge from the heat in the mountains.

The metaphysical or spiritual value of mountains will always be a very real and important social value. This almost universal primeval cultural fixation on mountain ranges and peaks can be used to help capture people's imagination as to the nature and extent of potential cataclysmic effects that could occur from global climate change. Many people believe that all species have a right to survive and prosper on this planet and feel that a healthy system of ecological preserves is an ethical responsibility for a civilized society. Therefore, mountainous landscapes cannot only be utilized to discover the early detection of the adverse effects of global climate change but can also be used as an inspirational focal point for capturing the attention of people as to the severity of the problem.

STEPS ECOSYSTEM MANAGERS SHOULD TAKE NOW TO ASSESS THE POTENTIAL IMPACTS OF CLIMATE CHANGE

Given the day-to-day priorities of most ecosystem managers, it is difficult to take insightful actions concerning possible future ecological problems related to global climate change. Waiting to take management action until all uncertainties associated with climate change are known is analogous to buying life insurance after you are in a coma. Small steps today could minimize loss of numerous sensitive species from the landscape in the future.

FRAMEWORK FOR SENSITIVITY ANALYSIS

What follows is a suggested strategy for ecosystem managers to begin the process of defining the influence of climate on ecosystem processes, biological diversity, cultural resources and social values. The intent is to suggest a proceedure to solicit expert advice using a minimal investment of time and funds. It is suggested that a panel of experts (hereafter referred to as the Panel), be formed to conduct a sensitivity analysis of the natural and cultural resource base. Participants would represent the scientific, community, business, and resource management interests of the landscape setting of concern. The suggested methodogy is generally extracted from that presented by Berish et al. in Chapter 7 of this volume, which deals with the conduct of regional resource assessments. What is suggested here is a simplified version of that process.

> *Goals and measurement end points.* The Panel should define goals for the sensitivity analysis that are restricted enough to provide direction to the effort. Vague goals for such an expansive topic will dilute the limited resources available to conduct the analysis. It is also important to define the form that the end products will take and how they are anticipated to be used. For instance, will the analysis be used to define a framework for a long-term monitoring strategy? Unfortunately, the state of the art of global change modeling is not precise at a subcontinental regional scale, making it extremely difficult to establish appropriate "end points" (USGCRP 1995).

Political and ecological boundaries. The physical extent of the study region should be defined by interested stakeholders, the physical distribution of the natural resources of value, the spatial reference of available data, and political considerations. The IPCC (1992) suggests that administrative units, (state to community); physiographic units (definition of physical and biotic landscape); ecological units (definition of ecoregion, forest type, and habitat); climatic zones (weather and climatic information) and sensitive subunits (treelines, wildlife corridors, etc.) be defined at a minimum.

Definition of valued resources. Another important perspective to aid in focusing the analysis is to define the valued resources of concern. This process should include a wide range of stakeholder interests in the natural and cultural resources and socioeconomic consider- ations of communities within the region. All of these are components of the quality of life and sense of place.

Assessment of data needs and availability and quality assurance. Data sufficient to answer stakeholder questions must be assembled, checked for uncertainty, and maintained. At a minimum, resources such as designated protected areas; areas dedicated to renewable resource management, such as forestry and range land; nonrenewable resource extractions, such as the mining of minerals; population centers and projected growth patterns; major point sources of pollution; and any landscape level disturbance that might be relevant, such as large scale fire, flooding, insect infestations, or the range of invasions by alien plants and animals should be identified. Conscientious data management and archiving are also extremely important as well.

Defining time frames and potential climate scenarios. Defining a time frame is useful to project the potential influence of confounding factors such as population growth, land use conversion, pollution emissions and resource utilization. On the other hand, projecting time frames for long-term trends in climate change is highly speculative and imprecise, making it very difficult to define specific time frames for the analysis.

It is important to look at historic records of extreme weather events such as hurricanes and torrential rains on one end of the spectrum to periods of extreme drought on the other. Records of impacts on natural and cultural resources during these events help place the dynamics of extreme weather in perspective. It is important to use one or two scenarios of climate change when conducting the analysis.

PRELIMINARY NATURAL RESOURCE SENSITIVITY ANALYSIS

The first step an ecosystem manager might take is to conduct a preliminary sensitivity analysis to identify those ecosystem elements most likely placed in jeopardy by rapid climate change. Such an analysis can be conducted even without knowledge of how the climate will ultimately change.

The first step in conducting a natural resource sensitivity analysis is to assemble relevant information and convene the Panel of experts to identify potentially vulnerable natural resources based on the information assembled for the resource area and their firsthand knowledge of the area. The goal of this exercise is to define a scenario of probable sensitivity due to significant long-term variance in climate. It is necessary to conduct the analysis in the context of one or more climate scenarios, such as warmer temperatures, drier conditions, and more extreme storm events, such as hurricanes or tornadoes. What is important is to identify resources that are particularly dependent upon or relevant to the dynamics of climate.

A good place to start is to characterize the prevailing weather patterns that bring moisture into the region and define those topographic features that tend to differentiate moisture patterns on the landscape. Most mountain ranges tend to experience much more wet deposition on the windward than the leeward side as the moisture is drawn from the atmosphere as it rises over the mountain range. These fundamental dynamics provide a basic force influencing the regional pattern of mesic vs. xeric habitats supporting species with very different tolerances to climate change.

Water resources are key. Natural lakes provide a cumulative indicator of the hydrologic impacts of climate change over the contributory system of watersheds. Lake volume, height, and chemistry are valuable indicators of climate influence. Wetlands and bogs replete with species that are dependent on or tolerant to extreme moisture conditions are key indicator communities. The edges of these systems are good places to observe early response to climate change. Alpine and subalpine vegetation should be reviewed as obvious places to detect the effects of climate change. Ecotones between vegetation types are another vulnerable area that should be identified as possible locations for sensitive species. The differentiation between dominant conifers vs. deciduous species in temperate forests is an example. These are also places where potential for species migration might first be detected as climate changes.

Wildlife sensitivity to climate change can be reflected by shifts in habitat that support predator/prey relationships. Small mammals can be useful indicators. They are more easily sampled than larger mammals and can be very sensitive to soil temperature and moisture gradients.

The analysis of identifying sensitive ecosystems and species should be conducted in the context of exacerbating circumstances which could affect the reaction of species to climate change, such as major disturbances. Many landscapes have a very high degree of disturbance related to fires and human activity such as agriculture, logging, and settlements. These areas will likely have different species composition and disturbance and therefore may react differently to climate change than undisturbed natural areas.

Rare and endangered species are important to consider in the context of sensitivity analysis of global climate change. Their range is usually restricted, and therefore they may be more vulnerable to climate change by being susceptible to isolation and a narrow gene pool which could restrict their capability to respond to subtle changes in their habitat. On the other end of the spectrum, alien species should also be monitored in case they experience more suitable habitat or reduced competition with the advent of climate change and therefore extend their range of invasion more deeply, creating altered ecosystems in previously undisturbed areas.

Once the broad-brush dynamics of moisture and temperature gradients reflecting slope, aspect, and elevation are generally defined and the sensitive ecosystems, such as alpine and subalpine meadows, tree lines and ecotones between plant communities are defined, the assembled specialists can then turn their attention to examination of the species assembled within these sensitive areas and attempt to choose potential bioindicators that would be particularly vulnerable to the influences of climate change within the context of these sensitive areas.

At the end of the process, one would anticipate that the team of experts assembled would have identified sites within sensitive ecosystems in the mountainous landscape that would be appropriate for monitoring the effects of global climate change as well as compiling a list of species residing in these areas that might provide good bioindicators of climate change and merit further study and analysis.

Preliminary Human Resource Sensitivity Analysis

The ecosystem manager should initiate a human resource analysis companion to the natural resource sensitivity analysis. In many mountainous landscapes around the world there reside indigenous people who have depended on the natural resources of mountain environments for their sustenance for countless generations. These people represent cultural resources of great importance. Their cultures demonstrate a mechanism by which man and nature can coexist within a sometimes marginal yet operable scale of time tested dimensions. The Man and the Biosphere Program of the United Nations Educational Scientific and Cultural Organization (UNESCO) is a global conservation program to demonstrate how man and nature can coexist (Gregg, Chapter 2 in this volume). The lessons that can be learned from indigenous populations living in mountain environments are quite significant. These people have learned how to live on what nature offers. Additionally, they are a wealth of information concerning ethnobiology or how humans can use native plants and animals to serve their needs. The pharmaceutical industry is funding research all over the world to gather information from what Anglo-Saxons would call "witch doctors and medicine men," or shamen, people who administer medicinal herbs and spices, for health and spiritual reasons, to indigenous people. The rituals performed and the utilization of plant and animal materials in a variety of ways provide a synthesis of generations of discovery of the value of native materials to humans. Therefore, it is very important and instructive for managers to document how indigenous populations use the natural resources.

This information can then be cross-referenced with the natural resource sensitivity analysis to identify potential conflicts that might occur with the advent of global climate change. This allows an opportunity to anticipate potential conflicts before they emerge, either due to changing availability of water or loss of species within ecosystems on which native people depend. For instance, seasonal grazing in high country that requires access to a traditional water supply may become in jeopardy. Alternative seasonal grazing lands might have to be established so as not to greatly impact the agricultural activities of the indigenous people. Such analysis should not be limited just to populations living within the mountainous landscapes of primary concern. A water supply dependent on mountain watersheds that experiences significant reduced production could provide significant regional hardship. Water-related recreation and tourism in the region could be adversely impacted as well. A degraded ecosystem or loss of chrismatic or game species from the landscape could discourage visitation.

Strategic Plan for Species Conservation

The sensitivity analysis should provide the ecosystem manager with reasonable insight into which species are most at risk, given the potential for climate change. In some of the worst case scenarios that have been expressed concerning climate change, there may be such dramatic adverse impacts on ecosystems that it becomes impractical to focus on single species management as has been done traditionally. The Endangered Species Act of 1973 (Public Law No. 93-205), which is focused on single species, is currently very controversial and is under considerable political pressure to be rescinded, in part because of the social conflicts associated with protecting single species. Ideally, species protection should be directed at the community level and associated habitat. Consideration should be given to managing protected areas of a viable enough size and configuration to sustain a large percentage of the plant and animal communities that occur naturally within the ecosystem supporting the species of concern. Nevertheless, there will always be an interest in preserving individual threatened species. The advent of climate change over the long term will make the job of single species protection daunting indeed.

An important consideration to take into account when evaluating the viability of an endangered species population and its vulnerability to climate change is to define a critical population size and distribution necessary to guarantee its survival. Determining the role that genetic variability contributes to the viability of the population is also important. Habitat characteristics relevant to climate should be characterized, such as soil moisture and temperature relationships, average number of frost-free days during the growing season, temporal distribution of precipitation, slope and aspect affecting orientation to the sun, dependency on water supply, and so forth. The relationship of these climate-related factors to the life history of the species is also important. How might climate change, for instance, affect the propagation process? Invariably, such an exercise will generate more questions than answers. The greatest value in going through such a thought process is in defining needs for research and monitoring for the species of concern. This analysis is also useful if it is determined that there is a need to establish a new population of the species at another, less vulnerable location. Habitat manipulation to protect species is becoming a more accepted practice. In some circumstances, for instance, it may be appropriate to store seeds, establish a seed orchard, or reintroduce an extirpated species.

MONITORING FOR CLIMATE CHANGE AND ITS EFFECTS

A critical step toward developing a global climate change program is to design a conceptual framework for a monitoring program that is oriented toward identifying early signs of global climate change and the biological, geophysical, and chemical responses to that change. In many mountain protected areas, there has been a long history of research already ongoing which can be applied in the design for such a program. Ideally, the monitoring program should be devised at different scales of resolution, including landscape, research watersheds, community, and species levels.

The heart of the conceptual monitoring program for climate change ideally should be a series of paired research watersheds designed to represent typical landscape patterns oriented to the windward and leeward sides of the prevailing weather patterns. Some minimal monitoring activity should be maintained by staff working for the managing agency. By tracking a few key parameters systematically for the long term, scientists would be attracted to the area for further study. Typical kinds of basic monitoring information that would be appropriate at the watershed level include flow rates, temperature, and chemistry for first or second order streams; meteorology, such as wet deposition, temperature, and humidity; glacier and snow field depth and position; and maps of vegetation distribution on the watershed. The research watersheds should be chosen in part because they represent areas considered to be sensitive to climate change as described in the previously mentioned sensitivity analysis.

The community level component of the monitoring strategy would be to establish study sites at the selected sensitive areas within the research watersheds. Vegetation plots, small mammal plots, and so forth, should be established to document population dynamics for vulnerable ecosystems. Monitoring activities on first, second, and third order stream segments should be established and populations of macroinvertebrates and fish described. The Smithsonian Institute has established protocols for a system of nested plots which can be utilized to document various biological realms in the study of biodiversity (Comiskey et al. 1995). Species populations identified from the sensitivity analysis should be described so that their dynamics can be monitored over time. Factors which should be considered in establishing a monitoring program at the species level include presence and absence, distribution, abundance, production, mortality, and phonology.

As mentioned earlier, rare and/or endangered species are important to consider in the context of sensitivity analysis of global climate change and therefore are important to consider as part of a monitoring program. On the other end of the spectrum, exotic species should also be monitored in case they experience more suitable habitat with the advent of climate change and therefore extend their range of invasion, creating altered ecosystems in previously undisturbed areas.

CASE EXAMPLE: GREAT SMOKY MOUNTAINS NATIONAL PARK

The circumstances at Great Smoky Mountains National Park (hereafter referred to as the Park) are used to illustrate the type of actions that ecosystem managers in the southern Appalachian mountains might take now to prepare for the eventuality of climate change predicted with a relatively high degree of certainty to occur within the first half of the next century (IPCC 1995). The suggestions offered are purposefully modest to illustrate the kinds of actions possible even in the current environment of limited funding availability and the downsizing of the role of the federal government.

The suggestion is to request that the Southern Appalachian Man And Biosphere Cooperative (SAMAB) convene a distinguished panel of scientists, community planners and leaders, and ecosystem managers from the SAMAB region to conduct and publish a sensitivity analysis concerning climate change in the southern Appalachians in general and the Park in particular. The findings of the group might be published in a special addition of an appropriate scientific journal.

GOALS AND MEASUREMENT ENDPOINTS

Suggested goals of the Park climate change sensitivity analysis are as follows.

- Identify ecosystems, ecotones and species that are particularly sensitive to climatic conditions
- Evaluate the effectiveness of the Park's resource monitoring program to detect ecosystem dynamics associated with climate change
- Provide information that can be used by the Park for education concerning climate change
- Devise a research and monitoring agenda concerning climate change in the Park and southern Appalachians

POLITICAL AND ECOLOGICAL BOUNDARIES

Suggested boundaries include at their narrowest the Park boundary and at their most expansive the the boundaries of the Southern Appalachian Assessment as described by Berish et al. in Chapter 7 of this volume. The ecological boundaries should be at their narrowest recognized forest types within major watersheds in the Park and at their most expansive in Bailey's designated Southern Blue Ridge Province (Bailey 1980).

DEFINITION OF VALUED RESOURCES*

The Panel assembled to conduct the sensitivity analysis would define resources of particular value. This exercise would provide focus for the analysis to insure its relevance to stakeholder interest in the natural and cultural resources of the region. Features to consider follow.

The ancient Appalachian mountain range in the eastern U.S. extends from Maine to Georgia, reaching its greatest elevation in the Southeast. In broad aspect, the topography consists of moderately sharp-crested, steep-side ridges separated by deep V-shaped valleys. Lesser ridges form radiating spurs from a central ridge line. This topography creats a very complex mozaic of temperature/moisture gradients harboring a multitude of resources sensitive to climatic conditions.

Many of the mountain ridges branch and subdivide, creating a complex of drainage systems with thousands of miles of fast-flowing clear mountain streams which are highly valued for recreation, aesthetics, and water supply. The Park contains 45 defined watersheds and over 3500 kilometers of streams (Parker and Pipes 1990). The water table tends to be near the surface in almost all regions. Pre-Cambrian metamorphic rocks consisting of gneisses, schists, and sedimentary rocks from the pre-Cambrian Ocoee series are predominant, while secondary rocks in the Appalachian valley are the youngest. Due to the rugged topography, the mountainous region is relatively sparsely populated.

Biological diversity at all levels (for example, genetic, species, and community levels) is high and many endemic species occur. Salamander fauna (*Plethodon* spp.) which tend to be climate sensitive is rich and locally diverse. The high mountains and rich array of microclimates promote, support, and harbor this diversity, and understanding the importance of this mountain mass to regional diversity in times of climate change is critical. The deeply dissected landscape present at the southern end of the Appalachian chain provided a refuge for a host of temperate and boreal species during the Pleistocene glaciation period. This has resulted in a rich vegetation mosaic

---

* All information concerning biological diversity in the Park was verified by personal communication with Keith Langdon, at Great Smoky Mountains National Park.

comprised of more than 2500 species of plants (including 130 species of trees) and over 2200 cryptogamous taxa (Randolph et al. in Chapter 4, in this volume). Over 30% of the Park's forests are high in virgin attributes (Pyle 1985). Areas which were farmed or logged have been recovering for varying periods of time and therefore represent a wide range of successional stages. Deciduous broadleaf and evergreen coniferous forests predominate, but treeless grass and heath balds, open wet meadows, and cliff communities occur as well. Vegetation changes continuously with elevation, slope, aspect, and topographic position.

Fourteen major forest types are currently recognized within the region. On mesic sites, low or midelevation cove hardwood (mixed mesophytic) and hemlock–hardwood forests grade, with increasing elevation, into northern hardwoods and finally, at about 1500 m, into spruce–fir. On a gradient from mesic to xeric, the cove hardwoods are replaced by mixed oak, xeric oak, and oak–pine. Heath balds represent the xeric extreme at the upper elevation and are dominated by ericaceous shrubs. Perhaps the most notable forest types are the cove hardwood and the spruce–fir. Cove hardwoods may contain upward of 20 different canopy species in the canopy at any one site. Diversity is present in the understory as well. A single 0.1-ha plot may support in excess of 50 vascular plant species throughout the year. Values of these protected terrestrial resources range from recreational and aesthetic to the provision of a refugium from which to conduct research to describe the ecosystem processes and functions associated with old and secondary growth eastern temperate forests.

The spruce–fir forest type occurring only at the highest elevations is of particular concern. This forest contains the largest contiguous block of virgin red spruce (*Picea rubens*) remaining on the Earth. Fully 75% of all southern Appalachian spruce–fir occurs within the boundaries of the Park (Nicholas et al., Chapter 21 in this volume). Additionally, grass balds, ridges, cliffs, and landslide scars within these high elevation forests are habitat for rare regional endemics (White and Sutter, Chapter 18 in this volume). Fifteen plants in the Park are listed as candidates for federal protection as threatened or endangered species. Moreover, 150 species are recognized as rare enough to be of managerial concern. A similar number of bryophytes, lichens, and fungi are also considered rare at the regional, national, or global level. The diversity of fauna includes at least 67 native mammal species. This biological diversity is internationally recognized, as the Park was designated a charter member International Biosphere Reserve in 1976 and in 1984 was designated a World Heritage Site. Within the context of this complex topography and rich biological diversity, the National Park Service manages a variety of facilities and services for an unprecedented number of visitors.

ASSESSMENT OF DATA NEEDS

As described by Berish et al. in Chapter 7 of this volume, an enormous database has been assembled for the Southern Appalachian Assessment. In addition, the Park has extensive natural resource databases as a legacy of the long history of research and resource monitoring having been carried out there (Smith et al., Chapter 8 in this volume). The panel should review the compilation of databases and suggest which ones are most relevant to the sensitivity analysis and what critical information needs remain unfullfilled.

TIME FRAME AND CLIMATE SCENARIOS

It is very difficult to set a time frame for the climate sensitivity analysis. What is more easily accomplished is to review past climate and stream flow records and define periods of extreme weather conditions. These periods can provide time frame indicators to look for environmental responses to the extreme weather conditions. From 1985 through 1988, an unprecedented period of extreme drought occurred in the southern Appalachian highlands. Environmental response to this extreme climatic condition provides some poignant insight into the potential impacts from a dramatically changed climate.

PRELIMINARY NATURAL RESOURCE SENSITIVITY ANALYSIS

Climatic changes from global warming that could have particularly adverse effects in the Park include changes in the range of annual and seasonal temperatures and alterations in the quantity and timing of precipitation. As experienced during the last decade, there could be a dramatic increase in seasonal drought and extreme weather events. With a worst-case scenario of significant climate change, there ultimately could be dramatic shifts in the structure and composition of forest communities and their position on the landscape, an overall reduction in native biodiversity and nutrient availability, periodic increases in fire frequency and intensity, and an increase in the invasion of exotic pests and pathogens. One dramatic example from the recent past of response to an abnormal climate condition was during the height of the drought of 1988 when there were a record 41 fires in the Park (Records at Great Smoky Mountains National Park).

Climate change may act alone or in combination with other agents to compound stress on the ecosystem. This is especially true for the high elevation spruce-fir ecosystem which is anticipated to be in particular jeopardy in the event of significant climate change (Delcourt and Delcourt, in press). Fraser fir (*Abies fraseri*) is endemic to the southern Appalachians and its survivorship is in question due to the infestation of the exotic insect pest the balsam wooly adelgid (*Adelges piceae*) (Nicholas et al., Chapter 21 in this volume). Dramatic climate change may provide a compounding stress which could threaten the survival of the species in the wild. A 5-year study of crown condition during drought during the 1980s showed a decline of healthy crowns on red spruce trees from 85% in 1985 to 50% in 1989. The crown condition is speculated to have been exacerbated by the impacts of air pollution loading in addition to the drought conditions (Nicholas and Zedaker 1990).

In addition, most of the 95 southern Appalachian endemic or near endemic vascular plants are found on high elevation north-facing slopes. Extreme changes in temperature and moisture might eliminate them from the Park and, in some cases, might lead to extinction. The region's nonvascular plant flora, such as mosses and liverworts, are closely tied to moisture-laden environments, with individual plants existing in very specific microcosms. In the Park there are 428 species of bryophytes, with 175 of those considered rare at the global, national, and park level.

Aquatic organisms can be extremely sensitive to changes in flow rates and temperature regimens. The native brook trout (*Salvelinus fontinalis*) has been a subject of research activity for many years in the southern Appalachian bioregion (Guffey et al., Chapter 12 in this volume). The southern Appalachian mountains represent the southern extent of the brook trout range. Due to historic land use conversion and an aggressive stocking program for nonnative rainbow trout, the distribution of the native salmonid species within the region has been narrowly restricted to headwater streams draining the region's forested mountain watersheds. It was observed during the extreme drought period in 1988 that as the stream flow decreased and the water temperature increased, there was mortality of some mature salmonids in the rivers of the Park. Reduced reproduction of salmonids resulted in the elimination of an entire age class. The fisheries in the lower order streams were more vulnerable to this phenomenon (Steve Moore, *pers. commun.*). Because the extent and condition of this aquatic habitat may be directly affected by changes in precipitation and temperature, the status of brook trout streams may serve as a key indicator of regional ecosystem response to climate change.

The region's streams might also be subject to critical changes in chemical composition as a consequence of altered patterns of elemental flux in watersheds. The streams are very low in ionic strength and poorly buffered against climate influenced chemical change. Results of the Integrated Forest Study of the late 1980s showed that high elevation areas of the Park receive some of the highest rates of atmospheric deposition in North America (Lindberg and Lovett 1992; Johnson and Lindberg 1992). Acidification of streams has been demonstrated (Mitchel and Lindberg 1992.). Changes in climate may exacerbate this existing environmental problem. It has been observed that episodic stream water acidification occurs when accumulated acidic materials are washed from watersheds into streams during periods of elevated runoff. The magnitude of this fact evidently depends on the length of time over which sulfur and other atmospherically transported materials have been deposited. Data collected in the Park suggest that nitrification of soils can also add to the accumulation of acidity during protracted dry periods. Thus, when climatic change results in greater frequency and intensity of drought or more extreme cycling between wet and dry conditions, then periods of episodic acidification and coincident stress on aquatic organisms may be more frequent or extreme. The significance of the potential for a synergistic effect of climate change and atmospheric pollution in this case is largely due to the present at-risk status of the resource.

Air stagnation events during summer months of 1988 was extreme, resulting in significant buildup of small particulate matter such as sulfates in the atmosphere, contributing to uniform regional haze. This condition resulted, at times, in a 90% reduction in visibility during the height of the visitor season (NPS 1996). In addition, some of the highest concentrations of ground level ozone pollution ever recorded occurred during the summer of 1988.

PRELIMINARY CULTURAL RESOURCE SENSITIVITY ANALYSIS

With over 50% of the nation's population living within a one-day drive, the Park is reported to have the highest visitation of any national park in the nation. Many of the recreational activities in the Park could be greatly impacted by climate change. For instance, visitors to the back country enjoyed drier conditions for their hiking experience during the drought period of 1985 to 1988, but many complained of the lack of water supply in the high elevation, where a series of springs usually provides adequate water year round for such use (Jim Renfro, *pers. commun.*). If the climate dries out and gets warmer as a result of global climate change patterns, the lack of water supply in the high elevations of the back country may become a cronic problem. Water-based recreational activities likely to be adversely impacted include fishing, tubing, and swimming.

Another example of potential impacts relates to the possible impairment of scenic views in a park where sightseeing is the most frequently cited recreational activity (Peine and Renfro 1988). If climate change results in increased air stagnation, a result would be an increase in the incidence of extreme air pollution events resulting in the virtual elimination of scenic views from the Park during those periods.

If climate change results in an increase in fires, insect infestation, and pathogens in the forests; the visual appeal of the Park could be diminished as well, thereby greatly detracting from the quality of the visitor experience. The most important feature of the Park to visitors is a healthy environment. The perception that a pristine environment is being protected in perpetuity is an extremely important social value (Ross et al. 1986). Research has shown that visibility of long range views is a primary component of the quality of visitor experience (Ross et al. 1986).

With the potential for more extreme and violent weather patterns predicted with global warming, the implications concerning operating costs for the Park, the nation's most visited national park, are quite serious. For example, on March 13, 1994, the Park experienced the "snow blizzard of the century." On October 5, 1995, the Park experienced the "flood of the century." The cost of repair of the damage from these two catastrophic events incurred by the federal government to date exceeds $10 million dollars. Even now, damage to back country trails has not been fully repaired (Bob Miller, pers. commun.). Clearly, the economic implications of climate change on park operations are quite significant.

STRATEGIC PLAN FOR SPECIES CONSERVATION

There are several active programs in the Park concerning conservation of rare species or species reintroduction. It would be useful for the climate change Panel to review Park programs for species of concern for possible influence associated with climate change. An obvious focal point is the high elevation spruce–fir ecosystem, vulnerable to climate change while harboring several species of concern (Nicholas et al., Chapter 21 in this volume). The Fraiser Fir is an obvious choice of a species of concern which is endemic to the southern Appalachian highlands and vulnerable to climate change.

MONITORING FOR CLIMATE CHANGE AND ITS EFFECTS

The Panel conducting the sensitivity analysis should review the ongoing Park resource monitoring program and suggest any adjustments necessary to make it more relevant to detecting effects of climate change (Smith et al., Chapter 8 in this volume). One strategy to facilitate coordinated research and monitoring would be to establish paired research watersheds on the Tennessee and North Carolina sides of the Park, which reflect very different orientation to major air masses that converge on the mountains. The University of Tennessee/Oak Ridge alliance could be recruited to sponsor a research watershed on the Tennessee side of the Park and Western North Carolina University Coweeta Hydrologic Laboratory for a paired watershed on the North Carolina side of the Park. Precedence has been established for such a collaboration on the Noland Divide research watershed site near Clingman's Dome in the Park. Ideally, the experimental watersheds designated as international biosphere reserves at Oak Ridge, TN and Cowetta, NC would be incorporated into this program as well. Periodic interpretation of ongoing research and monitoring activities could serve to provide refereed bench marks on relationships among climatological parameters and biological, physical, chemical, sociological, and economic response.

CONCLUSION

There are several steps which ecosystem managers should take to prepare for what could be a cataclysmic change in the natural and cultural resources within the next 50 to 100 years. Preparing now for such a vaguely defined event, which may be occurring now or not until the distant future, requires a great deal of courage and vision. Most ecosystem managers are desperately trying to respond to daily crises in the operation of their forests and parks without the luxury to plan far into the vaguely defined future. Ecosystem managers have difficulty planning 6 months in advance, let alone adhering to annual and 5-year plans. What is suggested here is that ecosystem managers take a leadership role in defining an early response to the potential effects of global climate change. Such a perspective has not been considered necessary by managers of protected areas. But with the advent of global climate change, now is the critical time to take action. Ignoring this pervasive concern in the near term may result in losing the option to effectively deal with it in the long term. Some scientists predict that if no action is taken now, within 50 years a greenhouse effect may be to such a state that it is irreversible, no matter what adjustments are made by societies to suppress emissions of precursor gases (USGCRP: 1995).

This situation is understandible but extremely unfortunate, considering how urgent the need is to get a reasonable fix on the magnitude of the potential implications associated with this phenomenon. Once the anticipated global climate change can be predicted with greater accuracy, the social resolve to deal effectively with the issue will most likely increase considerably. Never has society been faced with such a daunting and pervasive challenge that requires a faith in science, a commitment to focused public opinion, and an action plan that begins the process of controlling the greenhouse gases that are contributing to this phenomenon. The industrialized world is supposedly the most informed and therefore in a position to establish the political resolve to take the steps necessary to control these emissions. Managers of ecosystems, particularly those in mountainous protected areas can play a critical role in this process by providing the equivalent of that old American adage of the canary in the coal mine. This saying came from the times when coal miners took canaries down deep into the ground where they were mining minerals as an effective means to evaluate the quality of the air. If the canary died, they knew it was time to get out of that mine as quickly as possible. If the organisms in mountain protected areas show signs of stress due to climate change and these phenomena are systematically documented in the southern Appalachians and elsewhere, then alarm and concern may be raised throughout the industrialized world. Therefore, ecosystem managers share a great responsibility and opportunity to provide leadership on this critical issue.

Risk Assessment Report Card for Ecosystem Managers

Applying the *Risk Assessment Report Card for Ecosystem Managers* to a concept as obscure and controversial as climate change illustrates the degree of risk associated with enlightened management practice. In the case study of managing for climate change in the Park, the report card is as follows:

Vision: F

Little attention has been given to this issue in resources management, science or education.

Resource risk: D–

High elevation forests and streams are potentially at significant risk, especially the spruce–fir ecosystem.

Socioeconomic conflicts: D–

Potentially quite significant if there are higher incidents of air stagnation and less moisture adversely impacting water based recreation and municipal water supply.

Procedural protocols: C–

The Park's resource inventory and monitoring system will provide baseline information but the program was not designed specifically to test for sensitivity to climate change. The Southern Appalachian Assessment database is very helpful for baseline information.

Scientific validity: C–

Regional models do not yet predict with confidance a future climate change scenario for the southern Appalachians.

Legal jeopardy: A

There is none at this point.

Public support: C–

The concern is greater during periods of drought and unseasonally high temperatures.

Adequate funding: D–

Scientists with research interest in climate change are not being actively recruited to set up research watersheds, etc. in the Park.

Policy president: B

The Federal government continues to support programs in research, education and reduction of greenhouse gas emissions.

Transferability: A

A program developed in the Park could be transferable to other montane biosphere reserves around the world. It could be designed to maximize transferability from the beginning.

Global climate change clearly represents the greatest challenge facing ecosystem managers throughout the world. All bets are off about the future of the natural environment within the next century. The seriousness of the issue is masked by uncertainty and lack of clear evidence that elements of the biological, social and economic dimensions of ecosystems are in serious jeopardy. Enlightened ecosystem managers have a special responsibility to bring this serious problem to the forefront of human consciousness. International biosphere reserves should be linked together as living laboratories dedicated, in part, to document stressors on the environment induced by climate change. Given all the demands on ecosystem managers to effectively deal with pervasive problems that are much more obvious, it is particularly challenging to dedicate limited resources to prepare an information base to support making appropriate decisions today to prepare for this inevitable perturbation of unprecedented proportion on the natural, cultural, and economic environment of the earth.

REFERENCES

Bale, J.S. 1991. Insects at low temperatures: a predictable relationship? *Functional Biol.* 5:291–298.

Bailey, R.G. 1980. Description of the Ecoregions of the United States. U.S. Dept. of Agriculture Misc. Publ. 1391, Washington, D.C. iv + 77 pp. illustr. [including colored folding map, scale = 1:7,500,000].

Bailey, R.G. 1989. Locating sites for monitoring predicted effects of land management, proc. Global Natural Resource Monitoring and Assessments, September 24 to 30; Venice, Italy; American Society of Photogrammetry and Remote Sensing, Bethesda, MD; iii + 1495 pp. illustr. (p. 919–925).

Bailey, R.G. 1990. Design of ecological networks for monitoring global change. *Environ. Conserv.*

Barry., R.G. 1981. *Mountain Weather and Climate.* Methuen, New York.

Barry, R.G. 1990. Changes in mountain climate and glacio-hydrological responses. *Mountain Res. Dev.* 10:161–170.

Brubaker, L.J. 1986. Responses of tree populations to climatic change. *Vegetation* 67:119–130.

Canaday, B.B. and R.W. Fonda. 1974. The influence of subalpine snowbanks on vegetation pattern, production, and phenology. *Bull. Torrey Bot. Club* 101:340–350.

Climate Alert. 1988. NASA scientist testifies greenhouse warming has begun. Vol. 1(3):12.

Comiskey, J.A., G.E. Ayzanoa, and F. Dallmeier. 1995. A data management system for monitoring forest dynamics, *J. Trop. Sci.* 7(3): 419–427.

Committee on Earth Sciences. 1991. Our changing plant; the FY 1991 U.S. Global Change Research Program, a report. U.S. Geological Survey, Reston, VA.

Cusbach, U. and R. Cess. 1990. Processes and Modeling. Houghton, J.T. et al., Eds. *Climate Change: the IPCC Scientific Assessment.* Cambridge University Press: New York. 195–242.

Davis, M.B. 1983. Holocene vegetational history of the eastern United States. Pp. 166–181 in Wright, H.E., Ed. *Late-Qurternary Environments of the United States,* Vol. 2, *The Holocene.* University of Minnesota Press, Minneapolis.

David, M.B. 1989. Lags in vegetation response to greenhouse warming. *Climatic Change* 15:75–82.

Delcourt, P.A. and H.R. Delcourt. In press. Conservation of biology in light of the quaternary paleoecological record: should the focus be on species, ecosystems or landscapes? In W.J. Platt and R.K. Peet, Eds., special edition entitled: Use of Ecological Concepts in Conservation Biology: Lessons from the Southeastern Ecosystems, *Ecological Applications.*

Dullimier, F. 1992. Long Term Monitoring of Biodiversity in Tropical Rain Forests: Methods for Establishing Permanent Plots. MAB Digest II. MAB UNESCO Paris, 22 pp.

EPA. 1989. Smith, J. and D. Tirpak, Eds. Potential Impacts of Global Climate Change on the United States. Washington, D.C. [US EPA-230-05--89-050].

Folland, C.K. et al. 1990. Observed climate variations and change. Pp. 195–242 in Houghton, J.T. et al.;, Eds. *Climate Change: the IPCC Scientific Assessment.* Cambridge University Press: New York.

Fonda, R.W. 1976. Ecology of alpine timberline in Olympic National Park; Proc. Conf. Scientific Research in National Parks 1:209–212.

Graham, A. 1972. Outline and historical recognition of floristic affinities between Asia and North America; Graham, A., Ed. *Floristics and Paleofloristics of Asia and Eastern North America.* Elsevier: New York, 1972; pp. 1–6.

Graham, R.W. and E.C. Grim. 1990. Effects of global climate change on the patterns of terrestrial biological communities. *Trends Ecol. Evol.* 5:269–322.

Graumlich, L.J. 1991. Subalpine tree growth, climate, and increasing CO_2: an assessment of recent growth trends, *Ecology* 72:1–11.

Hamilton, L.S. 1991. Philippine storm disaster and logging. The wrong villain? *World Mountain Network Newsl.* 4:11.

Hammer, R.D. 1991. Landforms, soils, and forest growth: identification and integration with geographic information systems. Mengel, D.L., and D.T. Tews, Eds.; Proc. Symp. Ecological Land Classification: Applications to Identify the Productive Potential of Southern Forests. USDA Forest Service, SE Forest Experiment Station, Gen. Tech. Rep. SE-68, pp. 121–130.

Hansen, J. and S. Lebedeff. 1988. Global surface air temperatures: update through 1987. *J. Geophys. Res. Lett.* 15:323–326.

Houghton, R.A. and G.M. Woodwell. 1989. Global climatic change. *Sci. Am.* 260:36–44.

Intergovernmental Panel on Climate Change (IPCC). 1992. 1992 APCC Supplement: Scientific assessment of climte change. World Meteorological Organization. Geneva,Switzerland. pp. 4–6.

Johnson, D.W. and S.E. Lindberg, Eds. 1992. *Atmospheric Deposition and Forest Nutrient Cycling*. Springer-Verlag: New York. 707 pp.

Katz, R.W. 1988. Statistics of climate change: implications for scenario development, Pp. 95–112 in Glantz, M.H., Ed. *Societal Responses to Regional Climatic Change*. Westview, Boulder.

Keyfitz, N. 1991. Population growth can prevent the development that would slow population growth. Pp. 38–77 in Mathews, J.R., Ed. *Preserving the Global Environment*, W.W. Norton: New York.

LaMarch, V.C., D.A. Graybill, H.C. Fritts, and M.R. Rose, 1984. Increasing atmospheric carbon dioxide: tree ring evidence for growth enhancement in natural vegetation. *Science* 225:1019–1021.

Leggette, Jeremy, Ed. 1992. *Climate Change and the Financial Sector: The Emerging Threat — the Solar Solution*. Gerling Ahademie Verlag: Munich 212 pp.

Likens, G.E., C.T. Driscoll, and D.C. Buso. 1996. Long term effects of acid rain: response and recovery of a forest ecosystem. *Science:* 272(5259), 244.

Lindberg, S.E. and G.M.Lovett. 1992. *Atmospheric Environ.* 26A, 1477–1492.

Martinka, C.J. 1991. Conserving the natural integrity of mountain parks: Lessons from Glacier National Park, Montana. Paper for Peaks, Parks, and People — An International Consultation on Protected Mountain Environments, Environment and Policy Institute Workshop, Hawaii Volcanoes National Park, October 27 to November 2.

McCracken, G., C.R. Parker, and S.Z. Guffey. 1991. Genetic differentiation between hatchery stock and native brook trout in Great Smoky Mountains National Park. Trans. American Fisheries Society. Vol. 122, pp. 533–542.

McNeely, J.A. 1990. Climate change and biological diversity: policy implications. Pp. 406–429 in Boer, M.M. and De Groot, Eds. *Landscape-Ecological Impact of Climate Change*. IOS Press: Amsterdam.

Mitchell, J.F.B. 1990. Equilibrium climate change. Pp. 139–173 in Houghton, J.T. et al., Eds. *Climate Change: the IPCC Scientific Assessment*. Cambridge University Press.

National Park Service. 1996. IMPROVE data set at Great Smoky Mountains National Park, Gatlinburg, TN 37738.

Newmark, W.D. 1985. Legal and biotic boundaries of western North American national parks: a problem of congruence. *Biol. Conserv.* 33:97–208.

Nickolas, N.S. and S.M. Zedaker. 1990. Forest decline and regeneration success of the Great Smoky Mountains spruce–fir. In E.S. Smith, Ed., Abstr. First Annu. Southern Appalachian Man and Biosphere Conf. Tennessee Valley Authority, Norris, TN 68 pp.

Overpeck, J.T., D. Rind, and R. Goldberg, 1990. Climate-induced changes in forest disturbance and vegetation. *Nature* 343:51–53.

Parker, S.P., Ed. 1982. *Synopsis and Classification of Living Organisms*. McGraw-Hill: New York.

Parker, C.R. and D.W. Pipes. 1990. Watersheds of Great Smoky Mountains National Park: A Geophysical Information System Analysis. SER-91/01. 126 pp.

Pastor, J. and W.M. Post. 1986. Influence of climate, soil moisture, and succession on forest carbon and nitrogen cycles. *Biogeochemistry* 2:3–27.

Peters, R.L. and J.D.S. Darling. 1985. The greenhouse effect and nature reserves: global warming would diminish biological diversity by causing extinction among reserve species. *BioScience* 35:707–717.

Peterson, D.L., A. Woodward, and E.G. Schreiner. 1991. Assessing the response of high elevation ecosystems to a changing environment: consequences for management. Paper for Peaks, Parks, and People — An International consultation on Protected Mountain Environments; Environment and Policy Institute Workshop, Hawaii Volcanoes National Park, October 27 to November 2.

Peine, J.D. and J.R. Renfro, 1988. Visitor use patterns at Great Smoky Mountains National Park. USDI National Park Service Southeast Region. Res./Resour. Manage. Rep. SER-90. Atlanta, GA. 93 pp.

Peine, J.D., C. Martinka, and D. Peterson, [Letter to Jim Thorsell]. Located at the World Conservation Union, Avenue der Mont-Blanc, CH-1196 Gland, Switzerland. December 5, 1991.

Pyle, C. 1985. Vegetation disturbance history of Great Smoky Mountains National Park: an analysis of archival maps and records. National Park Service — Southeast Region. Res./Resour. Manage. Rep. SER-77. 69 pp.

Price, M.F. and J.R. Haslett. 1991. Complexities of climate change in the mountains. Paper for Peaks, Parks, and People — An International consultation on Protected Mountain Environments; Environment and Policy Institute Workshop, Hawaii Volcanoes National Park, October 27 to November 2.

Renfro, J.R., J.D. Peine, R.L. Van Cleave, D.J. Overton, and J.D. Absher, n.d. Backpacker use patterns at Great Smoky Mountains National Park. Unpublished report.

Response Strategies Working Group. 1990. Formulation of response strategies. WMO/UNEP Intergovernmental Panel on Climate Change, Geneva.

Retzer, J.L. 1974. Alpine soils. Pp. 771–802 in Ives, J.D. and Barry, R.G., Eds. Arctic and Alpine Environments. Methuen: London.

Root, T.L. and S.H. Schneider. 1993. Can large-scale climate models be linked with multiscale ecological studies? *Conserv. Biol.* 7:256–270.

Ross, D.M., W.C. Malm, and R.J. Loomis. 1986. An examination of the relative importance of park attributes at several national parks. Paper presented at the Air Pollution Control Assoc. Int. Conf., Visibility Protection Research and Policy Aspects Grand Teton National Park, WY.

Rowe, J.S. 1984. Understanding forest landscapes: what you conceive is what you get. The Leslie L. Schaffer Lectureship in Forest Science, October 25, 1984, University of British Columbia, Vancouver.

Rowe, J.S. 1991. Forests as landscape ecosystems: implications for their regionalization and classification. In Mengel, D.L. and Tews, D.T., Eds. Proc. Symp. Ecological Land Classification: Applications to Identify the Productive Potential of Southern Forests. USDA Forest Service SE Forest Experiment Station. *Gen. Tech. Rep.* SE-68.

Schneider, S.H. 1989. *Global Warming: Are We Entering the Greenhouse Century?* Sierra Club Books: San Francisco, CA.

Soule, M.E. 1991. Conservation tactics for a constant crisis. *Science* 253:744–750.

Strain, B.R. and Cure, J.D., Eds. 1985. Direct effects of increasing carbon dioxide on vegetation. Office of Energy Research, U.S. Dept. of Energy, Rep. DOE/ER 0238.

Strzepek, K.M. and J.B. Smith, Eds. 1995. *As Climate Changes: International Impacts and Implications.* Cambridge University Press: New York.

Thorsell, J. and Harrison, J. 1991. National parks and nature reserves in the mountain regions of the world. Background paper for Parks, Peaks, and People — An International Consultation on Protected Areas in Mountain Environments, Hawaii Volcanoes National Park, October 27 to November 2.

Titus, J.C., R. Park, S. Leatherman, R. Weggel, M. Greene, P. Maugel, M. Trechan, S. Brown, C. Gaunt and G. Yohe. 1991. Greenhouse effect and sea level rise: the cost of holding back the sea. *Coastal Manage.* 19:3

Titus, J.G. 1992. The cost of climate change to the United States. Pp. 385–409 in *Global Climate Change: Implications, Challenges and Mitigation Measures.* S.K. Majundar, K.S. Kalkstein, B. Yarnal, E.W. Miller and I.M. Rossenfield, Eds., Pennsylvania Academy of Science, Philadelphia.

Tranquillini, W. 1979 *Physiological Ecology of the Alpine Timberline.* Springer-Verlag: New York.

Udvardy, M.D.F. 1975. A classification of the biogeographical provinces of the world. Occas. Pap. 18, International Union for Conservation of Nature and Natural Resources, Morges, Switzerland. 48 pp.

U.S. Global Change Research Program: Subcommittee on Global Change Research (GCRP). 1995. Forum on Global Change Modeling. Rep. 95-01, No. 13. May 1995. http//www.gcrio.org.

Watson, et al. 1990. Greenhouse gasses and aerosols. Pp. 243–259 in Houghton, J.T. et al., Eds. *Climate Change: the IPCC Scientific Assessment.* Cambridge University Press.

Wells, P.V. 1983. Paleobiogeography of monane islands in the Great Basin since the last glaciopluvial. *Ecol. Monogr.* 53:241–282.

World Resources Institute. 1989. *Changing Climate: a Guide to the Greenhouse Effect.* 1709 New York Avenue, NW, Washington, D.C.

20 Ecosystem Stabilization and Restoration: The Clinch–Powell River Basin Initiative

Dennis H. Yankee, John D. Peine, Jack D. Tuberville, and Donald W. Gowan

CONTENTS

The Role of Restoration Ecology in Ecosystem Management 418
    The Clinch–Powell River Basin Case Study 418
Building Partnerships to Address Concerns 419
    The Nature Conservancy 419
    Tennessee Valley Authority 419
        Clean Water Initiative 419
        River Action Teams 419
    U.S. Fish and Wildlife Service 420
    U.S. Environmental Protection Agency 420
Inventory of Natural Resources 420
    Topography 421
    Land Cover/Use 421
    Caves 421
    Water Resources 421
        River System 421
        Aquatic Species 422
        Water Quality 422
    A Process to Inventory Natural Resources Applied in the Case Study Area 422
Inventory of Social Values and Resource Utilization 423
    Historical Context 423
    Social Context 423
    Agriculture 424
    Mining 424
    Forestry 424
    A Process for Defining Stakeholder Interests Applied in Case Study Area 424
Inventory of Environmental Stressors 425
    Water Quality Monitoring for the Clinch–Powell Initiative 425
Integrated Risk Assessment 425
    EPA Risk Assessment 425
    TVA River Action Teams 426
Mitigation Activities 426
    Riparian Restoration Program 426
    Community Activities and Education 427

0-57444-053-5/99/$0.00+$.50

    Compatible Economic Development 427
    Cave Registry Program 427
    Research on Threatened Organisms..427
  Strategies to Sustain Momentum..428
  Conclusion..428
    Risk Assessment Report Card for Ecosystem Managers 428
  References 430

THE ROLE OF RESTORATION ECOLOGY IN ECOSYSTEM MANAGEMENT

As we approach the 21st Century, the role of restoration ecology in ecosystem management will dramatically increase. One could argue that most issues facing natural resource managers today include an element of ecosystem stabilization or restoration. The topic will undoubtedly become a major international industry as society struggles to sustain and/or restore ecosystem processes necessary to support communities of native biological diversity, not the least of which includes *Homo sapiens.*

Restoration ecology implies a process of direct intervention by humans to restore *natural* ecosystem processes or desired populations of depleted or extirpated species. Landscape conditions at the broader scales affect the sustainability of valued components (Golley 1993). Therefore, ecosystem restoration must consider processes at the landscape scale. The need for restoration tends to be in response to a multitude of direct and indirect adverse impacts that are human induced. There are layers of complexity, e.g., understanding the ecosystem as it was before disturbance, the degree and nature of adverse impacts from various human induced perturbations, the various and usually divergent stakeholder values and interests in the resources, and the labyrinth of institutions and policies relevant to the issues of concern. Options for restoration should be judged as to their potential to succeed in attaining the envisioned goals, their potential adverse impacts to the ecosystem, their cost effectiveness, and the probability that actions taken will sustain the desired results over the long term.

In reality, it is nearly impossible to completely restore the natural biodiversity and ecosystem processes of a highly disturbed ecosystem. The restoration of large rivers to a pristine state is incompatible with present human population levels (Welcomme 1989). However, the manager can play a key role of recreating or stabilizing basic elements of ecosystems so as to allow an opportunity for naturally occurring ecosystem processes and functions to continue and for the ecosystem to evolve in as natural a state as practical.

Restoration ecology is invariably a long, arduous process with little guarantee that the steps taken will obtain the desired result. This chapter is organized according to a set of basic steps suggested to be taken in the process of ecosystem restoration. Each situation is unique and it is difficult to generalize, but a key role of the ecosystem manager is to provide leadership by encouraging that a comprehensive strategy be pursued such as the process suggested below.

THE CLINCH–POWELL RIVER BASIN CASE STUDY

The case study discussed in this chapter provides a poignant example of several ecosystem restoration programs that are truly works in progress. The river basin is a *hot spot* of concern in the southern Appalachian region, primarily due to the concentration of rare and threatened aquatic and cave species and the uniqueness of the natural area as the last remaining large scale free flowing headwater segment of the Tennessee river system. Extensive collaboration among various levels of government, nongovernment organizations, and private land owners is demonstrated. The enormity of the area of concern and complexity of the tasks at hand typify ecosystem restoration initiatives. The relatively small scale of the mitigation activities is illustrative of a real world ongoing process driven by the resolve of individuals searching for a realistic, effective means to solve the problems at hand. The job of saving this remaining biodiversity is made more difficult, due to the fact that very little of the land is managed by the government. Only 12% of the river basin is managed by the federal government. Cooperators must thus rely heavily on partnering with community members in order to effect change. In reality, there is no single restoration program in progress in the Clinch–Powell. Several entities are at work, each with its own mandate and interests. It should be emphasized that differences in interests will persist. The goal of restoration activity is not to eliminate differences but to minimize them and to work within them where goals are the same or complementary. Another hurdle that needs to be mentioned is the differing views of the states involved, as well as the lines that identify regions set up by the various federal agencies involved in these activities. Therefore, we not only have different federal agencies, but different regions of those agencies involved in different parts of the watershed.

BUILDING PARTNERSHIPS TO ADDRESS CONCERNS

A suggested first step in the ecosystem restoration process for the ecosystem manager is to build effective partnerships among stakeholders with vested interests in the ecosystem of concern. The solution of these problems must involve knowledge of human perceptions and desires (Ludwig et al. 1993). This is the most important role of the manager, who can leave much of the technical details to appropriate specialists. *The ecosystem manager must lead the effort, articulate the vision of anticipated result, and sustain enthusiasm and attention over the long term.* Ecosystem management is all about building partnerships among stakeholder interests in the landscape of concern. Land managing agencies; communities and industries using and/or polluting the natural resources; individuals; all families and organizations sharing social values related to the natural environment share either complimentary or conflicting interests in the ecosystem of concern. The greatest strength of the Clinch–Powell River Basin case study is the diversity of partnerships and the nature of the tasks currently being pursued. Discussion of the roles assumed by the river basin cooperative partners follows.

THE NATURE CONSERVANCY

Recognizing the importance of the Clinch–Powell River Basin, the Nature Conservancy has targeted the watershed as part of its "Last Great Places: An Alliance for People and the Environment" program. In 1990, the Virginia Chapter of the Nature Conservancy opened a field office in Abingdon to bring hands-on protection to the area. Four staff members are now enacting a comprehensive conservation strategy for the region that includes land acquisition, research, economic development, and community outreach. In the last 2 years, the Conservancy has launched several innovative conservation programs in the area. One of the Conservancy's priority areas is Big Cedar Creek. Their approach is to identify the most critically endangered species, then do projects based on their proximity to the endangered fauna. Although their strategies, and even goals, may be somewhat different from other cooperators, there is considerable area of common ground.

TENNESSEE VALLEY AUTHORITY

Clean Water Initiative

Recognizing the impact that land activities upstream have on reservoir water quality, the Tennessee Valley Authority (TVA) implemented a program called the Clean Water Initiative (CWI). Part of the CWI was the establishment of River Action Teams. These teams interact with the communities in the watersheds and offer advice, education, labor, and funding to help prevent or negate water resources problems. TVA's Clean Water Initiative is an indirect outgrowth of the Lake Improvement Plan. Adopted in 1991, the plan committed TVA to improving water quality in the tail waters of their hydroelectric plants and to keeping lake elevations up longer in certain of the tributary reservoirs in order to improve recreational uses. This new emphasis on recreation and especially on water quality soon led to the realization that the entire watershed was contributing to the water quality problem and that a watershed approach was needed to protect and improve the waters of the Valley. This led to TVA's commitment to make the Tennessee River the cleanest and most productive commercial river system in the U.S. Since TVA has no regulatory or enforcement authority, its approach to the challenge was to work cooperatively with other agencies and local groups to effect water quality improvements. This led to the creation of the CWI.

Goals of the CWI were designed to support one of TVA's overall goals, to be environmentally responsible in its management of the Valley's water resources. The CWI goals are to improve beneficial uses of water resources and to transfer responsibility to sustain these improvements to the user public by the year 2015.

River Action Teams

These teams are the facilitating agents of the CWI. They are interdisciplinary teams of 4 to 6 TVA employees usually including biologists, communications and education specialists, and environmental scientists or engineers. They focus on a single watershed and take a holistic, cooperative approach to water quality solutions. Teams set their own priorities and develop their own projects, based on their knowledge of the watershed and interests of local citizens. Business plans are prepared each year to organize work and make sure it aligns with CWI and TVA goals. Priority hydrologic units in the Clinch–Powell river basin currently assigned River Action Teams include the Guest River, Big Cedar Creek, North Fork Powell, and Norris Reservoir.

River Action Teams are supported by teams of TVA experts in various fields: the River Action Teams call on these experts to help solve specific problems and to provide technical advice and review. The River Action Teams make decisions and implement their own projects. In addition, there are Stream/Reservoir Teams that provide field crews to conduct stream and reservoir aquatic data collection upon request. Finally, there is the Aquatic Plant and Vector Team, which handles specialized work of aquatic plant and mosquito control Valley-wide.

U.S. FISH AND WILDLIFE SERVICE

The Fish and Wildlife Service has been involved in activities to conserve aquatic biodiversity of the Clinch–Powell Basin since the 1970s. In 1994 the Service opened the Southwestern Virginia Field Office in Abingdon, Virginia, to more fully and directly address endangered species recovery. Currently, the Service is working to stabilize the rapidly declining aquatic fauna through on-going contaminant studies, habitat restoration projects, and public education and outreach efforts (Hylton 1997).

U.S. ENVIRONMENTAL PROTECTION AGENCY

The U.S. Environmental Protection Agency is active in regulating point source pollutant discharges throughout the watershed. However, many of the problems in the Clinch–Powell are not point source related. With reduced amounts of funding for restoration work, there is a real need to focus in on the key problems. The EPA is currently funding a pilot ecological risk assessment in the watershed. The goal of this study is to determine the risk to aquatic resources from various stressors. This will allow for better application of restoration funds.

INVENTORY OF NATURAL RESOURCES

The second step in the suggested ecosystem restoration process is to gain an adequate understanding of the ecosystem in question: what are its key components and their interrelated processes, functions,

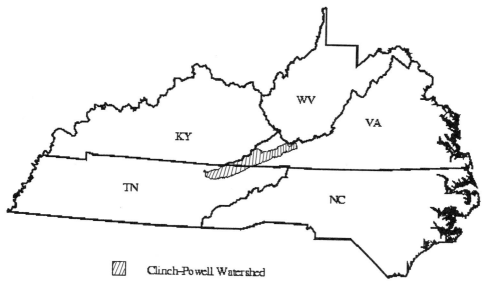

FIGURE 1   The Clinch–Powell Watershed.

and associated biological communities? In the Clinch–Powell River Basin case study described below, the central theme is the aquatic organisms associated with this free-flowing river system.

TOPOGRAPHY

The Clinch–Powell River Basin (Figure 1) drains approximately 286,470 ha in Tennessee and 437,317 hectares in Virginia (Soil Conservation Service 1992). The majority (95%) of the study area is located in the Appalachian Valley and Ridge physiographic province, and the remaining 5% is located on the Cumberland Plateau. Elevations in the Valley and Ridge province range from 225 to 300 meters on the valley floor to over 1000 meters on the ridges. The topography of the basin is characterized by mountains with steep valleys and ridges running from northeast to the southwest. Elevations on the Cumberland Plateau range from 300 meters to 1300 meters and the terrain is generally very rugged with steep slopes (Smalley 1984).

LAND COVER/USE

The predominant land cover is forest with over 50% so covered. Pasture and hay fields account for approximately 35% and cropland accounts for roughly 10% of the land use. Major annual crops include tobacco, hay and corn (Soil Conservation Service 1992).

CAVES

The karst region of the Appalachian Ridge and Valley portion of the Clinch–Powell Basin harbors a labyrinth of caves and underground streams supports two kinds of endangered bats, some fifty globally rare cave organisms and several unique natural communities.

WATER RESOURCES

The Clinch–Powell case study is particularly illustrative of the principles of ecosystem restoration because the focus is on the river system which serves as a conduit for the adverse effects of various forms of disturbance to the entire ecosystem. Freshwater mussels require exceptionally clean and fast-flowing waters to stay healthy. Therefore, fresh water mussel species and their host fish species of concern tend to be particularly vulnerable to chemical pollutants and changes in aquatic habitat effected by flow rate, sedimentation and shoreline land cover/use.

River System

The Clinch and Powell rivers are the only remaining ecologically intact headwater rivers of the Tennessee river system. Draining approximately 724,000 ha, the system begins in the mountains of southwest Virginia and empties into the Tennessee River near Knoxville. Ninety-five percent of the watershed falls in the Appalachian Ridge and Valley physiographic province, while the remaining 5% is on the Cumberland Plateau. The Powell River begins in Wise County, VA and flows 120 miles, where it enters the Clinch River as an impounded component of Norris Lake. Major tributaries to the Powell include the South and [N]Forth Fork Powell Rivers. The Clinch River begins in Tazewell County, VA, and flows for approximately 200 miles before reaching Norris Lake. Major tributaries to the Clinch include the Little River, Big Cedar Creek (80 mi^2 est.), Guest River (100), Copper Creek (130), and the North Fork Clinch River (90) (EPA 1997).

Aquatic Species

Originally representing arguably one of the greatest sources of species diversity of fresh water mussels anywhere in the world, the Tennessee river system has been largely impounded and polluted, leaving the Clinch–Powell watershed the last biological refuge for this vestige of aquatic diversity. Waterways of the Tennessee River Valley once supported nearly 60 species of mussels, which have dwindled to around 40 today. Among the last free-flowing sections of this once expansive system, these rivers are the sole remaining sanctuary for a group of freshwater mussels living nowhere else on earth. Inhabited by 28 mussel and 7 fish species listed as threatened or endangered, the river system contains one of the most diverse assemblages of freshwater mussel and fish species in North America (Shute 1994; Ahlstedt 1984). Recent fish and mussel surveys indicate that most rare species in this region continue to decline (Angermeir and Smogor 1993) (see also Ahlstedt and Tuberville 1997).

Water Quality

A 1991 report, *Understanding Surface Water Quality Trends in Southwestern Virginia* reviewed much of the current and historic water quality data for the Clinch and Powell watersheds (Zipper et al. 1991). While this report indicated improvements in water quality trends in the Clinch and Powell Rivers, the report also demonstrated high median fecal coliform concentrations at three of the four monitoring stations on the Powell River and increasing filterable residue values over time (a measure of total dissolved solids) in some regions of the Powell River and Guest River. The 1994 Virginia Water Quality 305b Assessment Report (EPA 1994) identified violations of the fecal coliform standards at various sites in the Guest River and Stock Creek, tributaries to the Clinch River. Of 109 highly impaired stream miles, 80 were attributed to elevated toxic metals instream. Recent fish and benthos surveys by the Tennessee Valley Authority have demonstrated potential water quality problems in several areas, especially in the headwater areas of the Powell and Guest Rivers, where abandoned mine acid drainage may be contributing to poor water quality (EPA 1997)

A PROCESS TO INVENTORY NATURAL RESOURCES APPLIED IN THE CASE STUDY AREA

The TVA River Action Teams, as part of the CWI, engage technical experts in selected fields of aquatic biology, hydrology, and water chemistry as well as water resource and land use management. Using this cadre of expertise, a screening assessment is made of the health of the river-related ecosystem.

In cases of insufficient information, field reconnaissance is conducted. This usually includes among other things, calculating an Index of Biotic Integrity (IBI) for the fishery (Kan et al. 1986) and characterization of aquatic insects as well. The assessment is a screening level activity and includes existing as well as newly acquired reconnaissance data. Although these reconnaissance level assessments do not prove any cause and effect relationships of water quality problems, they do provide clues to potential sources. Assessment of ecological condition, then, includes available information on aquatic resources and new reconnaissance data collected by the Teams. This provides a screening level of information on ecological resources.

INVENTORY OF SOCIAL VALUES AND RESOURCE UTILIZATION

This third step in the suggested ecosystem restoration process is the one most ignored and least understood, even though it is arguably the most important. Since human activity has invariably created the crisis requiring restoration ecology, it stands to reason that significant time and energy should be devoted to gaining an understanding of the dynamics of the socioeconomic activity that led to the problem. Only by obtaining a clear understanding of stakeholder values and interests in the natural resources in the ecosystem of concern, will viable, socially acceptable solutions be crafted to solve the perceived problems.

HISTORICAL CONTEXT

The Clinch–Powell watershed is rich in both human and natural history. The Wilderness Road that led 18th century settlers to the Kentucky country passed through both the Clinch and the Powell valleys on its way to Cumberland Gap, where Daniel Boone blazed a trail in 1750. During these early days the population was made up mostly of small farm owners and hunters. Up until the time of the Civil War, the Clinch–Powell remained somewhat isolated, though not to the extent sometimes described. By the late 19th century the industrial revolution sent railroads pushing up the valleys in search of cargoes of coal and timber. During the ensuing years land, coal and timber interests bought up much of the land in the watershed (Kincaid 1947).

In the coalfields, which lie principally along the Cumberland Plateau, much of the land and resources are still owned by large corporations, often controlled from outside the area. Terrain is difficult and does have some isolating effect. However, whatever isolation may still exist is rapidly being eliminated by the road-building activities supported in part by the Appalachian Regional Commission (Whisnant 1980). A more acute problem caused by terrain is related to sewage treatment systems. The mountainous terrain, especially in the coalfields, can increase dramatically the cost of installing sewage treatment systems; and, since the coalfields tend to be the most economically stressed, few funds are available to pay for such improvements. Therefore, a bacterial contamination is a widespread problem in the area, especially where homes are not served by local or regional systems.

Today coal is currently in decline in terms of production levels; and automation is causing even greater declines in employment (Hibbard 1990). As jobs become more scarce, fears of further environmental regulation (with its sometimes concomitant loss of jobs) increases. Added to this is the fact that wages in the coal industry are some of the highest available. For instance, in Lee and Wise counties, Virginia, weekly wages for coal work are higher than for any other category of employment (1994 data, VA Employment Commission). Thus, when environmental restoration is viewed as a trade of jobs for environmental improvements, it is a trade few wage-earners are willing to make.

SOCIAL CONTEXT

Today the economy of the region is driven primarily by coal, agriculture, and logging, in that order (EPA 1997). While the population of the state of Virginia grew nearly 16% from 1980 to 1990, the population within the Clinch–Powell River Basin declined by 10%. Thus the Clinch–Powell watershed remains essentially rural, the largest town being Lafollette, TN, with a population of about 7000. Few towns have more than 5000 residents. Limited population has probably been a major factor in maintaining biological diversity in the watershed. Now local and regional organizations are trying to capitalize on this by promoting tourism in the area. If done well, both economic development and resource protection can be achieved. However, abandoned mine lands continue to create erosion and sedimentation problems, over grazing and streambank erosion contribute to the problems, and timber harvest looms as an additional erosion threat. How well these threats and opportunities are handled will have a large bearing on how successful watershed restoration programs will be.

AGRICULTURE

Agriculture is a critical component of the social and economic fiber of the region. With nearly 35% of the watershed devoted to agriculture production in an area severely constrained by topography, agriculture land uses dominate the flood plain and lowland areas of the watershed. Primary income-generating commodities are beef and tobacco (Soil Conservaton Service [USCS] 1992). Livestock concentrations, overgrazing and tobacco production in these areas pose serious risks to sensitive aquatic and subterranean resources.

In the agricultural areas of the watershed, such as Tazewell, Russell and lower Lee counties, Virginia, and the Powell River Valley in Tennessee, land ownership tends to be more equitable. While income levels are still relatively low, wealth tends to be more evenly distributed and local economies more diverse.

MINING

More than 40% of the coal production from the state of Virginia occurs within the five counties encompassing the river basin; the coal-producing area comprises only 20% of the watershed. In recent years there has been a decline in coal production, which is due to economic forces, that is, coal prices. Reserves are sufficient for about 50 to 100 years depending on future demand, price and technology. On the technology side, production per unit of human labor has been increasing for decades (Hibbard 1990).

So, even while production was increasing, employment has been on the decline. When coal production itself declines, the loss in jobs is even worse. This decline is prompting fundamental economic changes, namely, unemployment and all the ills that go along with it. Unemployment ranks highest in the state in the coal-producing counties in the region. About half of the mining is surface mining.

FORESTRY

The watershed was intensively logged in the late 18th and early 19th and early 20th centuries. The last 15 years have seen a resurgence of the forest industry in the river basin. While growth of the timber industry is helping offset the loss of mining jobs, inappropriate logging practice can impair sensitive karst and aquatic resources, degrade water quality, and potentially pose other environmental threats.

A PROCESS FOR DEFINING STAKEHOLDER INTERESTS APPLIED IN THE CASE STUDY AREA

The TVA River Action Teams assess public interest and support for conservation practices. This is accomplished through public meetings, telephone surveys, and analysis of other available information, such as the number of organized, environmentally oriented groups within the watershed, and noting the number of protection/restoration actions taken. Another indicator is the positions local officials have taken on environmental issues such as solid waste and sewage treatment. Such information is periodically updated.

INVENTORY OF ENVIRONMENTAL STRESSORS

The fourth step in the suggested ecosystem restoration process is to define key stressors adversely effecting the ecosystem of concern. They need to be defined as precisely as possible as to their origin and nature and severity of adverse impact on the ecosystem, acting individually as well as collaboratively with other stressors. The greater the precision of understanding the nature of the problems caused by the stressors, the greater the likelihood that the chosen mitigating measures will provide maximum benefit and expectations will be met. A key to success is to limit the number of unknowns associated with the issue of concern. This is where research plays a central role in the process. Invariably there is a need to conduct applied and in some cases basic research to discover the cause/effect relationships of perturbations on the *natural* ecosystem. For the case study presented here, discussion will be limited to stressors to the rivers and their associate aquatic organisms.

WATER QUALITY MONITORING FOR THE CLINCH–POWELL INITIATIVE

On the Guest River, monitoring at 13 sites by TVA is underway for hydrologic parameters and water quality including fecal coliform, metals, and sediment. A U.S. Geologic Survey (USGS) National Water Quality Assessment Program (NAWQA) study unit is also monitoring water quality in the Clinch–Powell as part of their Upper Tennessee River Basin study. On a more local scale, the Nature Conservancy in conjunction with Virginia Tech is monitoring water quality in select subwatersheds within the Clinch–Powell.

INTEGRATED RISK ASSESSMENT

The fifth step in the suggested ecosystem restoration process is possibly the most critical in determining the potential for success of the endeavor. In order to making the most informed management decisions within the context of an imperfect real world of incomplete information, an integrated risk assessment should be conducted. This is the step of the process where counsel with experts is most helpful and involvement by key stakeholders is of paramount importance. The principles for the conduct of risk assessment are alluded to in Chapter 7 (Berish et al.) on the conduct of environmental assessments. Actions taken in the Clinch–Powell River Basin case study concerning risk assessment are described below.

EPA RISK ASSESSMENT

Due to its importance as a center aquatic biodiversity, the Clinch River Basin was selected by the USEPA as one of five Watershed Ecological Risk Assessment Case Study areas. The diversity of urban, rural, and industrial land uses within the watershed necessitates a broad approach to risk assessment. The philosophical framework behind the risk analyses proposed were developed with the intent of providing meaningful direction to resource managers and the public involved in this watershed. The over-arching goal of this assessment is to establish and maintain the unique native biological qualities of the Clinch–Powell watershed surface and subsurface aquatic ecosystem (EPA 1997). The EPA hopes to achieve this goal by giving resource managers information which will allow them to focus effort and resources in areas where they will do the most good.

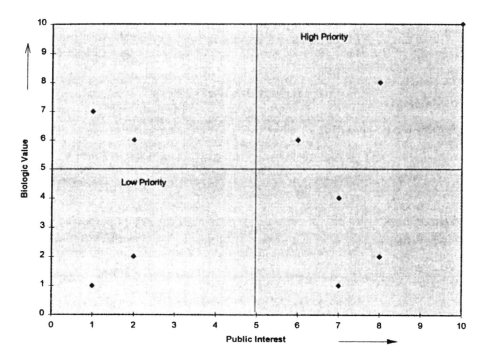

FIGURE 2   Public interest in relation to biologic value.

TVA River Action Teams

The results of the analysis of public interest and commitment to natural resource stewardship and the assessment of environmental health are compared in order to aid in targeting watersheds for work by the River Action Teams. Based on the assessment information, teams develop a matrix of condition vs. public interest as displayed in Figure 2. Those hydrologic units where resource need is great and public interest is high are assigned a high priority for action (upper right quadrant) by the TVA Teams. This group of high priority hydrologic units is then further evaluated and prioritized, based on many factors, including the nature of the suspected problem, the feasibility to mitigate the problem cost effectively, sources of support, etc. By using the USGS hydrologic units, the River Action Teams can communicate and coordinate more effectively with other agencies, since hydrologic units are the basic unit of consideration for many federal and state agencies.

MITIGATION ACTIVITIES

This sixth step in the ecosystem restoration process is the operative stage. Mitigation actions chosen should demonstrate clear progress toward solving the defined problems in order to sustain stakeholder interest in the process. Small yet meaningful actions are best at first as a means to build partnership involvement and to not get ahead of the learning curve. The following is a list of mitigation programs from the Clinch–Powell, which may be characterized as being at an achievable, modest scale with substantial stakeholder participation at the local level; and for the most part, voluntary rather than regulatory in nature.

Riparian Restoration Program

Teaming up with the U.S. Fish and Wildlife Service and local residents, the Nature Conservancy is working with farmers to build fences to keep cattle out of the river, find alternative sources of water for the cattle, and restore native vegetation to the stream banks. All this is done at no up front cost to the farmer.

COMMUNITY ACTIVITIES AND EDUCATION

Public interest in cleaning up unregulated dumps led TVA to begin assisting local communities in accomplishing this task. Initially, TVA considered this as simply a type of litter cleanup — nice to do, but not much effect on water quality. Locals convinced them there was some potential for surface and groundwater contamination, depending on what was lurking within the dump. TVA assisted in a few clean-ups and the public support for it was tremendous. They will be doing some each year for the foreseeable future. The nice thing about them is that a dump clean-up can be done in one to a few days, there is an immediate obvious result, and the public appreciates it. Typically, TVA supplies some heavy equipment, and the community or the state department of transportation or the local solid waste management provider supplies some services as well. Locals help by picking up some of the small refuse the equipment misses, and they help assure the dump doesn't get reborn later.

Realizing that education and public involvement are key ingredients in successfully dealing with water quality issues, many activities to inform and involve the public have occurred. One such activitiy is a Stream Walk. TVA in conjunction with the Natural Resources Conservation Service (NRCS), local Resource Conservation Districts (RC&D), community groups, colleges, and even Americorps hold these walks periodically. The purpose is to get as many people as possible surveying the stream in question, looking for anything that might impact water quality. The results of these surveys are collected and used in planning remediation activities. Another activity is the Kids in the Creek program. Again involving many federal, state, and local groups, this program takes school-age kids and gets them into the stream and lets them help with biological surveys. Both of these activities not only gather important data but serve to educate people on the importance of good water quality, factors that affect water quality, and also perhaps spark an interest in science and environmental issues.

COMPATIBLE ECONOMIC DEVELOPMENT

A central tenet of the "Last Great Places" initiative of the Nature Conservancy is the need to protect nature while allowing human communities to grow and prosper. The Conservancy is working with local planners, government agencies, and private industry to find environmentally sound approaches to such issues as sewage treatment, timber harvesting, coal mining, and airport and prison location.

CAVE REGISTRY PROGRAM

Working with the Cave Conservancy of the Virginias, Virginia Cave Board, and Virginia Department of Conservation and Recreation, the Nature Conservancy has registered for protection 13 biologically significant caves. Registry is a voluntary agreement by landowners to protect caves on their property providing a first layer of protection for eleven globally rare species, including the endangered gray bat.

RESEARCH ON THREATENED ORGANISMS

The Virginia Chapter of the Nature Conservancy is working with several universities to determine the habitat needs of the Clinch Valley's endangered plants and animals. Some of the research seeks to understand the complete life histories of endangered organisms and develop innovative methods to aid in the recovery of these species. Other research aims to quantify threats to the health of rivers and caves and find ways to reduce these threats without hurting the region's economy.

STRATEGIES TO SUSTAIN MOMENTUM

A seventh and final step is undoubtedly the most difficult to follow, and that is to maintain momentum for the project over the long term. Ecosystem restoration is invariably a long-term proposition that will likely span a longer period of time than the attention given to it by the original architects of the project. It is human nature to have the greatest ownership and enthusiasm for that which is self created. Inheriting someone else's *crusade* rarely leads to as much diligence and enthusiasm as that of the initiator. However, strategies such as those suggested below, can be built at the outset of a restoration program, which will hopefully institutionalize commitment to the project over the long term.

- Constantly work to expand shareholder involvement in the program. Spend energy and resources to identify additional vested interests and ways to involve individuals and institutions representing the invariably disparate array of stakeholder interests.
- As the program progresses, always seek ways to build and reinforce the institutional framework that is coordinating the program. Establishing permanent line item budgets is a key to instilling institutional commitment. Making it more feasible to conduct field work by providing support facilities and services will encourage long-term commitment. Examples of appropriate support are providing staff to coordinate logistical details, field laboratory facilities, housing, and transportation.
- Maintain a natural resource monitoring and evaluation program for the long term to document changes in ecosystem dynamics and the presence and influence of agents of perturbations.
- Build awareness of the program goals and benefits at a regional and local level.
- Build regulatory capability at the appropriate level of government to control the agents of perturbations and to manage the recovering ecosystem to the degree necessary to sustain the recovery process.

Since the Clinch–Powell River Basin is so sparsely populated, there are relatively few community and county facilities and service resources available to support the recovery program. Although much of the burden of recovery has been carried by federal and state government agencies and non government organizations, local people have been engaged at the early stages of the process. Ultimately, the burden of maintenance of ecosystem stewardship will shift to local authority and leadership.

CONCLUSIONS

This rapidly growing dimension of ecosystem management is unique in every case of application. Restoration has traditionally been attempted at a species or community scale. The Clinch–Powell case study exemplifies attempts to approach ecosystem stabilization and recovery at a more expansive landscape scale. The risk is much greater with an expanded scale of application but such shift in direction is ultimately mandatory if balance between protecting the *natural* environmental and accommodation of human needs and desires is to ever be achieved.

RISK ASSESSMENT REPORT CARD FOR ECOSYSTEM MANAGERS

Applying the *Risk Assessment Report Card for Ecosystem Managers,* as referenced in other chapters in Section III of this volume, to ecosystem restoration is particularly relevant because of the high level of risk to managers associated with the difficult choices associated with this topic. For the Clinch–Powell River Basin case study, the report card for managers is judged to be as follows.

Vision: B+

The uniqueness of the resources of the river basin and the importance of the species at risk provide a clear vision of the need for ecosystem restoration.

Resource risk: D

The aquatic and cave organisms are reported to be at great risk due to adjacent land uses.

Socioeconomic conflicts: B

Local land owners and communities have been engaged in the process of defining problems and projects to mitigate them. There has been significant effort to avoid having to make choices between conflicting resource values for the mitigation strategies.

Procedural protocols: C–

The mitigation procedures applied are unique to the circumstances of each hydrologic unit, so there has as yet not been the development of standard procedures for the program.

Scientific validity: B

Scientists are being consulted on a regular basis to insure that procedures for the conduct of resource condition assessment and mitigation techniques are scientifically valid. In some cases, research is underway to study the relative importance of various stressors and the critical components of habitat for species of concern.

Legal jeopardy: B

To this point, the mitigation activates have been on a voluntary basis, thereby suggesting minimal legal innuendo.

Public support: A

The mitigation measures are being pursued where there is considerable public support for such activity.

Adequate funding: C

The enormity of the targeted landscape area and the multiplicity of the sources of the perturbation suggests that a considerably higher funding level would greatly enhance the potential for success of the program.

Policy president: A

The Clean Water Initiative is being conducted by TVA throughout the entire Tennessee River Valley. The other agencies involved are supported by precedents of long standing program initiatives.

Transferability: B

Though the specifics of the recovery actions are unique to each hydrologic unit, there is significant commonalty as to the transferability of the principles applied.

Add this dimension of ecosystem management to the list that requires considerable insight to the future, courage to take risks, and the ability to sell the idea at the conceptual level to an army of skeptics and nay-sayers.

REFERENCES

Ahlstedt, S.A. 1984. Twentieth century changes in the freshwater mussel fauna of the Clinch River. *Walkerana* 5(13):73–122.

Ahlstedt, S.A. and J. Tuberville, 1997. Quantitative Reassessment of the Mussel Fauna in the Clinch and Powell Rivers, Tennessee and Virginia. *In press.*

Angermeir, P.L. and R.A. Smogor. 1993. Final report: Assessment of biological integrity as a tool in the recovery of rare aquatic species. Virginia Department of Game and Inland Fisheries, Richmond, VA.

EPA (Environmental Protection Agency). 1997. Clinch Valley Watershed Ecological Risk Assessment. Unpublished plan.

Golley, F.B. 1989. Landscape ecology and biological conservation. Landscape Ecol. 2:201–202.

Hibbard, W.R. 1990. *Virginia Coal: An Abridged History.* Virginia Center for Coal and Energy. Virginia Polytechnic Institute and State University: Blacksburg, VA.

Hylton, R. 1997. Personal communication.

Kan, J.R., K.D. Fausch, P.L. Angermier, P.R. Yont, and I.J. Schlosser. 1986. Assessing biological integrity in running waters: a method and its rationale. Spec. Publ. No. 5. Illinois Natural History survey, Champaign.

Kincaid, Robert L., 1947. *The Wilderness Road*, Bobbs-Merrill, Indianapolis.

Ludwig, D., R. Hillborn, and C. Walters. 1993. Uncertainty, resource exploitation and conservation: Lessons from history. *Science* 260, 17–36.

Shute, P. 1992. Little Known Fishes in the Clinch/Powell Watershed. Clinch Powell Rivers Bi-State Conference: Programs and Abstracts. 56–59.

Smalley, G.W. 1984. *Classification and evaluation of forest sites in the Cumberland mountains.* USDA Forest Service, Southern Forest Experiment Station, New Orleans, LA. *Gen. Tech. Rep. SO-50.*

Soil Conservation Service. 1992. *Plan of work: Clinch–Powell river basin study, Tennessee.* August 1992.

Stanford, J.A., J.V., Ward, W.J., Liss, C.A., Frissell, R.N., Williams, J.A., Lichatowich, and C.C. Coutant. 1996. A general protocol for restoration of regulated rivers. *Regulated Rivers: Res. Manage.,* 12: 391–413.

Virginia Department of Environmental Quality. 1994. Virginia water quality assessment 1994. Information Bull. No. 597. Richmond.

Welcomme, R.L. 1989. Floodplain fisheries management. In J. A. Gore, Ed., *Alternatives in Regulated River Management.* CRC Press: Boca Raton, FL.

Whisnant, David E. 1980. *Modernizing the Mountaineer — People, Power and Planning in Appalachia.* Appalachian Consortium Press: Boone, N. C. 296 pp.

Zipper, C.E., G.I., Holtzman, S., Rheem, and G.K. Evanylo. 1991. *Understanding Surface Water Quality Trends in Southwestern Virginia.* Virginia Polytechnic Institute and State University, Water Resources Research Center, Blacksburg, VA. 91–1:61.

21 Threatened Ecosystem: High Elevation Spruce–Fir Forest

Niki Stephanie Nicholas, Christopher Eagar, and John D. Peine

CONTENTS

Introduction 431
A Framework to Develop an Ecosystem Management Plan 432
    Resource Inventory...432
    Ecosystem *A Priori* Condition...432
    Stressors Inventory 432
    Risk Assessment..433
    Resource Monitoring..433
    Plan for Stabilization or Recovery...433
    Implementation Strategy 434
    Framework Summary 434
    Spruce–Fir Case Study...434
Southern Appalachian Spruce–Fir Ecosystem...435
    Natural History and Ecological Characteristics 435
    Current Stressors: a System Under Siege...437
        Effects of Logging..437
        Balsam Woolly Adelgid..438
        Air Pollution 441
        Climate Change 444
    Ownership and Use Patterns...444
    Current and Potential Management Initiatives 446
    Research Needs to Support Management Decision Making.......................................448
    A Call for Action 448
Risk Assessment Report Card 449
Acknowledgments...451
References 451

INTRODUCTION

The subject of dealing with threatened ecosystems is purposefully placed as the last two chapters in Section III of this book which describes various dimensions of ecosystem management. The preceeding Chapter 20, by Yankee et al., describes an ecosystem stabilization and restoration project for the Clinch and Powell River Basin. In that case, the subject is a restoration work in progress at a watershed scale. In contrast, this chapter focuses on a highly threatened forest ecosystem which does not have a comprehensive strategy for management nor mitigation of the numerous stressors that are impacting this unique resource.

Ecosystems can deteriorate due to a myriad of compounding adverse effects. As a result, the ecosystem manager is invariably faced with multiple challenges such as those collectively addressed in various chapters in this section of the book. In this chapter we first present a framework for managers to follow in developing an ecosystem management plan which integrates many of the principles presented in this book. Use of this framework will allow the manager to determine (1) how the ecosystem in question functioned in an *a priori* undisturbed state, (2) how much change has taken place in the system, (3) what are the principle causes of the decline of the system, and (4) what is the potential to stabilize and/or restore portions or all of the ecosystem. Steps in the process leading to answers to these key questions are followed by a detailed discussion of the status of the southern Appalachian spruce–fir forest.

A FRAMEWORK TO DEVELOP AN ECOSYSTEM MANAGEMENT PLAN

RESOURCE INVENTORY

Invariably, action to mitigate ecosystem decline will be done with incertitude and is likely to involve considerable cost, which will devert limited funds and personnel from other natural resources management activities. Managers concerned with a threatened ecosystem must be able to answer questions such as, "what is at stake," "why is a particular action important to protect/restore this system," and "why does it deserve special attention compared to other critically needed actions?" A well documented inventory of ecosystem attributes and their condition within the ecosystem of concern and to a lesser degree within the regional context can help provide answers to these questions. In addition, the resource inventory will provide a critical bench mark from which to judge the degree of change occurring in the system.

ECOSYSTEM *A PRIORI* CONDITION

Often, the ecosystem of concern has experienced extensive degradation before the problem is recognized. As a result, the degraded ecosystem is likely to be in a considerable state of flux in response to a variety of stressors. It will be necessary to estimate with some level of confidence the condition of the system prior to the beginning of degradation in order to establish a baseline of presumed normalcy. Such an analysis provides a "measuring stick" as to the degree of degradation of the system and provides insight into the scope of the restoration task.

A panel of scientists, resource managers, and other individuals representing a variety of complimentary disciplines and backgrounds and knowledgeable about the ecosystem of concern can provide an integrated perspective describing the *a priori* condition of the ecosystem, including its biodiversity and the probable naturally occurring dynamics of ecosystem processes and functions. Often such assessments are conducted only by scientists; however, it is important to not limit the panel to only scientists. Other individuals such as environmental historians, old-time local residents, and stakeholders with different viewpoints should be involved. Too often, exclusion of all but a few select scientists will result in a set of panel findings that have little credibility or acceptance by the resource managers that have to eventually implement the restoration process. Determining ecosystem *a priori* condition will be fraught with uncertainty, but such limitations should not detract from the importance of this exercise.

STRESSORS INVENTORY

Along with natural resources, it is important to inventory perturbations within the ecosystem of concern and the nature and degree of their adverse influences. Invariably, the human induced adverse effects are numerous, even in remote areas. They may be the result of direct manipulation of the environment or a result of human activity outside of the ecosystem of concern. Cause and effect relationships between stressors and ecosystem components or processes can be difficult to determine because often multiple disturbance agents are interacting with the interconnected biological and physical systems in complex and subtle ways. The key point is to think holistically in describing the litany of causes that could be responsible for change of the natural ecosystem condition.

The Southern Appalachian Assessment described in Chapter 7 by Berish et al. is an example of such an exercise to define an ecosystem, identify key stressors, and describe quantifiable adverse effects. This effort was conducted at an extremely broad landscape scale, which is not typical of the scale or resolution of most ecosystem restoration activities. The issue of tropospheric ozone pollution as described by Peine et al. in Chapter 15 and Berish et al. in Chapter 7, is illustrative of the difficulty of defining a cause–effect relationship between a human induced perturbation and a biological response.

RISK ASSESSMENT

A risk assessment is the most critical step to understanding the relative influence of the different disturbance agents that contribute to the decline of the ecosystem of concern. This step provides the rationale for the stabilization or restoration strategy. It is often very difficult to define the degree of influence of various perturbations. It is even more problematic to estimate the cumulative effect of the various perturbations on the ecosystem of concern. Naturally functioning ecosystems are extremely complex and the collective reaction of the many system components to a variety of perturbations is the central question that must be addressed in order to devise a successful restoration strategy. The more obvious the cause–effect relationship, the greater the potential to define the risk and predict the potential benefit of related mitigation strategies.

As discussed by Peine and Berish in Chapter 19, the process of conducting a risk assessment may raise more questions than it provides answers, but the end result will provide a basis for making more informed decisions as to the most appropriate actions to begin mitigation procedures. The risk assessment may focus on rare and endangered species, the status of dominant plant species which provide key habitat or food resources, sensitive ecotones, indicator species for biodiversity, and/or parameters reflecting key ecosystem processes such as soil nutrient status, stream chemistry, or turbidity.

RESOURCE MONITORING

In order to gather data as to the rate and extent of change occurring in the ecosystem, a resource monitoring system must be established. In addition to detecting and evaluating adverse change, such an initiative is critical to measure any progress achieved through mitigation activities and provides a mechanism to judge the effectiveness of the program. As described by Smith et al. in Chapter 8, and Peine and Berish in Chapter 19, the monitoring program should be targeted to indicator species and ecotones identified in the risk assessment to be at risk due to the circumstances afforded by the litany of perturbations present. It should be a balanced measurement among indicator species, the biological community as a whole and parameters which characterize key ecosystem processes and functions. It is also useful to design into the process the ability to make comparative measurements of key parameters within the regional context of the ecosystem of concern and, if practical, with a similar ecosystem more representative of *a priori* conditions.

PLAN FOR STABILIZATION OR RECOVERY

This is the "bottom line" step in the process, formulating a realistic strategy for mitigation with the goal to stabilize and/or restore the ecosystem of concern. The plan should be based on the insight gained from the previous steps of the process. It should provide a holistic perspective, suggesting a series of steps that begin with addressing the most critical threats that can be controlled and then leading to a sequence of recovery phases that will provide a foundation from which naturally occurring ecosystem processes can adequately function and begin the stabilization and restoration process. The plan goals and sequence of steps should be realistic and have a reasonable probability of success in achieving the desired biological and/or geophysical response.

IMPLEMENTATION STRATEGY

This is the action part of the process. It will require periodic review to insure that the stabilization and recovery plan is being followed and that adequate response is being achieved. There will invariably be numerous decisions requiring best judgment given incomplete information; such is the inherent nature of this work. Ecosystem stabilization/recovery requires a long-term commitment by a succession of ecosystem managers willing to share a vision over the long term. It will also be necessary to incorporate a feedback loop in the process in order to modify the plan if necessary.

FRAMEWORK SUMMARY

After studying this proposed framework, it will be obvious to resource managers that a comprehensive plan of resource inventory, analysis of the *a priori* ecosystem condition, inventory of stressors, risk assessment, monitoring of resources, system recovery planning, and implementation will be expensive and very time consuming if done as a step by step process. Several of the steps can and should be done simultaneously. However, haphazard short-cutting of the process may result in inadequate information and not achieving the goal of ecosystem stabilization.

SPRUCE–FIR CASE STUDY

The high elevation coniferous forest dominated by spruce (*Picea*) and fir (*Abies*) species provides one of the most striking ecological features of the mountains of the eastern U.S. and adjacent Canada. Spruce–fir forests are found on the upper slopes of the Adirondack and Appalachian mountains, over an area that extends from northern Quebec and New Brunswick south to western North Carolina. On the very highest of the southern Appalachian mountains is a forest that is much like ones found in Canada, dark and cool, dominated by spruce and fir trees. Southern Appalachian spruce–fir is a distinct variant of the spruce–fir forest type and is dominated by red spruce (*Picea rubens* Sarg.), a species found in all eastern spruce–fir forests, and Fraser fir (*Abies fraseri* (Pursh.) Poir.), a southern Appalachian endemic. Because of the geography and topography of the region, spruce–fir forests in the south are isolated from related northern forests and occur as a series of relic island-like stands on high elevation mountains in southwest Virginia, western North Carolina, and eastern Tennessee (see Figure 1).

The high elevation spruce–fir forests in the southern Appalachians provide an ideal case study to illustrate the difficulty in managing a threatened ecosystem. The ecosystem is a remnant of what was at one time a much larger forest system that dominated the southern Appalachian highlands and adjacent Piedmont during the last ice age. In Great Smoky Mountains National Park, the spruce–fir forests are a major destination of visitors. The spruce–fir forest has been called the crown jewel ecosystem of one of the nation's crown jewel national parks. The ecosystem is largely located on national forests and parks, so federal land managers have primary responsibility for stewardship. The forest has been heavily degraded by a wide array of disturbances ranging from highly destructive land use practices to exotic pests. While this forest has been the focus of research for over 20 years, there has never been a comprehensive interagency resources inventory, risk assessment, holistic monitoring strategy, management plan, or strategy to sustain the gene pool of the unique, isolated populations found on mountain peaks throughout the region. No systematic process has been followed to determine if mitigation measures are warranted, and the few piecemeal mitigation measures implemented to date have not been particularly effective.

Due to a combination of factors this unique ecosystem is deteriorating. In previous chapters in this book, there has been a description of a number of threats or stressors to natural communities. In this chapter we provide a detailed examination of the spruce–fir ecosystem in the southern

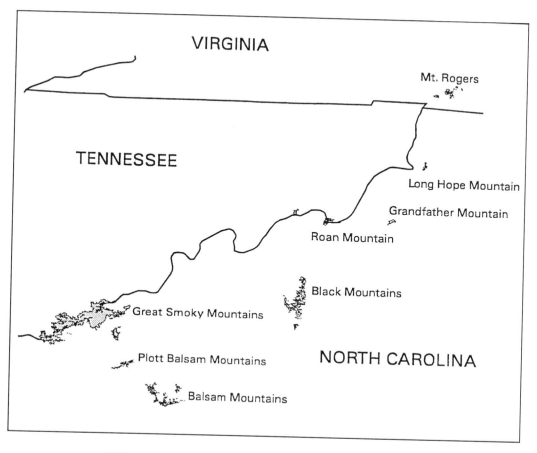

FIGURE 1 Location of southern Appalachian spruce–fir forests.

Appalachians with a focus on how different stressors such as logging, exotic insect infestations, and air pollution may interact at a whole system level. This chapter will first describe the forest system and community dynamics. Understanding these characteristics is critical for understanding how stressors impact the system. After describing the primary stressors that currently impact the ecosystem, we consider ownership and use patterns and discuss current and potential management initiatives that could result in maintenance and/or protection of some of the critical habitat. Finally, a brief consideration for research needs that remain in order to support management decision making will be made and will be followed by a Risk Assessment Report Card that is applicable to the spruce–fir forest.

SOUTHERN APPALACHIAN SPRUCE–FIR ECOSYSTEM

NATURAL HISTORY AND ECOLOGICAL CHARACTERISTICS

Traditionally spruce–fir forests in the eastern U.S. have been described in three different geographically based groupings. The southern Appalachian spruce–fir is usually recognized from Mount Rogers, VA, and southward, and the northern Appalachian ecosystem starts in New York and extends to the north and east. The mid-Appalachian spruce (and very little fir) forests are in the Allegheny Mountains region, primarily in West Virginia and Pennsylvania.

Available habitat for spruce–fir forests is not uniformly distributed along the Appalachians. Appropriate elevations are absent south of 35°N. The southern Appalachians reach their widest (43 miles) and most topographically diverse point at about 35°30'N and have several peaks above 6500 ft elevation. Northward into central and northern Virginia and West Virginia, the Appalachian Mountains become narrower and reach lower maximum elevations (ca. 3280 ft) (White and Cogbill 1992).

These unique southern forests are, in part, a result of their climate. Annual precipitation in the Great Smoky Mountains National Park averages 75 to 90 inches (Shanks 1954; Stephens 1969; Pauley 1989) near the ecosystem's current lower elevation limits (~5000 ft). Precipitation generally exceeds evapotranspiration (Shanks 1954), although some moisture stress occurs in summer and early fall (Pauley 1989). Average monthly temperatures (near 5000 ft elevation) range from a low in January of 29°F to a high in July of 63°F. The growing season is short, generally lasting 100 to 150 days (White and Cogbill 1992). Often enveloped in clouds, the spruce–fir zone has high humidity, high precipitation, and increased moisture inputs via interception of clouds. During winter, rime ice has the potential to cause severe damage to trees above the cloud base (White and Cogbill 1992).

Soils are a function of climate, vegetation, the underlying parent material (bedrock), and the weathering process. Spruce-fir soils are typically naturally acidic due to a combination of high leaching rates (affecting exchangeable acidity) and large accumulations of organic matter. Soils of red spruce–fir forests typically have thick organic horizons, ranging from 2 to 6 inches (Joslin et al. 1992). Depending on parent material and slope characteristics, these organic layers are underlain by (1) shallow bedrock (usually Histosols), (2) thick dark A horizons, rich in organic matter and formed in loamy or fine-textured soils (usually Inceptisols), and (3) ashy gray sands over dark sandy loams (Spodosols) (Kelly and Mays 1989; Joslin et al. 1992)

In order to determine what this forest was like in terms of species composition and forest stocking before logging and other recent stressors were introduced, we can use Whittaker's (1956) study of the vegetation of the Great Smoky Mountains. He found that within undisturbed southern spruce–fir forests species distribution tended to follow an elevation gradient: spruce forest from 4500 to 5400 ft, spruce–fir from 5400 to 6200 ft, and mostly fir forest above 6200 ft. On lower or drier slopes, red spruce shared dominance with yellow birch (*Betula lutea* Mich. f.). At the highest elevations, Fraser fir was often the sole dominant, and mountain-ash (*Sorbus americana* Marshall) often was the only other canopy tree present. Along with the change in composition from low to high elevations and increasing exposure to winds, there was a change in forest structure: average tree size and height decreased as elevation and exposure increased.

Although the overstory tree species composition is simple, the southern spruce–fir ecosystem is rich in rare and endemic (natural populations of species found nowhere else in the world) animal and plant species (White 1984). White and Renfro (1984) list 46 vascular plant species as characteristic of this forest type with 12 of those species found only in southern Appalachian spruce–fir.

The ecosystem's dominant overstory species are highly shade tolerant and adapted to cool, moist environments. Forest stand dynamics in this system are quite different than in other forest types of the southeast. Trees grow slowly and both spruce and fir saplings can remain in the understory for up to 50 years before being released and growing into canopy trees. White et al. (1985) suggest that, while lightning fire and debris avalanches occur, for the most part, small canopy gaps (0.05 acre) dominate the natural disturbance regime of old growth southern Appalachian spruce–fir forests.

Tree replacement patterns are unpredictable from gap size and age, but the gap events are important for species interactions. Most research done on stand dynamics of spruce–fir forests has been carried out on old-growth spruce–fir or almost pure fir stands. However, much of the southern Appalachian spruce–fir areas now consist of young second-growth forests or stands severely impacted by recent insect infestations; stand dynamics in these forests may not be dominated by small canopy gaps. It has been suggested that stand dynamics in second-growth or disturbed forests may occur at a more rapid pace and larger scale than in old-growth stands (Nicholas and Zedaker 1989).

CURRENT STRESSORS: A SYSTEM UNDER SIEGE

Prior to the beginning of the 20th century, there was relatively little human activity in the southern spruce–fir, primarily because of a lack of accessibility and little perceived economic value to European–American man. The climate was harsh, the topography was steep. However, during the last 100 years there have been a number of intense and long lasting disturbances introduced to the ecosystem. We will consider the most well documented: logging, the introduction of the balsam woolly adelgid (*Adelges piceae* Ratz.), and atmospheric deposition. We will also briefly consider the trends and potential impacts of climate change.

Effects of Logging

Land use history and the types of disturbances associated with this past use in the southern Appalachian spruce–fir varies widely by mountain cluster. Because of the inaccessibility of the high mountain forests, the spruce–fir was frequented by few European descendants and/or native Americans before the construction of logging railroads in the late 1800s. Some livestock grazing occurred, and fires occasionally spread into the spruce–fir forest from lower elevation deciduous forest cleared for farming or were accidentally set by hunters (Pyle and Schafale 1988). Overall, pre-logging human disturbances to the spruce–fir were minor compared to the damage caused to the deciduous forests downslope.

When the red spruce forests of the northeast became scarce, logging companies moved south. Substantial logging operations of spruce in the southern Appalachians began in 1905 on Mount Rogers in Virginia and quickly spread south, with the most inaccessible area, the Great Smoky Mountains, cut last (Pyle and Schafale 1988). Logging in the rugged southern Appalachians required large capital outlays for railroad construction and land or timber right purchases. In order to profit, large scale mechanized operations were necessary. Most of the wood was used for paper manufacturing, and home, automobile, and airplane construction. Aided by portable sawmills, logging railroads, and use of cable logging, access to extremely steep slopes was possible.

Why log such steep slopes? The focus of the logging activity was on red spruce. Fraser fir wood qualities are dismal for most wood use purposes. Red spruce is a strong, light-weight wood. Virgin spruce stands contained very high wood volumes. Korstain (1937) described 300-year-old spruce with a diameters of 60 inches and heights of 150 feet. Today we cannot find spruce near that size. Estimates of timber removed from logging operations on the northern slopes of Mt. Rogers range from 50,000 to 100,000 board feet per acre. Other southern Appalachian spruce harvests averaged 15,000 to 20,000 board feet per acre (Pyle and Schafale 1988).

Regeneration back to a spruce–fir forest did not always occur after logging. A tremendous amount of damage was done by clear cutting on very steep slopes. Logging often destroyed the advance spruce and fir regeneration and disturbed much of the soil organic layer. Accidental as well as purposefully set slash fires followed the logging, which often burned the soil organic layer. The high elevations are areas of high rainfall, and substantial soil erosion occurred as a consequence of the impacts of logging. The most exploitative phase of logging ended about 1930, but many areas had already received lasting impacts, with clear-cutting removing the overstory, fire killing most remaining understory and seedlings, and erosion removing much of the shallow soil. These conditions favored replacement of red spruce and Fraser fir with hardwoods such as fire cherry (*Prunus pennsylvanica* L. f.) and yellow birch, since both spruce and fir seeds require organic layers to germinate.

Although estimates vary, the southern spruce–fir forest may occupy less than half of its former area because of failed regeneration due to site degradation following logging (Pyle 1984). The Smokies spruce–fir forests were the least logged, and Pyle estimates that only 25% of the original forest was converted to hardwood forests after the logging era. Of the remaining 50,000 acres of spruce–fir in the Smokies, approximately 80% was never logged (Pyle and Schafale 1988). In total

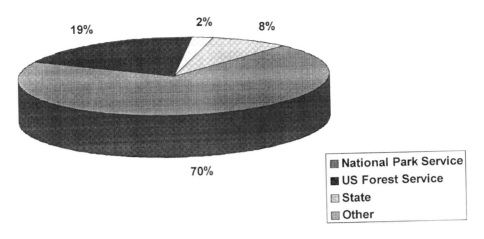

FIGURE 2 Distribution of southern Appalachian acreage by land ownership.

there are less than 70,000 acres of spruce–fir (and spruce–hardwood) forest left in the southern Appalachians (Groton 1985; Dull et al. 1988; Hermann 1996) scattered across eight mountain clusters including Mt.Rogers/Whitetop, VA; Long Hope Mountain; Grandfather Mountain, NC; Roan Mountain, TN; the Black Mountains, NC; the Balsam Mountains, NC; the Plott Balsam Mountains, NC; and the Great Smoky Mountains, TN and NC (see Figure 1). The forest type is restricted to upper reaches of mountains with elevations greater than 4800 ft. Logging records indicate that red spruce used to be found at elevations lower than 4000 ft where the biggest trees were found, but now we are unlikely to find spruce below 5000 ft elevation in areas that were logged.

During the 1920s public outrage flared over the widescale logging and resulting erosion of the southern mountain peaks. The Black Mountains were a popular tourist attraction and were also at the headwaters of the City of Asheville's water supply. Water quality had seriously deteriorated as soil erosion increased. In response to public outcry, federal- and state-level plans were laid to begin purchasing high elevation lands, with the result that more than 90% of southern spruce–fir is currently in the public trust (Figure 2). In response to pressure from the public, some logging operations began planting seedlings in the Black Mountains area where the worst and most visible damage was done. In order to aid in the reforestation of the denuded land, the U.S. Forest Service initiated planting experiments using red spruce, Fraser fir, and 18 nonnative high latitude or high elevation conifer species (Minckler 1940). Only red spruce, Fraser fir, and Norway spruce (*Picea abies* [L.] Karst) survived after 30 years.

BALSAM WOOLLY ADELGID

Tragically, it was those well-intentioned planting experiments that brought another major disturbance factor to the Southern Appalachian spruce–fir: the balsam woolly adelgid. This tiny wingless insect is a native of Caucasia in central Europe and is a pest of all true fir (*Abies* spp.) species. In the 19th century it spread throughout Europe (Balch 1952) feeding on silver fir (*Abies alba* Mill.), while causing only minor damage (Eagar 1984). The adelgid arrived in the U.S. around 1908 in Maine and spread primarily through wind dissemination to northeastern populations of balsam fir (*Abies balsamea* [L]. Mill.). The insect is suspected to have arrived in the southern mountains on silver fir planting stock purchased from northeastern nurseries to be used in reforestation efforts (Nicholas et al. 1992a).

Mature Fraser fir is highly susceptible to adelgid attack, with death occurring in 2 to 7 years (Amman and Speers 1965). Unfortunately, Fraser fir has proven to be the most susceptible of all true fir species, due to an inherent lack of resistance to the adelgid and the tendency for adelgid populations elsewhere in North America to be controlled by much more severe winters than are experienced in the southern Appalachians.

The balsam woolly adelgid is a small wingless insect whose North American populations are entirely female and parthenogenic (Balch 1952; Eagar 1984). An adult lays an average of 100 eggs and as many as 250 (Balch 1952; Amman and Speers 1965). At least two generations of the insect are produced per year in North America (Balch 1952); under more temperate conditions in the south three generations are not uncommon, and as many as four generations per year have been observed (Amman and Speers 1965). Dissemination of the insects within a stand and throughout the southern Appalachians is passive, with wind being the primary vector. Only eggs and first instars can be transported, since all other stages are attached to the host fir.

The insect feeds at bark fissures and lenticels, which allows access to parenchyma cells within the bark. During stylet insertion, the balsam woolly adelgid injects salivary compounds that aid in feeding and, in Fraser fir, prevents initiation of wound response within the bark tissue which would protect the tree from damage (Eagar 1985). Due to the lack of a wound response, the insect secretions diffuse into the stem of the tree. Adelgid saliva contains substances that act like plant growth hormones and alter growth patterns within the fir stem, producing early heartwood formation in sapwood tissues of the stem. Water and mineral transport up the stem is significantly reduced as a result of the formation of premature heartwood. The resulting water stress reduces photosynthesis to near zero. Drought can hasten death of the tree. Smooth-barked young stems are impervious to stem attack (Eagar 1985), but may suffer from crown infestations resulting in symptoms such as increased stem taper. In general, seedlings and small saplings are not as easily killed by the adelgid as mature fir and do not appear to support reproducing adults. Advanced regeneration under predominately fir overstory can be highly damaged due to the large source of insects from above. The seedlings and small saplings that escape damage provide a starting point for recovery.

Without development of environmentally acceptable control methods for the balsam woolly adelgid, we can expect a cycle of limited recovery followed by reinfestation. Eagar (1984) suggested that the amount of time to complete one cycle would be on the order of 35 to 60 years. Factors that will influence the timing of fir mortality within a given location include development of bark characteristics that allow sufficient adelgid feeding sites to support a large population of insects and a source of insects to colonize the stand. With fewer overstory fir than in the pre-adelgid forests, the number of eggs and first instars available for dissemination will be greatly reduced in the future compared to the initial wave of infestation. The probability of a wind blown adelgid landing on fir tree will be considerably lower since there will be few canopy fir trees.

While the adelgid was not actually discovered in the south until the mid-1950s, it had already spread throughout the Black Mountains (Nagel 1959) where the planting experiments were carried out . The Black Mountains are centrally located in the southern Appalachians, which allowed the balsam woolly adelgid to spread to Fraser fir communities throughout the southern Appalachians. The insect reached the Great Smoky Mountains by the late 1950s, and slowly moved west, reaching all parts of the Smokies spruce–fir forest by 1980, first infesting lower elevations and moving upslope. There is evidence that the adelgid reached the Mt. Rogers area by 1962, where both mature and immature fir trees appear to have an increased wound response resistance to the insect compared to Fraser fir growing at other locations; mortality has consequently been slower (Eagar 1984).

For the most part, there was no effort to control adelgid infestations in the southern Appalachians and, with the exception of the Fraser fir population on Mt. Rogers, VA, extensive mortality to Fraser fir occurred within 10 to 25 years after the infestation began in a particular area. In North America there are no known diseases or parasites of the adelgid (Balch 1952; Eagar 1984). Native and introduced predators of the insect have had little effect (Amman 1970). The limited human-applied control efforts to reduce the spread of the adelgid have largely failed. The first infested trees detected in the Great Smoky Mountains were cut down in order to slow the rate of infestation, but these sanitation cuts were soon discontinued when it was discovered that eggs and insects were detached during the felling and could be carried a considerable distance by wind (Ciesla et al. 1965; Lambert and Ciesla 1966).

Various insecticides were tested after the discovery of the adelgid in the southern mountains, and lindane emulsion was found to be the most effective (Amman and Speers 1965). Since the adelgid is a stem-feeder and is not found on the tree's foliage, aerial application techniques do not work, and highly labor intensive spraying of the entire bole by hand is required. Stands in a 200-ft wide margin on either side of access roads were sprayed with lindane annually during the 1960s and 1970s at Mount Mitchell and during the 1970s at Roan Mountain (Eagar 1984). This practice was discontinued due to cost as well as concerns about lindane's toxicity and persistence in the soil. Lindane is no longer available for this type of use. A less toxic alternative, potassium oleate soap, is currently applied annually to stands in the Great Smoky Mountains National Park around the parking lot at Clingmans Dome and the trail to the observation tower and roadside-accessible stands on Balsam Mountain (Eagar 1984; K. Johnson, pers. commun.). Currently, the program is budgeted on an annual basis and may be eliminated in the near future because of cost, limited effectiveness (only one generation of the insect is impacted), and the negative reaction of Park visitors when Clingmans Dome is closed for a week for treatment (K. Johnson, pers. commun.).

The impact of the balsam woolly adelgid on spruce–fir forests of the southern Appalachians has been severe and extensive resulting in dramatic change to this ecosystems. Because very few overstory tree species other than Fraser fir grow at the highest elevations, forest biomass has seriously declined (Busing et al. 1988; Busing and Clebsch 1988). In a comparison of data collected in the Great Smoky Mountains in 1986 with measurements from 40 years earlier by Oosting and Billings (1951), Nicholas et al. (1992a) found that live overstory densities and volume in 1986 decreased with increasing elevation, similar to 1946 patterns. However, live fir values were only 9 to 62% of those recorded earlier. When recently dead standing fir trees were added into the calculations for the 1986 data set, fir levels of the two studies compared fairly well. Red spruce volumes increased slightly over the 40 years.

The removal of mature Fraser fir from the canopy has profound implications for the spruce–fir ecosystem. Both red spruce and Fraser fir are highly shade tolerant species which are adapted to dark, moist closed canopy forests. Spruce is exposed to greater amounts of wind and sun with detrimental effects to tree growth and forest stand density (Nicholas et al. 1992a). Conversely, on sites with a low proportion of fir, spruce may respond positively to the loss of competing fir (Zedaker et al. 1988; Reams et al. 1993). Other, mostly understory, tree species have been reported to increase in numbers over time since canopy opening, including: striped maple (*Acer pensylvanicum* L.), mountain holly (*Ilex ambigua* var. *montana* [T. & G.] Ahles), mountain maple (*Acer spictum* Lam.), serviceberry (*Amelanchier arborea* var. *laevis* [Wiegand] Ahles), yellow birch, and fire cherry (DeSelm and Boner 1984). Responses of mountain-ash to increased exposure are uncertain because of significant mortality rates observed in the late 1980s. Damage from the exotic mountain-ash sawfly (*Pristiphora geniculata* Hartig) is thought to be an important factor (Nicholas 1992).

Understory changes due to adelgid infestation have also been quantified. Shrubs species which have increased markedly in density or cover in adelgid-disturbed areas include elderberry (*Sambucus pubens* L.), blueberry (*Vaccinium erythrocarpum* Michaux), hobblebush (*Viburnum alnifolium* Marshall), and especially blackberry (*Rubus canadensis* L.) (Boner 1979; DeSelm and Boner 1984). Blackberry density, in fact, was found to increase up to nearly tenfold in areas that had experienced increased canopy openness for up to 2 decades (Boner 1979), most likely invading in response to increased levels of light and temperature (Busing et al. 1988). Nonvascular species changes have also been observed. Nearly half of the 203 known bryophyte (mosses or lichens) species found in the Great Smoky Mountains spruce–fir forests are predicted to be affected by changes in the canopy (Harmon et al. 1983; Smith 1984), including the rock gnome lichen (*Gymnoderma lineare* [Evans] Yoshimura and Sharp), recently federally listed as an endangered species (USDI Fish and Wildlife Service 1995a).

In addition to impacts on vegetation, relative abundances of animals are affected by the demise of Fraser fir. Changes in canopy structure and fir seed availability may adversely impact some species of birds and mammals (Hall 1989). Increased amounts of sunlight reaching the forest floor may desiccate and warm the soil of this normally moist and cool habitat, possibly impacting species of salamander and invertebrates which live within the forest floor (Matthews and Echternacht 1984). One example is the recently listed federally endangered spruce–fir moss spider (*Microhexura montivaga*) which is probably declining because of desiccation of the moss mats in which it lives (USDI Fish and Wildlife Service 1995b). Other currently (March 1997) federally listed threatened and/or endangered animals found in the southern Appalachian spruce–fir ecosystem include the Carolina Northern Flying Squirrel *Glaucomys sabrinus coloratus*) and the Virginia Northern Flying Squirrel (*Glaucomys sabrinus fuscus*).

Ecosystem processes, such as nutrient cycling, energy flow, and disturbance regimes, may also be affected. Decomposition processes, and thus the rate of nutrient cycling, may accelerate with greater soil warming. Large amounts of dead fir wood provide greater nutrient influxes over the short term. Such changes in ecosystem processes are not limited to the spruce–fir forests, but could impact water quality since these high elevation sites are found at the headwaters of important southern Appalachian watersheds, including 17 of 28 major watersheds in the Great Smoky Mountains National Park (Pyle 1984) and the Asheville, NC municipal watershed. Increased export of nutrients as well as mobilization of potentially toxic elements like aluminum from soils could occur due to alterations in biogeochemical cycling and large inputs of organic matter. Potential changes in disturbance regimes include the greater possibility of windthrow and ice storm damage discussed above and increased probability of fire. Adelgid-caused mortality has increased fuel loading in spruce–fir forests with greater fuel levels occurring at higher elevations (Nicholas and White 1985).

AIR POLLUTION

Scientists and resource managers are concerned about more than just former logging practices and exotic species invasions. Reports of red spruce decline in the northern Appalachians (Siccama et al. 1982) have been attributed by many to air pollution and has raised concerns over the possible impact of air pollution on southern Appalachians spruce–fir forests (White 1984). Two forms of pollution that are of greatest concern are ozone and acidic deposition of sulfur and nitrogen compounds (commonly called acid rain).

Higher elevation sites tend to have greater atmospheric concentrations of ozone because much of the time they remain above the surface boundary layer where ozone depletion occurs at night (Mueller 1994). Based on the effects of ozone on other conifer species, such as loblolly pine (*Pinus taeda* L.), researchers were concerned that ozone may damage spruce chlorophyll or the wax layer on needle surfaces, thus leading to reductions in photosynthesis, decreased cold tolerance, and/or alterations in carbon allocation. Laboratory and field studies, however, have shown that red spruce and Fraser fir are not very sensitive to ambient ozone concentrations and do not support ozone as a casual factor in spruce decline (Tseng et al. 1987; DeHayes et al. 1991; McLaughlin and Kohut 1992; Thornton et al. 1994).

Acidic deposition to a forest occurs via three main pathways: (1) precipitation or wet deposition, where material is dissolved in rain or snow, (2) dry deposition, involving direct deposition of gases and particles (aerosols) to any surface, and (3) cloud water deposition, involving material dissolved in cloud droplets, which is deposited when cloud or fog droplets are intercepted by vegetation including forest canopies.

The high elevations of the Southern Appalachians are exposed to some of the highest pollution loadings of any place in North America due to the high amounts of precipitation, frequent immersion in clouds, and the large surface area of the needle-leafed spruce and fir trees which are efficient collectors of cloud moisture. Cloud water has higher concentrations of sulfate and nitrate than precipitation and an average pH of 3.5 (precipitation only is 4.2). Consequently, cloud water deposition increases the amount of acidic deposition by 2 to 4 times compared to low elevations that receive only rain (Mohnen 1992). Both direct canopy exposure and deposition to the forest floor have important implications for forest health.

The combination of high levels of acidic deposition and the inherent characteristics of the soil in southern Appalachian spruce–fir ecosystems has raised concern about changes in nutrient availability and potential effects on forest health. The soils of spruce–fir ecosystems are naturally acidic and low in essential base cations. Consequently, the high levels of acidic deposition received by these systems leaches base cations from the soil and increases the amount of aluminum in solution (Johnson and Fernandez 1992). Soil solution aluminum concentrations similar to those observed in spruce–fir forests inhibit the uptake of calcium and magnesium in controlled studies with spruce seedlings (Thornton et al. 1987). Application of calcium to spruce saplings growing in the Smokies caused an increase in growth and positive physiological effects in the foliage compared to untreated controls (Joslin and Wolfe 1994; Van Miegroet et al. 1993).

Another forest nutrition problem associated with acid deposition concerns is "nitrogen saturation" or nitrogen availability in excess of plant demand. Several pieces of evidence support a hypothesis of nitrogen saturation in spruce–fir ecosystems: soil solution nitrate levels are high throughout the year, leaching losses are in balance with atmospheric nitrogen deposition rates, and trees have not responded to experimental nitrogen fertilization (Johnson et al. 1991; Van Miegroet et al. 1992; Joslin and Wolfe 1994). The old-growth forests of the Smokies have characteristics that contribute to nitrogen saturation, including a large supply of nitrogen in the soil, which favors microbial conversion of nitrogen to more available forms, and low tree growth rates which reduces the demand for nitrogen. The recent death of Fraser fir also may contribute to reduced demand for nitrogen and, by opening the canopy, cause soil temperature to rise which would increase microbial conversion of nitrogen. Stream exports of nitrate from spruce–fir forests in the Great Smoky Mountains are high throughout the year and contribute to chronic and episodic stream acidification and overall reduction of water quality (Flum and Nodvin 1995; Nodvin et al. 1995).

Laboratory and field experiments which excluded ambient cloud water from red spruce seedlings and branches of mature trees have suggested several processes by which atmospheric deposition may contribute to spruce decline. Controlled laboratory exposure to ambient cloud water reduced spruce seedling foliar concentrations of essential plant nutrients such as calcium, magnesium, and zinc. Three possible mechanisms have been suggested: foliar leaching by acidic cloud water, soil leaching and reduced nutrient uptake, and interference of essential base cation uptake by elevated soil aluminum levels (Thornton et al. 1994). Ambient cloud deposition has been found to reduce the cold tolerance of red spruce seedlings (DeHayes et al. 1991) and mature branches (Vann et al. 1992) in the Northeast but not in the southern Appalachians. Results of acid deposition studies have also produced contradictory results regarding effects on photosynthesis and growth (Kohut et al. 1990; Patton et al. 1991; McLaughlin et al. 1993; Thornton et al. 1994).

Red spruce is the only tree species in the U.S. for which there is consensus agreement that acidic deposition contributes to decline conditions (Eagar and Adams 1992). However, some researchers are doubtful as to whether southern Appalachian red spruce is in significant decline (Busing et al. 1988; Nicholas et al. 1992a). Damage to red spruce is greater in the northeastern U.S. where harsh winter conditions interact with physiological changes caused by acidic deposition resulting in frequent winter injury of foliage, loss of needles from the crown, and significant mortality in some areas. While there were indications of spruce canopy deterioration in the southern Appalachians during the late 1980s, this coincided with a significant 3-year drought (Peart et al. 1992). Forest stands revisited 5 years after the original sampling ended showed only limited signs of recovery (Figure 3). Southern Appalachian populations of red spruce have not experienced tree mortality rates (Nicholas 1992; Busing and Pauley 1994) similar to those in the Northeast. However, as noted above, there are data that relate changes in nutritional status and physiological condition of red

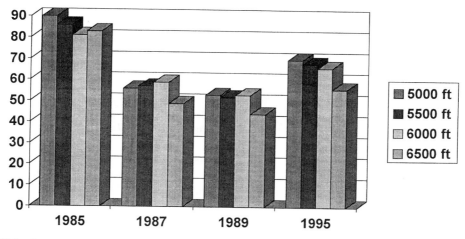

FIGURE 3 Proportion of red spruce with healthy crown condition in the Great Smoky Mountains between 1985 and 1995 at four different elevations.

spruce to exposure to acidic deposition Over time these subtle changes in red spruce could result in more obvious signs of adverse impact such as increased susceptibility to insects and pathogens, reduced growth, or poor response to changes in the environment.

There has been considerable debate over recent changes in the growth rate of red spruce in the southern Appalachians. The first tree-ring data for red spruce indicated a regional growth decline beginning about 1965 for trees growing above 6000 feet that could not be explained by climate variables or stand dynamics (McLaughlin et al. 1987). This study suggested that other environmental factors, including air pollution, might be causing these reductions in growth. More recent studies have suggested that the current decrease in incremental growth of southern red spruce is not greater than historical levels (Cook and Zedaker 1992) and have suggested that the 1965 downturn is part of a long-term, climate-induced cyclic fluctuation (Reams et al. 1993). In support of this idea, Reams et al. (1993) pointed out that a local maximum in radial growth preceded the 1960s decline and concurred with LeBlanc et al. (1992) that the recent downward trend is not unusual when compared with data from the past 200 years. McLaughlin et al. (1994) raised concerns about the Reams et al. (1993) study with regard to their method of analysis. The different analyses indicate that environmental and stand level factors that influence tree growth are complex and often subtle. Different methodological approaches also contribute to understanding tree-ring data. Even in the Northeast where there is consensus on how acidic deposition contributed to red spruce decline, there is still scientific debate about tree-ring data. Even the authors of this chapter have different interpretations of the tree-ring data from the Smokies.

It has been suggested that atmospheric deposition may weaken or predispose Fraser fir to attack by the balsam woolly adelgid. Hain and Arthur (1985) noted that silver fir in Germany seems to be highly pollution sensitive and that areas with greatest adelgid-induced mortality, the Great Smoky, the Black, and the Plott Balsam Mountains, are highest in elevation and closest to the major population centers of Knoxville, Tennessee, and Asheville, NC. However, since their study, other areas in the southern Appalachians have experienced high fir mortality. Hollingsworth and Hain (1991) and McLaughlin et al. (1997) proposed that calcium deficiencies in fir caused by leaching of basic cations from the soil may increase fir vulnerability to the adelgid. Since calcium may be limiting for cambial growth, infested trees lacking the nutrient may have a slower and less complete wound response to adelgid attack. This is currently only a hypothesis and there are no published data that support or reject this hypothesis.

Climate Change

Before we consider possible future potential impacts of climate change, we must first consider the past impacts of climate change and the forces that resulted in an isolated subalpine forest type left on the mountaintops of the southern Appalachians. Whittaker (1956) suggested that this ecosystem was a refuge sanctuary about 8000 to 4000 years ago when the climate was warmer than at present. Whittaker speculated that during this period, spruce–fir forests were restricted to elevations above 5600 to 5800 ft. As the climate cooled during the late Holocene (approximately 2000 years ago), the upward migration of spruce and fir stopped, and remaining populations were able to expand downslope (Delcourt and Delcourt 1984).

This theory may explain the absence of spruce–fir populations on some southern Appalachian mountains that seemingly have high enough elevation to currently support spruce–fir forests. Of the ten mountain areas that have peaks surpassing 5500 ft elevation, seven have a well-developed spruce–fir forest (Mount Rogers, VA; Grandfather Mountain, NC; Roan Mountain, TN and NC; the Black Mountains, NC; the Balsam Mountains, NC; the Plott Balsam Mountains, NC; and the Great Smoky Mountains, TN and NC), and an eighth has a well-developed spruce forest (Whitetop, VA). Both vascular plant and rare species diversity on these high peaks appears to be related to the size of the forest stand, or "island size" (amount of area above 5500 ft) (White 1984), which offers support to the idea that the southern Appalachian spruce–fir forest is a remnant of past climate change.

Given that the southern spruce–fir ecosystem exists at the edge of its geographic range, isolated on the tops of the highest mountains, the obvious question remains: how will this ecosystem respond to future climatic events? In fact, the spruce–fir forest of the southern Appalachian Mountains is geographically, topographically, and ecologically positioned to serve as a sensitive indicator of subtle changes in both chemical and physical climate.

Average global temperatures have been predicted to increase by 3 to 8°F over the next century because of the carbon dioxide-induced greenhouse effect (Root and Schneidner 1993). Expected regional-scale climatic changes could include modification of the seasonal distribution of air masses, locations of storm tracks, and intensification of frequency and strengths of storm events such as hurricanes (Schneider 1989). Regional projections of future greenhouse gas-induced climatic warming suggest that spruce–fir forests may become extinct in the southern Appalachians (Delcourt and Delcourt, in press). These predictions do suggest, however, that spruce forests would continue to be perpetuated in the mid-Appalachians. It has also been suggested that red spruce, Fraser fir, and associated species may be even more sensitive to predicted increases in global temperatures because of the possibility that atmospheric deposition may be causing slower growth rates at the highest elevations (McLaughlin et al. 1997). While these climatic change predictions are somewhat speculative and may sound alarmist, the take home message should also include the consideration that regional climate has always been dynamic and that the hypothesized greenhouse effect could accelerate the process.

Ownership and Use Patterns

By combining information from Dull et al. (1988), Groton (1985) and Herrman (1996), the southern Appalachian spruce–fir (and spruce–hardwood forest) covers about 79,000 acres. The map in Figure 1 demonstrates the disjunct nature of this forest type. The majority (65%) of the spruce–fir landbase is found in the state of North Carolina (Figure 4). The largest land owner is the National Park Service (Figure 2), with spruce–fir forest lands found in the Great Smoky Mountains National Park and along the highlands of the southern portion of the Blue Ridge Parkway. There is relatively little acreage of state held forest lands; however, all three states have state parks that hold some spruce–fir highlands. Almost 8% of the resource is privately held, including Grandfather Mountain and Long Hope Mountain.

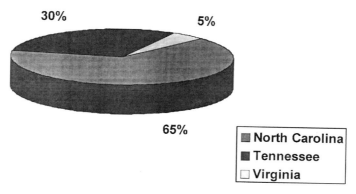

FIGURE 4 Distribution of southern Appalachian acreage by state.

The disjunct nature of the forest type across several different ownerships with different management objectives makes ecosystem level resource planning very difficult. To illustrate this point, we briefly examine Mt. Rogers/Whitetop Mountain as a case study. The Mt. Rogers/Whitetop Mountain area is the northernmost part of the southern Appalachian spruce–fir ecosystem. The Mt Rogers/Whitetop area makes up less than 2% of the entire ecosystem, yet is managed by two different land management agencies, including the USDA Forest Service as part of the Jefferson/George Washington National Forest and the Grayson Highlands State Park of Virginia. There are two disjunct parts to this forest: the spruce–fir on Mt. Rogers, and the spruce forest on Whitetop Mountain. They are separated by 3 miles of northern hardwoods forest and heavily grazed pasture, owned by the Forest Service, and State Route 58.

Both the state park and the U.S. Forest Service lands are managed for multiple types of recreation. Much of the Mount Rogers area was originally cleared by logging activity and nearby residents used the land for grazing cattle. Grazing leases in the highlands are still held and hikers to the area share the terrain with cattle. The U.S. Forest Service periodically (about every 3 years) burns a portion of the highlands in order to maintain grassy areas to facilitate a myriad of visitor uses including hunting, grazing, and aesthetics for some users. Visitors to the Mt. Rogers highlands describe the area as similar to Montana with open meadows, subalpine vegetation, and rock outcrops. The state park and U.S. Forest Service have fenced off areas that are burned and/or grazed from areas that are not to be disturbed; however, fences are not always well maintained and it is not at all unusual to encounter a small herd of cows in the midst of a spruce stand. Cattle are not the only nonnative grazers in these highlands. A herd of wild ponies, descendants of the Assateague Island, NC herd released in the Mt. Rogers area years ago, also roam the highlands. These ponies are managed by a local group, the Wilburn Ridge Pony Association, which has annual round-ups to sell some of the animals as pets each autumn. All of this activity takes place right up to the edge of designated national wilderness lands, since the Lewis Fork Wilderness Area covers the top of Mt. Rogers.

Much of the Cabin Ridge area of Mount Rogers can hardly be described as virgin forest, even before modern day recreational use. In the 1950s the area was developed as a Christmas tree farm and transplanted fir seedlings were set out in rows of 3 feet apart. When the tops of the trees were harvested, (every 5 years), the lowest branch was left to assume apical dominance. This resulted in trees with a sprouted appearance (Pyle and Shaffale 1988). Christmas tree cutting on Mt. Rogers officially ended in 1972. However, as mentioned earlier, Fraser fir on Mt. Rogers seem to be more resistant to the balsam woolly adelgid and fir from the Mt. Rogers area is much sought after in the Christmas tree industry. Illegal harvests of seedlings and seed have been reported over the years. The U.S. Forest Service has also been periodically pressured to give permission to commercial

Christmas tree growers to dig up Fraser fir seedlings and to harvest fir cones to replenish nursery stock. Fraser fir produces cones on the average of once every 2 to 4 years. Recent studies indicate that a very low percentage of seeds produced are viable (Nicholas et al. 1992b). Combine the pressures of grazing, fire, illegal tree cutting, the balsam woolly adelgid, and seedling and cone harvesting, and it would seem that Fraser fir and associated species on Mt. Rogers face enormous pressure. Given all the diverse user groups of the Mt. Rogers/Whitetop Mountain area, how can a land management agency accommodate everyone and keep a very fragile resource sustainable?

CURRENT AND POTENTIAL MANAGEMENT INITIATIVES

There has been very little management of spruce–fir forests in the south since these lands were purchased by federal and state agencies aside from protection of publicly held spruce–fir from logging and mining since becoming public lands. There were limited spray operations for control of the balsam woolly adelgid on Roan Mountain and Mt. Mitchell State Park in the early 1970s. Other than the routine burning of parts of Mt. Rogers National Recreation Area, and the National Park Service's limited spraying for adelgid in the Great Smoky Mountains, active land management in the spruce–fir is almost nonexistent today. As discussed earlier, most of the spruce–fir ecosystem is found on state and federal public lands. A small portion (8%) remains in private ownership. Most of the spruce–fir in private ownership is benignly managed for tourist or recreational use, such as Grandfather Mountain, a privately owned USMAB biosphere reserve with picnic areas, hiking trails, a nature museum, and enclosed animal habitats.

There was an effort in 1980 to develop a comprehensive research and management plan for Fraser fir that would address concerns and needs of the different land manager and land owners. The Southern Appalachian Research/Resource Management Cooperative (SARRMC) established a "Technical committee on Fraser Fir and the Balsam Woolly Aphid (Adelgid)" which held a workshop in May 1980. Participants included scientists, land managers from federal and state agencies, Christmas tree growers, and private land owners. The workshop addressed: (1) the status of Fraser fir, (2) short-term protection and management needs, and (3) long-term protection and management needs. A problem analysis with recommendations was produced and distributed by SARRMC to interested parties. Recommendations included:

1. Initiate and fund research on techniques to control the balsam woolly adelgid, understand changes in the ecosystem, fire hazard potential, Fraser fir genetics and relationship to resistance/tolerance, seed production areas and restoration techniques
2. Administer the research program through one office, designated to supervise research projects and to coordinate funding
3. Educate the public and disseminate information relative to Fraser fir and balsam woolly adelgid
4. Place Fraser fir seed in the National Seed Bank to preserve its germplasm
5. Sponsor an international symposium on Fraser fir and the balsam woolly adelgid

Unfortunately, the coordinated effort outlined in the report was never accomplished. Individual agencies used some recommendations from the workshop, mainly in the area of insect control and public education. The hard issues concerning restoration and genetic preservation were never addressed. Perhaps there might have been greater success at establishing a multiagency program if there had been more focus on management and less on a research agenda; however, at the time there were many information gaps that needed to be closed. Also funding was not available to implement the research agenda and no specific agency or individual assumed the leadership needed to develop the program.

Many of the issues and recommendations raised in the 1980 Problem Analysis are still highly relevant. Suggestions have been made recently to develop a more active management policy in the spruce–fir ecosystem, but as the SARRMC effort illustrates, one overriding question must be first asked. How can a comprehensive management plan be effective when the resource is spread in small disjunct parcels across three national forests (Cherokee, Jefferson–George Washington, and Pisgah); one national park (Great Smoky Mountains); a national parkway (Blue Ridge Parkway); three state parks (Grayson Highlands in Virginia; Mt. Mitchell State Park in North Carolina, Roan Mountain State Park), a city reservoir watershed (City of Asheville); and other landowners? Furthermore, a comprehensive management plan can only be effective if action strategies are developed for realistic goals. Several discrete goals have been suggested including:

1. Reduce negative impacts of the Christmas tree industry on Fraser fir native stands
2. Provide opportunities for the Christmas tree industry to continue to have a viable Fraser fir seed source
3. Provide gene pool protection for species in the southern Appalachian spruce–fir ecosystem that are threatened by loss of Fraser fir
4. Develop and implement restoration action plans to mitigate the impact of the multiple stressors on the system and in particular the loss of canopy Fraser fir trees
5. Continue to provide multiple forms of recreation opportunities for the public
6. Examine potentially varying land management scenarios if privately held lands change owners

The Fraser fir Christmas tree industry is an important small business opportunity for mountain landowners. Past practices of using public lands on Mt. Rogers or Roan Mountain as a primary seed source cannot be a long-term option either in terms of supply/demand aspects or impact on the ecosystem. The continued loss of Fraser fir reduces the seed output, while at the same time demand is increasing. Harvest of seed for Christmas tree production reduces the natural regeneration of Fraser fir that will provide the replacement stand for the trees being killed by the balsam woolly adelgid. This issue requires cooperation between state agencies and private individuals in terms of developing an approach that will protect an important economic opportunity and a unique forest ecosystem.

Long-term species viability is determined by its ability to adapt to the changing environment. Broad genetic characteristics will provide the best opportunity to adapt. The endemic nature of Fraser fir combined with its disjunct distribution in the southern Appalachians puts constraints on the genetic characteristics of Fraser fir. Loss of individuals or populations of Fraser fir, due to adelgid-caused mortality further reduces the genetic variability of the species. There are compelling reasons to initiate a program to protect the gene pool of Fraser fir as well as other species that may be at risk due to changes in habitat. There are also many questions to be answered in terms of how best to do this and how to insure the long-term success of the effort. There are also bioethical questions that need to be addressed. For example, since the Mt. Rogers Fraser fir appear to be less susceptible to adelgid attack than fir from other areas, should this gene pool be introduced to other areas in any type of restoration effort? Should there be a concerted effort to breed a more resistant Fraser fir?

The southern Appalachian spruce–fir ecosystem is currently in a state of significant flux. The combined impacts of balsam woolly adelgid infestations and changes in soil nutrient status due to acid deposition have the potential to permanently change inherent biophysical attributes which are fundamental to the sustainability of the spruce–fir ecosystem. In areas like Great Smoky Mountains National Park where there are large areas of virgin forests, the potential effects of these changes on the old-growth red spruce are a major consideration. In other areas that experienced past logging and where the forest is composed of mainly Fraser fir, there is the potential for conversion to hardwood forests. Some of the causes of stress, such as acid deposition, are beyond the direct control of land managers and are part of regional and national public policy. Other concerns can be managed at the local level, but would benefit from a multiagency and regional framework, while being sure to include private landowners in these discussions. A key consideration is to recognize the interactions that are present between both stressors and ecosystem response as well as the subtle, cumulative nature of the impacts.

The drastic changes due to the death of Fraser fir is having and will continue to have repercussions on recreation and visitor satisfaction. For visitors from the south and midwest, the high elevation spruce–fir forest has provided a unique opportunity to experience a setting common to the forest of northern New England or Canada. Observation towers are located on the two highest mountains in eastern North America: Clingmans Dome in the Smokies and Mt. Mitchell in the Black Mountains. These scenic vistas are now surrounded by large expanses of dead and dying fir trees. A similar scene is encountered along portions of the Blue Ridge Parkway. Backcountry users are exposed to a highly degraded forest and also pose a direct threat. The dead fir provide a very volatile fuel source (Nicholas and White 1985) and use of fire by backpackers or hikers could supply the source of ignition. Former nature trails that described the unique ecological characteristics of spruce–fir forests now provide public education on the balsam woolly adelgid. Managers face many challenges in meeting visitor needs and expectations and protecting the resource from other threats.

RESEARCH NEEDS TO SUPPORT MANAGEMENT DECISION MAKING

Southern Appalachian spruce–fir forests have received a considerable amount of scientific inquiry over the past 20 years, with the bulk of it focused on effects of air pollution. We are beginning to gain an understanding of impacts on red spruce. However, most of our understanding is based on short-term studies (3 to 5 years) of phenomena that occur over several decades, may produce subtle, long-term impacts, and are very complex. Additionally, most of the research has only studied the effects of a single stress factor at a time, but the ecosystem as a collection of parts and as a whole is exposed to multiple stresses. Changes caused by the two key stressors — past land use and exotic species — already have brought about profound changes in ecosystem structure and function. It is important for managers and policy makers to have better information on how these changes will impact other ecosystem components and processes besides red spruce and Fraser fir.

During the past 12 years we have had numerous discussions with the local supervisors at agencies charged with managing responsibility of the spruce–fir ecosystem. The structure of some of the land management agencies is such that they have little authority and/or funding to carry out research. In order for research to be useful to land managers, it must be focused on specific problems and the science must be credible. The scientific process does not often provide quick answers, and frequently managers become frustrated with the progress of research. There are often hard decisions that need to be made between spending money to gain more scientific understanding or investing that money in management activities. The main research needs identified by managers include:

- Effects of atmospheric deposition on soils and streams systems
- Effects of ozone exposure on species composition and stand dynamics
- Effects of cumulative, multiple stresses on forest structure and function
- Effects of cumulative, multiple stresses on animal populations and habitats
- Approaches for restoration of Fraser fir
- Impact of exotic plant species on genetic characteristics of native species
- Development and implementation of an integrated monitoring program

A CALL FOR ACTION

There are a number of reasons why there has been little progress in developing and implementing an ecosystem management strategy for the southern spruce–fir. It is often stated that we lack sufficient understanding of the problem and possible management options. As detailed in this chapter, there has been intensive research done on the deterioration of the southern Appalachian spruce–fir forest. The 1980 SARRMC workshop on Fraser fir and the balsam woolly adelgid made specific recommendations that were based on an extensive scientific understanding. The National Acid Precipitation Assessment Program's Forest Response Program and the Tennessee Valley Authority's Spruce–Fir Forest Inventory plot system (see Chapter 8 by Smith et al. for a discussion of both research and monitoring plot networks) have resulted in substantial baseline monitoring and forest health assessments. There can no longer be an excuse that "little is known or understood" about the ecosystem used in continuing to defer management activity.

Another reason there has been little active resource planning in this forest type is that different views are held by resource managers, scientists, and environmentalists on the relative influence of different stressors such as exotic species, air pollution, and past land use practices on the resulting forest deterioration. It is unlikely that there will ever be agreement among different groups about this issue. However, most involved individuals do agree that several stressors are involved in the short- and long-term deterioration of the system, and therefore, addressing just one problem will not "fix" the resource. Disagreement about causation should not stop strategy planning.

One of the most critical issues that has been an impediment to resource management planning for this ecosystem has been a lack of funding. Therefore, it is essential that a well-designed management strategy with multiple stakeholder input be advanced that includes detailed documentation of the deterioration of the ecosystem and of the significant public desire for resource preservation. Once a clearly articulated plan is developed, there is greater probability that funding authorities will be more interested in prioritizing and supporting spruce–fir activity plans.

The final issue is one of leadership. An individual, organization, or agency must be committed to developing a coordinated management plan and set the agenda for bringing the managers, stakeholders, and scientists together. One suggestion would be to use SAMAB as a mechanism for providing organizational assistance for a planning process that includes all the public land managers. Private land owners would be invited to the process. The role of SAMAB in protecting this ecosystem could be one of providing facilitation and coordination for a planning process, and encouraging communication among the diverse ownerships and stakeholders. Additional possible roles are coordinating continued monitoring and long-term research, and providing mechanisms for technology transfer and education. However, whether or not SAMAB can be useful for such a process is not the major issue. What must be done is a concentrated effort of the spruce–fir ecosystem land managers in coordinating discussion and developing strategy regarding this deteriorating resource.

RISK ASSESSMENT REPORT CARD

The highly threatened spruce–fir ecosystem provides an interesting case in which to apply the *Risk Assessment Report Card for ecosystem managers*, as referenced in other chapters in Section III of this volume. Like the Clinch–Powell restoration project described by Yankee et al. in Chapter 20, there is a high degree of uncertainty and risk associated with this threatened ecosystem issue. The report card on the spruce–fir ecosystem is judged to be as follows.

Vision: D

There has not been a comprehensive vision as to management responsibility for the ecosystem.

Resource risk: D

The ecosystem is already in a serious state of deterioration.

Socioeconomic conflicts: C

Loss of Fraser fir has significant ramifications for the local Christmas tree industry. Also, the scenic quality of the high elevation forests is highly degraded adversely impacting the quality of the visitor experience in Great Smoky Mountains National Park, along the Blue Ridge Parkway, in Mt. Mitchell State Park, and on other public lands.

Procedural protocols: C

There are protocols for some monitoring activity but none for potential mitigation activities.

Scientific validity: B+

The ecosystem has been *intensely* studied for the last 2 decades, providing detailed insight as to the condition of the forest and the impact of the various perturbations.

Legal jeopardy: A

No legal jeopardy is anticipated.

Public support: A+

The public has repeatedly expressed concern for the health of the spruce–fir forest.

Adequate funding: C

Significant funds for research have been expended. Little has been expended on management activities in recent years, but that may be partially due to a lack of an accepted mitigation activity plan.

Policy precedent: A

The emergence of ecosystem management as a formal accepted concept in public lands planning and the goals of SAMAB suggest a solid policy precedent for acting aggressively on this issue.

Transferability: C

The greatest potential for transferability is to learn from the process of evaluation of conditions, conduct of the risk assessment, and the development of a strategic plan for stabilization and/or recovery.

The current situation of the high elevation spruce–fir ecosystem presents a typical dilemma for ecosystem managers. Even though the ecosystem has been highly threatened for some 30 years, the disjunct nature of the forest type across several different land management agencies with different management objectives has resulted in little regional-level planning. The stewards of the ecosystem have not been able to collaborate in order to conduct a comprehensive risk assessment nor systematically explore potential mitigating activities. Furthermore, there has been some tendency for the media to characterize the causes of the deterioration of the forest in a simplistic and misleading manner, giving the public little understanding of the multiple stressors of the system and the complex task facing the resource managers.

On-site research and management action to date have been piecemeal. This may in part be due to the fact that the solution to the complex dilemma being faced is anything but obvious. As is repeatedly illustrated by other ecosystem management issues documented in other chapters in Section III of this volume, the primary role of those ultimately responsible for the stewardship of the threatened spruce–fir forest is to provide *leadership* to initiate a holistic and comprehensive management options analysis followed by mitigating actions. As of this writing, the issue has, for the most part, been left to scientists to euphemistically "study it to death."

ACKNOWLEDGMENTS

The writing of this chapter was, in part, supported by funding from the Environmental Protection Agency, the USDA Forest Service, and the Tennessee Valley Authority. This chapter has not been subject to any agency policy review and should not be construed to represent the policies of any of those agencies.

REFERENCES

Amman, G. D. 1970. Phenomena of *Adelges piceae* populations (Homoptera: Phylloxeridae) in North Carolina. *Ann. Entomol. Soc. Am.* 63: 1727–1734.

Amman, G. D. and C. F. Speers. 1965. Balsam woolly aphid in the southern Appalachians. *J. For.* 63: 18–20.

Balch, R. E. 1952. Studies of the balsam woolly aphid, *Adelges piceae* (Ratz.) and its effects on balsam fir, *Abies balsamea* (L.) Mill. Canadian Dept. of Agriculture Publ. 867.

Boner, R. R. 1979. Effects of Fraser Fir Death on Population Dynamics in Southern Appalachian Boreal Ecosystems. M. S. thesis, University of Tennessee, Knoxville. 105 pp.

Busing, R. T. and E. E. C. Clebsch. 1988. Fraser fir mortality and the dynamics of a Great Smoky Mountains fir-spruce stand. *Castanea* 53: 177–182.

Busing, R. T., E. E. C. Clebsch, C. C. Eagar, and E. F. Pauley. 1988. Two decades of change in a Great Smoky Mountains spruce–fir forest. *Bull. Torrey Bot. Club* 115: 25–31.

Busing, R. T. and E. F. Pauley. 1994. Mortality trends in a southern Appalachian red spruce population. *For. Ecol. Manage.* 64:41–45.

Ciesla, W. M., H. L. Lambert, and R. T. Franklin. 1965. Status of the balsam woolly aphid in North Carolina and Tennessee — 1964. USDA Forest Service, Div. S. & P.F., Zone 1, F.I. & D.C. Office, Asheville, NC. Rep. 65-1-1.

Cook and S. M. Zedaker. 1992. The dendrology of red spruce decline. In Eagar, C. and M . B. Adams, Eds., *Ecology and Decline of Red Spruce in the Eastern United States.* Springer-Verlag; New York.

Delcourt, H. R. and P. A Delcourt. 1984. Late-quaternary history of the spruce–fir ecosystem in the southern Appalachian mountain region. In White, P. S., Ed., The Southern Appalachian Spruce–Fir Ecosystem: Its Biology and Threats. USDI, National Park Service, Res./Resource Manage. Rep. SER-71, Atlanta, GA.

Delcourt, P. A. and H. R. Delcourt. In press. Conservation of biodiversity in light of the Quaternary paleo-ecological record: should the focus be on species, ecosystems, or landscapes? Ecological Applications.

DeHayes, D. H. F. C. Thornton, C. E. Waite, and M. A. Ingle. 1991. Ambient cloud deposition reduces cold tolerance of red spruce seedlings. *Can. J. For. Res.* 21:1292–1295.

DeSelm, H. R. and R. R. Boner. 1984. Understory changes in spruce–fir during the first 16–20 years following the death of fir. In P.S. White, Ed. The southern Appalachian spruce–fir ecosystem: its biology and threats. United States Department of the Interior, National Park Service, Res./Resource Manage. Rep. SER-71. 268 pp.

Dull, C. W., J. D. Ward, H. D. Brown, G. W. Ryan, W. H. Clerke, and R. J. Uhler. 1988. Evaluation of spruce and fir mortality in the southern Appalachian mountains. USDA Forest Service Southern Region R8-PR 13 October 1988. 92 pp.

Eagar, C. 1984. Review of the biology and ecology of the balsam woolly aphid in southern Appalachian spruce–fir forests. In P.S. White, Ed. The southern Appalachian spruce–fir ecosystem: its biology and threats. USDI, National Park Service, Res./Resource Manage. Rep. SER-71. 268 pp.

Eagar, C. 1985. Investigations of Balsam Woolly Aphid–Fraser Fir Interaction: Feeding Site Characteristics and Wound Response. Ph.D. dissertaion. University of Tennessee, Knoxville, TN 90 pp.

Eagar, C. and M . B. Adams, Eds. 1992. *Ecology and Decline of Red Spruce in the Eastern United States.* Springer-Verlag; New York.

Flum, T. and S. C. Nodvin. 1995. Factors affecting streamwater chemistry in the Great Smoky Mountains, U.S.A. *Water, Air Soil Pollut.* 85:1707–1712.

Groton, E. S. 1985. Assessment of spruce–fir forests in the southern Appalachian Mountains, Tennessee Valley Authority, Norris, TN.

Hain, F. P. and F. H. Arthur. 1985. The role of atmospheric deposition in the latitudal variation of Fraser fir mortality caused by the balsam woolly adelgid, *Adelges piceae* (Ratz.)(Hemip., Adelgidae): a hypothesis. *Zeitschrift Angew. Entomol.* 99:145–152.

Hall, G. S. 1989. Birds of the southern Appalachian subalpine forest. *Bird Conserv.* 3:101–107.

Harmon, M. E., S. P. Bratton, and P. S. White. 1983. Disturbance and vegetation response in relation to environmental gradients in the Great Smoky Mountains. *Vegetario* 55:129–139.

Hermann, K. A., Ed. 1996. The southern Appalachian Assessment GIS Data Base CD ROM Set. Southern Appalachian Man and the Biosphere Program. Norris, TN.

Hollingsworth, R. G. and F. P. Hain. 1991. Balsam woolly adelgid (Homoptera: Adelgidae) and spruce–fir decline in the southern Appalachians: assessing pest relevance in a damaged ecosystem. *Fl. Entomol.* 74: 179–187.

Johnson, D. W., H. Van Miegroet, S. E. Lindberg, R. B. Harrison, and D. E. Todd. 1991. Nutrient cycling in red spruce forests of the Great Smoky Mountains. *Can. J. For. Res.* 21:769–787.

Johnson, D. W. and I. J. Fernandez. 1992. Soil mediated effects of atmospheric deposition on eastern U.S. spruce–fir forests. In C. Eagar and M. B. Adams, Eds. *Ecology and Decline of Red Spruce in the Eastern United States.* Springer-Verlag: New York.

Joslin, J. D., J. M. Kelly, and H. van Miegroet. 1992. Soil chemistry and nutrition of North America Spruce–Fir stands: evidence for recent change. *J. Environ. Qual.* 21:12–30.

Joslin, J. D. and M. H. Wolfe. 1994. Foliar deficiencies of mature southern Appalachian red spruce determined by fertilizer trials. *Soil Sci. Soc. Am. J.* 58:1572–1579.

Kelly, J. M. and P. A. Mays. 1989. Root zone physical and chemical characteristics in southeastern spruce–fir stands. *Soil Sci. Soc. Am. J.* 53:1248–1255.

Kohut, R. A., J. A. Lawrence, R. G. Amundson, R. M. Raba, and J. J. Melkonian. 1990. Effects of ozone and acidic precipitation on the growth and photosynthsis of red spruce after two years of exposure. *Water, Air Soil Pollut.* 51:277–286.

Korstain, C. F. 1937. Perpetuation of spruce on cut-over and burned lands in the higher southern Appalachian mountains. *Ecol. Monog.* 7:125–167.

Lambert, H. L. and W. M. Ciesla. 1966. Status of the balsam woolly aphid in North Carolina and Tennessee — 1965. USDA Forest Service, Div. S. & P.F., Zone 1, F.I. & D.C. Office, Asheville, NC. Rep. 66-1-1.

LeBlanc, D. C., N. S. Nicholas, and S. M. Zedaker. 1992. Prevalance of individual-tree growth decline in red spruce populations of the southern Appalachian Mountains. *Can. J. For. Res.* 22:905–914.

Matthews, R. C. and A. C. Echternacht. 1984. Herpetofauna of the spruce–fir ecosystem in the southern Appalachian mountain region, with emphasis on Great Smoky Mountains National Park. In P.S. White, Ed. The southern Appalachian spruce–fir ecosystem: its biology and threats. USDI, National Park Service, Res./Resource Manage. Rep. SER-71. 268 pp.

McLaughlin, S. B., T. J. Blasing, and D. J. Downing. 1994. Two hundred year variation of southern red spruce redial growth as estimated by spectral analysis: comment. *Can. J. For. Res.* 24:2299–2304.

McLaughlin, S. B., D. J. Downing, T. J. Blasing, E. R. Cook, and H. S. Adams. 1987. An analysis of climate and competition as contributors to decline of red spruce in high elevation Appalachian forests of the Eastern United States. *Oecologia* 72:487–501.

McLaughlin, S. B., J. D. Joslin, A. Stone, R. Wimmer, and S. Wullschleger. 1997. The impact of global change on high elevation spruce/fir forests. In S. Fox and R. Mickler, Eds. *The Productivity and Sustainability of Southern Forest Ecosystems in a Changing Environment.* Springer-Verlag: New York.

McLaughlin, S. B. and R. J. Kohut. 1992.The effects of atmospheric deposition and ozone on carbon allocation and associated physiological processes in red spruce. In Eagar, C. and M . B. Adams, Eds., *Ecology and Decline of Red Spruce in the Eastern United States.* Springer-Verlag: New York. pp. 338–384.

McLaughlin, S. B., M. G. Tjoelker, and W. K. Roy. 1993. Acid deposition alters red spruce physiology: laboratory studies support field observations. *Can. J. For. Res.* 23: 380–386.

Minckler, L. S. 1940. Early planting experiments in the spruce–fir type of the southern Appalachians. *J. For.* 38:651–654.

Mohnen, V. A. 1992. Atmospheric deposition and pollution exposure of Eastern U.S. forests. In C. Eagar and M. B. Adams, Eds. *Ecology and Decline of Red Spruce in the Eastern United States.* Springer-Verlag: New York.

Mueller, S. F. 1994. Characterization of ambient ozone levels in the Great Smoky Mountains National Park. *J. Appl. Meteorol.* 33:465–472.

Nagel, W. P. 1959. Forest insect conditions in the Southeast during 1958. USDA Forest Service. Stn. Pap. SE-100.

Nicholas, N. S. 1992. Stand Structure, Growth, and Mortality in Southern Appalachian Spruce–Fir. Ph.D. dissertation, Virginia Polytechnic Institute and State University, Blacksburg, VA. 176 pp.

Nicholas, N. S. and P. S. White. 1985. The effect of balsam woolly aphid infestation on fuel levels in spruce–fir forests of Great Smoky Mountains National Park. USDI, National Park Service, Res./Resour. Manage. Rep. SER-74. 24 pp.

Nicholas, N. S. and S. M. Zedaker. 1989. Ice damage in spruce–fir forests of the Black Mountains, North Carolina. Can. J. For. Res. 19:1487–1491.

Nicholas, N. S., S. M. Zedaker, and C. Eagar. 1992a. A comparison of overstory community structure in three southern Appalachian spruce–fir forests. *Bull. Torrey Bot. Club* 119: 316–332.

Nicholas, N. S., S. M. Zedaker, and C. Eagar. 1992b. Seedling recruitment and stand regeneration in spruce–fir forests of the Great Smoky Mountains. *Bull. Torrey Bot. Club* 119: 289–299.

Nodvin, S. C., H. van Miegroet, S. E. Lindberg, N. S. Nicholas, and D. W. Johnson. 1995. Acid deposition, ecosystem process, and nitrogen saturation in a high elevation southern Appalachian watershed. *Water, Air, Soil Pollut.* 85:1647–1652.

Oosting, H. J. and W. D. Billings. 1951. A comparison of virgin spruce–fir forest in the northern and southern Appalachian system. *Ecology* 32: 84–103.

Patton, R. L., K. F. Jensen, and G. A. Schier. 1991. Responses of red spruce seedlings to ozone and acid deposition. *Can. J. For. Res.* 21: 1354–1359.

Pauley, E. F. 1989. Does *Rubus canadensis* Interfere with the Growth of Fraser Fir Seedlings?, M.S. thesis, University of Tennessee, Knoxville, TN. 113 pp.

Peart, D. R., N. S. Nicholas, S. M. Zedaker, M. M. Miller-Weeks, and T. G. Siccama. 1992. Condition and recent trends in high-elevation red spruce populations. In Eagar, C. and M. B. Adams, Eds., *Ecology and Decline of Red Spruce in the Eastern United States.* Springer-Verlag: New York.

Pyle, C. 1984. Pre-park disturbance in the spruce–fir forests of the Great Smoky Mountains National Park. In White, P. S., Ed., The Southern Appalachian Spruce–Fir Ecosystem: Its Biology and Threats. USDI, National Park Service, Res./Resour. Manage. Rep. SER-71, Atlanta, GA.

Pyle, C. and M. P. Schafale. 1988. Land use history of three spruce–fir forest sites in Southern Appalachia. *J. For. History* 32:4–21.

Reams, G., N. S. Nicholas, and S. M. Zedaker. 1993. Two hundred year variation of southern red spruce radial growth as estimated by spectral analysis. *Can. J. For. Res.* 23: 291–301.

Root, T. L. and S. H. Schneider. 1993. Can large-scale climatic models be linked with multiscale ecological studies? *Conserv. Biology* 7:256–270.

Schneider, S. H. 1989. The greenhouse effect: science and policy. *Science* 243:771–781.

Shanks, R. E. 1954. Climates of the Great Smoky Mountains. Ecology 35:354–361.

Siccama, T. G., M. Bliss, and H. W. Vogelmann. 1982. Decline of red spruce in the Green Mountains of Vermont. Bull. Torrey Bot. Club 109:162–168.

Smith, D. K. 1984. A status report on the bryophytes of the southern Appalachian spruce–fir forest, pp. 131–138. In P.S. White, Ed. The southern Appalachian spruce–fir ecosystem: its biology and threats. USDI, National Park Service, Res./Resour. Manage. Rep. SER-71. 268 pp.

Stephens, L. A. 1969. A Comparisons of Climatic Elements at Four Elevations in the Great Smoky Mountains National Park. M.S. thesis, University of Tennessee, Knoxville. 119 pp.

Thornton, F. C., M. Schaedle, and D. J. Raynal. 1987. Effect of aluminum on red spruce seedings in solution culture. *Environ. Exp.* Bot. 27:489–498.

Thornton, F. C. J. D. Joslin, P. A. Pier, H. Neufeld, J. R. Seiler, and J. D. Hutcherson. 1994. Cloudwater and ozone effects upon high elevation red spruce: a summary of study results from Whitetop Mountain, Virginia. *J. Environ. Qual.* 23:1158–1167.

Tseng, E. C., J. R. Seiler, and B. I. Chevone. 1987. Effects of ozone and water stress on greenhouse Fraser fir seedling growth and physiology. *Environ. Exp. Bot.* 28:37–41.

USDI Fish and Wildlife Service. 1995a. Endangered and threatened wildlife and plants; rock gnome lichen determined to be endangered. *Fed. Reg.,* 60:3657–3562.

USDI Fish and Wildlife Service. 1995b. Endangered and threatened wildlife and plants; spruce–fir moss spider determined to be endangered. *Fed. Reg.,* 60:6968–6974.

Vann, D. R., G. R. Strimbeck, and A. H. Johnson. 1992. Effects of ambient levels of airborne chemicals on freezing resistance of red spruce foliage. *For. Ecol. Manage.* 51:69–79.

Van Miegroet, H., D. W. Johnson, and D. W. Cole. 1992. Analysis of N cycles in polluted vs. unpolluted environments. Pp 199–202. In D. W. Johnson and S. E. Lindberg, Eds. *Atmospheric Deposition and Nutrient Cycling in Forest Ecosystems — a Synthesis of the Integrated Forest Study.* Springer-Verlag: New York.

Van Miegroet, H., D. W. Johnson, and D. E. Todd. 1993. Foliar response of red spruce saplings to fertilization with calcium and magnesium in the Great Smoky Mountains National Park. *Can. J. For. Res.* 23:89–95.

White, P. S. 1984. The southern Appalachian spruce–fir ecosystem: an introduction. In White, P. S., Ed., The Southern Appalachian spruce–fir ecosystem: its biology and threats. USDI, National Park Service, Res./Resour. Manage. Rep. SER-71, Atlanta, GA.

White, P. S. and L. A. Renfro. 1984. Vascular plants of southern Appalachian spruce–fir: annotated checklists arranged by geography, habitat, and growth form. In P.S. White, Ed. The southern Appalachian spruce–fir ecosystem: its biology and threats. USDI, National Park Service, Res./Resour. Manage. Rep. SER-71. 268 pp.

White, P. S., M. D. MacKenzie, and R. T. Busing. 1985. Natural disturbance and gap phase dynamics in southern Appalachian spruce–fir forests. *Can. J. For. Res.* 15:233–240.

White, P. S. and C. V. Cogbill. 1992. Spruce-fir forests of eastern North America. In Eagar, C. and M . B. Adams, Eds., *Ecology and Decline of Red Spruce in the Eastern United States.* Springer-Verlag: New York.

Whittaker, R. H. 1956. Vegetation of the Great Smoky Mountains. *Ecol. Monogr.* 26: 1–80.

Zedaker, S. M., N. S. Nicholas, C. Eagar, P. S. White, and T. E. Burk. 1988. Stand characteristics associated with potential decline of spruce–fir forests in the southern Appalachians. In Proc. U.S./F.R.G. research symposium: Effects of Atmospheric Pollution on the Spruce–Fir Forests of the Eastern United States and the Federal Republic of Germany. Oct. 19 to 23, 1987, Burlington, VT. USDA Forest Service, Northeastern Forest Experiment Station Gen. Techn. Rep. NE-120.

Section IV

The Future

22 The Role of Institutions in Ecosystem Management

David M. Ostermeier

CONTENTS

Introduction 457
An Institutional History 458
Institutional Framework 459
Case: Stakeholder Interactions in "Turn-of-the-Century" Commercial Timbering
in the Southern Appalachians 459
  Institutional Changes in the 20th Century 460
  More Institutional Change and Increased Complexity: Environmental Statutes
  of the 1970s 461
  An Institutional Dilemma 464
Designing Institutions for Sustainable Ecosystem Management 464
  Clarifying a Sustainable Vision 465
  Working toward Sustainability: Initiating and Facilitating Collaborative Processes...............466
  Structuring Involvement in Collaborative Natural Resource Management 467
    Gathering and Providing Information 467
    Providing Incentives and Disincentives 467
    Managing Monitoring Efficiency 468
    Conflict Resolution Mechanisms 468
    Nested Institutional Arrangements 469
Conclusions 470
References 472

INTRODUCTION

Throughout our globe there is a growing realization that as a species we have to redefine our relation with the natural world upon which both our sustenance and spirit depend. Sustainable development has emerged as a lingual logo by which we search in our struggle to live sustainably together. There is also an increasing realization that not only must we redefine our relation to the natural world, but we must also redefine our relation to each other. In fact, I will argue that it is only through change in how we relate with each other that we can bring our relationship with the natural world within sustainable parameters.

In the U.S., ecosystem management has emerged as a vision and paradigm for managing our landscapes. As McCormick (Chapter 1 in this volume) cites from Overby in this book, "Ecosystem management requires the maintenance of sustainable ecosystems while providing for a wider array of uses, values, products and services from the land to an increasingly diverse public." The focus of ecosystem management and sustainability is on *both* the desired conditions of ecological processes *and* the specific products of these processes. Many professional natural resource managers view ecosystem management as traditional science embodied within new language. Alternatively, ecosystem management advocates indicate that historical resource use is primarily focused on outputs, with limited attention to underlying ecological processes. It is the attention to the renewal of "broadly defined" ecological processes that differentiates this emerging view from traditional sustainable yield doctrines of forestry, wildlife and other renewable natural resources. Proponents of sustainability and ecosystem management foster choice and decision making whereby humans view — and act accordingly — our activities *within* an ecological and systemic context, not *outside* this context. This emphasis on process, on *how* we interact with each other and our environment, is in contrast to our historical homocentric view of resource use which has primarily been output oriented.

Given this contrast, a journey toward ecosystem management and sustainability is one of reform and change. More specifically, it is a journey of changing how we make decisions about how we should live with each other, with future generations, and with the reality of our interdependencies with our environment. "Ecosystem management calls for changes in how we approach nature, science, and politics. It requires that we ask ourselves what kind of society, and correspondingly, what kind of relationship with nature we want" (Cortner et al. 1995a:1).

Overriding goals of this chapter are to help the reader understand the critical role that institutions play in linking people and natural resources, and that institutional reform is needed if we are to progress toward sustainability. To these ends, a history of how institutions have linked people and natural resources is discussed. Institutional changes needed for sustainable living are then addressed and collaboration is introduced as a fundamental strategy for sustainable ecosystem use. The roles of institutional players — professional, multiple levels of government, private sector, citizen groups — are outlined, and various elements of structuring the involvement of people in sustainable processes are reviewed. Concluding points are then discussed and references are made to the Southern Appalachian Man and the Biosphere Cooperative (SAMAB).

It is argued that a sustainable path depends on developing strong, imaginative, decentralized institutions that are collaborative and foster open participation and deliberation. Throughout this chapter, examples of stakeholder interactions are discussed to illuminate the role institutions *have played*, *can play*, and *should play* in linking natural resources and society. Our story begins with an institutional history illustrating the evolution of institutions impacting natural resources. "As political and economic development occurs, societies tend to build complex institutional arrangements in a nested structure that enables them to use institutions at one level to modify institutions at other levels of organization" (Oakerson 1990: 44). For example, environmental interest groups and environmental management agencies are institutions that have grown from concern regarding the effects of unbridled market institutions. More recently, the "wise use" movement and associated legislative agendas and statutes, have grown in reaction to the regulatory nature of environmental management. It is argued that historical institutional arrangements have often marginalized or excluded the interests of at least some natural resource stakeholders. For example, historically environmental interests were often excluded from economic development decisions. Likewise, some environmental statutes, such as the Endangered Species Act, effectively exclude landowner interests. Such exclusion is inconsistent with sustainability, which demands participant inclusiveness and a wholistic, systemic view. It is suggested that working towards sustainability requires a search for congruence of socially determined goals including those of economic performance, social equity, and environmental integrity.

AN INSTITUTIONAL HISTORY

Any process of social change sinks deep into our institutions that govern how we live together. As suggested in their seminal treatment in *The Good Society*, Bellah et al. (1991) suggest that "we live through institutions" and it is only by changing how (process) we currently make decisions about living together that we can change how we live together. Oakerson (1990) suggests that institutions are "rule configurations" that any society erects over time to help its members make decisions about living together. Institutions capture, over time, the more dominant values of society or the values that dominate a society. Cortner et al. (1995b:2) suggest that "values of the past created the institutions of the present, and changing values will engender the institutions of the future."

INSTITUTIONAL FRAMEWORK

To better understand the effects of institutions and their history, an analytical framework has been adopted from work by Ostrom (1973, p. 89) and Oakerson (1990). Four principal elements of this framework are:

1. A *biophysical element* which represents the biophysical nature of the natural world and resources in question. This includes the technologies of use as well as the economic demand for such resources.

2. A *process and institutional element* whose collective arrangements mold the environment and form the incentives that determine how people will live and behave together. Institutions are important in governing any society, but it is how these institutions interact and function that drives citizen behavior. Institutional arrangements determine participant roles as well as rules that specify "who decides what in relation to whom, what is okay, and what is not." In addition to who is involved, the process of decision making is important — how open or closed decision making is, the degree to which reciprocity exits, and how conflicts are resolved.

3. A *behavioral choice element* which is the pattern of interactions among people in a community. Following Flora's typology (1994), a natural resource community is defined broadly to include community members who live in a certain geographic community and also "communities of interest" who are linked to natural resources through their values and interests.

4. An *outcomes or consequences element,* including multiple outcomes such as environmental, economic, political, and social.

The framework suggests that individuals respond to the incentives that they find in their environment and make choices based on the nature of the physical world which they face and the institutional arrangements or "rule configurations" that specify authority relationships. Outcomes result from the "interactions" of individuals as they make their choices — choices governed by the linkage between institutional arrangements and the reality of the physical world. In a democracy, a critical government role, then, is to see that existing institutional arrangements (mixtures of market, government and social/cultural institutions) reflect the general will of the populace, yet protect — to the extent foreseeable and possible — the rights of the individual. This is especially so in countries such as the U.S. which have very strong market institutions and relatively limited cultural institutions. Government institutions, at all levels, then become the primary institution of the "collective." "Institutions affect behavior by structuring the alternatives available to individuals and groups and by creating incentives and disincentives to choose one alternative over another." (Oakerson 1990, p. 47).

CASE: STAKEHOLDER INTERACTIONS IN "TURN-OF-THE-CENTURY" COMMERCIAL TIMBERING IN THE SOUTHERN APPALACHIANS

Before the 1880s, life in the southern Appalachians was characterized by "self-sufficient island communities divided by ridges and hills" (Eller 1979). Most farms were owner operated, and communities tended to concentrate along family lines. Due to the importance of family, early political activity was based on kinship ties. "Wealthier, landed families who controlled local businesses and provided political leadership formed a local elite" (USDA Forest Service). By the end of the 19th century, these more affluent families expanded their land holdings, and tenant farming comprised 30% of all farmers. At this same time, the development of the steam engine allowed the railroad to push into the area. This was immediately followed by outside investors from the northern U.S. and Britain, who purchased considerable land from traditional landholders. In the ensuing 2 decades powerful incentives emerged from the newly formed institutional mix dominated by absentee landownership and market institutions. The power of these incentives significantly changed life in the southern Appalachians. After profit and invested capital were recovered and after timber resources were depleted, incentives disappeared, and investors moved on. In their passing, the mining and milling jobs that had lured farmers from their land were gone. In the wake of outside forces, resources were depleted and degraded. Investments had primarily been in extraction technology and infrastructure, primarily railroads. These investments did little to raise the capabilities of the region's citizens, either individually via education or training or collectively by enhancing the collective abilities of communities. Unable to find work, many landless farmers moved back to lands they originally owned, but often as tenants.

In this case, the dominant institutional arrangements governing choice in the southern Appalachians before the 1880s were self-sufficient family communities. However, with the advent of the railroad, market institutions* "fit" the physical world of the timbered southern Appalachians so well that the incentives were very high for timber exploitation. In addition, because the primary "rule makers" within the market institution were from outside the area and had no long-term interest in the area, the costs of resource exploitation were simply transferred to the communities which remained. As Panayotou (1989, p. i) suggests, "The end result is an incentive structure which induces people to maximize their profits by appropriating other peoples' resources and shifting their own costs onto others, rather than by economizing on scarce resources and investing in enhancing their productivity."

It must be stressed that the power and incentives generated by the fit between physical/biological resources and the market institutions in the above case vastly dominated other institutions and the incentives other institutions might have generated. Stated another way, there was insufficient institutional capacity at that time to balance the incentives produced by the fit between the physical resource and the market institution. An institutional mechanism that empowered and rewarded local people to effectively represent their individual and communal interest was either weak or absent. Lacking such representation, the interests of local communities were externalized and dominated by interests of market institutions. This demonstrates the importance of *rule making*, in this case the process of decision making that determined "who decided what in relation to whom." The "patterns of interaction" that resulted from the institutional mix allowed outside forces and institutions to dominate the rules determining the future of communities in the southern Appalachians.**

INSTITUTIONAL CHANGES IN THE 20TH CENTURY

Over time and from lessons such as these, our society has been confronted with the problems associated with the unbridled market as a dominant institution. Simultaneously, by the early 20th century, forces had already developed to provide a departure from a traditional norm of limited government. The economy and Euro-American society of the U.S. were on the move via continued

* Market institutions included consumer and capital markets, producer institutions, and those developing technology.
** These conditions of a developing economy with rich natural resources are found in many areas of the developing world today, with the same institutional forces at work. Resulting strategies of export economies continue to disenfranchise local citizens and degrade natural resources (Bailey 1993).

westward expansion, including the theft,* exploitation and use of rich resources, and increased industrialization. A central government with a history of limited desire and ability to govern lacked the structure and institutional capacity to deal with the increasingly complex issues of the 20th century. "Its administrative capacity, revealed by two wars to be limited, was insufficient to handle more than the most rudimentary domestic responsibilities. New forms of government organization, more permanent, professional, and powerful, were clearly needed" (Chubb 1989, p. 13). Out of this era and with participants like Teddy Roosevelt, Woodrow Wilson and Grover Cleveland came the progressive movement and the beginning of a more "administrative and procedural state."

The progressive era, followed by the economic and environmental problems of the great depression, ushered in a number of now familiar public institutions: The U.S. Forest Service, the Soil Conservation Service, the National Park Service, and the Tennessee Valley Authority to name only a few. These institutions fostered a "gospel of efficiency" that stressed a more planned, efficient and "wise use" approach to our country's natural resources (Hays 1975). A wide variety of natural resources were affected, from forests to water resource planning. For example, this was the heyday of the development and promotion of the "river basin commission." After World War II, multiple state and federal institutions continued to evolve in concert with market institutions to form an increasingly complex and layered mix. This mix fit the rich natural, human and capital resources of the U.S. and provided both incentives and investments for continued economic growth.

The vast natural resources of this country and the "special interest" politics that was part of institutional environment noted in the southern Appalachian case described above continued to nurture resource degradation and human inequality well into this century.** Coal and mineral mining in the early 1900s pitted a favored industrial elite against the environment, labor, and local communities. By the late 1960s, and nurtured by a growing "activist" national culture (especially among younger citizens), reform increasingly grew in the public consciousness.

MORE INSTITUTIONAL CHANGE AND INCREASED COMPLEXITY: ENVIRONMENTAL STATUTES OF THE 1970s

By the late 1960s, Americans had become increasingly concerned with environmental issues. This concern was well founded. Rivers were catching on fire, air pollution was a major problem in industrial areas and in most large cities, and solid waste disposal was becoming problematic and expensive (Dunlap and Mertig 1992; Paehlke 1989). Of even more concern was pollutant toxicity and bioaccumulation, the buildup of toxic substances in food chains. There was also an increasing realization that due to the effects of human population and technologies, plant and animal species around the globe were disappearing at alarming rates. With this as background, Congress was in a "fix-it" mood in the late 1960s and early 1970s. Following in progressive footsteps imprinted early this century, Congress became a very activist body in the 1970 to 1972 period. Everyone was for the environment from Bobby Kennedy to Barry Goldwater, Richard Nixon, and Ed Muskie. Following civil rights reform and legislation, the U.S. Congress and presidency continued to centralize government power by passing a number of environmental statues calling for rules and procedures regarding environmental management. One of the more far reaching environmental statutes has been the Endangered Species Act, signed into law by President Nixon in 1973.

Known as the pit-bull of environmental legislation, the Endangered Species Act (ESA) has become a significant addition to the administrative and procedural state. Its very goal of "the

---

* Theft pertains to the "taking" of land and water resources from Native Americans, leaving them little or no access to traditional resources and traditional ways of life.

** Many contend that these same forces are active today, and even with significant recent (1960 to 1990) social and environmental legislation, social equity and resource degradation continue to be problematic. Others indicate that activist "special interest" politics are now a significant obstacle to needed growth in economic productivity.

recovery of endangered species to the point where their continued existence is no longer in doubt" (Tobin 1990, p. 235) was fostered by statutory language which dismissed economic and social considerations. Statutory intent, then, was not to balance — or find congruence between — environmental values and those of development but to represent or advocate for environmental values. In the 1970s and 1980s, the battle between environmental and economic "rights" became increasingly clear to implementing agencies, environmental interest groups, industrial associations and lobbyists, and U.S. courts. Traditional economic interests advocated their "rights" and lobbied agencies and politicians which had historically supported economic liberty (example, U.S. Department of Agriculture and Corps of Engineers). Simultaneously, environmental interests pushed for environmental integrity and advocated their "rights" by also lobbying politicians and newly designed agencies like the Environmental Protection Agency (EPA) or the Endangered Species Division of the U.S. Fish and Wildlife Service. Interest group activity became a key ingredient regarding the implementation of the ESA and other environmental legislation of the 1970 to 1990 period.

There are at least six important considerations regarding the decision making processes and institutional evolution that characterized the ESA and other natural resource policy making in the last 3 decades. First, decision making processes have grown increasingly adversarial, where participants have learned to "stand firm" on positions which they feel have been guaranteed by "rights" of either economic liberty or environmental integrity. These rights have not only become part of our social fabric, they are backed by either common or statutory law.

> Much of America's common-law tradition favors industrialization, economic expansion, and individual bargaining. With its roots flourishing in the last century, it takes the environment for granted, as if its capacity to endure were infinite. ... More recent environmental law is statutory — entrusting public agencies to generate a regulatory system for environmental improvement. It takes economic activity for granted, as if our capacity to endure regulation were infinite. (Flick 1994, p. 33).

Stated another way, "The discourse of divisiveness created by our current social, economic and political institutions threatens our capacity to address sustainability" (Shannon 1991, p. 42).

Second, natural resource decision making has become very procedural. Both administrative and judicial checks on administrative and legislative processes are very process oriented. "If you go through a good enough process of decision making, then the outcome is by definition okay" (Yaffee 1994, p. 195). However, in adversarial conflict resolution mechanisms, incentives are for participants to win. When such adversarial climates are combined with administrative procedures and regulations, adversaries will seek to uncover even the smallest of procedural deficiencies. "The adversarial, win–lose nature of judicial and administrative appeals promotes strong, one-sided argumentation from each of the affected parties, with very little incentive provided to think of creative solutions that bridge diverse interests" (Yaffee 1994, p. 196).

Third, natural resource and environmental decision making has become increasingly complex and technical. Of particular importance in this evolution has been the changing roles of those participating. Primary participants in both *representative* government and *representative* interest groups are the interest groups and government agencies themselves. The citizenry affected by decisions made by these participants have often not had direct access to the process of decision making. By empowering our institutions — governments, interest groups, and courts — we have taken the need to work out our conflicts away from communities and community members. We have asked and empowered our institutions to fix our problems of living sustainably together.

In so doing, two related problems have developed. First, one community value has been pitted against another, in an adversarial process where citizens do not directly struggle in seeking balance between *their* values of economic liberty and *their* values of environmental integrity (Boyte 1994). Without this opportunity for direct struggle, we will not develop the responsibility nor capability of collaborative living, a responsibility that is necessary to meet the reality of our complex world. Second, interests of unrepresented citizens or community groups are often "externalized" due to dominance of either market or governmental regulatory institutions. Such externalization occurs regardless of whether dominance is exercised by market institutions in far off places (banks, corporate boards) or by government institutions in far off places (state or federal capitals, regional government headquarters).

Fourth, this adversarial and interest group dominated decision making system has also become entrenched in our political system. Although legislative decision making often results in a "split in the stakes" among interest groups, rarely does the process result in creativity or new approaches. In both state and national legislatures, politicians and their staff are exceedingly busy, and this often results in significant interest group and/or agency input into legislation. "Generally, this means that interest groups and agencies control most input into Congressional decision making, and since these parties have few incentives to find creative crosscutting solutions to policy problems and often have heavily vested interests in particular kinds of solutions, new ideas are not often at the center of Congressional choices." (Yaffee 1994, p. 197).

In addition, entrenched politicians and institutional players (public and private) are resistant to reform because they are comfortable with the adversarial tools they have historically honed. Some interest groups have found that membership and press attention are positively correlated with adversarial behavior — Americans seemingly like a good fight. The discourse of divisiveness remains because historical players fear collaborative processes where the quest for power is not the issue and adversarial tools do not work. The Kettering Foundation has long attested that the problems of the "administrative and procedural state" are fundamentally problems of politics.

> From a nation of citizens, we have become a nation of clients. Many dynamics contribute to the erosion of responsible political participation, from mass communications and patterns of mobility to the emergence of the corporate economy and a consumer culture. But perhaps the least remarked and most central is this: civic relationships have become expert–client relations. In the process, public life has eroded from the fabric of America's civil society. Politics is not likely to improve in any substantial fashion until this pattern, itself, is challenged and changed. (Boyte 1994, p. 79).

Fifth, given the dominance of market and large government institutions, the development of local institutions has not only not been nurtured, it has often been preempted.

> External regulation of local interdependencies is an institutionally preempting, rather than institutionally enabling, policy regime. It deprives local communities of the discretion to be able to address their own problems, and, in the process, deprives them of the opportunity to develop productive local institutions. Instead it creates a relationship of dependency, in which individuals in the local community depend on external authorities to solve their problems for them. Local institutional capacity, instead of being developed over time, is stunted. (Oakerson 1990, p. 50).

Given the decentralized nature of natural resources, strong local institutions are necessary in working toward sustainability and ecosystem management. The generic issue of central dominance vs. decentralization is one of the most important governance issues of our times.

Sixth, our institutional arrangements have become ends in themselves. Whenever in our history we allow large institutional entities to dominate how we make choices as a society, the interests of citizens are not effectively represented — in fact, they are often marginalized. Institutions are systemic and self preservation is an ultimate force within any system. Institutions naturally place self-preservation above other interests, including the public interest (consider interest groups/associations, lobbyists, the media, the judiciary, corporations, legislative bodies, and the executive branch of government including administrative agencies). Large institutions have demonstrated their ability to assemble and use considerable power. The use of such power can reduce civic roles to that of consuming governance, or worse yet, a governance junky. Even more important, citizens lose — or never develop — both the ability and responsibility of representing their interests relative to those of other citizens.

An Institutional Dilemma

The foregoing discussion has focused on problems in the way our institutions, affecting the environment and natural resources, have developed. It is important to remember that this same institutional development has produced many benefits which are often taken for granted. In an attempt to improve our institutions such that they can better serve us, it is important not to forget some beneficial characteristics in the way they have evolved. These include the following:

- Strong market institutions which continue to deliver significant benefits to a multitude of citizens including consumers, producers, workers,* and holders of capital. The efficiency of our market system is unparalleled in delivering a flow of consumer products.**
- Strong governmental institutions which deliver significant natural resource and environmental benefits. Both the departments of Interior and Agriculture have numerous agencies which manage publically owned lands and others that preform important resource inventory and evaluation functions. Still others are a fundamental component of our research, education, and technical assistance systems. Environmentally, significant progress has been made in many areas of air, water, and land-based pollution abatement. Examples of these institutions include: U.S. Geological Survey, the Cooperative Agricultural Extension Service, the National Park Service, the U.S. Forest Service, the Agricultural Research Service, and the U.S. Fish and Wildlife Service.
- Strong nongovernmental organizations and interest groups whose activities include political advocacy and lobbying, provision of information and education, and more recently provision of technical and financial assistance toward various ends (conservation, free market principles, etc.). From Ducks Unlimited, the Sierra Club, and the Nature Conservancy to Professional Associations (Forestry, Wildlife, Soil/Water Conservation) and the National Forest and Paper Association, these organizations provide numerous benefits to their constituents and other members of society.

Our institutional dilemma is how to build upon the strengths of these institutions and yet address the above noted problems. It is argued here that our institutional problems are those of both design and implementation. These are discussed in the ensuing sections.

DESIGNING INSTITUTIONS FOR SUSTAINABLE ECOSYSTEM MANAGEMENT

From an institutional design standpoint, an important reality is that human existence is systemic — everything is related to everything else. We are bound within economic, social–ecological, and political systems that are increasingly connected and interdependent — a change in one affects the balance in all. Institutions must therefore be designed to meet this systemic reality; they must be designed so that there are strong incentives for all stakeholders to seek common ground regarding the use, allocation and management of natural resources (Cortner et al. 1995b; Ostrom 1995; Salwasser 1994; Fiorino 1989). This "congruence seeking" design criterion has a number of elements including: a strong participatory base (Fiorino 1989; Kemmis 1990), that fosters participatory equity; a strong participatory base that fosters *deliberative* participation (Stankey and Clark 1992; Mathews 1995); a need to be comprehensive and wholistic (Fiorino 1989; Mann and Plummer

---

* Many American workers continue to have working conditions and wages which allow them to enjoy a comfortable life style. However, there is significant evidence that the American worker is becoming increasingly disadvantaged. Considerable recent study and literature address this issue including: *Chaos or Community* (Sklar 1995); *The State of Working America, 1994-95* (Mishel and Bernstein 1995); and *The Work of Nations* (Reich, 1991).

** The author is not entering the debate about the value of a consumer society but rather is simply indicating that the market system is very efficient in providing consumer goods.

1995); and a need to be knowledgeable including efficient and effective use of science, as well as local and cultural wisdom (Cortner 1995b; Ostrom 1995). These characteristics are in sharp contrast to the past, where institutions produced strong incentives to fragment, divide, and conquer. In following Madison's view that our constitution must protect citizens from their meanest instincts, we have lost sight of Jefferson's views that our constitution must provide mechanisms to help us find common ground (Kemmis 1990). As a citizenry, we have become better at saying what we are against, as opposed to developing consensus regarding what we advocate.

CLARIFYING A SUSTAINABLE VISION

As natural resource communities (including both communities of place and communities of interest) meet to work toward common ground, resolve differences, and represent rights and interests, participants have three choices: walk away from the challenges and problems, fight about them, or negotiate (Fisher and Ury 1991). Walking away and not participating is not a viable option if we are to seek congruence and sustainable resource use. An absence of participation naturally occurs in situations of perceived "power-differentials." The fact that power-differentials persist and grow in our society (increasing concentrations of wealth, poverty, etc.), continues to signal the need to help "empower" those who tend to "walk away" because of a history of powerlessness. Associated with this are problems of apathy, "too busy to participate," and "let somebody else represent us." A sustainable path is an inclusive path, and reform is needed to foster inclusiveness and participation.

Likewise, fighting is not a viable option in working toward congruence and sustainable resource use. Much of our recent policy and political history has been adversarial, with individuals and groups pressing through the political process for a response to their (exclusively defined) needs. Alternatively, collaborative problem solving and deliberative negotiation strive to empower participants in seeking congruence so that they (inclusively) balance their collective interests. Participants in the latter include multiple stakeholders such as landowners, local communities, public agencies, the private sector, interest groups, and legislatures. In a collaborative ideal, lobbyists and the courts would play minor roles, and legislatures would play a role of responding to needs berthed from collaborative processes. A sustainable vision then requires a collaborative strategy.

A key characteristic of collaboration is that the interests of all parties are represented as effectively as possible. It is the right of each party to represent its interests; it is the responsibility of each party to help represent the interests of other parties. Winning is not part of the collaborative process; in fact, the process of collaboration attempts to transform relationships from "we–they" to "we–it", "it" being the differences that stakeholders have about natural resources. Collaboration is an attempt to change the relationship between players such that cooperation is the end result, not fighting. In this sense, collaboration means developing rules of social engagement that will preserve the ability of struggling and working together.

In research on collaboration, the Wilder Foundation defines collaboration as "a mutually beneficial and well-defined relationship entered into by two or more organizations to achieve common goals. The relationship includes a commitment to: a definition of mutual relationships and goals; a jointly developed structure and shared responsibility; mutual authority and accountability for success; and sharing of resources and rewards" (Winer and Ray 1994, p. 24). Although many collaborative efforts may differ somewhat from this definition, central points normally are: diverse and inclusive stakeholders working together to resolve differences and define common ground.

An important factor influencing the success of collaborative activities is the extent to which participants are able to develop trust and other forms of "social capital." Putnam (1993A) and others define social capital as "features of social organization, such as networks, norms and trust, that facilitate coordination and cooperation for mutual benefit." Social capital researchers (also see Flora and Flora 1993) have begun to discover correlation between lasting community development and the extent of social capital found in communities. For example, in a lengthy study of communities in Italy, Putnam (1993b) found that the quality of life, community development, and social capital were most developed where there have been *horizontal* forms of decision making and communication as found in collaborative processes. In his book on *Community and the Politics of Place*, Kemmis (1990, pp. 114–115) suggests two important elements of social capital.

> First is the indispensable element of trust. This is one of those civic virtues which people like Jefferson had argued were essential to public life, but which the procedural republic does not depend upon. Second is the fact that such civic virtues can only become a constitutive feature of public life in one way: through practice.

Hence, it is the investment in collaborative practice that enables groups to be innovative and creative in the future.

WORKING TOWARD SUSTAINABILITY: INITIATING AND FACILITATING COLLABORATIVE PROCESSES

Collaboration is not a naturally occurring phenomenon in our society nor is it nurtured by the collective arrangements of our institutions — in fact, it is often discouraged. In their work on fostering citizen participation in governance, the Kettering Foundation stresses a need to create "mediating institutions" where people can deliberate and collaborate (the Hardwood Group 1993). Facilitating collaborative processes and skill development is what Himmelman (1992) refers to as "collaborative betterment." Such betterment normally begins outside a community with public, private, or nonprofit institutions and is brought into the community. A common element of a variety of collaborative betterment activities is "structuring participant involvement in collaborative learning processes." A common intent is to raise the capacity of the community — normally at the grass roots level — to work collaboratively toward collective ends. A specific example is the "Environmental Issue Forums" supported by the North American Association for Environmental Education and the Kettering Foundation. In these forums, community members learn how to structure meetings and other deliberative activity so that all major interests are represented. A relatively rich and recent history of case study and literature on collaboration and collaborative processes is available for the interested reader (Potapchuk and Bailey 1994; Mattessich and Monsey 1992). Some common elements of such processes are summarized as follows:

1. Initiating and catalyzing. In a relatively independent and competitive society like ours, collaboration doesn't naturally happen. Initiating institutions and actors must often facilitate and help convene initial and on-going activity. Depending on the situation, an important goal of this initial activity is to invite, engage, and prepare people to collaborate. This is especially important for "new collaborators," as people often not only lack an understanding of the collaborative process, they also lack collaborative communication skills. Initial activity should also help those who have been in adversarial relations deal with issues of emotion and separate these from issues of substance. A collaborative principle is inclusiveness, and involvement of all affected parties should begin at the start of collaborative activity.
2. Sharing perspectives, diagnosing the situation, and collecting information. A major goal of this step is to identify and share perspectives and to strive for understanding. An open and participatory environment must be created where participants do not feel threatened, can drop their defenses, and share divergent world views. In such environments, people are free to understand and learn, and they are able to more effectively evaluate the situation from more than their perspective. Evaluating the situation, including gathering and/or deliberating about information, allows participants to more specifically isolate what they can agree on and what they have yet to resolve. It is important to help participants achieve a holistic view of the issues under consideration.

3. Articulating issues and underlying interests, and creating and evaluating issue options. It is critical to help participants fully develop and articulate their interests while simultaneously helping them to let go of their positions. By looking at underlying interests, common ground can often be found from which to build mutually supportive options. Moving from "positional bargaining" based on perceived "rights" to interest based negotiation is a fundamental element of the collaborative process. Working from interests, participants significantly increase their ability to bridge divergent world views.

4. Negotiating, making and reinforcing choices and agreements. Choices should be made on what participants want and what is possible. Objective criteria should be used and actions plans should specify who will do what, by when, and how. Monitoring is critical so that all parties understand how agreements are progressing.

STRUCTURING INVOLVEMENT IN COLLABORATIVE NATURAL RESOURCE MANAGEMENT

Implementing a collaborative strategy to work toward sustainability will require changes in various aspects of natural resource planning and management. Traditional approaches have fragmented activities whereby farmers, local communities, timber companies, wildlife agencies and other decision makers have often acted independently. A transition toward collaborative activities will require changes in how people and decision makers are involved in management functions such as information collection and monitoring. This section reports on a review of recent literature and case studies regarding stakeholder involvement in a more collaborative approach to natural resource use.

Gathering and Providing Information

Information and feedback help a community better understand the economic, equity, and environmental consequences of using and managing natural resources. Providing information regarding natural resource use has historically been the domain of state or federal agencies. Arguments favoring this domain have pointed to the following: trained professionals are necessary, given the complexity of what is being monitored; consistency and efficiency are higher, given a professional/trained work force; and bias is minimized with professional monitoring. Others suggest that professionals are also biased and can favor certain community values while suppressing others. In addition, because of the breadth and depth of needed information (economic performance, social impact, and ecological assessment), multiple groups should be involved, including agency professionals, interest groups, and community citizens (Berkes 1989; Hanna et al. 1995; Shannon 1991).

Even more important, however, is that gathering and developing information is significantly more than a technical or scientific task; it is part of a social process of seeking social congruence. The goal is not only development of information, but also nurturing community learning. If citizens don't trust and participate in such learning, it will simply not occur. In studying environmental risk assessment, as quoted by Fiorino (1989, p. 504) "citizens in a democratic society will eventually interfere with decisions in which they do not feel represented." Given a joint task of developing information and of building learning communities, institutional arrangements must facilitate this joint provision. This points to an important facilitation role for state or federal natural resource agencies, in addition to their historical technical role. Integrating facilitation roles and skills into natural resource agencies will not be easy and will require considerable leadership and skill development.

Providing Incentives and Disincentives

A very important governance question is how should the collective community provide positive and negative sanctions so that individual behavior is accountable to the collective good. Developing appropriate incentives and sanctions has also historically been the domain of the public, state, or federal agency. These institutions have the primary tools of public policy and sanction. However, as discussed earlier in this chapter, these institutions generally do not have a role of seeking congruence among competing natural resource values at the resource community level. In a sustainable approach using a collaborative and integrative strategy, critical tasks are not only to develop incentives and/or sanctions but to also foster compliance with those sanctions. By integrating compliance and sanctions, a collaborative strategy is necessary because all stakeholder interests must be represented, including those towards whom sanctions are directed. Jentoft (1989) suggests that compliance is related to the "legitimacy" of the incentive or sanction. "To be legitimate, the content of a regulation, the process by which it is made, the way it is implemented, and the effects of its distribution must be perceived as fair by resource users (Jentoft 1989, p. 143)." In her evaluation of institutional arrangements that have fostered sustainable resource use and endured long time periods, Ostrom (1995) found that "monitoring and sanctioning are undertaken primarily by the participants themselves." This is in contrast to the ESA example discussed earlier, where an outside institution assumes the sole responsibility for monitoring and compliance.

It must be stressed that the success of self-compliance is dependent on the extent to which congruence between economic performance, social equity, and ecological maintenance is actively and openly being sought. Jentoft (1989, p. 143) further suggests, "To be equitable, a resource management process must represent the range of user group interests and have a clear purpose and a transparent operation." It is not being suggested that self-compliance is the answer to all regulatory questions; nor is it suggested that regulatory agencies be eliminated. Rather, it is being suggested that providing sanctions and incentives is part of a larger holistic and inclusive community planning process. Such planning and action would include establishing consensus visions, strategies for vision attainment, and incentives/sanctions to help guide individual behavior.

Managing Monitoring Efficiency

Working towards a collective vision of sustainable resource use can be very resource intensive. If not efficiently managed, costs of providing various kinds of information can overwhelm resource use benefits. Collaborative processes involving multiple community interests have relatively high "up-front" costs compared to historical hierarchial processes. However, in a democracy the latter have increasingly had high "ex-post" or implementation costs. There is considerable evidence that the high costs of much of our command and control environmental regulation is due to an absence of seeking congruence between community values of economic performance, social equity, and environmental management (Alm 1992). In fact, these values have often been pitted against each other in very adversarial and costly battles (Fiorino 1989). As we begin to learn more about seeking congruence between community values, we can begin to learn more about increasing the efficiency of the following management or transaction costs: coordination costs between users, especially when ecological boundaries and property boundaries are not the same; monitoring and information gathering, including community learning; and developing appropriate incentives and sanctions.

Conflict Resolution Mechanisms

In a world of growing population, complexity and resource scarcity, conflicts will be increasingly common. When managed effectively, conflict often provides the innovation needed to address public issues (Cortner et al. 1995b). Accordingly, it will be critical to develop open and participatory procedures for resolving these conflicts. Considerable work has been done in alternative dispute conflict resolution that is consistent with the objective of seeking congruence and working toward sustainable resource use (Owen 1995; CORE 1994; Crowfoot and Wondolleck 1990). Of particular interest are examples that build conflict resolution mechanisms into efforts focusing on sustainable development. In 1992 and as a reaction to multiple decades of natural resource conflicts, the government of British Columbia established the Commission on Resources and Environment (CORE 1993–94). CORE's mandate has been to develop "for public and government consideration, a British Columbia-wide strategy for land use and related resource environmental management" (CORE 1994). CORE's role has been to change the way (process) decisions affecting natural resources are made so that they are:

- More *comprehensive* — all interrelations and tradeoffs regarding economic development and the use of natural resources are considered
- More *coordinated* — among the various functional government divisions (economic, environmental protection, etc.).
- More *democratic* — all interested stakeholders are participants in a "shared decision making process" of proactive planning and equitable dispute resolution.

At the heart of the shared decision making process (seeking congruence) is interest based, collaborative problem solving and dispute resolution. These collaborative arrangements have been institutionalized both by statutory law and in the way CORE's activities have been designed and implemented. As we learn from this and other cases, we will better understand the institutional arrangements necessary for effective conflict resolution and collaborative problem solving.

Nested Institutional Arrangements

In a society with multiple political and institutional layers (local, state, federal), the proper loci of authority regarding natural resource management have been historically debated. Case analysis by Hanna et al. (1995) of alternative institutional arrangements note various levels of user participation from passive receipt of government-imposed rules to active self-governance. In the use and management of decentralized natural resources, evidence suggests that a strong *local* level of participation and authority is consistent with sustained resource care and use. This does not suggest, however, an absence in participation from other levels, nor a necessarily weak federal authority. In a study of a number of property regimes associated with long-term sustained activity, Ostrom (1995) noted the importance of nested institutional arrangements with important functions being executed by various levels.

Although the presence or absence of a central authority is important, it is the *role* such central authorities play that is critical. When centralized institutions and authorities assume hierarchial-dominator roles, local authority is suppressed, even corrupted (Mathews 1995; Shannon 1991; Kemmis 1990; Oakerson 1990). This leads to a disinvestment in local social capital, since it fosters only vertical communication, rather than horizontal deliberation required in the development of improved social interaction. Alternatively, when centralized roles are informational and technical, and when they facilitate collaborative processes and skill development at the local level, multiple levels have the potential to work synergistically rather then competitively.

In executing a role of facilitating collaborative processes at the local level, central authorities must ensure that all communities of interests are effectively represented, be they geographic communities or communities of interest. It is important to remember that in working toward sustainable resource use, a goal is for communities of place and interest to seek congruence in how *they* will resolve the inevitable conflicts between economic performance, equity and ecological maintenance. "Thus, while large-scale governments are an essential part of the mix of governance, if these governments come to dominate decision making through massive funding of activities or the imposition of force, the effectiveness of local organizations is reduced substantially. On the other hand, the absence of supportive, large-scale institutional arrangements may be just as much a threat to the sustenance of biodiversity as the presence of preemptive, large-scale government agencies" (Ostrom 1995, pp. 42–43).

Quoting W. Ross Ashby's "Law of Requisite Variety,"* Ostrom (1995, p. 34) suggests that "Any governance system that is designed to regulate complex biological systems must have as much

---

* Ashby, a noted biologist, wrote *Design for a Brain: The Origin of Adaptive Behavior* in 1960.

variety in the actions that it can take as there exists in the systems being regulated." The lack of fit between complex, decentralized natural resource ecosystems and centralized influences (government or private sector) mandates that we reform and amend our institutional arrangements to ensure needed flexibility at the local or ecosystem level (Cortner et al. 1995b; Ostrom 1995; Shannon 1991). Common sense suggests that we tap into the energies, resources, and experience where they can be found or developed, be they local or otherwise.

CONCLUSIONS

The phrases "sustainable path" and "sustainable journey" have been implied numerous times in this chapter. These terms are used to underline the importance that: sustainability is a never ending social process that must involve affected individuals and groups. The intent of this chapter has been to help the reader better understand, from an institutional perspective, why past "paths" have not been sustainable and that reform will be difficult and require considerable time and commitment. Progress will necessitate considerable leadership and learning regarding how to organize people toward sustainable ends. Such organizing will require that we re-think and re-build our institutional approaches regarding natural resource use. This restructuring will require a collaborative approach and expanded stakeholder involvement. In addition, historical roles played by agencies, interest groups, the private sector, and other stakeholders will necessitate reform and change. Specific conclusions relevant to the Southern Appalachian Man and the Biosphere (SAMAB) cooperative, and other reform efforts, include:

1. A sustainable path must be grounded in a process whereby both communities of place and interest seek congruence among disparate values and interest in how natural resources are used and managed. Such interests should include those of economic performance, social equity and environmental integrity.
2. Because of the necessity of "social deliberation and determination", a sustainable path must employ collaborative strategies. Within collaboration — via the development of social capital — lies the seed bed for the innovation and creativity required to maintain a sustainable journey. Because of the necessity of collaboration, it is argued here that a sustainable path requires a collaborative ethic. Regarding issues of conflict in a democratic society, Willbern (1984) suggests an "ethic of compromise and social integration." He reminds us that Abraham Lincoln's classic formulation, "*with firmness in the right as God gives us to see the right*," seems to carry with it the implication that God may give someone else to see the "right" differently, and that they may also be firm. A collaborative ethic maintains the opportunity to examine our firmness as well as that of others; it also allows us to examine alternatives to such firmness. Through continued networking, a regional institution such as SAMAB can nurture such an ethic. Empirical evidence suggests that successful collaborative activity like the Southern Appalachian Assessment (SAMAB 1996), or recent efforts of the Foothills Land Conservancy*, are ways to build a collaborative ethic.
3. A sustainable path is one where collaborative activities are deliberative, comprehensive, and knowledge based. Deliberative discussions are not only at the core of democracy, they are at the core of interest based collaboration. Deliberation signifies the back and forth struggle that is required on a sustainable path. Because of the systemic reality of our world, collaboration must also be comprehensive and knowledge based. Comprehension not only implies diverse participant representation, but also that those participating must insist on a comprehensive and long-term view of the issues being deliberated.

---

* The Conservancy recently completed a collaborative public campaign to purchase nearly 5000 acres of forested wildlife habitat near the southern border of the Great Smoky National Park. Such collaboration ensures the maintenance of important wildlife corridors and habitat.

Knowledge based activities are not only those of applied science but also those encompassing local knowledge and wisdom. This opens opportunities for collaborative activity between scientists, technical and managerial experts, and local citizens.

4. Our market, government, interest group, judicial, political, and media institutions have evolved in ways whereby the resulting institutional chemistry promotes competitive, divisive, undemocratic, and exclusive behavior. Such behavior inhibits decision making that is collaborative, holistic, comprehensive and considers the long term. Ways of addressing this institutional dilemma include:

a. Legislative and policy reform

- Creating integrating institutions. As noted earlier in this chapter, in addressing issues of sustainability, British Columbia legislatively established the Commission of Resources and Environment (CORE). Their mandate is to change the way (process) decisions affecting natural resources are made so that they are more comprehensive, coordinated, and democratic. At the heart their decision making process is interest based, collaborative problem solving, and dispute resolution.

- Creating integrating policy. Such policy reform could take many forms, one of which could be to address current disincentives for collaboration. Possible examples include: tax incentives to foster private sector collaboration; policy to foster community wide natural resource planning including negotiated rule making to provide appropriate incentives and sanctions; and policy reform to foster intergovernmental and public–private alliance building.

b. Organizational and institutional reform — Significant reform potential lies in the hands of decision makers within both the public, private, interest group, and nonprofit sectors. Most promising are changes in *how* (process) agencies, companies and organizations are managed through a re-definition of roles and responsibilities. This redefinition must be consistent with the vision of sustainability (seeking congruence) and embody a collaborative ethic and strategy. Examples include:

- Natural resource and environmental management agencies would assume more significant roles of initiating and facilitating collaborative problem solving, proactive collaborative planning, and mediation. Simultaneously, such agencies would also sharpen their competitive advantage of collecting and managing technical information and making such information available to collaborative processes and participants. A collective result of these role changes would foster more decentralized and collaborative decision making. SAMAB has nurtured technical collaboration among agencies through efforts like the Southern Appalachian Assessment.

- Interest groups and nonprofits. There is significant opportunity for these groups to address the opportunities of what Himmelman calls "collaborative betterment" — building and developing collaborative capacity. Such capacity is needed if communities of place and interest, and the public and private sectors, are to join together on a sustainable journey. The nonprofit private sector, with its relative flexibility and ingenuity, is well positioned to promote and implement collaborative betterment activities. In its evolution, it is critical for SAMAB to expand its activities such that the interests of local communities and the private sector are more centrally integrated.

- Private sector companies and associated organizations. As with their governmental counterpart, the private sector can assume new roles of public–private facilitation. Because of the competitive nature of the private sector, assuming these new roles will require leadership. Such leadership can be fostered by public policy change to mitigate competitive losses resulting from collaborative alliances. It is important, however, that the private sector develop leadership in this area such that there is public support for policy change to foster public–private alliances.

5. Natural resources and associated communities are decentralized and are biologically, economically and socially unique. Because of this, localized decision making should be fostered. However, a collaborative strategy will require important, and sometimes significant, state and/or federal roles and responsibilities. Where collaboration and diverse citizen involvement has not been part of local decision making, it will be important that state and/or federal institutions, in association with non-profit institutions, work to build local collaborative capacity through collaborative betterment activities. Southern Appalachian communities have not had a strong history of collaborative local decision making involving diverse citizen participation. Working with federal and state governments, collaborative betterment will be an important role for SAMAB to play in the future. A specific example of such collaborative betterment would be the involvement of local communities in monitoring and assessment activities. Such involvement will not only help introduce important aspects of "local knowledge" into monitoring and assessment procedures, it will also result in developing skills and collaborative abilities of citizens as well as agency professionals. Such involvement could also lead to the inclusion of new variables in assessment and planning procedures, helping to make overall activities more collaborative and rewarding.

6. Last, a sustainable path is one that emphasizes both the use of, and investments in, various forms of capital, including economic capital — to ensure appropriate economic growth; environmental capital — to maintain environmental and ecological capacity; social capital — to provide the necessary learning and innovation to live well together; and human capital — to encourage the development of mind, body, and spirit, all important elements of sustainable living.

> *No man is an island entire of self.*
> *He is part of the continent, a piece of the main*
> *Therefore, seek not to know for whom the bell tolls,*
> *It tolls for thee.*

John Donne

REFERENCES

Alm, A.L. 1992. A need for new approaches: command-and-control is no longer a cure-all. *EPA J.*, 18:2.

Bailey, N.A. 1993. Foreign direct investment and environmental protection in the Third World. *Trade and The Environment: Law, Economics and Policy.*, D. Zaelke, P. Orbuch, and R.F. Housman, Eds. Center for International Environmental Law. Island Press: Washington, D.C. p. 133–143.

Bellah, R.N. and R. Madsen, W.M. Sullivan, A.Swidler, and S.M. Tipton. 1992. *The Good Society*. Vintage Books: New York.

Berkes, F. Ed. 1989., *Common Property Resources: Ecology and Community-Based Sustainable Development*. Belhaven Press: London.

Boyte, H. C. 1994. Reinventing citizenship. *Kettering Rev.*:Winter, 78–87. Kettering Foundation, Dayton, OH.

Chubb, J.E. and P.E. Peterson. 1989. American political institutions and the problem of governance. In *Can The Government Govern*, J.E. Chubb and P.E. Peterson, Eds. The Brookings Institution: Washington, D.C.

CORE 1994. *The Provincial Land Use Strategy. Vol. 1.A. Sustainability Act for British Columbia*. Seventh Floor, 1802 Douglas Street, Victoria, B.C. V8V1x4, Canada.

CORE *1993–94 Annual Report*. Victoria, B.C.

Cortner, H.J. and M.G. Wallace, S.Burke, M.A. Moote, and M.A. Shannon. 1995a. Research agenda for designing administrative structures that ensure sustainability. Invited paper, International Union of Forestry Research Organization, meeting in Tampere, Finland. August. Forest Policy and Forestry Administration Working Group.

Cortner, H.J. and M.A. Shannon, M.G. Wallace, S. Burke, and M.A. Moote. 1995b. Institutional barriers and incentives for ecosystem management — a problem analysis. Water Resources Research Center Issue Paper 16, College of Agriculture, University of Arizona, Tucsom.

Crowfoot, J. and J.M. Wondolleck. 1990. *Environmental Disputes, Community Involvement in Conflict Resolution*. Island Press: Washington, D.C.

Dunlap, R.E. and A.G. Mertig. 1992. The evolution of the U.S. environmental movement from 1970 to 1990: an overview. *American Environmentalism: The U.S. Environmental Movement, 1970–1990*. R.E. Dunlap and A.G. Mertig, Eds. Taylor % Francis: London.

Eller, R.D. 1979. Land and family: an historical view of preindustrial Appalachia. *Appalachian J.* 6 (Winter):84–86.

Fiorino, D.J. 1989. Environmental risk and democratic process: a critical review. *Columbia J. Environ. Law* 14: 501–547.

Fisher, R. W. Ury. 1991. *Getting to Yes, Negotiating Agreement Without Giving In*. 2nd ed. Penguin: New York.

Flick, W.A. 1994. Changing times, forest owners and the law. *J. For.* May: 30–33.

Flora, C.B. 1994. Sustainable Agriculture and Sustainable Communities: Social Capital in the Great Plains and Corn Belt. Unpublished ms, Dept. of Sociology, Iowa State University, Ames.

Flora, C.B. and J.L. Flora. 1993. Entrepreneurial social infrastructure: a necessary ingredient. *Ann. Am. Acad. Pol. Soc. Sci.* 529 (Sept.):48–58.

Hanna, S., C. Folke, and K.-G. Maler. 1995. Property rights and environmental resources. *Property Rights and the Environment: Social and Ecological Issues*. S. Hanna and M. Munasinghe, Eds. Beijer International Institute of Ecological Economics and The World Bank (copyright), Washington, D.C.

Hardwood Group. 1993. *Meaningful Chaos: How People Form Relationships with Public Concerns*. Dayton, OH: The Kettering Foundation.

Hays, S.P. 1975. *Conservation and the Gospel of Efficiency,* 2nd ed. Atheneum: New York.

Himmelman, A.T. 1992. Communities working collaboratively for a change. The Himmelman Consulting Group. Minneapolis, MN.

Jentoft, S. 1989. Fisheries co-management: delegating government responsibility to fishermen's organizations. *Mar. Pol.* 13(2):137–154.

Kemmis, D. 1990. *Community and the Politics of Place*. University of Oklahoma Press: Norman.

Mann, C.C. and M.L. Plummer. 1995. *Noah's Choice, The Future of Endangered Species*. Alfred A. Knopf: New York.

Mathews, D. 1995. *Politics for People: Finding a Responsible Public Voice*. University of Illinois Press: Champaign.

Mattessich, P.W. and B.R. Monsey. 1992. *Collaboration: What Makes It Work: A Review of Research Literature on Factors Influencing Successful Collaboration*. Wilder Research Center, Wilder Foundation: St. Paul, MN.

McCloskey, M. 1988. Debating the problems that underlie pollution control problems. *Environ. Law Reporter*, News and Analysis:18 ELF 10413–10418.

Mishel, L. and J. Bernstein. 1995. *The State of Working America, 1994–95*. Economic Policy Institute: Washington, D.C.

Moe, T.M. The politics of bureaucratic structure. *Can The Government Govern.*, J.E. Chubb and P.E. Peterson, Eds. The Brookings Institution: Washington, D.C.

Oakerson, R.J. 1990. Institutional diversity and rural development in America: An institutionalist's approach to rural studies. In *National Rural Studies Committee: a Proceedings*. E. Castle and B. Baldwin, Eds. Western Rural Development Center at Oregon State University, Corvallis.

Ostrom, E. 1995. Designing complexity to govern complexity. *Property Rights and The Environment: Social and Ecological Issues*. S. Hanna and M. Munasinghe, Eds. Beijer International Institute of Ecological Economics and The World Bank (copyright), Washington, D.C.

Ostrom, V. 1973 and 1989. *The Intellectual Crisis in American Public Administration*. University of Alabama Press: Tuscaloosa.

Owen, S.. 1995. Will interest negotiation produce a truce in B.C.'s "forest wars"? *Consensus* 27, July. MIT–Harvard Public Disputes Program: Cambridge, MA.

Paehlke, R.C. 1989. *Environmentalism and The Future of Progressive Politics*. Yale University Press: New Haven, CT.

Panayotou, T. 1989. *The Economics of Environmental Degradation: Problems, Causes and Responses.* Prepared for U.S. Agency for International Development under CAER Task Order #3. Harvard Institute for International Development, Cambridge, MA.

Potapchuk, W. and M.A. Bailey. 1994. *Building The Collaborative Community: A Selective Bibliography for Community Leaders.* Program for Community Problem Solving, Washington, D.C.

Putnam, R.D. 1993a. The prosperous community: social capital and public life. *Am. Prospect* 13:35–42.

Putnam, R.D. 1993b. *Making Democracy Work: Civic Traditions in Modern Italy.* Princeton University Press: Princeton, NJ.

Reich, R.B. 1991. *The Work of Nations*, Alfred Knopf: New York.

Salwasser, H. 1994. Ecosystem management: can it sustain diversity and productivity? *J. For.* 92 (8):6–10.

SAMAB. 1996. *The Southern Appalachian Assessment, Summary Report.* USDA Report R8-TP 25. Washington, D.C.

Shannon, M.A. 1991. Is American society organized to sustain forest ecosystems? *Proc. 1991 National Society of American Foresters Convention.* Bethesda, MD.

Sklar, H. 1995. *Chaos or Community*, South End Press

Stankey, G.H. and R.N Clark. 1992. *Social Aspects of New Perspectives in Forestry: a Problem Analysis.* Grey Towers Press: Milford, PA.

Tobin, R.J. 1990. *The Expendable Future: U.S. Politics and the Protection of Biological Diversity.* Duke University Press: Durham, N.C.

USDA Forest Service. *Mountaineers and Rangers: A History of Federal Forest Management in the Southern Appalachians, 1900–81.* Publ. No. FS-380. USDA, Forest Service, Washington, D.C.

Winer, M. and K. Ray. 1994. *Collaboration Handbook: Creating, Sustaining and Enjoying the Journey.* Amherst H. Wilder Foundation: St. Paul, MN.

Willbern, Y. 1984. Types and levels of public morality. *Public Admin. Rev.* March/April:102–108.

Yaffee, S.L. 1994. *The Wisdom of the Spotted Owl: Policy Lessons for a New Century.* Island Press: Washington, D.C.

23 Moving to an Operational Level: A Call for Leadership from the Southern Appalachian Man and Biosphere Cooperative

John D. Peine

CONTENTS

Defining Principles Through Practice..475
The Resource Assessment as a Basis for Vision...476
Shifting to an Operative Level...476
Suggested SAMAB Ecosystem Management Initiatives 477
    Assessment 477
    Environmental Monitoring 477
    Neotropical Migratory Songbirds 478
    Black Bear...478
    Red Wolf...478
    Brook Trout...479
    Wild Boar 479
    Gypsy Moth...479
    Air Quality..480
    Fire Management...480
    Managing Growth for Sustainable Communities 480
    Grassy Balds...480
    Climate Change 481
    Clinch–Powell River Basin 481
    Spruce–Fir Forest 481
A Call to Action for SAMAB 482

DEFINING PRINCIPLES THROUGH PRACTICE

As expressed by Bruce Babbitt in the Forward and McCormick in Chapter 1 of this volume, the definition of the terms "ecosystem management" and "sustainable development" suggest a paradigm that is key to survival for global biological and human cultural diversity in the next millennium. As documented by Gregg in Chapter 2, Hinote in Chapter 5, and Peine et al. in Chapter 6, the UNESCO-MAB program of the international network of 329 biosphere reserves in 83 countries provides an ideal venue to advance these principles. The application of these principles in real terms at biosphere reserves will ultimately sharpen their definition, drawing the philosophical rhetoric into more precise perspective. The case study programs presented in Section III of this volume provide real world examples of the multiple dimensions of ecosystem management. These examples of the application of principles help clarify the concepts and inform those associated with ecosystem management of the range of techniques employed to address these invariably difficult problems.

The southern Appalachian case study used to illustrate the principles and practices of ecosystem management is a work in progress. As documented in this volume, much has been accomplished to bring an integrated regional perspective to ecosystem management in the southern Appalachian highlands. The challenge is to continue to build momentum and acceptance of these principles in the adverse climate of shrinking budgets and greater responsibility. Such a climate of constraint heightens the necessity to form partnerships to achieve goals. Ecosystem managers can no longer afford the luxury of insular activity and tunnel vision while addressing the multidimensional problems facing the natural environment. This is true not only for the southern Appalachians but elsewhere in the U.S. and throughout the world. The necessity for cooperation breeds innovation, compromise, and consensus. Such is the nature of the opportunity for ecosystem management in the southern Appalachians.

THE RESOURCE ASSESSMENT AS A BASIS FOR VISION

As described by Berish et al. in Chapter 7 of this volume, the Southern Appalachian Assessment has provided an unprecedented regional perspective on the status of natural and cultural resources and their value and use by society. Adverse impacts on these resources were described and risks to them assessed. This extraordinary accomplishment provides the membership of the Southern Appalachian Man And Biosphere Cooperative (SAMAB) with an opportunity to identify "hot spots" of concern for the environment and establish priority programs to address the defined problems. As a result, SAMAB is at a threshold of opportunity to become more proactive in the arena of effectively targeted programs concerning ecosystem management.

SHIFTING TO AN OPERATIVE LEVEL

By seizing the momentum afforded by the Southern Appalachian Assessment, SAMAB is in an extraordinary position to strengthen its already recognized leadership in the field of ecosystem management. As documented by Hinote in Chapter 5 of this volume, SAMAB has provided its most effective leadership to this point as a facilitator of meetings on critical issues and the provision of education material. Ultimately, SAMAB needs to become more of an activist at the operational level concerning key issues related to ecosystem management. Stakeholders associated with on-the-ground programs need to feel comfortable turning to SAMAB to facilitate integration of programs and exploration of new ideas.

This would constitute a significant shift in the orientation of SAMAB toward influencing ecosystem management at the operational level. The cornerstone for practicing ecosystem management at a comprehensive level in the southern Appalachian highlands is to integrate specific resource management programs across institutional boundaries at an operational level. Section III of this volume includes numerous specific suggestions for such integration of practice. This chapter is devoted to the presentation of specific suggestions of how SAMAB might become more engaged in specific ecosystem management issues at an operational level.

As discussed by Ostermeier in Chapter 22, a key to social acceptance of the principles of ecosystem management lies with change and reform of the institutions either directly or indirectly engaged in the business of natural resources management, social services, and economic development. The SAMAB cooperative is in a unique position to facilitate institutional reform of the practice of ecosystem management with its multitude of human dimensions to embrace more holistic perspectives that require implementation at a landscape scale that ranges beyond traditional organizational boundaries of authority.

SUGGESTED SAMAB ECOSYSTEM MANAGEMENT INITIATIVES

All of the suggested SAMAB program initiatives listed below represent resource issues that critically need a landscape level, interagency vision; specific operational focus; and most of all, leadership — all of which SAMAB has the potential to deliver. In fact SAMAB *must* assume this leadership role in order for the principles of ecosystem management for sustainability to be fully implemented and the potential benefits fully realized in the southern Appalachians. Most of the following suggestions do not require a major new influx of funds into the region, but rather a shift in perspective among those individuals and institutions directly and indirectly involved in relevant research, resource management and/or education programs. The essence of the suggestions that follow reflect a dramatic shift in institutional perspective as called for by Ostermeier in Chapter 22.

ASSESSMENT

As described by Berish et al. in Chapter 7, the Southern Appalachian Assessment is undoubtedly the most significant accomplishment of the SAMAB cooperative. Now that the data are assembled and documented, the reports written, and basic conclusions drawn, there has been an understandable lull in assessment related work by the SAMAB family. There is a need to follow up the work completed with additional analysis directed to help SAMAB set priorities for the future. Rest assured that the database assembled is being used. As of December 1996, requests had been received by SAMAB from 35 universities for copies of the data (Hermann, pers. commun.). Since that time, the data have become available on the Internet.

It is suggested that SAMAB sponsor the conduct of a peer-reviewed *integrated* assessment of the resource conditions to refine and specify the general conclusions drawn from the original effort. A truly integrated analysis of the data describing the resources and their conditions was beyond the scope of the original assessment effort. This is a logical next step of what should be an *ongoing* assessment process. Such an analysis would most likely yield a list of "hot spots" of environmental concern in the region. It is imperative that SAMAB actively seek to build ecosystem management programs around lessons learned from this assessment process.

Another key role for SAMAB is to establish a procedure to maintain and add to the database assembled for the Southern Appalachian Assessment. This responsibility can be parceled out to various federal agency partners with vested interests in particular information. At this writing, no such procedures exist. It is critical that this information updating process become institutionalized very soon, before a collective disenfranchisement sets in among the SAMAB partners. Unfortunately, it is human nature to discount previous work, particularly databases that are a bit old. Again, yet another call for enlightened leadership from SAMAB.

ENVIRONMENTAL MONITORING

As described by Smith et al. in Chapter 8, there are numerous natural resource monitoring initiatives ongoing in the southern Appalachians. In most cases, there is minimal integration of these programs at any landscape scale. The most significant lost opportunity for collaboration for monitoring and research is among the cluster of designated international biosphere reserves in the region. As discussed by Gregg in Chapter 2 and Hinote in Chapter 5, one of the envisioned objectives for the southern Appalachian cluster of biosphere reserves was to foster collaboration among scientists engaged in research and monitoring activities being pursued at the three originally designated biosphere reserves of Coweeta Hydrologic Laboratory, Oak Ridge National Environmental Research Park, and Great Smoky Mountains National Park.

There is an obvious need for SAMAB to provide leadership to foster collaboration among the research and monitoring programs associated with these three biosphere reserves. A suggested initial strategy is for SAMAB to sponsor periodic meetings of scientists working at the three biosphere reserves and encourage presentations and research papers comparing trend information among the three research sites. SAMAB could publish proceedings from these meetings or coordinate the editing of the contributed papers for a special edition of an appropriate scientific journal.

In addition, SAMAB should convene a panel of scientists and ecosystem managers to review the recommendations for research and monitoring from the Southern Appalachian Assessment (Berish et al. Chapter 7) and make recommendations to member institutions concerning how to improve and integrate the disparate ongoing resource-monitoring programs to maximize their relevance to resource issues of greatest concern in the region.

NEOTROPICAL MIGRATORY SONGBIRDS

As described by Simons et al. in Chapter 9, the southern Appalachian highlands are a key refugium for neotropical migratory birds in the eastern U.S. The populations of several species of these birds utilizing mature temperate forests during their breeding cycle are in steep decline. Ongoing research in the southern Appalachians is directed in part toward defining the relationship of forest age structure and species composition to habitat utilization by these declining avian species. Collaboration among scientists has been loosely structured but complementary. Preliminary results suggest a need to conduct an overall assessment of forest stands on public lands in the southern Appalachians as to their relative importance as breeding habitat for declining avian species. SAMAB in conjunction with Partners In Flight should establish a regional task force of scientists and managers of public lands to examine national park and forest management practices including fire management within the context of the known habitat requirements for the targeted species in decline. The task force should identify critical habitat on federal lands and suggest management guidelines necessary to sustain the critical characteristics of the habitat condition such as continuous canopy cover, etc. Data collected for the Southern Appalachian Assessment are relevant to such analyses.

BLACK BEAR

As described by Clark and Pelton in Chapter 10, there has been a long history of collaboration concerning research and management of black bear (*Ursus americanus*) in the southern Appalachian highlands. Researchers studying bear biology and movement patterns began the collaborative efforts which have grown to include wildlife biologists dealing with hunting regulations and problem bears. Public education programs and bear-proof solid waste receptacles introduced into the Great Smoky Mountains National Park in the mid 1970s have been recently expanded to adjacent communities.

As expressed by the authors, there is a need for further cooperation among bear biologists and public land managing agencies. The Southern Appalachian Black Bear Study Group needs to be integrated into the regional context of ecosystem management. Broader issues such as land development creating urban/suburban barriers to movement patterns and habitat fragmentation, the reintroduction of fire into the landscape creating food sources, and the adverse impact from gypsy moth defoliation on mast crop production are of particular concern.

The SAMAB organization should invite the Southern Appalachian Black Bear Study Group to become a standing policy committee of SAMAB so that the researchers and wildlife biologists from federal and state agencies and universities coordinating policy, research, and management of this charismatic species symbolizing wilderness values in the southern Appalachians can do so in a broader context of regional ecosystem management.

RED WOLF

As described by Lucash et al. in Chapter 11, the next major phase of the arduous red wolf (*Canis rufus*) repatriation process into the southern Appalachian landscape is to release more families and extend the geographic range of release sites. This significantly escalates the level of effort, but is a critical step toward establishing a viable naturally reproducing population in the wild. This phase will require cooperation among the U.S. Forest Service, National Park Service, U.S. Fish and Wildlife Service, and appropriate state wildlife agencies. SAMAB should establish a region-wide task force on red wolf repatriation to lay the groundwork for coordinating the required policy, public education, identification of release sites, and the monitoring of released animals.

BROOK TROUT

As described by Guffey et al. in Chapter 12, the natural range of the native brook trout (*Salvelinus fontinalis*) has been reduced by 79%. The isolated populations that remain tend to be in high elevation first- and second-order streams which may be susceptible to acidification from air pollution. Studies have shown genetic variation among major watersheds harboring the species. The threat of widespread defoliation from gypsy moth threatens aquatic habitat as well. The greatest threat may ultimately be from climate change, which may result in dramatic changes in rainfall quantity, distribution, and intensity, all of which have the potential to greatly alter the aquatic environment of first- and second-order streams. This compilation of stressors over the long-term places this species at considerable potential risk. There is currently no regionally based strategy for protecting the genetic diversity of this species by major watershed within the context of these stressors, which in combination could conceivably greatly exacerbate the individual risks they represent.

Again, SAMAB could demonstrate leadership in the conservation of this native trout by assembling a task force of appropriate experts to formulate an integrated, interagency management, research and monitoring plan to insure protection of the gene pool and the expansion of the range into higher order, lower elevation streams within the major watersheds.

WILD BOAR

As documented by Peine et al. in Chapter 13, there are very different resource management agendas being practiced in the southern Appalachian landscape concerning the European wild boar (*Sus scrofa*). The National Park Service is working diligently to extirpate this exotic species from Great Smoky Mountains National Park, while state wildlife agencies are releasing these animals as a game species in nearby hunting areas. This is another case where SAMAB might provide enlightened leadership by convening a task force to review various feral hog management practices and make policy recommendations to reconcile, to the degree practical, the disparate management practices.

GYPSY MOTH

As described by Schlarbaum et al. in Chapter 14, gypsy moth (*Lymantria dispar*) defoliation may be the greatest immediate threat to the natural resources of the southern Appalachians. Experience from repeated infestation in the northern reaches of the southern Appalachians provides valuable insight as to probable impacts as the infestation moves south. The central conundrum being faced in the south is the potential conflicting management policies between the U.S. Forest Service and the National Park Service. The Forest Service's primary interests are to control the infestation in high value timber whereas the Park Service does not want to lose other invertebrate species from the landscape in the process of spraying insecticides to control the gypsy moth infestations. It is conceivable that Park Service lands may become reservoirs of insects adjacent to national forests where resource managers are likely to be engaged in active control programs. The SAMAB organization should establish a standing policy committee to keep up to date on the status of infestation and developments in control technology, review agency policy concerning control of the forest pest, and develop a strategic plan for coordinated policy and control activity years prior to heavy infestation. Lessons learned from such a program might serve as a model to coordinate activity for other pests and pathogens.

AIR QUALITY

As documented by Peine et al. in Chapter 15, the efforts of the Southern Appalachian Mountains Initiative (SAMI) are comprehensive in addressing the pervasive problem of air pollution in the southern Appalachians. This is probably the most complex issue threatening natural resources in the region. This is a case where the leadership of SAMAB resulted in specific action. A carefully planned and orchestrated workshop sponsored by SAMAB was held among various stakeholders concerning air quality in the region to devise a strategy to more comprehensively address the problem. That meeting led to collaboration among scientists dealing with air quality to create a multimillion dollar research and monitoring initiative and the establishment of an institution (SAMI) designed to comprehensively address the issue from the disparate perspectives of the various stakeholders. When the work of the SAMI team is complete, SAMAB will remain in place over the long term and at that time may need to again provide leadership in this important issue.

FIRE MANAGEMENT

As documented by Buckner and Turrill in Chapter 16, as an agent of disturbance, fire in the landscape increases species richness and diversity. Federal land managers are now recognizing the role of fire in sustaining heterogeneity in the forested landscape. Unfortunately, no fires have yet been allowed to burn as "prescribed burns" in Great Smoky Mountains National Park, but plans are being made to do so. The U.S. Forest Service has conducted six prescribed fires in four national forests in the southern Appalachian region to study yellow pine regeneration. None of the these fires have been at the watershed scale as recommended by the authors in order to sustain a remnant gradation of plant communities reflecting the influence of natural and human set fires on the landscape. The authors warn that seed sources for some of these fire-dependent species are becoming scarce, thus reducing the effectiveness of reintroducing fire for their benefit.

There is a need to coordinate fire policy among the federal land managing agencies in the southern Appalachians. SAMAB should establish a standing policy committee to coordinate an effort to evaluate the role of fire on a regional scale and select specific watersheds most suitable for large scale prescription fires to sustain fire-dependent and enhanced species at a viable scale.

MANAGING GROWTH FOR SUSTAINABLE COMMUNITIES

As described by Peine in Chapter 17, ecosystem management is inextricably tied to the human dimension. This is an intuitively obvious statement, but not universally recognized by those charged with ecosystem stewardship. Understandably, natural resource managers focus primarily within the boundaries of their management units where they have jurisdiction and are more familiar with the issues of concern. They tend to tread lightly into the "outside" world, even though the source of many resource perturbations occur outside their land management boundaries. As awareness grows of the pervasive influence of human activity, even within the most remote reaches of protected natural areas, the need for ecosystem managers to become more involved in a regional perspective is becoming readily apparent.

As documented by Peine in Chapter 17 and Hinote in Chapter 5, the SAMAB cooperative has begun a community outreach program to support efforts to manage growth following the principles of sustainable development reported by McCormick in Chapter 1 of this volume. As experience is gained, SAMAB should expand this program to serve more communities through a limited sharing of resources and expertise represented by the federal and state agency member institutions. Such a program might be entitled the "SAMAB Partnership for Sustainable Communities."

GRASSY BALDS

As documented by White and Sutter in Chapter 18, the grassy balds are important and unique historic landscape features scattered throughout the Blue Ridge Province of the southern Appalachian highlands. They are of particular social and cultural as well as botanical value. Currently, only a small percentage of the balds is being maintained through active management. Associated rare and endemic species are dispersed geographically throughout the range of grassy balds in the region. The authors call for an assessment of the species distribution throughout the compilation of grassy balds and a cost–benefit analysis of bald restoration and maintenance throughout the southern Appalachian region. Such an analysis and maintenance agreement should be coordinated through a SAMAB coordinated initiative.

CLIMATE CHANGE

As described by Peine and Berish in Chapter 19, global climate change may become the greatest threat to natural resources in the southern Appalachians. The high elevation spruce–fir forest ecosystem, elements of which are endemic to the southern Appalachians, will likely be in particular jeopardy. Federal land managers are the stewards for almost the entirety of this ecosystem that is a relic of past ice ages. High elevation streams are also particularly vulnerable.

As suggested by Peine and Berish, there is a series of actions that ecosystem managers should initiate now to actively address the potential long-term consequences of climate change. The SAMAB cooperative should convene a task force of experts to conduct a sensitivity analysis to climate change, utilizing data from the Southern Appalachian Assessment. Their assignment should include an evaluation of the adequacy of existing resource monitoring activity in the region to detect sensitivity to climate change. Suggestions should be made for the integration of monitoring activity at the U.S. Forest Service operated Cowetta, National Biological Service operated Noland Divide, and the Oak Ridge National Laboratory-operated Walker Branch research watersheds in the southern Appalachians to detect climate-sensitive ecosystem dynamics.

CLINCH–POWELL RIVER BASIN

As described by Yankee et al. in Chapter 20, probably the most important "hot spot" of environmental concern throughout the entire southern Appalachian region is the Clinch–Powell River Basin. This area has been identified as a "hot spot" of concern for endangered species at the national level, as well. The innovative resource stabilization and restoration efforts underway consist of a loosely configured set of initiatives by several institutions, both governmental and not-for-profit.

There is a need to solicit more help and improve coordination of service so as to maximize the collaborative benefits from energies expended. Again, SAMAB is ideally suited to act as a facilitator for such an endeavor. This situation points out yet another example of a critical need for leadership from the SAMAB cooperative.

SPRUCE–FIR FOREST

As described by Nicholas et al. in Chapter 21, the high elevation forests are experiencing a variety of stressors, such as exotic pests and pathogens, airborne pollution, and wind shear as a result of a more open canopy, which collectively have had devastating effects. Climate change may become the greatest potential threat of all. Several endemic species of concern in the southern Appalachians including the Frasier fir (*Abies fraseri*) are found in this relic boreal forest. This is arguably the most threatened forest community occurring in the southern Appalachian highlands.

There is no comprehensive strategy in place to sustain this very important ecosystem. Studies of various components of the ecosystem such as small mammals, lichens, birds and herbaceous plants and trees have been conducted, but no comprehensive monitoring program or risk assessment has been conducted. There has been minimal research to describe potential genetic diversity among the isolated stands of Frasier fir, nor are there any programs to preserve the integrity of the gene pool from these isolated populations.

SAMAB should take the initiative to form an interdisciplinary, interagency task force to draft a comprehensive management, research, and resource monitoring plan for this "crown jewel" ecosystem that is at the heart of one of the country's "crown jewel" national parks.

A CALL TO ACTION FOR SAMAB

Will these recommended actions by SAMAB require a huge amount of effort and energy beyond that which staff members of the member institutions already expend? Not necessarily! The above agenda implies that the front line specialists working day to day in ecosystem management related programs need to recognize the value in utilizing SAMAB as a tool for institutional coordination for program development and integration of operation. This will result in SAMAB brokering a higher level of influence on the polices and actions associated with operative programs in ecosystem management. This level must be achieved if SAMAB is to reach its true potential as a facilitator of integrated ecosystem management.

Hopefully, the process, the struggle, the setbacks, and triumphs cumulatively result in steady progress made in the southern Appalachians to embrace the principles of ecosystem management for sustainability and that this journey, joined by so many dedicated souls as documented in this volume, will serve as inspiration to those working in the other 328 international biosphere reserves throughout the world. As has occurred in the southern Appalachians, the UNESCO Man And Biosphere program continues to inspire leadership by example to encourage the principles of sustainability in the utilization of indigenous natural and cultural heritage throughout the world, the only planet we have to depend on for our collective survival.

Index

Index

A

Abies alba Mill. (silver fir), 438

Abies balsamea (L.) (Balsam fir), 438

Abies fraseri (Fraser fir), 134, 178, 294, 299–300, 390, 409, 431–454, see also Spruce–fir forest management

Abrams Creek park, 50

Absolute priority principle, 10

Acadian flycatcher *(Empidomax virescens),* 197, 202

Acceptance
   international, 10, see also Public attitudes
   national, 10–11
   regional and local, 11–12

Acer pensylvanicum (striped maple), 440

Acer rubrum (red maple, swamp maple), 273

Acer saccharum (sugar maple), 275, 295

Acer spictum Lam. (mountain maple), 440

Acidic deposition, 133, 137, 149–150, see also Air quality
   in spruce–fir forest, 441–442

Action Plan for Biosphere Reserves, 30, 87

Adelgid, woolly *(Adelges* spp.), 134, 178, 294, 296, 409, 438–441, 443

Adverse effects of environmental monitoring, 169

African-Americans, as early inhabitants, 76–77

Agriculture
   Cades Cove park development and, 50–52
   of Clinch–Powell River Basin, 424
   economic importance of, 77–78
   forest clearing and, 74
   grassy balds and, 384
   land clearing by burning in, 333

Ailanthus altissima (Tree-of-Heaven), 298–299

Air pollution, 142–154, 308–310, see also Acidic deposition; Air quality
   in spruce–fir forest, 441–443

Air quality, 94, 142–154, 160, 179–180, 307–327, see also Climate change; Environmental monitoring
   air as natural resource, 307
   key research findings overview, 312–313
   management initiatives for, 480
   new source permit application reviews, 322–324
      conclusions, 324–326
      Eastman Chemical Company, 323
      number and nature of, 322–323
      Tenn Luttrell, 323–324
   principles of management, 308–312
      defining air quality-related values, 310–311
      integration of management functions, 311–312
      manager responsibilities, 308
      modeling of pollutant transport, 310
      monitoring, 310, see also Air pollution; Environmental monitoring

pollution source inventorying, 309–310
pollution source regulation, 308–309
research support, 311
in Russia versus U.S., 111
Southern Appalachian Mountains Initiative (SAMI), 313–322
   effectiveness of decision-making process, 317–322
   federal land manager perspective, 318–319
   industry perspective, 321–322
   state regulator perspective, 317–318
   volunteer perspective, 319–321
   organizational structure and function, 313–317

Air Quality-Related Values (ARQVs), 310–311

Albizzia julibrissin Durazz. (mimosa), 298–299

Alkes alkes (European elk), 102

Allegheny chinquipin *(Castanea pumila* Mill.), 293

Amelanchier arborea (serviceberry), 212, 440

American beech *(Fagus grandifolia),* 178, 210, 273, 297

American Land Sovereignty Protection Act, 36

Amphibians, 75

Andrews Bald, 388–389

Angelica triguinata, 275

Animal species (fauna), 127–128, see also individual species
   public attitudes and, 132

Anthracnose, dogwood *(Discula destructiva* Redlin), 133–134, 295–296

Anthropogenic impact, 8–9, see also Land use patterns

APHIS (USDA Animal and Plant Health Inspection Service), 300–301

Appalachian Mountains, 68

Appalachian Plateaus, geological features of, 66–68, 121

Appalachian Task Force, 113

Appalachian Trail Conference, 113

Aquatic habitat, 138–141

Aquatic monitoring, 178–179

Aquatic resources, 76

Aquatic species, of Clinch–Powell River Basin, 422

Aquatic team environmental assessment, 136–142, 160

Aquila chrysaetos (golden eagle), 120

Architectural guidelines, for Pittman Center, 367–368

Areas
   core, 25
   managed use (buffer zones), 25
   transition (cooperative), 25–26

Armellaria mellea (beech fungus), 275, 297

Ash, mountain *(Sorbus americana* Marshall), 436

Assateague Island pony importation, 445

Assessment
   community-defined endpoints for, 119–120
   environmental, 117–166, see also Environmental assessment

environmental impact (EIA), 118
    resource management, 118
Atlantic Coastal Plain Biosphere Reserve, 31
Atmospheric team environmental assessment, 142–154
    acidic deposition, 149–150
    ozone exposure
        forest tree response to, 148
        at ground level, 143–149
    pollutants of concern, 142
    questions developed for, 161
    visibility, 150–154
Attitudes, public, 132, 207, 222, 241–244, 287, 304, 325,
        372, 392, 450
    toward fire, 334–336
Automobiles, 50, 57
Avens
    bent *(Geum geniculatum),* 390
    mountain *(Geum radiotum),* 390
Azalea, see *Rhododendron*

B

Backman's Warbler, 190
Badgeworth Nature Reserve (Great Britain), 391
Bait station monitoring, 177
Balsam fir *(Abies balsamea* [L.]), 438
Balsam wooly adelgid *(Adelges piceae),* 134, 178, 294, 409,
        438–441, 443
Barberry, Japanese *(Berberis thunbergii),* 298–299
Bay-breasted warbler, 190
Bear, black *(Ursus americanus),* see Black bear *(Ursus
        americanus)*
Beaver *(Castor canadensis),* 102, 228
Beech
    American *(Fagus grandifolia),* 178, 210, 273, 297
    gray *(Fagus grandifolia),* 275
Beech bark disease complex, 297
Beetle, pine bark *(Dendroctonous frontalis),* 329
Bent avens *(Geum geniculatum),* 390
Berberis thunbergii (Japanese barberry), 298–299
Betula alleghaniensis (yellow birch), 178, 440
Betula lutea (yellow birch), 436, 437, 440
Big South Fork National River and Recreation Area, 215
Big Thicket National Preserve (Texas), 31
Biodiversity, 34
    fire as agent of, 339–340
    grassy bald case study, 375–395
        background, 375–377
        conclusions, 391–392
        distribution and floristics, 377–380
        distribution of flora and significance, 390–391
        management, 387–390
        origin, maintenance, and succession, 380–387
BioMon database, 36
Biosphere Reserve Integrated Monitoring Program (BRIM),
        36
Biosphere reserves
    cluster concept of, 85–87, 105
    conceptual evolution of, 81–83
    early development of regional, 85–88
    map of U.S., 27

multiple functions of, 84–85
origins and definition of concept, 24–26
Prisko-Terrasny Reserve (Russia) and Great Smoky
        Mountains National Park compared, 99–114, see
        also individual locations
    management, 106–108
    national government and, 104–106
    overview, 100–104
    priority setting, 108
    programs and partnerships, 108–114
    in Russia, 100–101
    Southern Appalachian Mountain, 88–97, see also
        Southern Appalachian Man and the Biosphere
        Program (SAMAB)
    UNESCO criteria for, 25–26
    in United States, 101
Birch
    yellow *(Betula alleghaniensis),* 178
    yellow *(Betula lutea),* 436, 437, 440
Birds, 75
    neotropical migratory song, 187–208, see also
        Neotropical migratory song birds
    translocation of, 226–227
Bison
    American *(Bison bison),* 120, 210
    European *(Bonasus capreolus),* 102, 109
Bison antiques, 42
Bittersweet, oriental *(Celastrus orbiculatus* Thunb.),
        298–299
Black bear *(Ursus americanus),* 112, 177, 209–223, 229,
        276
    ecology of, 211–213
    ecosystem management and conservation, 215–221
    habitat available for, 128–131
    historical perspective, 209–210
    management initiatives for, 478
    North American distribution of, 211
    Operation Smoky and, 53–54
    public policy and, 53–54
    research and management, 214–215
Blackberry *(Rubus* spp.), 210, 212, 275, 383, 440
Blackburnian warbler *(Dendroica fusca),* 187, 202
Black-capped vireo, 190
Black cherry *(Prunus serotina),* 48, 212, 295
Black mountains, see Spruce–fir forest
Black oak *(Quercus velutina),* 210
Black-throated blue warbler, 197, 202
Black-throated green warbler *(Dendroica virens),* 197,
        202
Black warbler, 187
Blueberry *(Vaccinium* spp.), 210, 212, 383, 387, 388, 440
Blue Ridge, geological features of, 66
Bluets, mountain *(Houstonia purpurea* var. *montana),* 390
Boar, European wild *(Sus scrofa),* see European wild boar
        (Sus scrofa)
Bobwhite *(Colinus virginianus),* 210
Bonasa umbelous (ruffed grouse), 276
Bonasus capreolus (European bison), 102, 109
Boundaries, 7–8
Breeding Bird Survey, 190–193, 194, 203–206, see also
        Neotropical migratory songbirds

Brook trout, southern Appalachian *(Salvelinus fontinalis)*, 220, 247–265, 409
  biological diversity and, 259
  distinctiveness and impact of stocking, 252–257
    effects of hybridization on, 255–256
    genetic inventory, 254–255
    heterogeneity among native populations, 256–257
    molecular population genetics, 253–254
    native and local history, 252–253
    stocking history, 252
  European wild hog and, 276
  history and status of, 248–252
    decline in Southern Appalchians, 249–250
    distribution, 248–249
    in Great Smoky Mountains National Park, 250
    rainbow trout effect on, 250–252
  management and conservation of, 257–258
  management initiatives for, 479
  monitoring of, 177–178
Brown-headed cowbird *(Molothrus ater)*, 192, 197
Buffer zones (managed use areas), 25
Bulls Island wolf release, 226, see also Red wolf *(Canis rufus)*
Butternut canker *(Sirococcus clavigignenti-juglandacerum)*, 134, 294–295
Butternut canker *(Sirococcus clavigignent-juglandacerum)*, 134
Butternut *(Juglans cinera)*, 134, 294–295

C

Cades Cove
  agriculture and, 50–51
  geological cross section of, 68
  historic landscape management at, 51–52
  wolf release in, 229–235, see also Red wolf *(Canis rufus)*
Calvinist doctrine, 45–46
Canadian Man and the Biosphere Program (MAB), see under Man and the Biosphere Program (MAB)
Canis latrans (coyote), 225, 240
Canis lupis (timber wolf), 120
Canis rufus (red wolf), see Red wolf *(Canis rufus)*
Canker, butternut *(Sirococcus clavigignenti-juglandacerum)*, 134, 294–295
Cape Romain wolf release, 226, see also Red wolf *(Canis rufus)*
Capreolus capreolus (Siberian roe deer), 102
Carbon dioxide, 398–399, see also Climate change
Cardaria nutans L. (musk thistle), 298–299
Carex misera (wretched sedge), 390
Carolina hemlock *(Tsuga caroliniana)*, 134
Carolina Northen flying squirrel *(Glaucomys sabrinus coloratus)*, 441
Carolinian–South Atlantic Biosphere Reserve, 31–32
Carya spp., 54, 210, 273
Castanea dentata (American chestnut), 74, 135, 210, 293–294, see also Pathogen and pest control
Castanea pumila Mill. (Allegheny chinquapin), 293
Castor canadensis (beaver), 102, 228
Cat, European pole *(Mustele putoris)*, 102
Cattle ranching, 47–48

Caves, 36, 37, 421, 427
Celastrus orbiculatus Thunb. (oriental bittersweet), 298–299
Central California Coast Biosphere Reserve, 37
Cerulean carbler, 220
Cervus canadensis (American elk), 120, 210, 227
Chattooga River Project, 12
Chenopodium spp., 42
Cherokee Nation
  European settlers' interaction with, 45–47
  historical land use practices of, 43–44
Cherry
  black *(Prunus serotina)*, 48, 212, 295
  fire *(Prunus pensylvanica)*, 178, 437
Chestnut, American *(Castanea dentata)*, 74, 135, 210, 293–294, see also Pathogen and pest control
Chestnut blight *(Cryphonectria parasitica)*, 74, 135, 210, 293–294, see also Pathogen and pest control
Chestnut oak *(Quercus prinus* Lamb.), 338
Chestnut-sided warbler *(Dendroica pennsylvanica)*, 187, 202
Chihuahuan Desert, 37
Chipmunk *(Tamias striatus)*, 276
Christmas tree growing, 446
Cinquefoil *(Potentilla rumex)*, 275, 379
Civilian Conservation Corps (CCC), 50, 103, 173, 334
Claytonia virginica (spring beauty), 54, 275
Clean Air Act, 150, 152, 300, 314, 318, 321–322, see also Air pollution; Air quality
Clean Water Initiative of Tennessee Valley Authority (TVA), 419–420
Clemson University, 214
Climate, 69–71
  fire behavior and, 336–337
  of grassy balds, 382
Climate change, 397–415
  ecological and social effects of, 398–399
  Great Smoky Mountains National Park case example, 406–412
    assessment of data needs, 408
    definition of valued resources, 407–408
    goals and measurement endpoints, 407
    monitoring, 411
    political and ecological boundaries, 407
    sensitivity analysis, 409–410
    preliminary cultural resource, 410
    preliminary natural resource, 409–410
    strategic plan for species conservation, 410–411
    time frame and climate scenarios, 408
  historical background, 397–398
  immediate assessment steps for, 402–406
    monitoring, 406
    sensitivity analysis, 402–405
    framework for, 402–403
    preliminary human, 404–405
    preliminary natural resources, 403–404
    strategic plan for species conservation, 405–406
  management initiatives for, 481
  mountainous areas in detection of, 399–402
  spruce–fir forest and, 444
Clinch–Powell River Basin Initiative, 417–438
  environmental stressor inventory, 425

integrated risk assessment, 425–426
  EPA, 425
  TVA, 426
management initiatives for, 481
mitigation activities, 426–427
  cave registry program, 427
  community activities and education, 427
  compatible economic development, 427
  riparian restoration, 426–427
  threatened organisms research, 427
natural resources inventory, 420–423
  aquatic species, 422
  caves, 421
  land cover/land use, 421
  river system, 422
  topography, 421
  water quality, 422
  water resources, 421–422
partnership participants in, 419–420
  Nature Conservancy, 419
  Tennessee Valley Authority (TVA), 419–420
  Clean Water Initiative, 419–420
  river action teams, 420
  U.S. EPA, 420
  U.S. Fish and Wildlife Service, 420
significance of, 419
social values and resource utilization inventory, 423–425
  agriculture, 424
  forestry, 424
  historical context, 423
  mining, 424
  social context, 423–424
strategies to sustain momentum, 428
Clingmans Dome, 180, 209
Cluster concept of biosphere reserves, 85–87, 105
Colinus virginianus (bobwhite), 210
Colorado Rockies, 37
Commission on Sustainable Development, 34–35
Communities, resource-dependent, 154–155, 157
Community-defined endpoints for assessment, 119–120
Community involvement, 123, see also Stakeholder involvement
Community planning
  for Pittman Center, 362–363
  programs for, 93
Community vision statement, for Pittman Center, 366
Congaree Swamp National Monument (South Carolina), 31
Conopholis americanus (squawroot), 212
Conservation biology, SAMAB program in, 93–94
Consortium for Study of Man's Relationship with the Environment, 28
Continuous Forest Inventory program, 173
Convention of Biological Diversity, 34
Cooperative (transition) areas, 25–26
Copper Hill Region, sulfur dioxide pollution in, 142
Core areas, 25
Cornus florida (flowering dogwood), 133–134, 295–296
Cornus spp., fungus-resistant, 295–296
Cottontail rabbit *(Sylvilagus floridanus)*, 210, 229
Cove Mountain ozone monitoring program, 180

Cowbird, brown-headed *(Molothrus ater)*, 192, 202
Coweeta Biosphere Reserve, 32
Coweeta Hydrologic Laboratory, 12, 174
Coyote *(Canis latrans)*, 225, 240
Creating Successful Communities (Mantell et al.), 366–367
Cryphonectria parasitica (chestnut blight), 74, 135, 293–294, see also Pathogen and pest control
Cryptococcus fagisuga (beech scale), 297
Cultural aspects
  of climate change, 401
  of fire, 331–332
  of hunting, 212
  of Southern Appalachian region, 78
Cultural programs, 94
Cultural resources, climate change and, 410
Cumberland Plateau, 68

D

Danthonia compressa (mountain oat grass), 275
Databases, 35–36
Data management, 169–170
Deer
  environmental impact of, 55
  Siberian roe *(Capreolus capreolus)*, 102
  white-tailed *(Odocoileus virginianus)*, 132, 177, 210, 228, 229, 241
Dendroctonus frontalis (pine bark beetle), 329
Dendroica fusca (Blackburnian warbler), 187, 202
Dendroica pennsylvanica (chestnut-sided warbler), 187, 202
Dendroica virens (black-throated green warbler), 197, 202
Diagnostic versus triage approach, 122
Dioscorea batatas (wild yam), 273
Discula destructiva Redlin (dogwood anthracnose), 133–134, 295–296
Dogwood anthracnose (*Discula destructiva* Redlin), 133–134, 295–296
Dogwood *(Cornus* sp.)
  flowering *(Cornus florida)*, 133–134, 295–296
  fungus-resistant, 295–296
Drosera rotundifolia (roundleaved sundew), 388

E

Eagle, golden *(Aquila chrysaetos)*, 120
Eastern hemlock *(Tsuga canadensis)*, 134, 296, see also Woolly adelgid
Eastern kingbird *(Tyrannus tyrannus)*, 187–188, 202, see also Neotropical migratory song birds
Eastern Workshop on Black Bear Management, 216
Eastman Chemical permit application, 323
Economic development, bear conservation and, 217
Economic trends, 156–158
Economy, of Southern Appalachian region, 77–78
Ecosystem management
  acceptance of, 10–12
    international, 10
    national (U.S.), 10–11
    regional and local, 11–12
  definition of, 118

environmental assessment in, 117–165
    approach, 118–120
    conceptual basis, 117–118
    conclusions drawn from, 158–159
    questions used in, 160–162
    Southern Appalachia as focus of, 120–121
    Southern Appalachian regional program, 121–123
    Southern Appalachian results, 123–158, see also
        Environmental assessment
fire and, 336–340, see also Fire management
goals of, 118
for high-elevation spruce–fir forest, 431–454, see also
    Spruce–fir forest management
historical origins of, 9
historical precedents and evolution of, 41–58
    Cherokee Nation (1540–1750), 43–44
    early European settlement of U.S. (1750–1838),
        45–47
    future challenges, 57
    large-scale resource extraction period (1838–1930),
        47–49
    national park creation period (1930–1975), 49–52
    natural cultural resources protection period
        (1976–1987), 52–55
    prehistoric period, 42–43
    sustainable development approach (1988 to present),
        55–57
holistic principles of, 6–8
institutions and, 457–474
    changes in 20th century, 460–461
    conclusions about, 470–472
    design of, 464–470
    dilemma of, 464
    environmental statutes of 1970s, 461–463
    history of, 458–459
    "turn-of-the-century" commercial timbering, 459–460
plan framework for, 432–435
    ecosystem *a priori* condition, 432
    framework summary, 434
    implementation strategy, 434
    resource inventory, 432
    resource monitoring, 433
    risk assessment, 433
    spruce–fir forest case study, 434–435
    stablization/recovery plan, 433–434
    stressors inventory, 432–433
principles of, 12–19
    ecological basis for, 17
    eight central principles of, 16
    implementation implications of, 18–19
    key, 15
    main components of ecosystem approach, 15
    management implications of, 17
    overview of, 14
    seven pillars of, 16
    for U.S. Forest Service management planning process,
        18
    for vision of "sustainable United States," 13–14
as related to sustainability, 6–9
relevance of Southern Appalachian region, 79

SAMAB program in, 94
    sustainable development as defined in, 3–6
Ecosystem stabilization and restoration, Clinch–Powell
    River Basin Initiative, 417–430, see also
        Clinch–Powell River Basin Initiative
Ecotonal hypothesis, of grassy balds, 382–383
Education
    environmental professional, 94
    public, 94, 228, 344
Eicotyles tajacu (collared peccary), 271
Elderberry *(Sambucus pubens)*, 440
Elk
    American *(Cervus canadensis)*, 120, 210, 227
    European *(Alkes alkes)*, 102
EMAP (Environmental Monitoring and Assessment
    Program), 120, 180–181
Empidomax virescens (Acadian flycatcher), 197, 202
Employment, 156
Endangered and threatened species, 127–128, 131, 241
    climate change and, 404
    exotic pest impact on, 299–300
    Jones middle-tooth snail *(Mesodon jonesianus)*, 276
    salamander *(Plethodon* spp.), 195, 276, 407, 441
    of spruce-fir forest, 441
Engel, Ronald, 37
Enlightenment, philosophy of, 46
Entomophaga maimaiga (gypsy moth fungus), 297
Environmental assessment, 117–165
    approach to, 118–120
    conceptual basis of, 117–118
    framework for, 432–435
    land area by ecological sections, 127
    management initiatives for, 477
    regional program for Southern Appalachia, 121–123
    results for Southern Appalachia, 123–158
        aquatic team, 136–142
        atmospheric team, 142–154
        acidic deposition, 149–150
        ozone exposure and forest tree response, 148
        ozone exposure at ground level, 143–149
        pollutants of concern, 142
        visibility, 150–154
        human use patterns at landscape scale, 124–127
        social, cultural, and economic team, 154–158
        terrestrial team, 127–135
    Southern Appalachia as focus of, 120–121
    triage versus diagnostic approach to, 122
    types of, 118–119
Environmental education
    Russia versus U.S., 113–114
    SAMAB program in, 94
Environmental impact assessment (EIA), 6, 9, 118
Environmental integrity, 118
Environmental monitoring
    adverse effects of, 169
    definitions and objectives of, 167–170
    future role of SAMAB in, 183–184
    history in Southern Appalachia, 173–174
    management initiatives for, 477
    overview of current, 174–176

within administrative areas, 174–175
    intensive-site monitoring, 174
    regional-scale, 175–176
regional framework for, 170–173
in Russia versus U.S., 110
SAMAB program in, 93
scale in, 171–173
specific programs in Southern Appalachia, 176–183
    Forest Health Monitoring program, 180–182
    Great Smoky Mountains National Park, 176–180
    risk assessment report card for, 182–183
Environmental Monitoring and Assessment Program
    (EMAP), 120, 175–176, 180–181
Environmental policy, see Policy issues
Environmental Protection Agency (EPA), 179
    EMAP Assessment Framework of, 120, 175–176,
      180–181
    environmental risk as identified by, 120
    particulate standards of, 314
    Watershed Assessment Approach of, 119
Environmental risk, 120
Environmental stressors, interrelationship of, 126–127
Ermine, European *(Mustela erminea),* 102
Erythronium americanum (fawn lily), 275
Eureopean bison *(Bonasus capreolus),* 102, 109
European ermine *(Mustela erminea),* 102
European gypsy moth *(Lymantria dispar* L.), 134–135, 214,
    296–297, 479
European hare *(Lepus timidus),* 102
European pole cat *(Mustela putoris),* 102
European settlement period (1750–1838), 45–47, 76, 77, 384
European wild boar *(Sus scrofa),* 54, 102, 228
    conclusions about, 286–287
    control program, 276–280
    Great Smoky Mountains National Park setting, 271
    home range and activity patterns, 272–274
    management initiatives for, 479
    monitoring, 280–281
    origin and invasion by, 271–272
    policy mandates, 271
    population models, 282–283
    recommendations, 283–286
    reproductive characteristics, 274
    resource impacts, 275–276
Exotic pest damage, 293–297, see also Pathogen and pest
    control
Exotic plant invasions, 297–299
Exotic species
    European wild boar *(Sus scrofa),* 267, 271–290
      conclusions about, 286–287
      control program, 276–280
      Great Smoky Mountains National Park setting, 271
      home range and activity patterns, 272–274
      monitoring, 280–281
      origin and invasion by, 271–272
      policy mandates, 271
      population models, 282–283
      recommendations, 283–286
      reproductive characteristics, 274
      resource impacts, 275–276
    principles of control of, 268–271

F

Fagus grandifolia (American beech), 178, 210, 273, 297
Fagus grandifolia (gray beech), 275
Falcon, peregrine *(Falco peregrinus),* 109, 227
Falco peregrinus (peregrine falcon), 109, 227
Fauna (animal species), 127–128
Fawn lily *(Erythronium americanum),* 275
Feather grass *(Stipa joannis),* 102
Felis concolor (mountain lion), 120, 227
Fern, Virginia chain *(Woodwardia virginica),* 177
Fir
    balsam *(Abies balsamea* [L.]), 438
    Fraser *(Abies fraseri),* 134, 178, 294, 299–300, 390, 409,
      see also Spruce–fir forest
    silver *(Abies alba* Mill.), 438
Fire, in grassy bald formation, 383–384
Fire cherry *(Prunus pensylvanica),* 178, 437
Fire management, 329–347
    by Cherokee Nation, 44
    conclusions about, 342–345
    by early Europeans, 46
    ecological principles of, 336–340
      biodiversity, 339–340
      fire behavior, 336–337
      soil effects of fire, 337
      vegetative response, 337–339
    fire history in Southern Appalachia, 330–336
      federal policy and, 334–336
      historic setting, 333–334
      prehistoric setting, 330–332
    initiatives for, 480
    options for, 340–342
    public policy and, 55
Fish community monitoring, 178–179, see also Brook trout
Fisher *(Martes pennanti),* 227
Flowering dogwood *(Cornus florida),* 133–134, 295–296
Flycatcher, Acadian *(Epidomas virescens),* 197, 202
Forest Health Monitoring Demonstration, 180–182
Forest Health Monitoring Program, 93, 180–182
Forestry management programs, 93
Forests, 72–74, see also Vegetation and individual species
    high-elevation spruce–fir, 431–454
      ecosystem characteristics, 435–448
      ecosystem management plan, 432–435
    old-growth versus second-growth as bird habitat,
      198–202
    successional class and forest type group, 75
    types of, 128
Forest tree ozone response, 149
Fox *(Vulpes vulpes),* 102
Fraser fir *(Abies fraseri),* 134, 178, 294, 299–300, 390, 409,
    431–454, see also Spruce–fir forest management
Fraxinus spp. (ash), 48
Fritillaria ruthenica (Russian hazel), 102
Funding, 27–28, 207
    for air quality management, 325
    for black bear program, 222
    for climate change programs, 412
    for Clinch–Powell River Basin Initiative, 430
    for European wild boar *(Sus scrofa)* control, 287

for fire management, 344
for grassy balds, 393
for pathogen and pest control, 304
for Pittman Center, 372
for red wolf repatriation, 244
in Russia as compared with U.S., 107–108
of Southern Appalachian Man and the Biosphere Program
    (SAMAB), 95–97
for spruce–fir forest management, 450
Future challenges, 57, 475–482
FutureScape Plan, see Pittman Center

G

Gateway communities, 358–360
Gatlinburg, TN bird survey, 195
Gaylussacia spp. (huckleberry), 212
Genetic conservation, 301–302
Geographic Information System (GIS), 121
Geological history, 68–69
George Washington National Forest, 214
Geum geniculatum (bent avens), 390
Geum radiotum (mountain avens), 390
Ginseng, 45, 46
Glaucomys sabrinus coloratus (Carolina Northern flying
    squirrel), 227, 441
Glaucomys sabrinus fuscus (Virginia Northen Flying
    squirrel), 441
Glyceria nubigenu (Smoky Mountain manna grass),
    390
Golden-cheeked warbler, 190
Golden eagle *(Aquila chrysaetos),* 120
Grass
    Japanese *(Microstegium vimineum* [Trin.] A. Camus),
        298–299
    Johnson *(Solanum halepense* [L.] Pers.), 298–299
    Smoky Mountain manna *(Glyceria nubigenu),* 390
Grassy balds, 375–395
    community diversity by Nature Conservancy
        classification, 380
    distribution and floristics, 377–380
    general discussion, 375–377
    management initiatives for, 480–481
    management of
        Gregory and Andrews Balds (Great Smoky Mountains
            National Park), 388–389
        Roan Mountain, 389–390
    origin, maintenance, and succession of, 380–387
        climate and topography, 382
        ecotonal hypothesis of, 382–383
        European settlement and, 384–385
        fire in, 383–384
        geology and soils, 382
        Native Americans in, 384
        native grasses, 383
        natural factors possibly explaining, 383
        rate of succession in, 386–387
        summary of knowledge about, 385–386
    vascular plants in Southern Appalachian
        common, 380
        rare and endemic, 381

Gray beech *(Fagus grandifolia),* 275, 297, see also
    American beech
Gray's lily *(Lilium grayi),* 390
Great Smoky Mountains, Cherokee tradition and, 43–44
Great Smoky Mountains National Park, 32, see also Pittman
    Center
    bear population of, 213
    bear research in, 214
    brook trout in, 250, see also Brook trout, southern
        Appalachian
    climate change case example, 406–412
        assessment of data needs, 408
        definition of valued resources, 407–408
        goals and measurement endpoints, 407
        monitoring, 411
        political and ecological boundaries, 407
        sensitivity analysis, 409–410
        preliminary cultural resource, 410
        preliminary natural resource, 409–410
        strategic plan for species conservation, 410–411
        time frame and climate scenarios, 408
    as compared with Priosko-Terrasny Reserve (Russia),
        99–114
    creation of, 49–52
    environmental monitoring in, 176–180
        of air quality, 179–180
        aquatic and watershed, 178–179
        landscape-scale, 179
        problems and rationale for, 176–177
        risk assessment report card for, 182–184
        at species level, 177–179
        of vegetation, 179
    exotic plant pest invasions in, 297–299
    government policy and, 105–106
    grassy bald management in, 388–389
    history of, 103–104
    management of
        budget and funding of, 108
        mission of, 106–107
        organization of, 107
    as part of national system, 101
    permit applications, 322–324
        Eastman Chemical, 323
        number and nature of, 322–323
        Tenn Luttrell, 323–324
    priority setting for, 108
    programs and partnerships of
        air quality monitoring, 111
        black bear management, 112
        environmental education, 113–114
        native brook trout management, 112
        research and monitoring, 110–111
        species reintroduction, 109–110
        strategic planning for wilderness management,
            112–113
    regional and national acceptance and, 12
    significance to forest bird species of, 190
    timber industry in, 48
    tourism in, 48–49
    wolf release in, 228–235, see also Red wolf *(Canis rufus)*
Greenhouse warming, see Climate change

Gregory Bald, 388–389
Groundhog *(Marmota monax)*, 229
Grouse, ruffed *(Bonasa umbelous)*, 276
Guest River, see Clinch–Powell River Basin Initiative
Gymnoderma lineare (rock gnome lichen), 299, 440
Gypsy moth, European *(Lymantria dispar* L.), 134–135,
    214, 296–297, 479

H

Habitat
    aquatic, 138–141
    classification of, 131
    exotic pest impact on, 299–300
    for neotropical migratory birds, 190–192
    nesting productivity and, 201
    old-growth versus second-growth forests as, 198–202
Hard mast surveys, 177
Hare, European *(Lepus timidus)*, 102
Hawaiian Islands Biosphere Reserve, 31
Hazel, Russian *(Fritillaria ruthenica)*, 102
Helianthus spp., 42
Hemlock *(Tsuga* spp.)
    Carolina *(Tsuga caroliniana)*, 134
    Eastern *(Tsuga canadensis)*, 134, 296
    fungus-resistant, 296
Hemlock woolly adelgid *(Adelges tsugae)*, 134, 296
Hickory *(Carya* spp.), 210, 273, see also Butternut
Highways, 52
Historical development
    of ecosystem management, 9, 41–58
        Cherokee Nation (1540–1750), 43–44
        early European settlement of U.S. (1750–1838), 45–47
        future challenges, 57
        large-scale resource extraction period (1838–1930),
          47–49
        national park creation period (1930–1975), 49–52
        natural cultural resources protection period
          (1976–1987), 52–55
        prehistoric period, 42–43
        sustainable development approach (1988 to present),
          55–57
    of fire and fire management, 330–336
    of land use ethics, 41–58
    of Man and the Biosphere Program (MAB), 23–40
    of U.S. MAB, 27–32
Historic authenticity, 52
Historic landscapes, Cades Cove park as, 51–52
Hobblebush *(Viburnum alnifolium)*, 440
Hog, European wild *(Sus scrofa)*, see European wild boar
    (Sus scrofa)
Holistic principles of management, 6–8
Holly, mountain *(Ilex ambigua* var. *montana)*, 440
Honeysuckle, Japanese *(Lonicera japonica* Thumb.),
    298–299
Hooded warbler *(Wilsonnia citrina)*, 201
Hoopers Bald, 271–272
Houstonia purpurea var. *montana* (mountain bluets), 390
Huckleberry *(Gaylussacia* spp.), 212
Human population growth, bear conservation and, 216–217
Human use patterns, 124–127, see also U-index

Hunting, 219
    as cultural value, 212
    regulation of bear, 214–215
Hylocichia mustelina (wood thrust), 192, 197, 201

I

Ilex ambigua var. *montana* (mountain holly), 440
Indicator species, see also Neotropical song birds and
    individual species
    black bear *(Ursus americanus)*, 195, 209–223
        ecology of, 211–213
        ecosystem management and conservation, 215–221
        historical perspective, 209–210
        research and management, 214–215
    neotropical migratory songbirds, 187–208
        ecosystem approach to conservation of, 193–201
        evidence and causes of decline in, 190–193
        Great Smoky Mountains National Park versus
          Cherokee National Forest, 203
        high priority southern species of, 203
        importance of in Southern Appalachians, 190
        in old-growth versus second-growth forests, 198–201
        research insights about, 201–203
        research programs in Southern Appalachians, 194
        Roaring Fork Study census results for, 196
        species codes used for, 197
        types and distribution, 187–189
    salamander *(Plethodon* spp.), 195, 276, 407, 441
    tall milkweed, 180
Information, access to, 35–36
Institutions and ecosystem management, 457–474
    changes in 20th century, 460–461
    conclusions about, 470–472
    design of, 464–470
        clarification of vision, 465–466
        initiating and facilitating collaboration, 466–467
        restructuring of involvement, 467–470
        conflict resolution, 468–469
        incentives and disincentives, 467–468
        information gathering and dissemination, 467
        monitoring, 468
        nested institutional arrangements, 469–470
    dilemma of, 464
    environmental statutes of 1970s, 461–463
    history of, 458–459
    "turn-of-the-century" commercial timbering, 459–460
Interagency Ecosystem Management Task Force, 34
Interagency Monitoring of Protected Visual Environments
    (IMPROVE), 179–180
International acceptance, 10
International Association for Ecology (INTECOL), 10
International Biological Program (IBP), 10
International Biosphere Reserve Congress, 29–30
Internet, 35–36
Inventory
    natural resources, 356
    resources for Pittman Center, 364–365
Isolated population management, southern Appalachian
    brook trout as example of, 247–265, see also Brook
    trout, southern Appalachian *(Salvelinus fontinalis)*

J

Japanese barberry *(Berberis thunbergii)*, 298–299
Japanese grass *(Microstegium vimineum* [Trin.] A. Camus), 298–299
Japanese honeysuckle *(Lonicera japonica* Thumb.), 298–299
Japanese walnut *(Juglans sieboldi* var. *cordiformis)*, 295
Jefferson, Thomas, 46
Johnson grass *(Solanum halepense* [L.] Pers.), 298–299
Jones middle-tooth snail *(Mesodon jonesianus)*, 276
Joyce Kilmer Forest and Wilderness Area, 44
Juglans cinera (butternut), 134, 294–295
Juglans sieboldi var. *cordiformis* (Japanese walnut), 295

K

Kalmia latifolia (mountan laurel), 212, 338
Kentucky warbler *(Oporonis formosus)*, 190, 202
Key principles of ecosystem management, 15
Kirtland's warbler, 190
Komandorsky Biosphere Reserve (Russia), 100
Konza Prairie Research Natural Area (Kansas), 28
Kudzu *(Pueraria montana)*, 298–299

L

Laissez faire economics, 47–49
Land, cultural values and, 78–79
Land Between the Lakes Area, 36
Land cover type, 128, 130, see also Plant species(flora); Vegetation
Landscape ecology/landscape monitoring programs, 93
Landscape heterogeneity, fire and, 339–340
Land use ethic, as term, 41–42
Land use management, See Ecosystem management
Land use patterns, 124–127
    bear conservation and, 216–217
    bird populations and, 195
    in spruce–fir forest, 444–446
Land-use planning
    principles for sustainable development, 355–357
    for sustainable tourism, 349–374, see also Tourism
Last Great Places program of Nature Conservancy, 419
Laurel, mountain *(Kalmia latifolia)*, 338
Laws, hunting, 214–215
Legal jeopardy
    of air quality management, 325
    of black bear program, 222
    of climate change programs, 412
    of Clinch–Powell River Basin Initiative, 430
    of European wild boar *(Sus scrofa)* control, 287
    of fire management, 344
    of grassy balds, 392
    of neotropical migratory bird program, 207
    of pest and pathogen control, 304
    of Pittman Center, 372
    of red wolf repatriation, 244
    of spruce–fir forest management, 450
Lepus timidus (European hare), 102
Lichen, rock gnome *(Gymnoderma lineare)*, 299, 440

Lightning, 330–331, 343, see also Fire management
Ligustrum vulgare L. (privet), 298–299
Lilium grayi (Gray's lily), 390
Lily
    fawn *(Erythronium americanum)*, 275
    Gray's *(Lilium grayi)*, 390
Lindane, in adelgid control, 440
Lion, mountain *(Felis concolor)*, 120, 227
Liriodendron tulipifera (tulip poplar), 44, 48
Little River Lumber Company, 48
Livestock depredation, 229, 235–239
Local acceptance, 11–12
Locke, John, 46, 49
Lodgepole pine *(Pinus contort* Dougl.), 339
Logging, 48, 424, 437–438
Lonicera japonica Thunb. (Japanese honeysuckle), 298–299
Look Rock visibility monitoring program, 179–180, 312–313
Lutra canadensis (river otter), 109
Lutra lutra (otter), 102
Lymantria dispar L. (European gypsy moth), 134–135, 214, 296–297

M

MAB (Man and the Biosphere Program), see Man and the Biosphere Program (MAB)
Mammoth Cave Area, 36, 37
Managed use areas (buffer zones), 25
Management implications, 18–19
Management initiatives, 475–482
Man and the Biosphere Program (MAB), 10, see also Biopshere reserves
    budget and funding of, 27–28
    Canadian, 32, 33
    future directions for, 37–38
    historical evolution of, 23–40, 81–83
        conservation/development harmonization (1990s), 33–37
        designated site establishment (1974–), 26–29
        multisite linkages (1981–1989), 29–33
    Mexican, 33, 37
    Southern Appalachian (SAMAB), establishment of, 32–33
    UNESCO
        origins of, 23–24, 81–82
        U.S. political climate and, 30
    U.S.
        historical evolution of, 27–32, 82–83
        information access in, 35–36
        local partnerships in, 36–37
        political climate and, 30
        restructuring of, 33–37
        strategic framework for, 35
Man in nature/man vs. nature, 8
Maple
    mountain *(Acer spictum* Lam.), 440
    red *(Acer rubrum)*, 273
    striped *(Acer pensylvanicum)*, 440
    sugar *(Acer saccharum)*, 275, 295
Marmota monax (groundhog), 229

Martes martes (pine martin), 102
Martes pennanti (fisher), 227
Meleagris gallopavo (wild turkey), 132, 276
Mephitis mephitis (skunk), 229
Mesodon jonesianus (Jones middle-tooth snail), 276
Microbexura montivaga (spruce-fir spider), 299, 441
Microstegium vimineum [Trin.] A. Camus (Japanese grass),
     298–299
Milkweed, tall, 180
Mimosa (*Albizzia julibrissin* Durazz.), 298–299
Mining, 78, 423, 424
Mississippian culture, 42–43
Molothrus ater (brown-headed cowbird), 192, 197
Monitoring, see Air quality monitoring; Environmental
     monitoring
Monumenting, 169
Moth, European gypsy (*Lymantria dispar* L.), 134–135, 214,
     296–297, 461
Mountain-ash sawfly (*Pristiphora geniculata* Hartig),
     440
Mountain ash (*Sorbus americana* Marshall), 436
Mountain avens *(Geum radiotum)*, 390
Mountain bluets (*Houstonia purpurea* var. *montana)*, 390
Mountain holly *(Ilex ambigua* var. *montana)*, 440
Mountain laurel (*Kalmia latifolia)*, 212, 338
Mountain lion *(Felis concolor)*, 120, 227
Mountain maple (*Acer spictum* Lam.), 440
Mountain oat grass (*Danthonia compressa)*, 275, 379
Mountain pine (*Pinus pungens* Lam.), 337, 339
Mountain Protected Areas, 401
Mountain rattlesnake root *(Premanthes roanensis)*, 390
Mt Rogers, see Spruce–fir forest management
Multiflora rose (*Rosa multiflora* Thun. ex Murray), 298–299
Musk thistle (*Cardaria nutans* L.), 298–299
Mustela erminea (European ermine), 102
Mustela putoris (European pole cat), 102

N

Nantahala National Forest, 216, 271–272
National acceptance, 10–11
National Acid Precipitation Assessment Program (NAPAP),
     122, 173–174, 180
National Ambient Air Quality Standards (NAAQS), 300, see
     also Air quality
National Atmospheric Deposition Program, 175
National Environmental Policy Act of 1969 (NEPA), 24, 52
National Parks, see also specific parks
National parks, creation period (1930–1975), 49–52
National Park Service, 29, 30, 176, 179, 317–318
     in black bear management, 54, 214
     exotic species policy, 271
     regional programs of, 88
     reorganization of, 105
     threat list of, 52–53
     wild hog control by, 54
     wolf release and, 227
National Weather Service, 312
Native Americans
     as early inhabitants, 76–77
     fire practices of, 331–334

grassy balds and, 384
     historical land use practices of, 43–44
Naturalness concept, 330
Natural resources inventory, 356
Nature Conservancy, 28, 74
     community classification of, 179, 379
     Last Great Places program of, 419
     Roan Mountain bald and, 389–390
Neotropical migratory songbirds, 187–208
     ecosystem approach to conservation of, 193–201
     evidence and causes of decline in, 190–193
     Great Smoky Mountains National Park versus Cherokee
          National Forest, 203
     high priority southern species of, 203
     importance of in Southern Appalachians, 190
     management initiatives for, 478
     in old-growth versus second-growth forests,
          198–201
     research insights about, 201–203
     research programs in Southern Appalachians, 194
     Roaring Fork Study census results for, 196
     species codes used for, 197
     types and distribution, 187–189
New Jersey Pinelands National Reserve, 31
Nitrogen oxide, 180
Nitrogen saturation, 442
Nixon–Breshnev agreement, 24
Noland Divide Research Watershed, 178–179, 312
North Carolina State University, 214
North Carolina Vegetation Survey, 179
North Carolina Wildlife Resources Commission, 216
Northern flying squirrel (*Glaucomys sabrinus coloratus)*,
     227, 441
Northern Parula, 202
Notorus baileyi (Smoky Mountain mad tom), 109, 220, 228

O

Oak (*Quercus* spp.), 54, 273
     black *(Quercus velutina)*, 210
     chestnut *(Quercus prinus* Lamb.), 338
     scarlet *(Quercus coccinea)*, 338
     white *(Quercus alba)*, 210
Oak Ridge National Environmental Research Park, 32, 174
Oak Ridge National Laboratory, 179
Oak Ridge Reservation, 11, 12, 175
Oat grass, mountain *(Danthonia compressa)*, 275, 379
Odocoileus virginianus (white-tailed deer), 132, 177, 210,
     228, 229, 241
Office of Management and Budget (OMB), 29
Office of Science and Technology, 29
Oncorhynus mykiss (rainbow trout), 249, 250–252, see also
          Brook trout, southern Appalachian
Operation Smoky, 53–54, 215
Oporonis formosus (Kentucky warbler), 190, 202
Opportunities and constraints mapping, 356–357
Orchid, purple fringeless (*Plantathera peramoena)*, 177
Oriental bitterweset (*Celastrus orbiculatus* Thunb.),
     298–299
Otter, river *(Lutra canadensis)*, 109
Otter *(Lutra lutra)*, 102

Outer Banks, 31–32, see also Carolinian–South Atlantic
  Biosphere Reserve
Ovenbird *(Steurus aurocapillus)*, 202
Ozone
  ground level, 180
  ground-level, 143–149
  in spruce–fir forest, 441

P

Parks, national, see National Parks and specific parks
Partners in Flight program, 192, 201
Pathogen and pest control, 291–296, see also individual
  species
  damage examples in Southern Appalachian ecosystems,
    293–297
    American chestnut blight *(Cryphonectria parasitica)*,
      293–294
    beech bark disease complex, 297
    butternut canker, 134, 294–295
    dogwood anthracnose, 133–134, 295–296
    gypsy moth, 134–135, 214, 296–297
    pear thrips, 295
    woolly adelgid
      balsam, 134, 178, 294, 409, 438–441, 443
      hemlock, 134, 296
  forest ecosystem change and stress and, 292–293
  habitat impact of exotic pests, 299–300
  invasions in Great Smoky Mountains National Park,
    297–299
  management, 300–303
    ecosystem protection, 300–301
    genetic conservation, 301–302
    pest control, 302–303
    reintroduction protocols, 303
    species reintroduction, 301
Paulownia tomentosa (Thunb), 298
Pear thrip *(Taeniothrips inconsequens)*, 295
Peccary species, 271
Peer review, 169
Peregrine falcon *(Falco peregrinus)*, 109, 227
Performance zoning, 369–370
Periwinkle *(Vinca minor)*, 298–299
Pest control, 291–306, see also Pathogen and pest control
Phytolacca americana (pokeberry), 210
Picea rubens (red spruce), 178, 299–300, 431–454, see also
    Spruce–fir forest management
  acidic deposition and, 442–443
Picolides borealis (red-cockaded woodpecker), 227
Piedmont, geological features of, 66
Pine
  lodgepole *(Pinus contort* Dougl.), 339
  mountain *(Pinus pungens* Lam.), 337, 339
  pitch *(Pinus rigida* Mill.), 337
  Virginia *(Pinus virginiana* Mill.), 338–339
Pine bark beetle *(Dendroctonous frontalis)*, 329
Pinelands National Reserve (New Jersey), 31
Pine martin *(Martes martes)*, 102
Pinus contort Dougl. (lodgepole pine), 339
Pinus pungens Lam. (mountain pine), 337, 339
Pinus rigida Mill. (pitch pine), 337

Pinus virginiana Mill. (Virginia pine), 338–339
Piranga olivecea (scarlet tanager), 197, 202
Pisgah National Forest, 216, 219
Pitch pine *(Pinus rigida* Mill.), 337
Pittman Center, 362–371
  architectural guidelines for commercial structures,
    367–368
  community planning process, 362
  community vision statement, 366
  conceptual plan, 365
  history of community, 362
  implementation strategies, 371
  performance zoning, 369–370
  primary street design standards, 368–369
  proximity to National Park Service properties, 362
  public policy and, 56
  relation to Greenbrier Park entrance, 371
  resources inventory, 364
  ridge-top ordinance, 366
  river corridor ordinance, 368
  sign ordinances, 366–367
  sustainable development commission, 370
  visioning process, 363
Plantathera peramoena (purple fringeless orchid), 177
Plant data bases, 36
Plant protection programs, 94
Plants
  Cherokee tradition and, 43–44
  ozone effects on, 145–147
Plant species (flora), 128, 130, see also specific types and
    species
  of grassy balds, 380, 381, see also Grassy balds
  of spruce–fir forests, 436
Plethodon jordani (red-cheeked salamander), 276
Plethodon spp. (salamanders), 195, 276, 407, 441
Poaching, 215, 219
Pokeberry *(Phytolacca americana)*, 210
Policy issues, Man and the Biosphere Program (MAB)
    evolution, 23–40
Pollutants, see also Atmospheric team environmental
    assessment
  of concern, 142–143
  contribution to visibility by source, 155
  visibility and, 150–154
Poplar, tulip *Liriodendron tulipifera*, 44, 48
Potentilla rumex (cinquefoil), 275, 379
Precipitation, 71, see also Climate
Prehistoric land use practices, 42–43
Premanthes roanensis (mountain rattlesnake root), 390
Prevention of Significant Deterioration Program, 300–301,
    see also Air quality management
Princess tree *(Paulownia tomentosa [Thunb.])*, 298
Priority setting, in Russia as compared with U.S., 108
Prisko-Terrasny Reserve (Russia)
  as compared with Great Smoky Mountains National Park
    (U.S.), 99–114
  government and, 104–105
  history of, 102–103
  management, 106–108
  management of
    budget, 107–108

mission, 106
   organization, 107
as part of Russian national system, 100–104
priority setting, 108
programs and partnerships of, 108–114
   air quality, 111
   environmental education, 113–114
   research and monitoring, 110
   species reintroduction, 109
Pristiphora geniculata Hartig (mountain-ash sawfly), 440
Privet (*Ligustrum vulgare* L.), 298–299
Procyon lotor (raccoon), 229
Program design, 169
Prunus pensylvanica (fire cherry), 178, 437
Prunus serotina (black cherry), 48, 212, 295
Public attitudes, 132, 207, 222, 241–242, 244, 287, 304, 325, 372, 392, 450
Public information programs, 94
Public policy, in Russia as compared with U.S., 104–106
Pueraria montana (kudzu), 298–299
Purple fringeless orchid *(Plantathera peramoena)*, 177

Q

Quercus alba (white oak), 210
Quercus coccinea (scarlet oak), 338
Quercus prinus Lamb. (chestnut oak), 338
Quercus spp. (oak), 54, 212, 273
Quercus velutina (black oak), 210

R

Rabbit, cottontail *(Sylvilagus floridanus)*, 210, 229
Raccoon *(Procyon lotor)*, 229
Ragwort, Rugel's *(Rugelia nudicaulis)*, 390
Rainbow trout *(Oncorhynus mykiss)*, 249–252
Rattlesnake root, mountain *(Premanthes roanensis)*, 390
Recreational use, 158, 445, see also Tourism
Red-cheeked salamander *(Plethodon jordani)*, 276
Red-cockaded woodpecker *(Picolides borealis)*, 227
Red maple *(Acer rubrum)*, 273
Red spruce *(Picea rubens)*, 178, 299–300, 431–454, 442–443, see also Spruce–fir forest management
Red wolf *(Canis rufus)*, 109–110, 120, 177
   management initiatives for, 478–479
   public policy and, 56–57
   species repatriation of, 225–246
      in Great Smoky Mountains National Park, 228–236
      background of, 228–229
      experimental release, 229
      mortality in, 233
      subsequent release, 229–236
      historical background of, 225–226
      keys to future success in, 241–243
      livestock depredation and, 229, 235–239
      obstacles to, 239–241
      hybridization with coyotes, 240
      limited land base, 239
      small staff, 239–240
      taxonomic status, 240–241

   in Southern Appalachians, 227
   translocation ecology and, 226–227
Regional and local acceptance, 11–12
Regional context, importance of, 168
Regional Ecosystem Office, 119
Reporting, 170
Reptiles and amphibians, 75
Research
   air quality adverse effects, 311
   Russian versus U.S., 110
Research on Monitoring Techniques program, 182
Resource conflicts
   with air quality management, 325
   with black bear program, 222
   with fire management, 343
   with pest and pathogen control, 304
   with red wolf repatriation, 243
Resource-dependent communities, 154–155, 157
Resource extraction
   cattle ranching, 47–48
   timber, 48
   tourism, 48–49
Resource extraction period (1838–1930), 47–49
Resource inventory, 432
Resource management assessments, 118
Resource monitoring, 433
Rhododendron maximum, 275
Rhododendron spp., 212
   in grassy balds, 380
Riparian zone assessment, 140–142
Risk, environmental, 120
Risk assessment, 6, 168–169
Risk Assessment Report Cards
   for air quality management, 324–326
   for black bear program, 221–222
   for climate change, 412
   for Clinch–Powell River Basin Initiative, 428–429
   for European wild boar *(Sus scrofa)* control, 287
   for European wild hog program, 287–288
   for fire management, 343–345
   for grassy bald management, 392–393
   for Great Smoky Mountains National Park environmental monitoring, 182–184
   for neotropical migratory bird program, 206–207
   for pest and pathogen control, 303–304
   for Pittman Center, 372
   for red wolf repatriation, 243–244
   for spruce–fir forest management, 449–450
River corridor ordinance, for Pittman Center, 368
River otter *(Lutra canadensis)*, 109
Road development, 52
Roan Mountain grassy bald, 389–390
Robust Multiple Objective Linear Programming Model, 9
Rock gnome lichen *(Gymnoderma lineare)*, 299, 440
Rocky Mountains, 37
Rosa multiflora Thun. ex Murray, 298–299
Roundleaved sundew *(Drosera rotundifolia)*, 388
Rubus spp. (blackberry), 210, 212, 275, 383, 440
Rudbeckia spp., 42
Ruffed grouse *(Bonasa umbelous)*, 276
Rugelia nudicaulis (Rugel's ragwort), 390

Rural Environmental Planning for Sustainable Development (Sargent), 355–357
Russia
  biosphere reserves in, 100–101
  government policy in, 104–105
  Priosko-Terrasny Reserve as compared with Great Smoky Mountains National Park, 99–114, see also individual locations
Russian hazel *(Fritillaria ruthenica),* 102

S

Salamander *(Plethodon* spp.), 195, 276, 407, 441
  red-cheeked *(Plethodon jordani),* 276
SAMAB Associated Colleges and Universities, see also Southern Appalachian Man and the Biosphere Program (SAMAB)
SAMAB Associated Colleges and Universities (SAMAB Consortium), 92, 93
SAMAB Cooperative, 91, 92, see also Southern Appalachian Man and the Biosphere Program (SAMAB)
SAMAB-FHM Demonstration, 181–182
SAMAB Forest Health Monitoring Demonstration, 180–182
SAMAB Foundation, 92, 93, see also Southern Appalachian Man and the Biosphere Program (SAMAB)
Sambucus pubens (elderberry), 440
Sawfly, mountain-ash *(Pristiphora geniculata* Hartig), 440
Scale
  spatial, 171–173
  temporal, 172–173
Scarlet oak *(Quercus coccinea),* 338
Scarlet tanager *(Piranga olivecea),* 197, 202
Science Applications International Corporation (SAIC), 156–157
Scientific partnerships, 169
Scientific validity
  of air quality management, 325
  of black bear program, 222
  of climate change, 412
  of Clinch–Powell River Basin Initiative, 430
  of European wild boar *(Sus scrofa)* control, 287
  of fire management, 344
  of grassy balds, 392
  of neotropical migratory bird program, 207
  ofr Pittman Center, 372
  of pest and pathogen control, 304
  of red wolf repatriation, 244
  of spruce–fir forest management, 450
Scuires vulgaru (squirrel), 102, 276, 441
Sedge, wretched *(Carex misera),* 390
Sensitivity analysis, see under Climate change
Serviceberry *(Amelanchier arborea),* 212, 440
Seven pillars of ecosystem management, 16
Sevier County, TN, 360–381
Shenandoah National Park, 213, 214
Siberian roe deer *(Capreolus capreolus),* 102
Sign ordinances, 366–367
Silver fir *(Abies alba* Mill.), 438
Sirococcus clavigignent-juglandacerum (butternut canker), 134, 294–295
Skunk *(Mephitis mephitis),* 229

Smithsonian Institution, 406
Smithsonian/MAB Biodiversity Program, 36
Smokey Bear, 334–335, see also Fire management
Smoky Mountain mad tom *(Noturus baileyi),* 109, 220, 228
Smoky Mountain manna grass *(Glyceria nubigenu),* 390
Social, cultural, and economic environmental assessment, 154–158
  questions developed for, 161–162
Socioeconomic analysis, 315
Socioeconomic conflicts
  with black bear program, 222
  with climate change programs, 412
  with Clinch–Powell River Basin Initiative, 430
  with European wild boar *(Sus scrofa)* control, 287
  with fire management, 343–344
  with grassy balds, 392
  with neotropical migratory bird program, 206
  with pest and pathogen control, 304
  with Pittman Center, 372
  with red wolf repatriation, 244
  with spruce–fir forest management, 450
Soils, 71–72
  fire effects on, 337
  of grassy balds, 382
  of spruce–fir forests, 436
Solanum halepense (L.) Pers. (Johnson grass), 298–299
Sonoran Desert, 37
Sorbus americana Marshall (mountain ash), 436
Southern Appalachian Assessment, 64, 74, 117–166, 220, see also Environmental assessment
Southern Appalachian Biosphere Reserve, 88–89
Southern Appalachian Black Bear Study Group, 215–216, see also Black bear *(Ursus americnus)*
Southern Appalachian brook trout, 247–265, see also Brook trout, southern Appalachian *(Salvelinus fontinalis)*
Southern Appalachian Man and the Biosphere Program (SAMAB), see also SAMAB entries
  call for action in, 475–482
  components of, 88–89
  development strategies and public policy, 55–57
  environmental assessment by, 117–166
  environmental monitoring by, 167–185
  formation of program in, 89–91
  funding of, 95–97
  organization of, 91–93
  program areas of, 93–95
  regional and local acceptance of, 11–12
  toward operational level of, 475–482
Southern Appalachian Mountains Initiative (SAMI), 56, 313–322, see also under Air quality management
Southern Appalachian region, 63–80
  aquatic resources of, 76
  climate of, 69–71
  culture of, 78
  economy of, 77–78
  geographical description of, 63–65
  geological features of, 65–69, 121
    Appalachian Plateau, 66–68
    Blue Ridge, 66
    Piedmont, 66
    Valley and Ridge province, 66

geological history of, 68–69
people of
  early European settlers, 77
  early inhabitants, 76–77
physiogeographic provinces of, 65
relevance to ecosystem management of, 79
soils of, 71–72
vegetation of, 72–74
wildlife of, 74–75, see also individual species
Southern Oxidants Study (SOS), 175
Spatial scale, 171–173
Species
  endangered and threatened, see Endangered and
    threatened species and specific species
  indicator, see Indicator species and specific species
Species-level monitoring, 177–178
Species reintroduction
  of black bear, 215, see also Black bear (Ursus
    americanus)
  of exotic pests, 301
  of red wolf (Canis rufus), 225–246
    in Great Smoky Mountains National Park, 228–236
    historical background of, 225–226
    keys to future success in, 241–243
    livestock depredation and, 229, 235–239
    obstacles to, 239–241
    in Southern Appalachians, 227
    translocation ecology and, 226–227
  Russia versus U.S., 109
  species suitable for, 226–227
Spider, spruce-fir (Microbexura montivaga), 299, 441
Spring beauty (Claytonia virginica), 54, 275
Spruce, red (Picea rubens), 178, 299–300, 408, 431–434,
    442–443, see also Spruce–fir forest
Spruce–fir forest, 431–454
  ecosystem management plan framework, 432–435
    ecosystem a priori condition, 432
    framework summary, 434
    implementation strategy, 434
    resource inventory, 432
    resource monitoring, 433
    risk assessment, 433
    spruce–fir forest as case study, 434–435
    spruce–fir forest case study, 434–435
    stablization/recovery plan, 433–434
    stressors inventory, 432–433
  management initiatives for, 481–482
  Southern Appalachian spruce–fir ecosystem, 435–449
    call for action, 448–449
    current potential management initiatives, 446–448
    current stressors, 437–444
    air pollution, 441–443
    balsam wooly adelgid, 134, 178, 294, 409, 438–441,
      443
    climate change, 444
    logging effects, 437–438
    natural history and ecological characteristics, 435–436
    ownership and use patterns, 444–446
    research needs, 448
Spruce-fir spider (Microbexura montivaga), 299, 441
Squawroot (Conopholis americanus), 212

Squirrel
  Carolina Northen flying (Glaucomys sabrinus coloratus),
    441
  northern flying (Glaucomys sabrinus coloratus), 227,
    441
  Virginia Northen Flying (Glaucomys sabrinus fuscus),
    441
Squirrel (Scuires vulgaru), 102, 276
Stachys clinmannii, 275
Stakeholder involvement, 119, 123, 321–322, 355–356,
    424–425
Standard visual range, 150
Steurus aurocapillus (ovenbrid), 197, 202
Stipa joannis (feather grass), 102
Strategic Plan for U.S. Biosphere Program, 35
Street design standards, for Pittman Center, 368–369
Stressors, environmental, see Environmental stressors and
    specific stressors
Stressors inventory, 432
Striped maple (Acer pensylvanicum), 440
Strip mining, 78
Sugar maple (Acer saccharum), 275
Sulfur dioxide, 142–143, 149, 180, see also Acidic
    deposition; Air quality
Sundew, roundleaved (Drosera rotundifolia), 388
Sus domestica (domestic pig), 271
Sus scrofa, see European wild boar (Sus scrofa)
Sustainability, see also Sustainable development
  bear conservation and, 219
  as concept, 106
  management initiatives for, 480
Sustainable development
  as applied to tourism, 351–355
  definition of, 3–6
  dynamic model of, 5
  historical approaches to (1988 to present), 55–57
  land-use planning for tourism, 349–374, see also Tourism
  land-use planning principles for, 355–357
  management for, 5
  operational aspects of, 5–6
  public policy and, 55–57
  right of unborn to, 5
  SAMAB program in, 93
Sylvilagus floridanus (cottontail rabbit), 210, 229

T

Taeniothrips inconsequens (pear thrip), 295
Tall milkweed, 180
Tamiasciurus hudsonicus, 276
Tamias striatus (chipmunk), 276
Taxonomy, of red wolf (Canis rufus), 241
Tayasuidae (peccary) spp., 271
Temperature, 69–70, see also Climate change
Temporal scale, 172–173
Tennessee
  bear population of, 219
  bear reintroduction in, 215
Tennessee Valley Authority (TVA), 111, 173, 175, 182,
    312–313, see also TVA entries
  air pollution by, 142

Clean Water Initiative of, 419–420
River Action Teams of, 420, 426
Tenn Luttrell permit application, 323–324
Terrestrial team environmental assessment, 127–135
    questions developed for, 160
Thistle, musk (*Cardaria nutans* L.), 298–299
Threatened and endangered species, 127–128, 131, 241
    climate change and, 404
    exotic pest impact on, 299–300
    of spruce-fir forest, 441
Thrush, wood (*Hylocichia mustelina*), 192, 197, 201–203
Timber industry, 48
Timber wolf (*Canis lupis*), 120
Topography, 65, 120, see also Geological features
    of Clinch–Powell River Basin, 421
    of grassy balds, 382
Tourism, 349–374, 448
    climate change and, 402
    early, 48–49
    growth of industry, 350–351
    Pittman Center case study, 362–371
        architectural guidelines for commercial structures, 367–368
        community planning process, 362
        community vision statement, 366
        conceptual plan, 365
        history of community, 362
        implementation strategies, 371
        performance zoning, 369–370
        primary street design standards, 368–369
        proximity to National Park Service properties, 362
        relation to Greenbrier Park entrance, 371
        resources inventory, 364
        ridge-top ordinance, 366
        river corridor ordinance, 368
        sign ordinances, 366–367
        sustainable development commission, 370
        visioning process, 363
    public policy and, 56
    relationship to environment of, 351
    in Southern Appalachians, 357–361
        gateway communities, 358–360
        regional setting, 357–358
        Sevier County, TN, 360–381
    sustainable development applied to, 351–355
        alien species invasion, 352
        cultural sustainability, 353–354
        domestic animal intrusion, 352
        ecosystem sustainability, 352
        light and noise pollution, 353
        low impact services and facilities, 353
        pollutant introduction, 353
        social and economic sustainablity, 354–355
        wildlife conflicts, 353
Tranferability, of European wild boar (*Sus scrofa*) control, 287
Transferability
    of air quality management, 325
    of black bear program, 222
    of climate change programs, 412
    of Clinch–Powell River Basin Initiative, 430

    of fire management, 345
    of grassy balds, 393
    of neotropical migratory bird program, 207
    of pathogen and pest control, 304
    of Pittman Center case study, 372
    of red wolf repatriation, 244
    of spruce–fir forest management, 450
Transition (cooperative) areas, 25–26
Translocation ecology, 226–227
Tree-of-Heaven (*Ailanthus altissima*), 298–299
Tree species, see Forests and individual species
Triage versus diagnostic approach, 122
Tri-State Black Bear Study Group, 216
Trout
    rainbow (*Oncorhynus mykiss*), 249, 250–252
    southern Appalachian brook (*Salvelinus fontinalis*), 247–265, see also Brook trout, southern Appalachian (*Salvelinus fontinalis*)
Trout fisheries, 50
Tsuga canadensis (Eastern hemlock), 134, 296
Tsuga caroliniana (Carolina hemlock), 134
Tsuga spp. (hemlock), fungus-resistant, 296
Tulip poplar *Liriodendron tulipifera*, 44, 48
Turkey, wild (*Meleagris gallopavo*), 132, 276
TVA Countrywide Survey program, 173
Tyrannus tyrannus (Eastern kingbird), 187–188, 202

U

U-index (human use assessment), 124–126
U.N. Conference on Environment and Development (1992), 34
U.N. Environment Program, 29
Understanding Surface Water Quality in Southeastern Virginia, 422
UNESCO, 33, 81–83
    criteria for Biosphere reserves, 25–26
    Man and the Biosphere Program, see under Man and the Biosphere Program (MAB)
United States
    biosphere reserves in, 101
    Vision of Sustainable, 12–14
University of Georgia, 214
University of Michigan Biological Station, 28
University of Tennessee, 214
Uplands Field Research Laboratory, 106
Upper Three Runs Creek, 11
Ursa americanus (black bear), see Black bear (*Ursa americanus*)
U.S. Agency for International Development (USAID), 10
U.S. Department of Energy (DOE), 11
U.S. Environmental Protection Agency (EPA), 420, 425
U.S. Fish and Wildlife Service, 226, 227, 420
U.S. Forest Service (USFS), 18, 175, 182
    in black bear management, 214
    Continuous Forest Inventory program, 173
    fire policy of, 334–336
    Forest Inventory and Analysis program, 173
    policy reform in, 11
U.S. Man and the Biosphere Program (USMAB), see under Man and the Biosphere Program (MAB)

U.S. Office of Management and Budget (OMB), see Office of Management and Budget (OMB)
U.S. Office of Science and Technology, see Office of Science and Technology
USDA Animal and Plant Health Inspection Service (APHIS), 300–301
USDA Forest Service, 300–301
Utilitarianism, 49–50

V

Vaccinium spp., 210, 212, 383, 387, 388, 440
Valley and Ridge province, 66
Vegetation, 72–74, see also Grassy balds
  monitoring of, 179
  ozone effects on, 145–147
  response of to fire, 337–339
Viburnum alnifolium (hobblebush), 440
Vinca minor (periwinkle), 298–299
Viola spp., 275
Vireo, black-capped, 190
Virgina Polytechnic Institute and State University, 214
Virginia chain fern *(Woodwardia virginica)*, 177
Virginia Northern Flying squirrel *(Glaucomys sabrinus fuscus)*, 441
Virginia pine *(Pinus virginiana* Mill.), 338–339
Virgin Islands Biosphere Reserve, 30
Visibility, 150–154, 312
Visioning process, 355–356
  for Pittman Center, 363
Vulpes vulpes (fox), 102

W

Walker Branch Watershed, 174
Walnut, Japanese *(Juglans sieboldi* var. *cordiformis)*, 295
Warbler species, 187–188, 190, 197, see also Neotropical migratory songbirds
Water quality, 138–139
  of Clinch–Powell River Basin, 422
  monitoring of, 178–179
Water resources, 76, 137–138
Watershed Assessment Approach of Environmental Protection Agency (EPA), 119

Watershed monitoring, 178
Watersheds, 136–137
Wetlands programs, 93
White oak *(Quercus alba)*, 210
White-tailed deer *(Odocoileus virginianus)*, 132, 177, 210, 229, 241
White warbler, 187
Wilderness designation, 52
Wildlife, 74–75, see also individual species
Wild turkey *(Meleagris gallopavo)*, 132, 276
Wilsonnia citrina (hooded warbler), 201, 202
Wolf
  red *(Canis rufus)*, see Red wolf *(Canis rufus)*
  timber *(Canis lupis)*, 120
W126 value, 143–147, see also Acidic deposition
Woodland tradition, 42–43
Woodpecker, red-cockaded *(Picolides borealis)*, 227
Wood thrush *(Hylocichia mustelina)*, 192, 197, 201–203
Woodwardia virginica (Virginia chain fern), Virginia chain fern, 177
Woolly adelgid *(Adelges* spp.), 134, 178, 294, 296, 409, 438–441, 443
World Conservation Strategy, 29
World Conservation Union, 29
World Summit for Social Development, 34
World Wide Web, 35–36
World Wildlife Fund, 29
Worm-eating warbler, 197, 201
Wretched sedge *(Carex misera)*, 390

X

Xeric conditions, fire behavior and, 336–337

Y

Yam, wild *(Dioscorea batatas)*, 273
Yellow birch *(Betula alleghaniensis)*, 178
Yellow birch *(Betula lutea)*, 436, 437, 440
Yellowstone National Park, 339–340

Z

Zoning, performance, 369–370